Lecture Notes in Electrical Engineering

Volume 942

The book series *Lecture Notes in Electrical Engineering* (LNEE) publishes the latest developments in Electrical Engineering - quickly, informally and in high quality. While original research reported in proceedings and monographs has traditionally formed the core of LNEE, we also encourage authors to submit books devoted to supporting student education and professional training in the various fields and applications areas of electrical engineering. The series cover classical and emerging topics concerning:

- Communication Engineering, Information Theory and Networks
- Electronics Engineering and Microelectronics
- Signal, Image and Speech Processing
- Wireless and Mobile Communication
- Circuits and Systems
- Energy Systems, Power Electronics and Electrical Machines
- Electro-optical Engineering
- Instrumentation Engineering
- Avionics Engineering
- Control Systems
- Internet-of-Things and Cybersecurity
- Biomedical Devices, MEMS and NEMS

For general information about this book series, comments or suggestions, please contact leontina. dicecco@springer.com.

To submit a proposal or request further information, please contact the Publishing Editor in your country:

China

Jasmine Dou, Editor (jasmine.dou@springer.com)

India, Japan, Rest of Asia

Swati Meherishi, Editorial Director (Swati.Meherishi@springer.com)

Southeast Asia, Australia, New Zealand

Ramesh Nath Premnath, Editor (ramesh.premnath@springernature.com)

USA, Canada:

Michael Luby, Senior Editor (michael.luby@springer.com)

All other Countries:

Leontina Di Cecco, Senior Editor (leontina.dicecco@springer.com)

** This series is indexed by EI Compendex and Scopus databases. **

More information about this series at https://link.springer.com/bookseries/7818

Zhihong Qian · M. A. Jabbar ·
Xiaolong Li
Editors

Proceeding of 2021 International Conference on Wireless Communications, Networking and Applications

Set 2

 Springer

Editors
Zhihong Qian
College of Communication Engineering
Jilin University
Jilin, Jilin, China

M. A. Jabbar ⓘ
Department of AI & ML
Vardhaman College of Engineering
Hyderabad, Telangana, India

Xiaolong Li
College of Technology
Indiana State University
Terre Haute, IN, USA

ISSN 1876-1100   ISSN 1876-1119   (electronic)
Lecture Notes in Electrical Engineering
ISBN 978-981-19-2455-2   ISBN 978-981-19-2456-9   (eBook)
https://doi.org/10.1007/978-981-19-2456-9

This Springer imprint is published by the registered company Springer Nature Singapore Pte Ltd.
The registered company address is: 152 Beach Road, #21-01/04 Gateway East, Singapore 189721, Singapore

Preface

WCNA2021 [2021 International Conference on Wireless Communications, Networking and Applications] will be held on December 17–19, 2021, Berlin, Germany (virtual conference). Due to the COVID-19 situation and travel restriction, WCNA2021 has been converted into a virtual conference, which will be held via Tencent Meeting.

WCNA2021 hopes to provide an excellent international platform for all the invited speakers, authors, and participants. The conference enjoys a wide spread participation, and we sincerely wish that it would not only serve as an academic forum but also a good opportunity to establish business cooperation. Any paper and topic around wireless communications, networking, and applications would be warmly welcomed.

WCNA2021 proceeding tends to collect the most up-to-date, comprehensive, and worldwide state-of-the-art knowledge on wireless communications, networking, and applications. All the accepted papers have been submitted to strict peer review by 2–4 expert referees and selected based on originality, significance, and clarity for the purpose of the conference. The conference program is extremely rich, profound, and featuring high-impact presentations of selected papers and additional late-breaking contributions. We sincerely hope that the conference would not only show the participants a broad overview of the latest research results on related fields but also provide them with a significant platform for academic connection and exchange.

The technical program committee members have been working very hard to meet the deadline of review. The final conference program consists of 121 papers divided into six sessions. The proceedings would be published on Springer Book Series Lecture Notes in Electrical Engineering as a volume quickly, informally, and in high quality.

We would like to express our sincere gratitude to all the TPC members and organizers for their hard work, precious time and endeavor preparing for the conference. Our deepest thanks also go to the volunteers and staffs for their long-hours work and generosity they have given to the conference. Last but not least, we would like to thank each and every of the authors, speakers, and participants for their great contributions to the success of WCNA2021.

WCNA2021 Organizing Committee

Organization

Committees

Honor Chair

Patrick Siarry — Laboratoire Images, Signaux et Systèmes Intelligents, University Paris-Est Cré, Paris, France

General Chair

Zhihong Qian — College of Communication Engineering, Jilin University, China

Co-chairs

Isidoros Perikos — Computer Engineering and Informatics, University of Patras, Greece

Hongzhi Wang — Department of Computer Science and Technology, Harbin Institute of Technology, China

Hyunsung Kim — School of Computer Science, Kyungil University, Korea

Editor in Chief

Zhihong Qian — College of Communication Engineering, Jilin University, China

Co-editors

M. A. Jabbar   Head of the Department, Department of AI &ML, Vardhaman College of Engineering, Hyderabad, Telangana, India

Xiaolong Li   College of Technology, Indiana State University, USA

Sivaradje Gopalakrishnan   Electronics and Communication Engineering Department, Puducherry Technological University, Puducherry, India

Technical Program Committee

Qiang Cheng   University of Kentucky, USA

Noor Zaman Jhanjhi   School of Computing and IT, Taylor's University, Malaysia

Yilun Shang   Department of Computer and Information Sciences, Northumbria University, UK

Pascal Lorenz   University of Haute Alsace, University of Haute Alsace, France

Guillermo Escrivá-Escrivá   Department of Electrical Engineering, Universitat Politècnica de València, Spain

Surinder Singh   Department of Electronics and Communication Engineering, Sant Longowal Institute of Engineering and Technology, India

Pejman Goudarzi   Iran Telecom Research Center (ITRC), Iran

Antonio Muñoz   University of Malaga, Spain

Manuel J. Domínguez-Morales   University of Seville, Spain

Shamneesh Sharma   School of Computer Science & Engineering, Poornima University, India

K. Somasundaram   Amrita Vishwa Vidyapeetham, India

Daniela Litan   Deployment & Delivery (Oracle Technology Center), Oracle Developer, Romania

Artis Mednis   Institute of Electronics and Computer Science, University of Latvia, Latvia

Hari Mohan Srivastava   Department of Mathematics and Statistics, University of Victoria, Canada

Chang, Chao-Tsun   Department of Information Management, Hsiuping University of Science and Technology, Taiwan

Sumit Kushwaha   Department of Electronics Engineering, Kamla Nehru Institute of Technology, India

Bipan Hazarika   Department of Mathematics, Gauhati University, India

Petko Hristov Petkov   Technical University of Sofia, Bulgaria

Pankaj Bhambri   Department of Information Technology, I.K.G. Punjab Technical University, India

Aouatif Saad   National School of Applied Sciences, Ibn Tofail University, Morocco

Marek Blok   Telecommunications and Informatics, Gdańsk University of Technology, Poland

Phongsak Phakamach   College of Innovation Management, Rajamangala University of Technology Rattanakosin, Thailand

Mohammed Rashad Baker   Imam Ja'afar Al-Sadiq University, Iraq

Ahmad Fakharian   Islamic Azad University, Iran

Ezmerina Kotobelli   Department of Electronics and Telecommunication, Faculty of Information Technology, Polytechnic University of Tirana, Albania

Nikhil Marriwala   Electronics and Communication Engineering Department, Kurukshetra University, India

M. M. Kamruzzaman   Department of Computer and Information Science, Jouf University, KSA

Marco Listanti   Department of Electronic, Information and Telecommunications Engineering (DIET), University of Roma "La Sapienza," Italy

Ashraf A. M. Khalaf   Electrical Engineering (Electronics and Communications), Minia University, Egypt

Kidsanapong Puntsti   Department of Electronics and Telecommunication Engineering, Rajamangala University of Technology Isan (RMUTI), Thailand

Valerio Frascolla   Director of Research and Innovation at Inte, Intel Labs Germany, Germany

Babar Shah   College of Technological Innovation, Zayed University, Dubai

DijanaIlišević   Department for Planning and Construction of Wireless Transport network

Xilong Liu   Department of Information Science and Engineering, Yunnan University, Yunnan University, China

Suresh Kumar   Computer Science and Engineering, Manav Rachna International University, India

Sivaradje Gopalakrishnan   Electronics and Communication Engineering Department, Puducherry Technological University, India

Kanagachidambaresan   Vel Tech University, India

Sivaradje	Department of Electronics and Communication Engineering, Pondicherry Engineering College, India
A. K. Verma	CSED, Thapar Institute of Engg. and Technology, India
Kamran Arshad	Electrical Engineering, Ajman University, UAE
Gyu Myoung Lee	School of Computer Science and Mathematics, iverpool John Moores University, UK
Zeeshan Kaleem	COMSATS University Islamabad, Pakistan
Fathollah Bistouni	Department of Computer Engineering, Islamic Azad University, Iran
Sutanu Ghosh	Electronics and Communication Engg., India
Sachin Kumar	School of Electronic and Electrical Engineering, Kyungpook National University, South Korea
Anahid Robert Safavi	Wireless Network Algorithm Laboratory Huawei Sweden, Sweden
Hoang Trong Minh	Telecommunications Engineering, Telecommunications Engineering, Vietnam
Devendra Prasad	CSE, Chitkara University, India
Hari Shankar Singh	Electronics and Communication Engineering, India
Ashraf A. M. Khalaf	Faculty of Engineering, Minia University, Egypt
Hooman Hematkhah	Electrical and Electronics Engineering, Chamran University (SCU), Iran
Mani Zarei	Department of Computer Engineering, Tehran, Iran
Jibendu Sekhar Roy	School of Electronics Engineering, KIIT University, India
Luiz Felipe de Queiroz Silveira	Computer Engineering and Automation Department, Federal University of Rio Grande do Norte, Brazil
Alexandros-Apostolos A. Boulogeorgos	Digital Systems, University of Piraeus, Greece
Trong-Minh Hoang	Posts and Telecommunication Institute of Technology, Vietnam
Jagadeesha R. Bhat	Electronic Communication Engg., Indian Institute of Information Technology, India
Tapas Kumar Mishra	Computer Science and Engineering, SRM University, India
Zisis Tsiatsikas	Information and Communication Systems Engineering, University of the Aegean, Greece
Muge Erel-Ozcevık	Software Engineering Department, Manisa Celal Bayar University, Turkey
E. Prince Edward	Department of Instrumentation and Control Engineering, Sri Krishna Polytechnic College, India

Prem Chand Jain — School of Engineering, Shiv Nadar University, India

Vipin Balyan — Department of Electrical, Electronics and Computer Engineering, Cape Peninsula University of Technology, South Africa

Yiannis Koumpouros — Department of Public and Community Health, University of West Attica, Greece

Aizaz Chaudhry — Systems and Computer Engineering, Carleton University, Canada

Andry Sedelnikov — Department of Space Engineering, Samara National Research University, Russia

Alexei Shishkin — Faculty of Computational Mathematics and Cybernetics, Moscow State University, Russia

Sevenpri Candra — S.E., M.M., ASEAN Engg., BINUS University, Indonesia

Meisam Abdollahi — School of Electrical and Computer Engineering, University of Tehran, Iran

Sachin Kumar (Research Professor) — Kyungpook National University, South Korea

Thokozani Calvin Shongwe — Electrical Engineering Technology, University of Johannesburg, South Africa

Ganesh Khekare — Department of Computer Science and Engineering, Faculty of Engineering & Technology, Parul University, Vadodara, Gujrat, India

Nishu Gupta — ECE Department, Chandigarh University, Mohali, Punjab, India

Gürel Çam — Iskenderun Technical University, Turkey

Ceyhun Ozcelik — Muğla Sıtkı Koçman University, Turkey

Shuaishuai Feng — Wuhan University, China

W. Luo — School of Finance and Economics, Nanchang Institute of Technology, China

Y. Xie — Party School of CPC Yibin Municipal Committee, China

Thanh-Lam Nguyen — Lac Hong University, Vietnam

Nikola Djuric — University of Novi Sad, Serbia

Ricky J. Sethi — Fitchburg State University, USA

Domenico Suriano — Italian National Agency for new Technologies, Energy, and Environment, Italy

Igor Verner — Faculty of Education in Science and Technology Technion, Israel Institute of Technology, Israel

Nicolau Viorel — "Dunarea de Jos" University of Galati, Romania

Snježana Babić — Polytechnic of Rijeka, Rijeka, Croatia

Esmaeel Darezereshki	Department of Materials Engineering, Shahid Bahonar University, Kerman, Iran
Ali Rostami	University of Tabriz, Iran
Hui-Ming Wee	Department of Industrial and Systems Engineering, Chung Yuan Christian University, Taiwan
Yongyun Cho	Dept. Information and Communication Engineering, Sunchon National University, Sunchon, Korea
Lakhoua Mohamed Najeh	University of Cathage, Tunisia
M. Sohel Rahman	Bangladesh University of Engineering and Technology, Bangladesh
Khaled Habib	Materials Science and Photo-Electronics Lab., RE Program, EBR Center KISR, Kuwait
Seongah Chin	Sungkyul University, Korea
Ning Cai	School of Artificial Intelligence, Beijing University of Posts and Telecommunications, China
Zezhong Xu	Changzhou Institute of Technology, China
Saeed Hamood Ahmed Mohammed Alsamhi	MSCA SMART 4.0 FELLOW, AIT, Ireland
Lim Yong Kwan	Singapore University of Social Sciences, Singapore
Imran Memon	Zhejiang University, China
Anthony Kwame Morgan	Kwame Nkrumah University of Science and Technology, Ghanaian
Ali Asghar Anvary Rostamy	Tarbiat Modares University, Iran
Hasan Dincer	Istanbul Medipol University, Turkey
Prem Kumar Singh	Gandhi Institute of Technology and Management-Visakhapatnam, India
Dimitrios A. Karras	National and Kapodistrian University of Athens, Greece
Cun Li	Eindhoven University of Technology, Netherland
Natalia A. Serdyukova	Plekhanov Russian University of Economics, Russia
Sylwia Werbinska-Wojciechowska	Wroclaw University of Science and Technology, Poland
José Joaquim de Moura Ramos	University of A Coruña, Spain
Naveen Kumar Sharma	I.K.G. Punjab Technical University, India
Tu Ouyang	Case Western Reserve University, USA
Nabil El Fezazi	Sidi Mohammed Ben Abdellah University, Morocco
Pedro Alexandre Mogadouro do Couto	University of Trás-os-Montes e Alto Douro, Portugal

Sek Yong Wee	Universiti Teknikal Malaysia Melaka, Malaysia
Muhammad Junaid Majeed	AuditXPRT Technologies, SQA Engineer, Pakistan
Janusz Kacprzyk	Systems Research Institute, Polish Academy of Sciences, Poland
Cihan Aygün	Faculty of Sports Sciences, Eskişehir Technical University, Turkey
Ciortea Elisabeta Mihaela	"December 1, 1918" University of Alba Iulia, Romania
Mueen Uddin	University Brunei Darussalam, Negara Brunei Darussalam
Esingbemi Princewill Ebietomere	University of Benin, Benin City, Nigeria
Samaneh Mashhadi	Iran University of science and Technology, Iran
Maria Aparecida Medeiros Maciel	Federal University of Rio Grande do Norte, Brazil
Josefa Mula	Universitat Politècnica de València, Spain
Claudemir Duca Vasconcelos	Federal University of ABC (UFABC), Brazil
Katerina Kabassi	Head of the Department of Environment, Ionian University, Greece
Takfarinas Saber	School of Computer Science, University College Dublin, Ireland
Zain Anwar Ali	Beijing Normal University, China
Jan Kubicek	VSB-Technical University of Ostrava, Czech Republic
Amir Karbassi Yazdi	School of Management, Islamic Azad University, Iran
Sujata Dash	Dept. of Computer Science and Application, North Orissa University, India
Souidi Mohammed El Habib	Abbes Laghrour University, Algeria
Dalal Abdulmohsin Hammood	Middle Technical Education (MTU) Electrical Engineering Technical College, Iraq
Marco Velicogna	Institute of Legal Informatics and Judicial Systems, Italian National Research Council, Italy
Hamad Naeem	College of Computer Science Neijiang Normal University, China
Hamid Jazayeriy	Babol Noshirvani University of Technology, Iran
Rituraj Soni	Engineering College Bikaner, India

Qutaiba Abdullah Hasan   University of Tikrit, Iraq
  Alasad
Alexandra Cristina González   Universidad Técnica Particular de Loja,
  Eras   Department of Computer Science and
                                  Electronics, Ecuador
Falguni Roy   Noakhali Science and Technology University,
                                  Bangladesh
Ioan-Lucian Popa   Department of Computing, Mathematics,
                                  and Electronics, "1Decembrie 1918"
                                  University of Alba Iulia, Romania

Keynote Speakers

Advanced Architectures of Next Generation Wireless Networks

Pascal Lorenz

University of Haute-Alsace, France

Abstract. Internet Quality of Service (QoS) mechanisms are expected to enable wide spread use of real-time services. New standards and new communication architectures allowing guaranteed QoS services are now developed. We will cover the issues of QoS provisioning in heterogeneous networks, Internet access over 5G networks, and discusses most emerging technologies in the area of networks and telecommunications such as IoT, SDN, edge computing, and MEC networking. We will also present routing, security, and baseline architectures of the Internet working protocols and end-to-end traffic management issues.

Biography: Pascal Lorenz received his M.Sc. (1990) and Ph.D. (1994) from the University of Nancy, France. Between 1990 and 1995, he was a research engineer at WorldFIP Europe and at Alcatel-Alsthom. He is a professor at the University of Haute-Alsace, France, since 1995. His research interests include QoS, wireless networks, and high-speed networks. He is the author/co-author of three books, three patents, and 200 international publications in refereed journals and conferences. He was Technical Editor of the IEEE Communications Magazine Editorial Board (2000–2006), IEEE Networks Magazine since 2015, IEEE Transactions on Vehicular Technology since 2017, Chair of IEEE ComSoc France (2014–2020), Financial chair of IEEE France (2017–2022), Chair of Vertical Issues in Communication Systems Technical Committee Cluster (2008–2009), Chair of the Communications Systems Integration and Modeling Technical Committee (2003–2009), Chair of the Communications Software Technical Committee (2008–2010), and Chair of the Technical Committee on Information Infrastructure and Networking (2016–2017). He has served as Co-Program Chair of IEEE WCNC'2012 and ICC'2004, Executive Vice-Chair of ICC'2017, TPC Vice Chair of Globecom'2018, Panel sessions co-chair for Globecom'16, tutorial chair of VTC'2013 Spring and WCNC'2010, track chair of PIMRC'2012 and WCNC'2014, symposium Co-Chair at Globecom 2007–2011, Globecom'2019, ICC 2008–2010, ICC'2014 and '2016. He has served as Co-Guest Editor for special issues of IEEE Communications Magazine, Networks Magazine, Wireless Communications Magazine, Telecommunications Systems, and LNCS. He is an associate editor for International Journal of Communication Systems (IJCS-Wiley), Journal on Security and Communication Networks (SCN-Wiley) and International Journal of Business Data Communications and Networking, Journal of Network and Computer Applications (JNCA-Elsevier). He is a senior member of the IEEE, IARIA fellow, and member of many international program committees. He has organized many conferences, chaired several technical sessions, and gave tutorials at major international conferences. He was IEEE ComSoc Distinguished Lecturer Tour during 2013–2014.

Role of Machine Learning Techniques in Intrusion Detection System

M. A. Jabbar

Department of AI and ML, Vardhman College of Engineering, Hyderabad, Telangana, India

Abstract. Machine learning (ML) techniques are omnipresent and are widely used in various applications. ML is playing a vital role in many fields like health care, agriculture, finance, and in security. Intrusion detection system (IDS) plays a vital role in security architecture of many organizations. An IDS is primarily used for protection of network and information system. IDS monitor the operation of host or a network. Machine learning approaches have been used to increase the detection rate of IDS. Applying ML can result in low false alarm rate and high detection rate. This talk will discuss about how machine learning techniques are applied for host and network intrusion detection system.

Biography: Dr. M. A. JABBAR is Professor and Head of the Department AI&ML, Vardhaman College of Engineering, Hyderabad, Telangana, India. He obtained Doctor of Philosophy (Ph.D.) from JNTUH, Hyderabad, and Telangana, India. He has been teaching for more than 20 years. His research interests include artificial intelligence, big data analytics, bio-informatics, cyber-security, machine learning, attack graphs, and intrusion detection systems.

Academic Research

He published more than 50 papers in various journals and conferences. He served as a technical committee member for more than 70 international conferences. He has been Editor for 1st ICMLSC 2018, SOCPAR 2019, and ICMLSC 2020. He also has been involved in organizing international conference as an organizing chair, program committee chair, publication chair, and reviewer for SoCPaR, HIS, ISDA, IAS, WICT, NABIC, etc. He is Guest Editor for the Fusion of Internet of Things, AI, and Cloud Computing in Health Care: Opportunities and Challenges (Springer) Series, and Deep Learning in Biomedical and Health Informatics: Current Applications and Possibilities–CRC Press, Guest Editor for Emerging Technologies and Applications for a Smart and Sustainable World-Bentham science, Guest editor for Machine Learning Methods for Signal, Image and Speech Processing –River Publisher.

He is a senior member of IEEE and lifetime member in professional bodies like the Computer Society of India (CSI) and the Indian Science Congress Association (ISCA). He is serving as a chair, IEEE CS chapter Hyderabad Section. He is also serving as a member of Machine Intelligence Laboratory, USA (MIRLABS) and USERN, IRAN , Asia Pacific Institute of Science and Engineering (APISE) Hong Kong , Member in Internet Society (USA), USA , Member in Data Science Society, USA, Artificial Intelligence and Machine Learning Society of India (AIML), Bangalore.

He received best faculty researcher award from CSI Mumbai chapter and Fossee Labs IIT Bombay and recognized as an outstanding reviewer from Elsevier and received outstanding leadership award from IEEE Hyderabad Section. He published five patents (Indian) in machine learning and allied areas and published a book on "Heart Disease Data Classification using Data Mining Techniques," with LAP LAMBERT Academic publishing, Mauritius, in 2019.

Editorial works

1. Guest Editor: The Fusion of Internet of Things, AI, and Cloud Computing In Health Care: Opportunities and Challenges (Springer)
2. Guest Editor: Deep Learning in Biomedical and Health Informatics: Current Applications and Possibilities (CRC)
3. Guest Editor: Emerging Technologies and Applications for a Smart and Sustainable World-Bentham science
4. Guest Editor: Machine Learning Methods for Signal, Image, and Speech Processing-River Publisher
5. Guest Editor: The Fusion of Artificial Intelligence and Soft Computing Techniques for Cyber-Security-AAP–CRC Press
6. Guest Editor Special Issue on Web Data Security: Emerging Cyber-Defense Concepts and Challenges Journal of Cyber-Security and Mobility-River Publisher

Data Quality Management in the Network Age

Hongzhi Wang

Computer Science and Technology, Harbin Institute of Technology, China

Abstract. In the network age, data quality problems become more serious, and data cleaning is in great demand. However, data quality in the network age brings new technical challenges including the mixed errors, absence of knowledge, and computational difficulty. Facing the challenge of mixed errors, we discover the relationships among various types of errors and develop data cleaning algorithms for multiple errors. We also design data cleaning strategies with crowdsourcing, knowledge base as well as web search for the supplement of knowledge. For efficient and scalable data cleaning, we develop parallel data cleaning systems and efficient data cleaning algorithms. This talk will discuss the challenges of data quality in network age and give an overview of our solutions.

Biography: Hongzhi Wang is Professor, PHD supervisor, the head of massive data computing center and the vice dean of the honors school of Harbin Institute of Technology, the secretary general of ACM SIGMOD China, outstanding CCF member, a standing committee member CCF databases, and a member of CCF big data committee. Research fields include big data management and analysis, database systems, knowledge engineering, and data quality. He was "starring track" visiting professor at MSRA and postdoctoral fellow at University of California, Irvine. Prof. Wang has been PI for more than ten national or international projects including NSFC key project, NSFC projects, and national technical support project, and co-PI for more than ten national projects include 973 project, 863 project, and NSFC key projects. He also serves as a member of ACM Data Science Task Force. He has won first natural science prize of Heilongjiang Province, MOE technological First award, Microsoft Fellowship, IBM PHD Fellowship, and Chinese excellent database engineer. His publications include over 200 papers in the journals and conferences such as VLDB Journal, IEEE TKDE, VLDB, SIGMOD, ICDE, and SIGIR, six books and six book chapters. His PHD thesis was elected to be outstanding PHD dissertation of CCF and Harbin Institute of Technology. He severs as the reviewer of more than 20 international journal including VLDB Journal,

IEEE TKDE, and PC members of over 50 international conferences including SIGMOD 2022, VLDB 2021, KDD 2021, ICML 2021, NeurpIS 2020, ICDE 2020, etc. His papers were cited more than 2000 times. His personal website is http://homepage.hit.edu.cn/wang.

Networking-Towards Data Science

Ganesh Khekare

Department of Computer Science and Engineering, Faculty of Engineering and Technology, Parul University, Vadodara, Gujrat, India

Abstract. For communication, network is a must. Nowadays, networking is generating big data. To handle and process this huge amount of data, data science is required. Due to the increase in connectivity, interactions, social networking sites, platforms like YouTube, then invention of big data, fog computing, edge computing, Internet of Everything, etc., network transactions have been increased. Providing the best network flow graph is a challenge. Researchers are working on various data science techniques to overcome this. Node embedding concept is used to embed various complex networking graphs. To analyze different nodes and graphs for embedding, KarateClub library is used with Neo4j. Neo4j Graph data science library analyzes multigraphs networks in a better way. When network information is required in a fixed size vector, node embedding is used. This information is used in a downstream machine learning flow. Pyvis library is used to Visualize Interactive Network Graphs in Python. It provides a customization facility by which the network can be arranged for user requirements or to streamline the data flow. Researchers are also looking for interactive network graphs through data science algorithms that are capable of handling real-time scenarios. To draw Hive plots, the open-source Python package Hiveplotlib is available. The intelligible and visual probe of data generated through networking can be done smoothly by using Hive Plots. A data science algorithm viz., DeepWalk, is used to understand relationships in complex graph networks using Gensim, Networkx, and Python. Undirected and unweighted network visualization is also possible by using Mercator graph layout/embedding for a real-world complex network. Visualization of high dimensional network traffic data with 3D 360-degree animated scatter plots is the need. A huge research scope is there in networking using data science for the upcoming generations.

Biography: Dr. Ganesh Khekare is currently working as an Associate Professor in the department of Computer Science and Engineering at Parul University, Vadodara, Gujrat, India. He has done Ph.D. from Bhagwant University India. He pursued Master of Engineering from G H Raisoni College of Engineering, Nagpur, in the year 2013, and Bachelor of Engineering from Priyadarshini College of Engineering, Nagpur, in 2010. He has published more than 25 research articles in reputed international journals and conferences including Thomson Reuters, IGI Global, Inderscience, Springer, IEEE, Taylor and Francis, etc. He has published one patent and three copyrights. He guided more than 50 research as well as industry projects. His main research work focuses on data science, Internet of everything, machine learning, computer networks, artificial intelligence, intelligent transportation system, etc. He has more than 12 years of teaching and research experience. He is an active member of various professional bodies like ACM, ISTE, IEEE, IAENG, IFERP, IERD, etc.

Contents

Devices, Tools, and Techniques for WSN and Other
Wireless Networks

Internet of Things (Iot)

Internet of Things (Iot)

Research on Visualization of Power Grid Big Data

Jun Zhou[1,2(✉)], Lihe Tang[1,2], Songyuhao Shi[3], Wei Li[1,2], Pan Hu[1,2], and Feng Wang[1,2]

[1] NARI Group Corporation/State Grid Electric Power Research Institute, Nanjing 211106, China
453927489@qq.com
[2] NARI Information Communication Science and Technology Co. Ltd., Nanjing 210003, China
[3] University of Florida, Gainesville, FL 32611, USA

Abstract. With the constant improvement of power grid planning and management requirements and the gradual advancement of the urbanization process, the problems that need to be taken into account in the planning process are increasing, especially the demand for big data visualization of the power grid has increased sharply. About 80% of the information that humans obtain from the external environment comes from the visual system. A picture is worth a thousand words. A good visualization platform can monitor the overall operation of the power grid, which is convenient for analyzing and monitoring the operation of power supply companies to provide customers with high-quality services. The platform can complete the interactive simulation of different services, and can display the monitoring and analysis of the power grid through a rich visual interface, which is convenient for people to understand the real-time status of the power grid. This paper uses various advanced visualization technologies and data module algorithms at home and abroad to cooperate with the monitoring network to realize the visualization platform of power grid big data, promote the further development of power grid big data applications, and form a big data standard system for power big data technology research, product research and development, and pilot construction.

Keywords: Data visualization · Big data · Monitoring network

1 The Importance of Big Data Visual Analysis

Big data visualization analysis refers to the use of the user interface with information visualization and the human-computer interaction methods and technologies with analysis process while the automatic analysis and mining methods of big data are used to effectively integrate the computing power of the computer and the cognitive ability of the human in order to obtain insights into large-scale and complex data sets [7]. From the construction perspective of a smart grid visualization platform with big data structure, it is necessary to further consolidate and improve the optimization and design work of the computer visualization platform and the unified data interface of other sub-projects,

© The Author(s) 2022
Z. Qian et al. (Eds.): WCNA 2021, LNEE 942, pp. 511–517, 2022.
https://doi.org/10.1007/978-981-19-2456-9_52

so that the platform can play an active role in the storage and calculation of power big data and realize data analysis and control as well. Using intuitive visualization methods to display analysis results can effectively guide the operators to make scientific decisions, facilitate the realization of intelligent and visualization of electricity consumption, serve the company and related industries, and realize the intelligence of the production process.

2 System Construction and Realization

2.1 System Module

This paper designs four-layer system modules, which are:

The first layer is the collection and access of big data. Big data is a data collection with the main characteristics of large capacity, multiple types, fast access speed, and high application value. The characteristics of grid big data are shown in Fig. 1. We use sensors, smart devices, video surveillance equipment, audio communication equipment, mobile terminals and other information acquisition channels to collect data with a huge amount, scattered sources, and diverse formats.

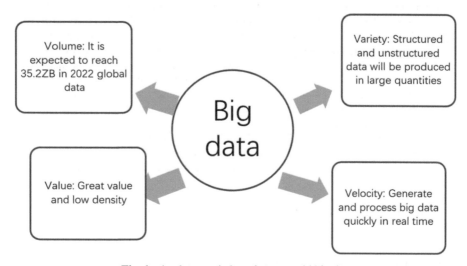

Fig. 1. 4v characteristics of power grid big data

The second layer is data storage. The storage technology used in this article is to use the current cloud storage technology to classify and store. The data types of big data are divided into structured and unstructured data. Sorting them into storage is conducive to subsequent efficient analysis and processing.

The third layer is data statistical analysis, mining, calculation, and management, providing security services with data backup and analysis services such as expert analysis and algorithm libraries. At present, methods such as feature extraction, data mining analysis, and statistical analysis of structured data have been widely used. For unstructured data, video, audio, and text are research hotspots, and intelligent analysis methods, such as machine learning, pattern recognition, and association analysis, are needed to achieve in-depth mining and multi-dimensional display of big data. Analyzing the data in the smart grid can help us to obtain information such as load and fault, which is helpful for the maintenance and operation of the power system, upgrading and updating. For example, the University of California, Los Angeles integrates the distribution of users, real-time electricity consumption information, temperature and weather and other information into a "electricity map", which can intuitively show the electricity consumption of people in each block and the power consumption of buildings, providing effective load data for the power sector.

The fourth layer is to integrate the information derived from various data algorithms such as classification, clustering, and association rules, and then visualize it graphically. Visualization is the use of graphics and images to describe complex data information. A reasonable and good visualization can make people have a more intuitive and three-dimensional understanding of data information. Each data item in the database is represented as a single graphic element and constitutes a data image, and the data is integrated, processed and analyzed according to different dimensions (time, space, etc.). The visualization of smart grid big data not only meets the needs of production and operation, but also meets the requirements of external support. Visualization can display the data status of power system production, operation, and operation as a whole and in an all-round way. When there is a special status or a warning status, it can be promptly and quickly discovered by operators and management personnel.

2.2   The Key to System Design

The third and fourth layers are the key modules of the big data visualization system. The key point of the third layer is the algorithm. This article does not use a single algorithm to apply to all modules, but uses the optimal and most suitable algorithm for this data module based on the conclusions drawn from the characteristics and needs of a certain data module. The algorithms to be used in this article include Hadoop, MapReduce, whole-process data processing, big data causal analysis algorithms, self-recommended adaptive full-life data, data set technology and hybrid computing technology. The key point of the fourth layer is to prepare to introduce advanced visualization technologies at home and abroad, including the latest network visualization, spatiotemporal data visualization, multi-dimensional data visualization, and WebGIS visualization (as shown in Fig. 2), and use these advanced technologies to build a visualization platform.

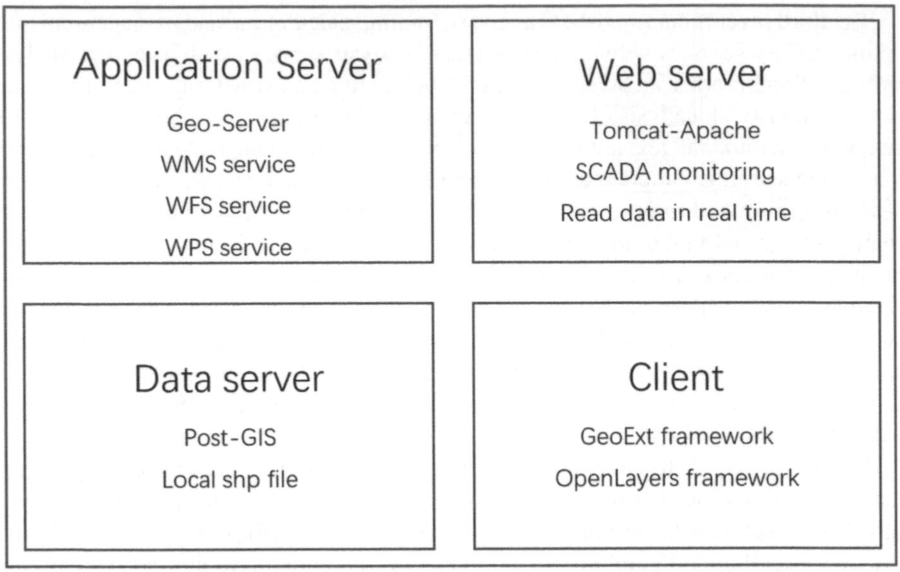

Application Server	Web server
Geo-Server WMS service WFS service WPS service	Tomcat-Apache SCADA monitoring Read data in real time
Data server	Client
Post-GIS Local shp file	GeoExt framework OpenLayers framework

Fig. 2. WebGIS visualization architecture diagram

2.3 Realization of System Function Module

The system has built five application function modules, including statistical analysis center, trend warning center, intelligent search center, panoramic image center, and visualization display center. The smart grid visualization platform is mainly based on the overall perspective, using big data technology architecture to carry out the overall construction, and to accommodate the grid status data. The content involved includes various data collections that appear in the process of power grid operation, maintenance and energy collection.

The massive data and specific cloud computing models provided by the smart grid big data information platform can provide more targeted guidance for the operation and development of the smart grid to a certain extent. As a result, the realization of the smart grid visualization platform based on the big data architecture can become an important field of future development. The existence of big data technology can not only implement advanced applications from the perspective of the field of intelligent scheduling, but also solve the problems in state detection and conduct a comprehensive analysis of power consumption. The functional system of the big data visualization monitoring system is shown in Fig. 3.

In the report technology, we use a Python-based multi-dimensional report platform, the main types of functions are: Overall template design: it can be selected from the existing template library, or can be customized according to needs; Statistical chart type selection: 6 types of statistical chart forms including line chart, scatter chart, and histogram are provided, which are conducive to the intuitive display of data; Chart

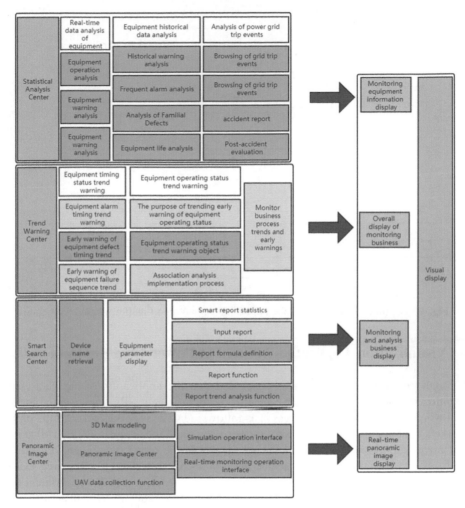

Fig. 3. The functional system of the big data visualization monitoring system

parameter setting: diversified operations can be performed, such as importing files and setting coordinates axis, add legend, add notes, etc. In the module composition of the automatic chart generation system, there are two major modules: template setting and chart generation, which cooperate with each other to support the operation of the platform [9] (Fig. 4).

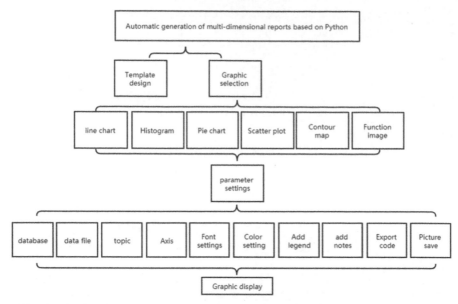

Fig. 4. The functional structure of a multi-dimensional report platform based on Python

3 Visualization Application Method of Power Grid Big Data

Through practical application, it is concluded that the application of big data visualization technology analysis in power grid big data is generally implemented in the following process. The main process is as follows: (1) The user puts forward the problems encountered in actual work, and clarifies the goal of the analysis. (2) By collecting and investigating the possible influencing factors of the target (equipment reliability, grid risk, etc.), analyze the data source and obtain relevant data. (3) Research factor classification attributes (such as time series, space, static, etc.). (4) Choose different big data visualization techniques for different types of factors (such as basic diagrams, network diagrams, tree diagrams, multidimensional diagrams, geographic diagrams, etc.). Variables of the same type can be put together for multi-dimensional analysis to realize the analysis of the degree of influence of potential factors on the target. (5) Through the feedback of the visualization results, continuously improve or replace the visualization technology to make the potential relationship or characteristics more obvious [4].

4 Prospect

Data visualization can show the potential connections between numbers more clearly. Through data mining and summarization of the massive data obtained by calculation, the essential connections within the data can be discovered and indirect indicators that can accurately represent the state of the system can be obtained. Finally, visualize it in the correct way. It can present a panoramic view of the development of the power grid system, thereby presenting the direction of changes in electricity-side data and economic development, and embodying the important role of the power industry in social and economic development.

Acknowledgements. This work is supported by the State Grid Corporation Science and Technology Project Funded "Key technology and product design research and development of power grid data pocket book" (1400-202040410A-0-0-00).

References

1. Keim, D., Konlhammer, J., Ellis, G., Mansmann, F.: Mastering the information age: solving problems with visual analytics. Goslar: Eruographics Association, pp. 1–168 (2010)
2. Wang, W., Hao, P., Song, L.: Application of big data visualization monitoring system in power grid centralized control operation and maintenance. Rural Power Gasification **10**(89), 39–40+59 (2021)
3. Shen, G., Li, L., Di, F., et al.: Data integration and visualization display of UHV power grid dispatching automation system. Autom. Electric Power Syst. **32**(23), 94–97 (2009)
4. Pan, Y., Hu, J., Zhu, Y.: Application research of WEB visualization technology in grid big data scenarios. Electric Power Big Data **21**(445), 8–12 (2019)
5. Leng, X., Chen, G., Bai, J., Zhang, J.: Overall design of big data analysis system for smart grid monitoring operation. Autom. Electric Power Syst. **42**(12), 22–25 (2018)
6. Yang, Y.: Application prospects of data visualization in operation monitoring. Smart Grid **8**(5), 457–464 (2018)
7. Ren, L., Du, Y., Ma, S., Zhang, X., Dai, G.: Overview of big data visual analysis. J. Softw. **25**(9), 1909 1936 (2014)
8. Wang, H., Zhou, Y., Zuo, C., Liu, Z.: Three-dimensional intelligent virtual operation inspection system for substation. **32**(4), 73–78 (2017)
9. Xin, H.: Research on automatic generation of library business report based on python. Comput. Knowl. Technol. **27**, 72–74 (2016)

Vehicle Collision Prediction Model on the Internet of Vehicles

Shenghua Qian$^{(\boxtimes)}$

Tianjin University of Finance and Economics Pearl River College, Tianjin 301811, China
qsh0709@163.com

Abstract. An active collision prediction model on the Internet of Vehicles is proposed. Through big data calculation on the cloud computing platform, the model predicts whether the vehicles may collide and the time of the collision, so the server actively sends warning signals to the vehicles that may collide. Firstly, the vehicle collision prediction model preprocesses the data set, and then constructs a new feature set through feature engineering. For the imbalance of the data set, which affects predictive results, SMOTE algorithm is proposed to generate new samples. Then, the LightGBM algorithm optimized by Bayesian parameters is used to predict the vehicle collision state. Finally, for the problem of low accuracy in predicting the collision time, the time prediction is transformed into a classification problem, and the Bayesian optimization K-means algorithm is used to predict the vehicle collision time. The experimental results prove that the vehicle collision prediction model proposed in this paper has better results.

Keywords: Vehicle collision prediction · Unbalanced data · SMOTE · LightGBM · K-means

1 Introduction

The safe driving of vehicles has always been an important research direction in the field of transportation. There are about 8 million traffic accidents every year, causing about 7 million injuries and about 1.3 million deaths. Traffic problems cause the global domestic productivity to drop by 2% [1, 2]. The annual cost of personal automobile transportation (excluding commercial and public transportation) in the United States is about 3 trillion US dollars, of which 40% of the cost comes from parking, vehicle collisions, etc. [2, 3]. The research on vehicle collision prediction is a important topic in the field of traffic safety.

Traditional vehicle collision prediction mainly relies on the equipment carried by the vehicle itself, generally including millimeter wave radar, sensors, and cameras. These equipment are used to perceive and recognize objects around the vehicle. Collect the information of surrounding objects for input, rely on its own algorithm to calculate, thereby judging whether the vehicle is in an emergency state [4]. The traditional method is based on the information collected by the single vehicle itself for early warning, which

© The Author(s) 2022
Z. Qian et al. (Eds.): WCNA 2021, LNEE 942, pp. 518–530, 2022.
https://doi.org/10.1007/978-981-19-2456-9_53

has certain limitations. In bad weather or harsh environmental conditions, the vehicle-mounted sensor may have errors in the collected information or errors that deviate from the real situation. These deviations are often unacceptable in real traffic scenarios.

The Internet of Vehicles provides a new direction for the development of automotive technology by integrating global positioning system technology, vehicle-to-vehicle communication technology, wireless communication and remote sensing technology [5]. At present, some scholars have conducted research on vehicle collision prediction based on the Internet of Vehicles. Gumaste et al. [6] used V2V (vehicle-to-vehicle) technology and GPS positioning technology to predict the potential collision position of the vehicle, generate the vehicle collision area, and design the vehicle collision avoidance system to control the movement of the vehicle to avoid collision. Sengupta et al. [7] proposed a cooperative collision avoidance system based on the acquired pose information of their own vehicle and neighboring vehicles, which used the collision time and collision distance to determine whether a collision occurred. Yang Lan et al. [8] constructed a highway collision warning model based on a vehicle-road collaboration environment. The simulation results show that the model can effectively warn the occurrence of rear-end collision and side collision accidents. X.H.XIANG et al. [9] use DSRC (Dedicated Short Range Communication) technology, based on the neural network, established a collision prediction model to solve the problem of high false alarm rate in the rear-end collision system and invalid early warning in emergency situations. C.M.HUANG et al. [10] proposed an ACCW (advanced vehicle collision warning) algorithm to correct the errors caused by speed and direction changes. The results show that ACCW algorithm has a higher early warning accuracy rate at intersections and curved roads.

By analyzing the existing vehicle collision prediction model, we proposed an active collision prediction model based on the Internet of Vehicles, using the algorithm combined with SMOTE (Synthetic Minority Oversampling Technique) and LightGBM (A Highly Efficient Gradient Boosting Decision Tree), and using big data calculations on the cloud computing platform to predict whether the vehicles may collide and the collision time. If a collision is predicted, proactively send an early warning signal to vehicles that may have a collision.

2   Background

2.1   Internet of Vehicles Platform Architecture

The Internet of Vehicles platform [11] mainly includes OBU(onboard unit) and mobile communication network. Vehicles are required to have the ability to broadcast and receive V2N (Vehicle to Network) messages, that is, the vehicles communicates with the cloud computing server, as shown in Fig. 1.

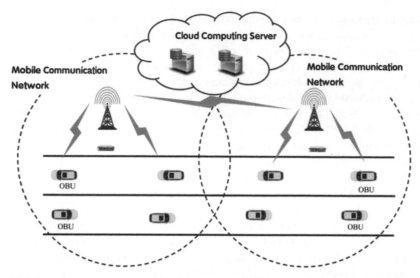

Fig. 1. Schematic diagram of communication network based on Internet of Vehicle

The OBU is carried by the vehicle and is equipped with a mobile communication network interface. The communication network base station can ensure a wide range of network coverage and ensure the communication between the vehicle and the cloud computing server. At the same time, the vehicle-mounted OBU can be connected to the surrounding vehicles that also carry the OBU. Each vehicle-mounted OBU has a unique electronic tag, and the vehicle can receive early warning information directly. The vehicle information will be uploaded to the database module of the cloud computing server in real time, and the data will be processed and calculated. The processed information will be fed back to the vehicle in real time.

2.2 Task Description

On the cloud computing server, real-time information of a large number of vehicles is obtained through the Internet of Vehicles, to identify whether the vehicle has a collision, and to predict the time of the collision. Therefore, the prediction model is divided into two layers, the first layer is to predict the state of vehicle collision, and the second layer model performs accurate time prediction of vehicle collision on the basis of the first layer.

The vehicle prediction model in our research mainly predicts vehicle collision state and collision time via a large amount of vehicle information obtained from the cloud computing server, and then verifies the proposed model. After completing the prediction, transmitting the signal to the vehicle in advance through the communication network for warning, which will no longer be the main focus of our research.

3   Methodology

3.1   Sampling

The problem of category imbalance often leads to large deviations in the model training results. Therefore, for the case where the number of samples in the positive and negative categories is relatively large, sampling techniques are generally used to add or delete the original data to build a new data set. Doing so can make the training results of the model more stable.

SMOTE. The SMOTE algorithm [12, 13] is to generate new samples by random linear interpolation between the minority samples and its neighbors to achieve the purpose of balancing the data set. The principle of the algorithm is as follows:

1) For each minority sample $X_i (i = 1,2,3,...,n)$, calculate the nearest neighbor M minority samples $(Y_1, Y_2, Y_3,..., Y_m)$ according to the Euclidean distance.
2) Several samples are randomly selected from the M nearest neighbor samples, and random linear interpolation is performed between each selected sample Y_j and the original sample X_i to generate a new sample S_{new}. The interpolation method is shown in Eq. (1), where rand (0,1) is expressed as a random number in the interval (0,1).

$$S_{new} = X_i + \text{rand}(0, 1) * (Y_j - X_i) \tag{1}$$

3) Add the newly generated samples to the original data set.

The SMOTE algorithm is an improved method of random oversampling, it is simple and effective, and avoids the problem of over-fitting.

3.2   LightGBM

LightGBM [14, 15] is a framework of GBDT (Gradient Boosting Decision Tree) based on decision tree algorithm. Compared with XGBoost (eXtreme Gradient Boosting) algorithm, it is faster and has lower memory usage.

An optimization of LightGBM based on Histogram, which is a decision tree algorithm, is to discretize continuous eigenvalues into K values and form a histogram with a width of K. When traversing the samples, the discrete value is used as an index to accumulate statistics in the histogram, and then the discrete value in the histogram is traversed to find the optimal split point.

Another optimization of LightGBM is to adopt a leaf-wise decision tree method with depth limitation. Different from the level-wise decision tree method, the leaf-wise method finds the leaf with the largest split gain from all the current leaves and then splits it, which can effectively improve the accuracy, while adding the maximum depth limit to prevent over-fitting.

The principle of LightGBM algorithm is to use the steepest descent method to take the value of the negative gradient of the loss function in the current model as the approximate value of the residual, and then fit a regression tree. After multiple rounds of iteration, the results of all regression trees are finally accumulated to get the final result. Different from the node splitting algorithm of GBDT and XGBoost, the feature is divided into buckets to construct a histogram and then the node splitting calculated. For each leaf node of the current model, it is necessary to traverse all the features to find the feature with the largest gain and its division value, so as to split the leaf node. The steps of node splitting are as follows:

1) Discrete feature value, divide the feature value of all samples into a certain *bin*.
2) A histogram is constructed for each feature, and the histogram stores the sum of the gradient of the samples in each *bin* and the number of samples.

Traverse all *bin*s, take the current *bin* as the split point, and accumulate the gradient sum S_L from the *bin* on the left to the current *bin* and the number of samples n_L. According to the total gradient sum S_p on the parent node and the total number of samples n_p, by using the histogram to make the difference, the gradient sum S_R of all *bin*s on the right and the number of samples n_R are obtained. As Eq. (2) calculate the gain value, take the maximum gain value in the traversal process, and take the feature and the feature value of *bin* at this time as the feature of node splitting and the value of the split feature.

$$gain = \frac{S_L^2}{n_L} + \frac{S_R^2}{n_R} - \frac{S_P^2}{n_p} \tag{2}$$

3.3 Prediction Model

Firstly, the predictive model in this paper preprocesses the data set, secondly, extracts features to build the training set, and then generates new samples through SMOTE algorithm, and adds them to the original training set to balance the data set, after that uses LightGBM algorithm on the new training set to train according to the features constructed by feature engineering, and finally establish SMOTE-LightGBM predictive model.

The prediction modeling process is shown in Fig. 2, and the specific implementation process is as follows:

1) Input data set D, and preprocess the data set, including clearing vacant values, deleting invalid data, and processing abnormal values to form a new data set D_1.
2) Feature engineering 1 selects new features to form a new data set D_2.
3) Apply SMOTE algorithm to the data set D_2 to synthesize new minority samples, and add them to the original data set to form a new data set D_3.
4) The LightGBM algorithm is used to train the new data set D_3, and the Bayesian algorithm is used to determine the best parameter combination for model optimization, and obtain the prediction model of the vehicle state.
5) In order to better complete the prediction of vehicle collision time in Feature Engineering 2 have revised the features from Feature Engineering 1, and the prediction of collision time is mainly for the collision vehicles, so the features of the collision vehicles form a new data set D4. The K-means algorithm is used to predict the collision time of the collision vehicle, and the final prediction model is obtained.

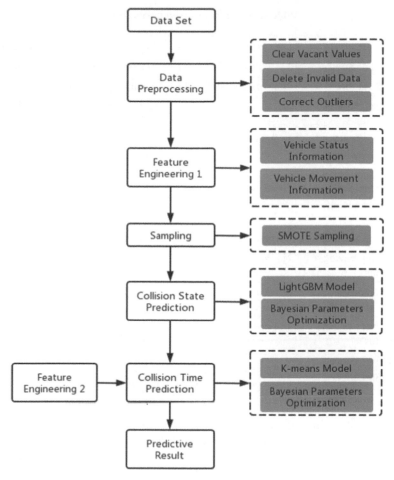

Fig. 2. Predictive model process

6) Test with the test set to verify the effect of the prediction model.

4   Experiments

4.1   Data Set

The data used in the predictive model comes from Internet of Vehicles of a Chinese automobile company. The data mainly includes vehicle state information and vehicle movement information. Each CSV file corresponds to a vehicle. The following Table 1 gives specific information of the vehicle.

Table 1. Vehicle information data format.

No.	Feature	Data example	No.	Feature	Data example
1	Vehicle number	1	11	Handbrake status	Handbrake up
2	Collect time	2020/8/30 6:59:14	12	Vehicle key status	Off
3	Accelerator pedal position	0	13	Low-voltage battery voltage	12.55
4	Battery pack negative relay status	Disconnect	14	Current vehicle gear status	Neutral gear
5	Battery pack positive relay status	Disconnect	15	Vehicle total current	0
6	Brake pedal status	No pedal	16	Vehicle total voltage	114.4
7	Driver leaving prompt	No Warning	17	Vehicle mileage	6738
8	Main driver's seat occupation status	Someone	18	Vehicle speed	0
9	Driver seat belt status	Not tied	19	Steering wheel angle	1.438
10	Driver demand torque value	0			

The data set is divided into training data and testing data: There are 120 CSV files for training data, each file contains 2–5 days of data, and the total number of data for each file is between 4324 and 114460. There are 90 CSV files for testing data, each file contains 1–4 days of data, and the total number of data for each file is between 3195 and 116899.

The data set has a label CSV file, which is a label file for collision prediction. "Vehicle number" is the vehicle number corresponding to the previous data file, "Label" column is the label information corresponding to the vehicle (1 means collision, 0 means no collision), and "Collect Time" column is the time when the vehicle collision occurred. The following Table 2 gives the label file format.

Table 2. Label file format.

Vehicle number	Label	Collect time
1	1	2020/8/30 21:36
2	0	
3	1	2020/8/12 8:36
4	0	
5	1	2021/1/6 16:24
…	…	…

The training data is trained with the previous data and label data, the test data is used to predict whether the test vehicle will collide, and the time of the collision, and the data in the test set labels is used for evaluation.

4.2 Data Preprocessing and Feature Engineering

First of all, the missing data, data redundancy, and abnormal data values are processed. The data is sorted according to "collect time", and then the preprocessed data extracts features.

Feature engineering 1, which is for vehicle collision state prediction, is mainly considered from two aspects: vehicle state information and movement information. The following Fig. 3 gives the operation of feature engineering 1 in predictive model process.

For the state information, the features such as "battery pack negative relay status", "brake pedal status", "main driver's seat occupancy status", "driver demand torque value", "handbrake status", "vehicle key status", "vehicle total current" and "vehicle total voltage" are selected. The most important is the construction of new features "if_off" and "if_on" in the start-stop state. When the relay changes from connection to disconnection, if_off gradually changes from −5 to −1, the rest of the time is 0.when the relay changes from disconnection to connection, if_on gradually changes from −1 to −5, and the rest of the time is 0.

For the vehicle motion information, three features such as "accelerator pedal position", "steering wheel angle" and "vehicle speed" are selected. The features such as "instantaneous acceleration", "local acceleration" and "speed difference" are newly constructed. Several important features like "accelerator pedal position", "vehicle speed" and "speed difference" are carried out for data bucketing. These new features have a strong correlation with collision labels, making subsequent sampling and model construction easier.

Feature Engineering 2 is to predict the time of vehicle collision, which construct the features "current instantaneous acceleration", "next instantaneous acceleration", "collision judgment", and "main driver's seat occupation status".

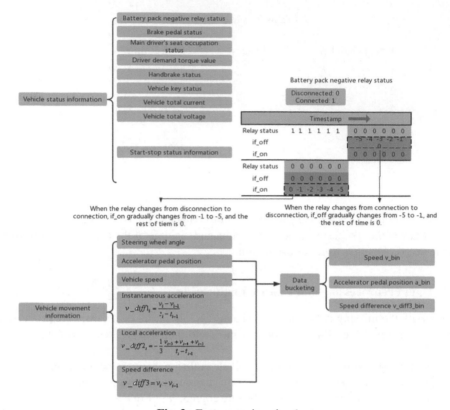

Fig. 3. Feature engineering 1

In order to convert the time prediction into a two-class model, add "time_label", and mark the time if_off $= -5$ in the data set label as 1, and the other time labels as 0. The following Fig. 4 gives the operation of feature engineering 2 in predictive model process.

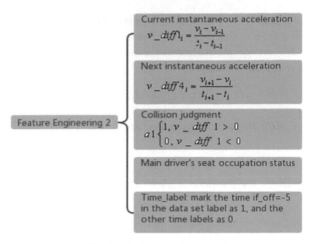

Fig. 4. Feature engineering 2

4.3 Sampling

Because the positive and negative samples of the vehicle collision label data are extremely unbalanced. Therefore, the SMOTE algorithm is used to oversample the small number of negative samples. After sampling, the number of positive and negative samples of the data is close, which improves the generalization ability of the model prediction.

4.4 Model Evaluation Index

Classification Evaluation. For the evaluation of the vehicle collision state results, the four basic indicators of the classification results are used: TP (true positive example), FP (false positive example), TN (true negative example), FN (false negative example). These four basic indicators are mainly used to measure the number of correct and incorrect classifications of positive and negative samples in the prediction results.

Precision represents the proportion of correct predictions by the model among all positive example by predicting, which is shown in Eq. (3).

$$P = \frac{TP}{TP + FP} \tag{3}$$

Recall rate represents the proportion of correct predictions by the model among all real positive example, which is shown in Eq. (4).

$$P = \frac{TP}{TP + FN} \tag{4}$$

F_1 can be regarded as a weighted average of precision P and recall R. Its maximum value is 1, and its minimum value is 0. F_1 is used as the evaluation index to predict the collision classification result, which is shown in Eq. (4).

$$F_1 = \frac{2 \cdot P \cdot R}{P + R} \tag{5}$$

Evaluation of Collision Time Prediction Results. The evaluation standard for predicting the collision time is the absolute difference MAE, which is shown in Eq. (6). Among them, abs is the function to calculate the absolute value, f is the predicted collision time, and y is the real collision time.

$$MAE = abs(f - y) \tag{6}$$

The difference *MAE* has a corresponding relationship with *Score*, as shown in the following Table 3.

Table 3. The corresponding relationship of MAE and Score.

MAE	Score	MAE	Score
0 s	10	Within 2 h	5
Within 10 s	9	Within 3 h	4
Within 1 min	8	Within 4 h	3
Within 10 min	7	Within 5 h	2
Within 1 h	6	Within 6 h	1

F_2 is the evaluation standard for evaluating of predicting collision time, which is shown in Eq. (7). Among them, *sum* is the function to calculate the sum value.

$$F_2 = \frac{sum(score)}{(\text{total number of samples}) \cdot 10} \tag{7}$$

Final Evaluation. The standard for comprehensive evaluation of vehicle collision state and collision time is Eq. (8).

$$F = \frac{F_1 + F_2}{2} \tag{8}$$

4.5   Experimental Results and Analysis

The experiment process is implemented using python, using a five-fold cross-validation method, and the final results Are averaged. In the prediction model in Fig. 1, after preprocessing and feature engineering of the data set, firstly, GBDT, XGBoost, and LightGBM algorithms are verified, and then LightGBM algorithm after SMOTE sampling operations is compared. These algorithms mainly predict the collision state of vehicles, and take the earliest time of the predicted collision as the result, and obtain the values of F_1, F_2 and F respectively. The experimental results are shown in the following Table 4.

Table 4. Experimental results.

Algorithm	F_1	F_2	F
GBDT	0.952	0.854	0.903
XGBoost	0.951	0.856	0.904
LightGBM	0.958	0.886	0.922
SMOTE + LightGBM	0.975	0.912	0.944
SMOTE + LightGBM + K-means	0.975	0.972	0.974

A single LightGBM model is better than other models in results, and the LightGBM model that uses sampling technology has three indicators higher than other models. The model results are the best, but it can also be seen that the prediction results for the vehicle collision time are not very good.

Since the prediction result of the vehicle collision time is not ideal, refer to the prediction model in Fig. 1, after predicting the vehicle collision state, perform feature engineering 2 again, convert the time prediction into a two-class model, and use the K-means algorithm to predict the collision time. The experimental results in Table 4 show that the best experimental results are obtained by using sampling, LightGBM algorithm to predict collision status, and K-means algorithm to predict collision time.

5 Conclusion

The vehicle collision prediction model is proposed in this paper, data preprocessing improves the data quality; sampling improves the accuracy of collision label prediction; Feature engineering and the LightGBM model improve the robustness of the model; the K-nearest neighbor model prediction time improves the collision time prediction accuracy. The running result of the whole model is stable, and the total running time of the data set code is only 60–90 s.

In the next step, we will optimize the model according to the importance of different features, perform more detailed processing of the feature space, and further improve the results of the model. In the current data, the vehicles that have collided are more obvious. Consider more types of collisions, it is necessary to increase the amount of data in the training set and the test set to enhance the generalization ability of the model.

References

1. Litman, T., Doherty, E.: Transportation Cost and Benefit Analysis II–Vehicle Costs. Victoria Transport Policy Institute (VTPI) (2015). http://www.vtpi.org. Accessed 2009
2. Contreras-Castillo, J., Zeadally, S., Guerrero-Ibañez, J.A.: Internet of vehicles: architecture, protocols, and security. IEEE Internet Things J. 5(5), 3701–3709 (2017)
3. Mai, A.: The internet of cars, spawning new business models (2012). https://www.slideshare.net/AndreasMai/12-1024scvgsmaciscoperspectivef

4. Jun, Z.: Study on Driving Behavior Recognition and Risk Assessment Based on Internet of Vehicle Data. University of Science and Technology of China (2020)
5. Xu, W., Zhou, H., Cheng, N., et al.: Internet of vehicles in big data era. IEEE/CAA J. Automatica Sinica **5**(1), 19–35 (2017)
6. Bhumkar, S.P., Deotare, V.V., Babar, R.V.: Intelligent car system for accident prevention using ARM-7. Int. J. Emerging Technol. Adv. Eng. **2**(4), 56–78 (2012)
7. Fallah, Y.P., Khandani, M.K.: Analysis of the coupling of communication network and safety application in cooperative collision warning systems. In: Proceedings of the ACM/IEEE Sixth International Conference on Cyber-Physical Systems, pp. 228–237 (2015)
8. Yang, L., Ma, J., Zhao, X., et al.: A vehicle collision warning model in expressway scenario based on vehicle infrastructure cooperation. J. Highway Transp. Res. Dev. **34**(9), 123–129 (2017)
9. Xiang, X.H., Qin, W.H., Xiang, B.F.: Research on a DSRC-based rear-end collision warning model. IEEE Trans. Intell. Transp. Syst. **15**(3), 1054–1065 (2014)
10. Huang, C.M., Lin, S.Y.: An advanced vehicle collision warning algorithm over the DSRC communication environment: an advanced vehicle collision warning algorithm. Camb. J. Educ. **43**(2), 696–702 (2013)
11. Santa, J., Pereniguez, F., Moragon, A., et al.: Vehicle-to-infrastructure messaging proposal based on CAM/DENM specifications. In: 2013 IFIP Wireless Days (WD). IEEE (2013)
12. Japkowicz, N., Stephen, S.: The class imbalance problem: a systematic study. Intell. Data Anal. **6**(5), 429–449 (2002)
13. Zeng, M., Zou, B., Wei, F., et al.: Effective prediction of three common diseases by combining SMOTE with Tomek links technique for imbalanced medical data. In: 2016 IEEE International Conference of Online Analysis and Computing Science (ICOACS), pp. 225–228. IEEE (2016)
14. Guo, L.K., Qi, M., Finley, T., et al.: LightGBM: a highly efficient gradient boosting decision tree. In: Proceedings of the 2017 Advances in Neural Information Processing Systems. California: NIPS, pp. 3146–3154 (2017)
15. Meng, Q., Ke, G., Wang, T., et al.: A communication-efficient parallel algorithm for decision tree. In: Advances in Neural Information Processing Systems, pp. 1279–1287(2016)

Design of Abnormal Self-identifying Asset Tracker Based on Embedded System

Xianguo Lu, Chenwei Feng(✉), Jiangnan Yuan, and Huazhi Ji

School of Opto-Electronic and Communication Engineering, Xiamen University of Technology, Xiamen, China
chevyphone@163.com

Abstract. This design was aimed at the requirements of the asset tracker's working time, abnormal identification, remote alarm, information prompt, using STM32 as the core MCU for data collection and processing, with a built-in GNSS wireless communication module, three-axis acceleration sensor, and other sensors, design and implement an asset tracker device that automatically recognizes and reports abnormalities. In this design, the GPS positioning information was processed, and the positioning accuracy of the device was improved. The acceleration sensor data was performed by the Kalman filter, which could effectively judge the movement of assets. The sleep-work-sleep work mode was adopted to reduce the device's power consumption and enhance the device's endurance. The test results showed that the device could reasonably identify the device's abnormal condition, quickly locate the device, and upload the device information to the server. Each working life could be applied to the tracking of all kinds of assets.

Keywords: Asset tracker · STM32 · Acceleration sensor · GPS · Kalman filter

1 Introduction

With the development of technology, the location tracking was integrated into our daily life. At present, the positioning tracker in the market has a more miniature asset tracker [1–3]. This tracker was positioned by Wi-Fi, Bluetooth, and GPS, with high positioning accuracy and was generally used to track keys, valuables, and pets. A wearable type real-time location tracker by GSM wireless communication technology [4–6], such trackers generally only used GPS for real-time positioned, with high power consumption, and used the elderly and children for location tracking. Traditional logistics tracked was generally based on warehouse storage for location tracked by online registration [7]. Based on the positioning and tracking function of the above tracker, this design was based on the embedded system [8, 9] and used the STM32 chip as the primary control MCU. A vehicle asset tracker was designed with Internet reminder, intelligent anomaly identification, precise location tracking, convenient disassembly, and use.

© The Author(s) 2022
Z. Qian et al. (Eds.): WCNA 2021, LNEE 942, pp. 531–542, 2022.
https://doi.org/10.1007/978-981-19-2456-9_54

2 The Overall Framework of the System

The research device receives the current environmental information through the peripheral sensor modules, including temperature and humidity information, light-sensing information, GPS information, network signal, and three-axis acceleration information. The data is then processed by MCU and packaged into an asset information package (AIP). The device accesses the Internet through the wireless communication module. Then the MCU packaged AIP was subscribed and published to the MQTT (Message Queuing Telemetry Transport) server through the MQTT. By subscribing to the same topic as MCU, the webserver can receive the AIP published by MCU, then parse, process, and store it. Finally, the device's current location and other related information were displayed on the web map. Figure 1 shows the overall block diagram of the system.

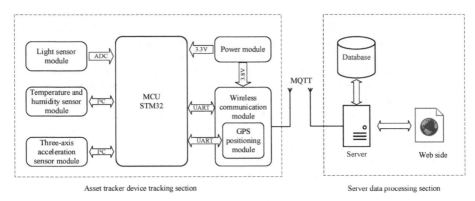

Fig. 1. Overall block diagram of the system

3 System Hardware Module Design

3.1 Overview of System Hardware

The hardware design of this research device was mainly composed of a light intensity sensor, temperature and humidity sensor, three-axis acceleration sensor, wireless communication module, GPS positioning module, power management control module and STM32F105 development board.

STM32 read the temperature and humidity sensor and three-axis acceleration information through I^2C communication mode, the wireless communication module and GPS positioning module communicate and control through UART port and I/O port, the light intensity and battery information were obtained by ADC sampling, and the indicator LED was controlled by I/O port.

3.2 Peripheral Hardware Circuit Design

Temperature and Humidity Sensor Module. This module used an SHTC3 temperature and humidity sensor to detect the temperature and humidity of the environment where the device was located. SHTC3 is a digital humidity and temperature sensor integrated with a complete sensor system. Figure 2 shows the circuit diagram of the temperature and humidity module.

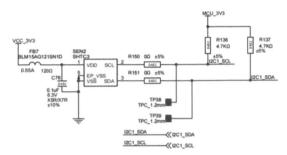

Fig. 2. Temperature and humidity module circuit diagram

Three-Axis Accelerometer Module. This module used LIS3DH three-axis linear accelerometer to collect the current three-axis acceleration information of the device. The LIS3DH has a dynamic user-selectable complete scale of ± 2g/ ± 4g/ ± 8g/ ± 16g, can measure acceleration at an output data rate of 1 Hz to 5.3 kHz, and has 6D/4D direction detection, free-fall detection, and motion detection. Figure 3 shows the circuit diagram of the three-axis accelerometer module.

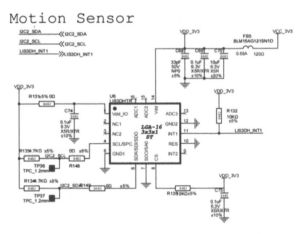

Fig. 3. Three-axis accelerometer module circuit diagram

Wireless Communication Module. This module used the BG95 module to access the wireless network and obtain GPS location information. BG95 is a series of multimode LTE-Cat M1/Cat-NB2/EGPRS modules with an integrated GNSS function developed by Quectel. Figure 4 shows the circuit diagram of the wireless communication module.

Module BG95

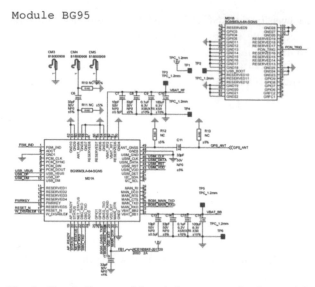

Fig. 4. Circuit diagram of the wireless communication module

4 System Software Design

The main body of the system software design was divided into four parts: system architecture design, sensor data processing algorithm, data transmission control, and web data processing.

4.1 System Architecture Design

The design architecture of the research software was that after the device is powered on, MCU initializes and self-tests each module, obtains the relevant information of the device, and uploads it to the server. After receiving the server's feedback, the device enters a dormant state and continues to monitor the status through each module. When the regular wake-up time arrived, or each module detected an abnormal state of the device, the device was awakened. It then enters the normal tracking process of hibernation-work-hibernation.

The data collected by this research device were detected and processed by the following program modules: light sensing data detection and processing, temperature and humidity data detection and processing, GPS positioning information acquisition, sensor data detection, and processing.

Program Design for Detection and Processing of Light-sensitive Data. The light-sensing data acquisition only needs to collect the current of the I/O port connected by the photosensitive sensor then compare it with the light characteristic curve of the sensor. The luminance of the current environment can be obtained.

STM32 collected the current of the light-sensitive sensor many times and calculated the average value i_{ls}. According to the optical characteristic curve, a light-sensitive abnormal threshold was i_{abn}. When $i_{ls} > i_{abn}$, the light perception is abnormal; otherwise, it is normal.

Program Design for Temperature and Humidity Data Detection and Processing.
The temperature and humidity data acquisition was written to the reading address of the SHTC3 device by MCU, and collected many times, and calculated that the average values of the current ambient temperature and humidity data were Tcur and Hcur, respectively, and determined the standard temperature threshold Tmin, Tmax and humidity threshold Hmax. When Tmin $\leq$ Tcur $\leq$ Tmax, the current ambient temperature is normal; otherwise, it is abnormal; when Hcur $\leq$ Hmax, the current ambient humidity is normal, and vice versa.

Program Design for Obtaining GPS Location Information. GPS positioning information was based on the BG95 module for transceiver and collection. Suppose N pieces of GPS information are obtained, each GPS information is expressed as $B_i = \{Lat_i, Lon_i\}$, $i = 1,..., N$, where Lat_i and Lon_i are latitude and longitude, respectively. At this time, taking B_1 as the initial point, the distance d_i between each point and point B_1 is calculated according to Eq. (1).

$$\text{haversin}(\frac{d}{R}) = \text{haversin}(Lat_2 - Lat_1) + \cos(Lat_1)\cos(Lat_2)\text{haversin}(|Lon_2 - Lon_1|)$$

(1)

where R is the radius of the earth, the average value is 6371 km, d is the distance between two positions, and *haversine* is Eq. (2),

$$\text{haversin}(\theta) = \sin^2(\frac{\theta}{2}) = \frac{1 - \cos(\theta)}{2}$$

(2)

Figure 5 shows the block diagram of the GPS information processing algorithm, where d_{fen} is the radius of the fence with B_1 as the center. The distance d_i relative to the B_1 point is calculated by Eqs. (1) and (2). Then through the comparison of d_i and d_{fen}, we can get the number n_s and n_m of the above location information inside and outside the fence. η is a static factor. By comparing the magnitude of n_s and $N*\eta$, we can judge whether the device is in a static state or a moving state.

$$C = \frac{\sum_{i=0}^{n_s} S_{n_s}}{n_s}$$

(3)

When the device is in a static state, the current position coordinate D_0 of the device can be calculated by Eq. (3). All the current position information is linearly fitted when the device moves and the linear equation $y = ax + b$ with B_1 as the coordinate origin is obtained. Then H_1, $H_{nm/2}$, and H_{nm} are substituted into the linear equation, and the position information D_1, D_2, and D_3 are obtained.

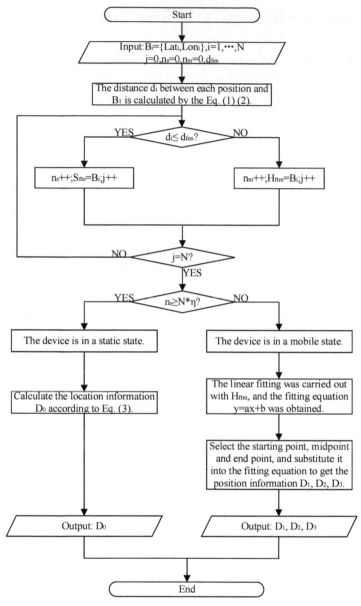

Fig. 5. Program block diagram of GPS information processing algorithm

4.2 Program Design for Acceleration Sensor Data Processing

In this study, Sensor data read three-axis data through the I^2C communication module. In order to be compatible with the characteristics of portable disassembly and assembly and at the same time achieve simple and effective judgment and recognition, the average acceleration a_{ave} was used to reduce the complexity of the three-axis vector operation.

In order to filter out the occasional acceleration fluctuation, the average acceleration state X_t was processed by Kalman filter [10]:

$$\begin{cases} \hat{X}_t^- = A\hat{X}_{t-1} + Bu_{t-1} \\ Z_t = H_t X_t + V_t \end{cases} \tag{4}$$

The formula: H_t is the unit matrix, V_t is the measurement noise with mean 0 and variance R, u_{t-1} is discrete white noise with mean 0 and variance Q, $\hat{X}_t^-$ is the a priori estimation of time, Z_t is the measured value of t-time.

From Eq. (4),

$$P_t^- = AP_{t-1}A^T + Q \tag{5}$$

$$K_t = P_t^- H^T (HP_t H^T + R)^{-1} \tag{6}$$

$$\hat{X}_t = \hat{X}_t^- + K_t(Z_t + H\hat{X}_t^-) \tag{7}$$

$$P_t = (I - K_t H)P_t^- \tag{8}$$

The formula: $\hat{X}_t$ is a posteriori estimate of t-time, P_t is a posteriori variance, P_t^- is a priori variance, K_t is Kalman gain of t-time.

Through the analysis and processing of the posterior estimated value $\hat{X}_t$, we can accurately judge whether the device is abnormal or not.

4.3 Program Design for Data Transmission Control

Data transmission was mainly based on the connection between the device and the server through the BG95 communication module, and the BG95 communication module connects to the network through 4G communication. The device SN number and IMEI number were used as the unique identification for the server to distinguish and register the device. Figure 6 shows the block diagram of the data transfer program.

4.4 Program Design for Web Data Processing

Web-side data processing was mainly operated by the *webServlet* class. Since the messages forwarded by the back-end server were mainly POST operations, the *doPost()* method was used in this class. The front-end web page used a JSP page and set up a form to determine that the parameter *sleepTime* that needed to be passed could be entered on the web page and then transferred to the background using *submit()*.

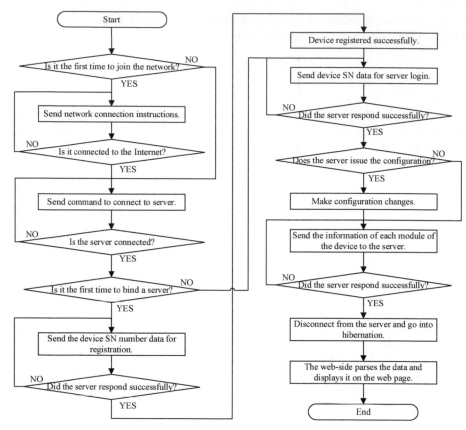

Fig. 6. Block diagram of the data transmission program

In data processing at the back end, the *Frameheader* of the data string was used to determine the information category, verify it, and separate the registration, login, device information, logout, and other information categories. Figure 7 shows the block diagram of the web-side data processing flow.

5   System Testing

Through the tested of the device, after the device was powered on, it enters the work cycle of self-tested, uploaded data—dormant—awaken—self-tested, and uploaded data. Figure 8 shows the simulation results of GPS information processing. It could be seen that the processed positioned coordinates coincide with the actual coordinates. That was, the research device could read the positioned information more accurately.

The Kalman filter processed the data collected by the acceleration sensor. Figure 9 shows the results of sensor data simulation. It could be seen that the filtered data could filter out most of the acceleration fluctuations, which was convenient for the device to identify the abnormal conditions.

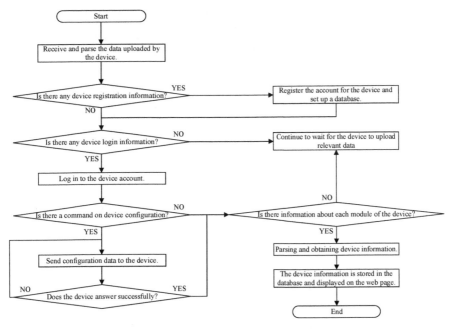

Fig. 7. Data processing flow chart on web-side

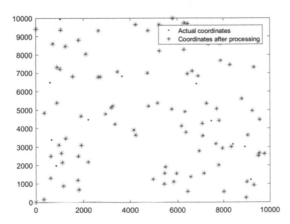

Fig. 8. Result of GPS simulation

The data relating to the device's working time and timing wake-up time was obtained through the power consumption test. Table 1 shows the battery test data. The working days of the device in this study were proportional to the wake-up time interval, and the maximum working time could be up to 170 days.

Through the above tests, this research device could be applied to all kinds of asset tracking.

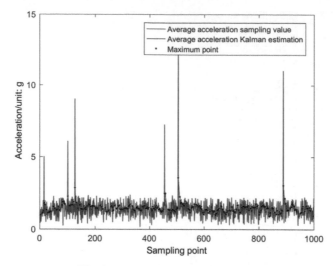

Fig. 9. Sensor data simulation results

Table 1. Battery test data

Wake-up time interval/h	Number of wake-ups per day/times	Device working days/days
1	24	≥24
2	12	≥42
6	4	≥86
12	2	≥116
24	1	≥140
Only dormancy	0	≥170

6 Conclusion

Integrating the specificity, scalability, reliability, and power consumption of embedded systems, STM32 was used as the data processing core MCU, and other functional modules were used to design and implement an asset tracker device that automatically recognizes and reports abnormalities.

The main advantage of this design was that the device could automatically identify according to the surrounding environment and movement of the asset, display the relevant data on the web page, and support users to remotely modify the dormancy time of the device according to the situation—the data processing method of the accelerometer provides convenience for device installation. The GPS information processing method improves positioning accuracy without Wi-Fi and Bluetooth assistance. Acceleration sensor data and GPS information processing methods are not complex; STM32 could carry out related processing. The ultra-long life span enables the device to be used in all kinds of asset tracking.

Acknowledgement. This work was supported by National Science Foundation of China (Grant No. 61801412), High-level Talent Project of Xiamen University of Technology (Grant No. YKJ17021R, and No. YKJ20013R), Scientific Research Climbing Project of Xiamen University of Technology (Grant No. XPDKT19006), and Education and Scientific Research of Young Teacher of Fujian province (Grant No. JAT190677, No. JAT200471, and No. JAT200479).

References

1. Hyunsung, K., et al.: Design of a low-power BLE5-based wearable device for tracking movements of football players. In: 2019 International SoC Design Conference (ISOCC), pp. 11–12. IEEE, Jeju (2019)
2. Bauyrzhan, K., Muhammad Fahad, F., Rana Muhammad, B., Atif, S.: A Wi-Fi tracking device printed directly on textile for wearable electronics applications. In: 2016 IEEE MTT-S International Microwave Symposium (IMS), pp. 1–4. IEEE, San Francisco (2016)
3. Hannah, A.S.A., Francis, K.O., Tan, S., George, K.A., Manasah, M.: Developing a bluetooth based tracking system for tracking devices using arduino. In: 2020 5th International Conference on Computing, Communication and Security (ICCCS), pp. 1–5. IEEE, Patna (2020)
4. Hind Abdalsalam, A.D.: Design and implementation of an accurate real time GPS tracking system. In: The Third International Conference on e-Technologies and Networks for Development (ICeND2014), pp. 183–188. IEEE, Bcirut (2014)
5. Fatima Nadhim, A., Ziad Saeed, M., Abdulrahman Ikram, S.: An economic tracking scheme for GPS-GSM based moving object tracking system. In: 2018 2nd International Conference for Engineering, Technology and Sciences of Al-Kitab (ICETS), pp. 28–32. IEEE, Karkuk (2018)
6. Padmanabhan, R., Pavithran, R., Shanawaz Mohammad, R., Bhuvaneshwari, P.T.V.: Real time implementation of hybrid personal tracking system for anomaly detection. In: 2016 Eighth International Conference on Advanced Computing (ICoAC), pp. 93–98. IEEE, Chennai (2017)
7. Lei, Y., Long-qing, Z.: Logistics tracking management system based on wireless sensor network. In: 2018 14th International Conference on Computational Intelligence and Security (CIS), pp. 473–475. IEEE, Hangzhou (2018)
8. Hao, T., Jian, S., Kai, L.: A smart low-consumption IoT framework for location tracking and its real application. In: 2016 6th International Conference on Electronics Information and Emergency Communication (ICEIEC), pp. 306–309. IEEE, Beijing (2016)
9. Rui, W., Shiyuan, Y.: The design of a rapid prototype platform for ARM based embedded system. IEEE Trans. Consum. Electron. **50**(2), 746–751 (2004)
10. Md Masud, R., Nazia, H., Md Mostafizur, R., Ahmed, A.: Position and velocity estimations of 2D-moving object using kalman filter: literature review. In: 2020 22nd International Conference on Advanced Communication Technology (ICACT), pp. 541–544. IEEE, Phoenix Park (2020)

Emotional Analysis and Application of Business Space Based on Digital Design

Yaying Wang[✉] and Jiahui Dou

College of Landscape Architecture, Beijing University of Agriculture, Beijing , China
aziwyy@126.com

Abstract. As social science and technology progressing, people pay more attention to themselves. Jewelry, whether as a daily design or exquisite art, deeply carries individual feeling. Commercial space design, as an important embodiment of tolerance and foil, could show its value and meet people's emotional needs. Based on jewelry store design, this paper studies the emotional design contained in digital commercial space to enrich the emotional experience in space design. Through the construction and design of jewelry store space, it can better convey the value and emotion of goods, and apply emotional elements to the layout, color and form of digital commercial space, so as to build a digital commercial space full of emotion and design [1].

Keywords: Digital space design · Emotional experience · Woman · Research background

1 Introduction

With the development of economic globalization and the outbreak of the epidemic in early 2020, with social progress and the rapid development of economy and culture, plain commercial exhibitions and sales can no longer meet people's pursuit of beauty and psychological and emotional needs. Therefore, the design of digital stores came into being, which can meet various needs of consumers [2]. For the design of digital commercial space, it is necessary to integrate and reasonably use the digital elements in the layout, framework, color and material of the space, coordinate and integrate the various elements, make the commercial space, goods and consumers operate and display as a whole, and design from the perspective of consumers, so as to make the space meet the emotional needs of consumers [3].

2 Analysis of Concept and Research

2.1 Analysis of Thematic Business Space

Design never comes out of nothing, it needs the people, environment and social background it serves as its cornerstone [4]. Design derives from different geographical environment and cultural background is different. When entering a commercial space, consumers would focus on the commodity itself, while the emotional space design could

© The Author(s) 2022
Z. Qian et al. (Eds.): WCNA 2021, LNEE 942, pp. 543–549, 2022.
https://doi.org/10.1007/978-981-19-2456-9_55

create an appropriate atmosphere, set off the products, and let the consumers entering the space with spiritual resonance and emotional comfort. Emotional design needs to impress customers through design and imperceptibly influence users' cognitive style of beauty. From the perspective of consumers, it could help consumers better understand products and services, which results in good interaction between consumers and enterprises. The emotional expression of digital commercial space is displayed through design, and the emotional experience of consumers is the ultimate goal. The space design for emotional experience is advanced, an important people-oriented way, and the exploration and creation of human emotional needs [5].

2.2 Comparative Analysis of Research

Taking jewelry as an example, data show that jewelry consumers in China are concentrated in middle and low-end jewelry; The ratio of male to female is about 4:6. It can be seen that the jewelry market is gradually diversified, but the main consumer is still women. (See Fig. 1, Fig. 2) Among all the samples collected in the questionnaire survey, the number of women filling in the questionnaire accounts for a large proportion, most of them are post-90s, and the samples are mainly middle-aged and young people. Based on the above, the emotional design of jewelry stores should take groups of mid-low income and age as the main targeted consumers, take female-friendly as the keynote of design, take active guidance and de-gender as the direction of design.

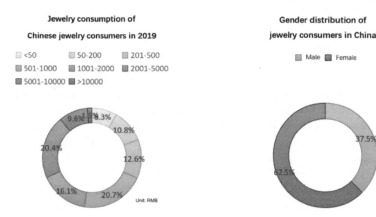

Fig. 1. Consumption amount of jewelry consumers.

Fig. 2. Gender distribution of jewelry consumers

2.3 Design Concepts

Female-Friendly Digital Design of Business Space. Space is not only the carrier of things within the philosophy, but also the intermediary of aesthetics [6]. Everything in the world can be classified as time and space. It is abstract in meaning and thought but actually exists. Architectural space for life is the most common scientific space. The volume, proportion and shape of buildings are the most direct visual science. Space design is gradually developing towards "feminization" [7]. Because the product consumer group is and will be dominated by women for a long time, people need to design the digital space from a female and female friendly perspective.

Women have more detailed requirements for the performance and sensory experience, and pay more attention to emotion and psychology. Due to women's special needs for space, their characteristics must also be reflected in the design. According to the hierarchy theory of needs, human needs from low to high are physiology, security, love and belonging, respect and self realization. The most fundamental is the realization of the desire for physiology and safety. For example, the vision should be wide, the action should be relatively simple and free, the road should be smooth, and so on.

All art must strive for beauty, and its form must be closely centered on its core and function. Current design should focus on "metaphysics", such as art, culture, fashion and style. Both "form obeys function" and "form follows experience" have their own functions [8]. Consumers' demand for space is not only limited by the function of space, but also pursues the experience that space could offer them. The design should be committed to mobilizing consumers' emotions towards life. When necessary, it needs to please women's mood and make them feel comfortable. It also needs to pay attention to the psychology and emotion in their emotional needs, so that women could integrate into the space and get a sense of belonging.

Conceptual Analysis-Spatial Emotional Design. Emotional experience is divided into three: instinct, behavior and reflection. With the development of modern society, corporate culture has changed from material culture to spiritual culture. Commercial space is a place to provide services or products that meet commercial requirements. Now it has evolved into a commercial trade network system that takes the world as a stage [9]. In centralization, commercial space has also changed from dynamic to specific. Because the commercial space is fixed, both parties to the transaction have certain requirements for the commercial space - have certain commercial facilities, and design and create culture and speciality. Space itself has special emotional characteristics, which can stimulate and meet people's emotional needs. This is because the psychological needs of users are everywhere in our life and work. Qualified spatial emotional design can give full play to this function and summarize people's emotions. And psychologically and physiologically, public design can be used to meet the needs of people in a specific space.

1) Instinct takes precedence over one's subjective consciousness and thought. The benchmark of human first impression is instinct. Humanized design will be highly praised by human beings, while instinctive design focuses on the first impression and the beauty of the appearance seen for the first time. Therefore, in order to get a good design that evokes human instincts, we must coordinate and unify external attributes (such as shape, material and color) to conform to the "aesthetic" standards of human beings, to integrate the most real and instinctive experiences into human feelings, to adjust the overall appearance and design, and to find a balance between all contradictions.

2) Behavior is mainly related to the user experience brought by design. Behavior is mainly related to the user experience brought by design - whether the function division included in the experience is scientific, whether the control system is clear and

whether human care is achieved. Good behavior can give consumers a sense of identity and produce pleasant and positive emotions while achieving their expected goals [10]. Interior design that lacks attention to behavior usually has a negative impact on consumers. Easy to use, it is a "considerate" humanized design and exquisite design that pay attention to details. This is the concept of "Empathy" advocated by design and science.

3) Reflection is related to the meaning of goods. It is affected by environment, culture, identity and identity. It is more complex and changes rapidly. The most important thing of reflective design is to help users establish their self-image and social status, so as to meet their emotional needs. Reflection exists in consciousness and higher-level feelings and emotions. Only this level can reflect the complete integration of thought and emotion (Figs. 3 and 4).

Fig. 3. Interactivity,

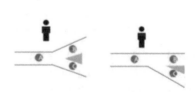

Fig. 4. Selectivity

The Relationship Between Concepts and Emotional Space. The focus of emotional design includes the following two aspects: (1) emotional stimulation and experience generated by the design (2) emotion and experience generated by users under specific use conditions. Resonance with space is the abstract expression of emotion in space design. This resonance of thought and emotion would not directly caused by any specific characteristics [11].

3   Design Schemes

3.1   Derivation of Space

Case design scheme is crystal, one of the main materials of jewelry, that is, the process of crystal development, collection, processing and wearing, and is presented in a narrative way. The design elements come from the NACA crystal cave in Mexico. Crystal usually develops in the harsh environment of high temperature and high pressure, and becomes shining after tens of thousands of years of precipitation. Crystal has been endowed with tenacity, purity and thoroughness from the very beginning. It symbolizes innocence, kindness, purity and unyielding. And the ancient Chinese also believed that crystal was the "Ice of the Millennium" and that crystal was full of energy, which covered the crystal with a sacred veil. When people give these beautiful words to things, they are expecting and confirming the quality of the things they own. From the subject to human beings, we will apply this idea to the theme of design. Comparing people to crystals growing under adversity and full of expectations and worship for the world, and they will eventually shine like crystals after the erosion of time and the challenge of adversity. The form of design will focus on using crystal cluster, crystal cave, mineral deposit and other elements, and a large number of common hexahedral biconical and rhombohedral crystals will be used as blocks in space design [12] (Figs. 5 and 6).

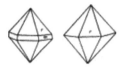

Fig. 5. Hexagonal biconical crystal.   **Fig. 6.** Rhombohedral crystal

It will take the process of people entering the crystal pit, discovering, excavating, cutting and inlaying as the narrative node in space, and show the process of people longing for light and finding treasure in the dark and adversity. Under site selection, the plane of the building would be deduced on the basis of the crystal shape to form a plane building (Figs. 7 and 8).

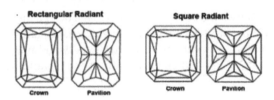

Fig. 7. Radine cutting

Fig. 8. The floor plan

3.2 Analysis of Colors in Spaces

In real life, architectural space is an integral part of personal activities. It must have enough privacy, security and let people comfortable. With the increasing awareness of gender equality, modern architecture reduces the gender differences in space and creates a homogeneous space, which is a breakthrough in balancing the relationship between men and women. Obviously, color is very important for shaping an overall commercial space. It maximizes the aesthetic experience and spiritual satisfaction. There is a certain distance between human vision and the real color of decoration, and the color of objects is usually scattered and colored in a large area. Therefore, the color in the space structure should be very pure and harmonious. The gray shadow can make the small space show a sense of intimacy and warmth. The emotional expression of color makes the commercial space more colorful and meaningful. The specific expression of color in business space can make customers emotional and stimulate their inner feelings. A good commercial space will undoubtedly let people move their sight, because color can quickly and effectively capture people's mind, and the emotional information conveyed by color used in buildings could resonate with more customers.

4   Conclusion

In the study of commercial space design, firstly, the design direction is put forward according to the background, comparing the research at home and abroad, analyzing the design condition and the users, then determining the location according to the users, combining with the design principle of space emotional design theory; Secondly, the theme of digital business design is integrated to describe the design concept of digital business related to jewelry stores. [13]Finally, make the design theme scheme, from the overall layout to local details, fully reflect the characteristics and significance of space emotional design, so that users can resonate emotionally and integrate into the space, which can be emotionally satisfied and fully reflect the significance of spatial digital design.

References

1. Won, Y.-T., Kim, H.-D., Lee, M., Noh, H.-W., Kwak, H.-S.: Emotion based real-time digital space design system for development virtual reality and game. J. Korean Soc. Comput. Game **14**, 14–18 (2008)
2. Scott, B.B.: The epistemic significance of emotional experience. Emotion Rev. **13**(2), 113–124
3. Yi, Q.: Research on Emotional Indoor Light Design. Changchun University of Technology, 201
4. Jiang, S.: Research on the Application of Color and Emotion in Commercial Space Design. Guangxi Normal University (2019)
5. Yang, Y.: Research on Business Space Design Based on Emotional Needs. Jingdezhen Ceramic University (2019)
6. Sun, L.: Research on the Application of Emotional Design Expressions in Office. Southwest Jiaotong University (2013)
7. Guo, Z.: Humanized Design of Female Business Space Based on Emotional Needs. Shandong Normal University (2017)
8. Light flight. Emotional Design of Urban Business Space. Beauty and Times (City Version) (08), pp. 61–62 (2018)

 9. Zhang, F.: Brief Analysis on Expression and Application of Emotional Design in Buildings. Building materials and decoration. (02), 113–114 (2020)
10. Luo, Z.: Humanized design in commercial space display design. Housing **36**, 105–132 (2019)
11. Xue, Y.: Gender-free commonality toilets in commercial buildings. China Real Estate **02**, 64–68 (2018)
12. Kim, Y.-J.: A Study on Digital Space Design Method and System using Virtual Reality. Archives of Design Research (2004)
13. Anne., K.: The Social Environment in Digital Space - DGPPN 2020]. Psychotherapie, Psychosomatik, medizinische Psychologie **70**(11) (2020)

Research on the Design of Community Residential Space from the Perspective of Digitization

Chenglin Gao[✉] and Shuo Tong

School of Landscape Architecture, Beijing University of Agriculture, Beijing, China
chenglingao@126.com

Abstract. The residential architecture in the process of urban digital development has become a living complex with real and virtual mirrors, in which people are the unity of connection between spatial environment, identity and living relationship. In this paper, the new value orientation of community residential design is analyzed by sorting out the meaning of community; within the design system of residential space, the intimacy and public consciousness of residents' neighborhood relationship is enhanced through spatial transition and cultivation of shared living space. The argument is developed from three levels: individual residents' self-reconstruction, residents' new behavioral decisions, and spatial behavioral output. Through a series of argumentation, the relationship between community and residential space planning and design is explored, and the data on the interaction between users, usage behavior and space usage of different households are statistically obtained. At the same time, this paper simulates and designs the community residential space module system based on this data and combined with the computer 3D model derivation. The residential block formed by the combination of the smallest modules, as the smallest residential unit, continues to form the design path of a sustainable residential system through the process of combination and deformation of space.

Keywords: Digital modeling · Community residential space design · Modular space · Lifestyle characteristics · Computer-aided diagnosis

1 The Development of Community Residential Space

1.1 The Evolution of the Connotation of Community

The development of community has its origins in Aristotle's "idea of the city-state community: perfectionism", in which the city-state is a community and all communities are established for a common good. This concept evolved through a series of connotations until the end of the twentieth century, when Western liberal theory emerged as both a reflection of community thought on the problems of real society and an extension of the Western rational cultural tradition, playing an important role in real social problems. Focusing on the value of community, which emphasizes the new vision of conceiving the

© The Author(s) 2022
Z. Qian et al. (Eds.): WCNA 2021, LNEE 942, pp. 550–559, 2022.
https://doi.org/10.1007/978-981-19-2456-9_56

state of complementary and harmonious coexistence between self and other, individual and family, and family and family, is an important ideological resource for enriching lifestyles and promoting neighborhood relations in the design context. Starting from spatial justice, American urban sociologist David Harvey proposes the theory of spatial squeeze, advocating that spatial community is a remedial strategy to safeguard citizens' basic rights and prevent urban spatial risks. Based on the ontology of residential community, it is concluded that any community practice is a spatial presence and invariably shapes the spatial layout of the community At the same time, if the community wants to form a warm and comfortable place in the process of spatial production and reproduction, it can only resort to a spatial effort practice oriented to solidarity and mutual benefit, and this place is also the third domain where the residents' material space and psychological space are transformed.

1.2 The Value of Community Residential Space

Influenced by the idea of community, the function of "connection" of residential space, the way of thinking and decision making of residents have also undergone important changes, which are caused by the increasing awareness of diverse life under the influence of information. Based on this, this paper understands community residential space as "spatial community" and "housing". In this paper, it is interpreted as a group of people living together under the conditions and goals of common residence.

Residential community can be understood through the form of community. The so-called residential community refers to a family group that is established in the same geographical, blood, action and neighborhood internal spatial environment, spontaneously interacts and has a certain sense of sharing; under the same lifestyle, it spontaneously interacts with its neighbors in the residential space and has a certain sense of sharing, thus generating an autonomous, interactive and united interaction relationship. In the design of community housing, considering the emergence of new family structures, it is first necessary to take into account the segmentation of target users, as well as the characteristics that influence the gradual change of modern Chinese family structures into smaller scale, structural nucleation and diversification of types.

1.3 Community Residential Space in Modern Context

In the modern context, the most consistent spatial forms of community residential design in China are the quadrangle dwellings of Beijing and the earth buildings of Fujian. These spatial forms are characterized by the public space as the center and the open entry space and the private living space as the enclosure, so that the public space in the center has a certain natural privacy and people spontaneously interact in it. In foreign countries, the main high-rise public housing in use and in line with the concept and spatial form of community residential space is Singapore, whose design is characterized by the following six points.

First, it has supporting infrastructure, such as transportation system, schools, stores and cleanliness and safety; second, it is planned comprehensively between the completion of the building, divided into three levels of new town, neighborhood and neighborhood; third, it needs to pass through the air street to enter the neighborhood common space; fourth, its design system that allows residents to participate; fifth, its use of apartment layout to deal with height difference, and supporting convenience stores, nursing homes and small plazas to provide a convenient way for the elderly to age in place; sixth, its introduction of eco-neighborhood models and neighborhood parks.

2 Digital Value of Community Residential Space

2.1 Residents' New Perception of Individual Self-reconstruction

With the advent of digitalization and informatization, one of the first results is the reawakening of man's perception of himself, what is the constituent essence of his existence. The current complete understanding includes three aspects, one is the physical person, that is, a real person with a body and weight, and belonging to a specific place at any given time; the second is the information person, who can process the input information in the behavioral environment on the basis of certain cognition and previous experience, and finally form behavioral decisions and output; the third is the cyber person, who lives in the cyber space as a disproportionate incarnation, but whose role is real.In particular, cyberspace has brought certain changes to the social construction of personal identity. Specifically, its transformation of individual life patterns that include beliefs, values and cognitive styles from modernism to postmodernism and the use of these as a symbol to complete the self-proof of human existence has led to an increased need for self-attribution in space.

2.2 New Behavioral Decision-making Model of Residents

The cognitive basis of human behavioral decision pattern represents the cognitive and processing ability of information formed in the brain and varies depending on the spatial environment, culture and family life of the person as the cognitive subject of the objective world. In addition, even with a certain cognitive base, the availability, accuracy and richness of information can produce different behavioral outcomes. The public environment in residential space plays an important role as the main activity place for residents in interaction. If a spatial environment suitable for interaction is built in a house, it is not only important to promote the establishment of good neighborhood relations among users, but also to enhance parent-child relationships. Behavioral information originates from the part of the objective environment that people perceive, i.e., the behavioral environment. Given that the human behavioral decision-making process can be generally described as "need-information search-information processing-behavioral decision selection-behavioral output and behavior", the human behavioral decision-making process is essentially a process of information flow.

2.3 New Types of Behavioral Output for Residents

In this study, the analysis of the output types of residents' behaviors is mainly based on the data statistics of the case study in the user analysis method. First, a representative sample of five households in Beijing was selected for analysis. By conducting in-home interviews and CCTV recording of modern household users, the living behavior and usage time records of modern households were summarized. At the same time, the location plan was recorded by camera (as Fig. 1) then the interactions of users, usage patterns and use space spaces in modern households at different stages were analyzed sequentially to derive the relatively public spaces in modern household living spaces. Finally, by integrating the relatively public space in the residence, a residence with multiple families sharing a common space is established to form a new public shared environment.

Fig. 1. CCTV settings record tracks

A week-long user analysis process was conducted for five representative households, during which the behavioral characteristics of users in their homes, the current status of usage problems, and the characteristics of different users' stage lifestyles were recorded from 7:00 to 22:00 every day, and this was used to derive the design requirements for the future residential space. The following is a description of the specific analysis process for one of the households.

By recording statistics, it can be concluded that the daily demand behavior of different families in the same type of space is as follows.

Based on the user behavior, it can be seen that family communication, parent-child play, and family work are the main events occurring in family interaction, while smart home, acting, and talking to oneself are the events occurring alone in children's lives. Combining the results of the questionnaire, household interviews, and CCTV observations, it can be analyzed that family interaction education and children's free growth are intertwined. Families that do not know each other are more likely to communicate and interact with each other spontaneously using children as a channel and emerge with a sense of sharing, more autonomy in the form of interaction, and more solidarity when problems arise. However, considering the small amount of public space in the existing residential form and the fact that most of the residential space has only access space outside the living space of each household, a community residential space with abundant public space was selected for the main users of two-child families.

Table 1. Cases analysis of family lifestyle characteristics in different time periods

Time	Behavior trajectory analysis from 7:00 A.m. to 22:00 P.m
1.Monday	
2.Tuesday	
3.Wednesday	
4.Thursday	
5.Friday	
6.Saturday	
7.Sunday	

(Legend behavior kind: ■Financial Study ■Interest ■Nature ■Electron ■Dialogue ■Unconscious ■Imagination ■Housework Amusement)

(Legend behavior kind: ■Financial Study ■Interest ■Nature ■Electron ■Dialogue ■Unconscious ■Imagination ■Housework Amusement)

Fig. 2. New types of behavioral output for resident

3   Digital Community Residential Space Design

3.1   Prototype of Spatial Design of Community Residential

In this study, based on the spatial forms of the traditional quadrangle dwellings of Beijing and the earth buildings of Fujian and combined with the modern design of the quadrangle dwellings and earth buildings, the prototype of community residential space is modeled to derive the spatial form of future communitarian residential design.

At the same time, by adjusting the composition of space in modern houses, appropriately reducing the area of public spaces such as kitchen, living room and dining room, a model centered on public space is established. The open entry space and the private living space are enclosed, so that the public space in the center has a certain natural privacy, allowing people to interact spontaneously in it. With regard to the process of building residential interiors, the spatial forms are combined and innovated on the basis of the living space required for the residence, creating a form of residence that guides people to communicate and enhances neighborhood relations.

3.2 Computerized 3D Space Modular Construction

Regarding the modular construction of computerized 3D space, firstly, the basic household area of 120 m^2 was calculated based on GB 50096–2011, which states that the core household of four people should have 30 m^2 of usable area per person. Given that there may be elderly people coming to take care of children at home from time to time, the area of 20 m^2 is increased and decomposed in modules of 1000mm*1000mm, resulting in 140 space modules, and the space modules are given functions to divide the living room, dining room, kitchen, master bedroom, second bedroom, children's room and bathroom. Then, according to the spatial forms and data application of traditional quadrangle dwellings of Beijing, traditional earth buildings, and modern quadrangle dwellings, the spatial forms that can accommodate four households are derived. The public space in each household is integrated to form a new public space in the center, in which a functional space with parent-child activities, reading and learning, viewing greenery, audio and video, and urban viewing platform is established; finally, the spatial system is integrated to leave a 1600mm passage, and the passage is given the functions of entry, stairwell and shared activity platform. Ultimately, the spatial form of community residential space is obtained.

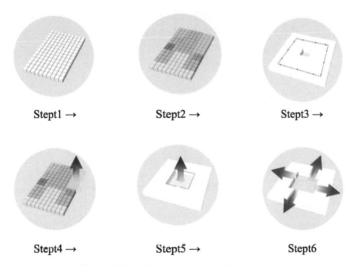

Stept1 →   Stept2 →   Stept3 →

Stept4 →   Stept5 →   Stept6

Fig. 3. Modular space generation process

3.3 New Residential space under the role of digitalization

The simulated residential space under the role of digitization is committed to establishing a unified body of space daily life module, family activity module and neighborhood interaction module, where the space module is divided into four levels: functional space, morphology, combination unit and shared space.

(Classification of spatial function modules: 1. Activity space, 2. Learning space, 3. Experience space, 4. Traffic space, 5. Emotional space, 6. Rest space, 7. Public communication space).

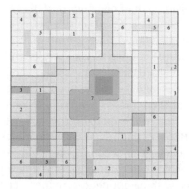

Fig. 4. Modular community residential space composition

The final space design drawing is as follows.

Fig. 5. Digital community residential space plan

Regarding the design of residential space under the role of digitalization, it is necessary to first construct functional components according to the overall demand ratio of space, and then retrieve the confidence of household design resources through the household type resource library; along with the gradual depth of space morphology and standardized drawing design and refine the space allocation into parent-child activities, reading and learning, viewing greenery, audio and video viewing and urban viewing platform functional space and give the wall parent-child interaction and neighborhood communication. In the process of establishing the combination unit, it is necessary to create a shared part set of transition space that needs to link public space, and finally form a complete residential monolithic design scheme.

Communication space1

Communication space2

Learning space

Activity space

Rest space1

Rest space2

Fig. 6. Digital community residential space renderings

4 Conclusions

The new period of social development has given rise to the increasing perfection of digital building technology, which brings more possibilities and ways of realization for residential space. On the basis of sorting out the evolution of the connotation of community and analyzing the new value orientation of community residential design, this paper provides constructive thoughts on the path of constructing a community of residential space through spatial transition and cultivation of shared space to enhance the intimacy and public awareness of residents' neighborhood relationship within the design system of residential space. At the same time, the modular space scheme in the form of community family life is proposed, the unity of daily life module, family activity module and neighborhood interaction module of residential space is formed through the construction of digital space model. The combination of computer 3D modeling and digital space design is used to realize the unification of indoor and outdoor residential environment. According to the residents' behavior, lifestyle and spatial interactions within the family, this paper analyzes and derives diverse living modules applicable to the modern community residential space and carries out three-dimensional spatial modular design, which facilitates the rapid transformation of design ideas to physical space construction and forms an integrated spatial design logic of "life-design-services". The above research, on the one hand, met the diversified needs of modern residents for residential space to a certain extent and promoted the sustainable development of residential space, and on the other hand, played an active role in enhancing the economic benefits of the residential construction industry.

References

1. Chen, M.P.: Community: the evolution of a sociological discourse. J. Nantong Univ. (Soc. Sci. Edition) **25**(01), 118–123 (2009)
2. Lu, Y.F.: The logical relationship between "communicative behavior theory" and "human destiny community." Kanto J. **04**, 14–20 (2016)
3. Xie, J.H.: Community in the context of modernity and the contemporary imagination of community. J. Yibin Univ. **11**(04), 1–8 (2021)
4. Sen, X.S., Lu, H.B., Shan, M.T.: Discussion on habitat community under landscape thinking. Constr. Technol. Dev. **48**(03), 85–87 (2021)
5. Jacklin-Jarvis, C.C.M.: "It's just houses": the role of community space in a new housing development in the digital era. Voluntary Sector Rev. **10**(01), 69–79 (2019)
6. Zeng, H.D.: Application of digital expression and prototype technology in residential design. Building Materials Decoration **9**(36), 99–100 (2018)
7. Jiao, K., Du, Z.L., Yang, X.: Research and practice of digital intelligent construction system in the whole process of Architecture. Civil Eng. Inf. Technol. **13**(02), 1–6 (2021)
8. Chen, Z.G., Ji, G.H.: Transformation and trend of digital architectural design characteristics driven by construction. Architect. **03**, 107–112 (2020)
9. Zhang, Y.P., Jiang, H., Wang, X.Y.: Research on future residential design based on the background of digital age. Housing Ind. **11**, 51–53 (2020)
10. Dara, C., Hachem-Vermette, C.: Evaluation of low-impact modular housing using energy optimization and life cycle analysis. Energy, Ecol. Environ. **4**(6), 286–299 (2019). https://doi.org/10.1007/s40974-019-00135-4
11. Ye, H.W., Zhou, C., Fan, Z.S.: Thinking and application of integrated digital construction of prefabricated buildings. J. Eng. Manage. **31**(05), 85–98 (2017)
12. Hong, L.: Energy simulation and integration at the early stage of architectural design. J. Asian Architect. Build. Eng. **19**(01), 16–29 (2020)
13. Sun, H., Fei, J.T., Xie, W.: Data value core: re architecture of architectural design methods in the digital background. Contemporary Archit. **03**, 32–34 (2021)
14. Pezhman, S., Bijan, S., Hamid, R.: Automated spatial design of multi-story modular buildings using a unified matrix method. Autom. Constr. **10**(82), 31–42 (2017)

Study on the Comparison and Selection of County Energy Internet Planning Schemes Based on Integrated Empowerment-Topsis Method

Qingkun Tan[1]([✉]), Jianbing Yin[2], Peng Wu[1], Hang Xu[2], Wei Tang[1], and Lin Chen[2]

[1] State Grid Energy Research Institute Co., LTD, Changping, Beijing 102209, China
tanqingkun@163.com
[2] State Grid HangZhou Power Supply Company, Hangzhou 310000, Zhejiang, China

Abstract. Energy Internet is an important way to solve current energy and environmental problems. It combines the planning of multi-energy systems such as electricity, natural gas, heat and transportation, combines energy conversion and utilization with comprehensive demand response, and integrates energy supply network planning with sources and loads. Energy hub planning is combined. Firstly, through the literature survey method and expert interview method to identify the factors that affect planning, and establish a factor index system. Secondly, in order to make the calculation results more meaningful, subjective and objective weighting are combined, and the expert scoring method and the entropy weight method are used to determine the weight of the factors at each stage. Finally, a calculation example is used to verify the rationality of the topsis method for county-level energy Internet collaborative planning. The results of the calculation example show that collaborative planning can avoid the shortcomings of single-subject planning, and the model has certain applicability.

Keywords: County energy internet planning · Influencing factor index system · Integrated weighting method · Topsis model

1 Introduction

Due to multiple connotations, and cross-domain characteristics, the concept of the Energy Internet covers towns, cities, provinces and the country. Therefore, its development evaluation also involves many levels and scopes, such as eco-city, development zone, and park. Due to differences across domains, evaluation often uses indicators of different dimensions, such as economic, environmental, and social dimensions, energy supply, transmission, transaction, demand and other dimensions [1], energy quality, safety and reliability, use and service, etc.; key Technology and innovation capabilities, etc.

In addition to primary energy coal, petroleum, and natural gas, county energy resources generally include renewable energy sources such as agricultural and forestry

© The Author(s) 2022
Z. Qian et al. (Eds.): WCNA 2021, LNEE 942, pp. 560–568, 2022.
https://doi.org/10.1007/978-981-19-2456-9_57

biomass, household waste, wind resources, light resources, and geothermal resources. Except for a few resource-based counties, most counties are short of fossil energy, but renewable resources such as biomass, wind resources, and light resources are abundant. Existing research on energy system planning mainly focuses on the location and capacity of energy station equipment. Multi-energy complementary forms include electro-thermal coupling [2], electrical coupling [3], and cooling-heat-electric coupling system [4]. Literature [5] constructed a combined cooling, heating and power system including wind turbines and photovoltaics, and carried out a multi-objective optimization study on the capacity of the key equipment of the micro-energy grid.

From the perspective of sustainability and practicality of the project, this paper combines the case with the method of literature survey and expert interview to identify the factors affecting the planning of the county energy Internet, and builds the topsis collaborative planning evaluation model based on each core stakeholder. The effectiveness of the constructed model is verified through case analysis, and the results of the calculation example show that the shortcomings of incomplete risk identification and excessively idealized collaborative planning of similar projects in existing research are avoided.

2   County Energy Internet Planning Impact Index System

2.1   County Energy Internet

Focus on the local utilization of clean energy in counties rich in renewable energy. The utilization of energy resources is shown in Fig. 1. Its resource utilization methods generally include: (1) agricultural and forestry biomass: It can be used for cooking,

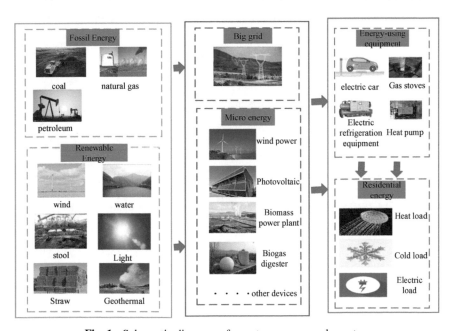

Fig. 1. Schematic diagram of county energy supply system

briquette fuel, gasification, power generation and heating. (2)Domestic waste: Domestic waste can be used to generate electricity. (3) Light: Light energy can convert into electric energy, and solar collector plate can convert light energy into heat energy. (4) Wind: Wind energy can be used to generate electricity. (5) Water: Water energy can be used to generate electricity. (6) Reclaimed water/geothermal: Reclaimed Water/geothermal can be used for heating (cold).

2.2 Index System

The terminal energy Internet focuses on flexibly interacting with users through the integration of heat, electricity, gas and other energy production, transmission, conversion, storage and other links, to enhance the coupling and complementarity between energy sources, to smooth the fluctuations caused by high-penetration renewable energy, and to improve the Renewable energy consumption capacity and users' energy quality. The main body of energy Internet construction is power grid, gas grid, heating network, etc. This paper studies two county energy Internet solutions. Plan 1 is a single gas network, heating network, and power grid planning, and Plan 2 is a joint planning of gas, heat, and power grids. According to the four dimensions of green development, smart empowerment, safety assurance, and value creation, determine the influencing factors of the county energy Internet under different planning schemes.

Table 1. Index system of county energy internet planning

Secondary indicators	Three-level indicators	Plan 1	Plan 2
Green development	Proportion of non-fossil energy in primary energy	64.38%	92.32%
	Renewable energy as a proportion of electricity generation	40%	50%
	Electricity accounts for the proportion of final energy consumption	59.68%	91.33%
	Energy consumption per unit GDP	0.26	0.23
	Typical daily load peak-valley difference rate	77.26%	50.23%
	Distributed power penetration rate (%)	100%	100%
	Distributed clean energy consumption rate (%)	100%	100%
Security	Power supply reliability rate	99.9315%	99.965%
	Average annual power outage time of households (hours)	13.40	3.07
	Power quality	99.826%	99.998%

(*continued*)

Table 1. (*continued*)

Secondary indicators	Three-level indicators	Plan 1	Plan 2
	Information security protection capability	95%	98%
Wisdom empowerment	Digital development index	20%	40%
	Electric vehicle charging pile vehicle ratio (%)	85.96	13.55
	Service radius of electric vehicle charging facilities (km)	3	1.5
Value creation	Universal service level	100%	100%
	Comprehensive energy service business development index	100%	100%
	Business model innovation index	2%	10%
	Customer service satisfaction (%)	90%	95%

3   Evaluation Model of Integrated Weighting-Topsis Method

3.1   No Quantitative Treatment of Indicators

The county-level energy Internet benefit impact index system established in this paper has the characteristics of multiple levels and multiple indicators. In order to facilitate comparative analysis, it is necessary to eliminate the difference in the unit dimensions of the evaluation indicators. Generally, the types of indicators generally have benefit type and cost type. Since the dimensions of different attributes may be different, in order to eliminate the influence of different dimensions on the decision-making results, the attribute indicators need to be dimensionless.

For benefit attributes, generally:

$$r_{ij} = \frac{a_{ij} - \min\limits_{i} a_{ij}}{\max\limits_{i} a_{ij} - \min\limits_{i} a_{ij}} \tag{1}$$

For cost attributes, generally:

$$r_{ij} = \frac{\max\limits_{i} a_{ij} - a_{ij}}{\max\limits_{i} a_{ij} - \min\limits_{i} a_{ij}} \tag{2}$$

The matrix $R = (r_{ij})_{m \times n}$ obtained by the above dimensionless processing, which is called the standardized decision matrix.

3.2   Differential Weighting Method

Entropy weight method is an objective weighting method, which mainly uses information entropy to calculate the entropy weight of each indicator according to the degree of variation of each indicator, and then corrects the weight of each indicator through entropy weight to obtain a more objective indicator weight.

Step 1: Calculate the bias coefficient α, β.

According to the basic idea of moment estimation theory, for each evaluation index, the expected value of subjective weight and the expected value of objective weight are respectively.

Step 2: Solve the optimal combination weight set.

Taking into account the different weighting coefficients of different indicators, in order to calculate the feasibility, the weighting coefficients of different indicators are defined as the same, and the objective function obtained is as follows:

$$\min H = \alpha \sum_{j=1}^{n} \sum_{s=1}^{p} \left(w_j - w_{sj} \right)^2 + \beta \sum_{j=1}^{n} \sum_{t=1}^{q} \left(w_j - w_{tj} \right)^2 \tag{3}$$

The constraint function is:

$$\sum_{j=1}^{n} w_j = 1$$
$$0 \leq w_j \leq 1, 1 \leq j \leq n \tag{4}$$

Step 3: Solve the trend optimal combination weight set.

The integrated weight also reflects the importance of the indicators. The weight results reflect the different importance of the indicators. Optimal objective function:

$$\min G = \alpha \sum_{j=1,k=1,k\neq j}^{n} \sum_{s=1}^{p} \left(\frac{w_j}{w_k} - \frac{w_{sj}}{w_k} \right)^2 + \beta \sum_{j=1,k=1,k\neq j}^{n} \sum_{t=1}^{q} \left(\frac{w_j}{w_k} - \frac{w_{tj}}{w_k} \right)^2 \tag{5}$$

The constraint function is:

$$s.t. \begin{cases} \sum_{j=1}^{n} w_j = 1 \\ \sum_{k=1}^{n} w_k = 1 \\ 0 \leq w_j \leq 1, 1 \leq j \leq n \\ 0 \leq w_k \leq 1, 1 \leq k \leq n \end{cases} \tag{6}$$

Step 4: Solve the integrated weight set.

At the same time, considering the two objective functions of the smallest deviation and the best trend, the two optimization objectives are treated equally, and the final multi-objective function is obtained:

$$\min Z = \frac{1}{2} \min H + \frac{1}{2} \min G \tag{7}$$

Obtain the index benchmark weight set W_j based on the optimal combination through the above formula. When selecting the evaluation method, it was decided to take a comprehensive evaluation based on the TOPSIS method. The specific formula is as follows shown:

$$y_i = \sum_{j=1}^{m} w_j(x_{ij} - x^*)^2 \tag{8}$$

Wherein y_i is the distance, x^* is the ideal point. The queuing indicator value is used to measure the distance from the negative ideal point. The larger the queuing indicator value, the better the queuing indicator value of this scheme:

$$c_i = \frac{y_i^-}{y_i^- + y_i^+} \tag{9}$$

4 Case Analysis

The initial matrix is standardized according to formula (1–2) to obtain a matrix. In order to avoid the subjectivity of experts' scoring, the entropy method is used to quantitatively obtain the weight of each core stakeholder of the county energy Internet that affects the benefits of the county energy Internet, as shown in Table 2:

Table 2. Index weights of influencing factors in county energy internet planning

Three-level indicators	Index label	AHP weight	Entropy weight	Combination weight
Proportion of non-fossil energy in primary energy	C1	0.0123	0.03351	0.02291
Renewable energy as a proportion of electricity generation	C2	0.0096	0.021871	0.015736
Electricity accounts for the proportion of final energy consumption	C3	0.0136	0.03156	0.02258
Energy consumption per unit GDP	C4	0.0213	0.006407	0.013854
Typical daily load peak-valley difference rate	C5	0.013	0.050843	0.031922

(continued)

Table 2. (*continued*)

Three-level indicators	Index label	AHP weight	Entropy weight	Combination weight
Distributed power penetration rate (%)	C6	0.0422	0.049504	0.045852
Distributed clean energy consumption rate (%)	C7	0.0252	0.03156	0.02838
Power supply reliability rate	C8	0.1155	0.022834	0.069167
Average annual power outage time of households (hours)	C9	0.0627	0.03156	0.04713
Power quality	C10	0.1351	0.027127	0.081114
Information security protection capability	C11	0.1948	0.059314	0.127057
Digital development index	C12	0.0139	0.003714	0.008807
Electric vehicle charging pile vehicle ratio (%)	C13	0.1258	0.016048	0.070924
Service radius of electric vehicle charging facilities (km)	C14	0.0638	0.08856	0.07618
Universal service level	C15	0.0446	0.335995	0.190298
Comprehensive energy service business development index	C16	0.0217	0.012277	0.016989
Business model innovation index	C17	0.0521	0.088765	0.070433
Customer service satisfaction (%)	C18	0.0328	0.08856	0.06068

Based on the above formula, the queuing indication value of each scheme can be calculated, as shown in Table 3:

Table 3. Item queuing indicator value

	Plan 1	Plan 2
Distance from positive ideal point	0. 549	0.225
Distance from negative ideal point	7.84	7.45
Queue indication value	0.94	0.97
Comprehensive sort number	2	1

5   Conclusions

This paper establishes an indicator system for the evaluation of county energy Internet development from four dimensions: green development, smart empowerment, safety assurance, and value creation, and uses structural entropy-factor analysis to verify the effectiveness of the indicators, and further constructs a variable weight function based on policy factors To determine the index variable weight, and use the model to evaluate the development of the energy Internet in a certain county. The evaluation result objectively reflects the development of the county energy Internet, verifies the validity of the model, and can be used for county energy Internet development evaluation.

Acknowledgments. This work was supported by the State Grid science and technology projects under Grant 5400-202119156A-0–0-00. (Research on Key Technologies of planning and design of county energy Internet for energy transition).

References

1. Zhao, J., Wang, Y., Wang, D., et al.: Research progress in energy internet: definition, indicator and research method. Proc. CSU-EPSA **30**(10), 1–14 (2018)
2. Muke, B., Wei, T., Cong, W., et al.: Optimal planning based on integrated thermal-electric power flow for user-side micro energy station and its integrating network. Electric Power Autom. Equipment **37**(6), 84–93 (2017)
3. Jun, W., Wei, G., Shuai, L., et al.: Coordinated planning of multi-district integrated energy system combining heating network model. Autom. Electric Power Syst. **40**(15),17–24 (2016)
4. Shao, C.C., Wang, X.F., et al.: Integrated planning of electricity and natural gas transportation systems forenhancing the power grid resilience. IEEE Trans. Power Syst. **32**(6), 4418–4429 (2017)
5. Liu, W., Wang, D., Yu, X., et al.: Multi-objective planning of micro energy network considering P2G-based storage system and renewable energy integration. Autom. Electric Power Syst. **42**(16), 11–20, 72 (2018)

Study on the Analysis Method of Ship Surf-Riding/Broaching Based on Maneuvering Equations

Baoji Zhang[(⊠)] and Lupeng Fu

College of Ocean Science and Engineering,
Shanghai Maritime University, Shanghai 201306, China
bjzhang@shmtu.edu.cn

Abstract. In order to understand the mechanism of the surf-riding/broaching profoundly, the four- degree- of-freedom(4DOF) maneuvering equation (surge, sway, yaw and roll) is simplified to a one- degree-of-freedom (1DOF) equation, and the fourth-order Runge-Kutta method is used to integrate a 1DOF surge equation in the time domain to analyze the two motion states of the ship during the surging and surf-riding. The critical Froude number is calculated using the Melnikov method. Taking a fishing boat as an example, the ship's surf-riding/broaching phenomenon is simulated under the condition of wavelength-to-ship-length ratio and wave steepness, 1 and 1/10 respectively, providing technical support for the formulation of the second generation intact stability criteria.

Keywords: Surf-riding/broaching · Maneuvering equations · Melnikov method · Second generation intact stability

1 Introduction

A ship will subject to a large surging moment due to the broaching phenomenon caused by surf-riding. The centrifugal force generated by serious yaw motion can lead to ships capsizes, especially for small vessels or high-speed vessels. Surf-riding is a condition in which a ship is captured by a wave in advance at a wave speed under conditions of waves or wake waves. Broaching is the violent shaking motion of the ship. Even if the maximum rudder angle is reversed, the heading phenomenon cannot be changed. Under normal circumstances, surf-riding is a prerequisite for broaching. The stability assessment method of surf-riding /broaching is divided into three levels, the safety margin is from high to low, and the judgment method is from simple to complex [1]. The third level needs to be directly evaluated, and there is no standardized conclusion. In recent years, domestic and foreign scholars have carried out various studies on the surf-riding/broaching. Spyrou [2] also conducted a nonlinear dynamic analysis for ship broaching. Umeda et al. [3] attempted to develop a more consistent mathematical model for capsizing associated with surf-riding/broaching in following and quartering waves by taking most of the second-order terms of the waves into account. Yu et al. [4] used

© The Author(s) 2022
Z. Qian et al. (Eds.): WCNA 2021, LNEE 942, pp. 569–575, 2022.
https://doi.org/10.1007/978-981-19-2456-9_58

the wave theory to calculate the surge force, utilized Melnikov method to predict the threshold value of surf-riding and used numerical analysis to solve the thrust and drag equilibrium equations, and the calculation program of the second-generation weakness of surf-riding/ broaching is developed. Chu [5] determined the surf-riding phenomenon by constructing a new Melnikov function of surge system to calculate the first-order zero threshold value. On the basis of summarizing the previous research results, based on the 4DOF maneuvering equation, this paper focuses on the surf-riding and surge of the ship in the following and quartering seas by simplifying it into 1DOF maneuvering equations. Then, the Melnikov method is used to calculate the critical Froude number required in the second level criteria and plot the ship's velocity and displacement phase diagrams. The study presented in this paper can lay a theoretical foundation for the direct calculation of the intact stability of the second generation.

2   The Maneuvering Equations for 4DOF

The 4DOF maneuvering equations of the ship can be expressed as [6]:

$$\dot{\xi} = \{u \cos \chi - v \sin \chi - c\} \tag{1}$$

$$\dot{u} = \{T(u; n) - R(u) + X_W(\xi_G, \chi)\} / (m + m_x) \tag{2}$$

$$\dot{v} = \left\{ \begin{array}{l} -(m + m_x)ur + Y_v(u; n)v + Y_r(u; n)r + Y_\phi(u)\phi \\ +Y_\delta(u; n)\delta + Y_w(\xi_G, \chi) \end{array} \right\} \bigg/ (m + m_y) \tag{3}$$

$$\dot{\chi} = r \tag{4}$$

$$\dot{r} = \left\{ \begin{array}{l} N_v(u; n)v + N_r(u; n)r + N_\phi(u)\phi \\ +N_\delta(u; n)\delta + N_w(\xi_G, \chi) \end{array} \right\} \bigg/ (I_{zz} + J_{zz}) \tag{5}$$

$$\dot{\phi} = p \tag{6}$$

$$\dot{p} = \left\{ \begin{array}{l} m_x Z_H ur + K_v(u; n)r + K_r(u; n)r + K_\phi(u)\phi \\ +K_\delta(u; n)\delta + K_w(\xi_G, \chi) - mgGZ(\phi) \end{array} \right\} \bigg/ (I_{xx} + J_{xx}) \tag{7}$$

$$\dot{\delta} = \{-\delta - (\chi - \chi_c)\} / TE \tag{8}$$

where X_w and Y_w are wave force, N_w and K_w are wave moments, ξ_G is the longitudinal coordinate of the ship center of gravity. u is the speed of surge, v is the speed of sway, N is the movement of yaw, K is the movement of roll, the superscripts of u, v, N, K are hydrodynamic coefficients except for the wave force. χ is the heading angle, χc is the designed heading angle, r is the speed of yaw, φ is the angle of roll, p is the speed of roll, δ is the rudder angle. There is a dot on the letter that represents the first derivative of time. T is the thrust, R is the resistance, n is the propeller speed, c is the wave velocity. m and mx, my represent the hull mass and additional mass, respectively, I and J are the moment of inertia and the additional moment of inertia, respectively, Z_H is the center of the sway force, g is the acceleration of gravity, GZ is the restorative arm, TE is the constant of steering gear set as 0.63.

3   The Analysis of Hull Form Data and 1DOF Model

A fishing boat is selected within this study. The basic parameters of the ship hull shown in Table 1.

Table 1. The general properties of a fishing boat

Length between perpendiculars/L_{pp}	34.5 m
Breadth/B	7.60 m
Draft/d	2.65 m
Block coefficient C_B	0.597
Wake fraction/ω	0.156
Thrust reduction/t_p	0.142
Propeller diameter/D_p	2.60 m

The wave condition used in this study is as follow: Wave steepness $h/\lambda = 1/10$, Wavelength $\lambda = 34.5$ m. By reading a large amount of literatures, surf-riding always occurs when the wavelength λ is close to the ship length L. Therefore, $\lambda/L = 1$ is selected as the wave condition within this study and the wave steepness is set as 1/10 based on existing literature.

Since the wave condition calculated in this section is completely random and without tailgating, the heading angle is equal to zero. Then, the Eq. (1) to Eq. (8) will be simplified as follows. First, the predetermined heading χ_c, the steering angle χ and the rudder angle δ are set as zero. The ship has no sway force when sailing along a straight line. Without considering the capsizing, the yaw moment can also be ignored. Therefore, the simplified equations can be written as:

$$\dot{\xi}_G = \{u - c\} \tag{9}$$

$$\dot{u} = \{T(u; n) - R(u) + X_w(\xi_G, \chi)\}/(m + m_x) \tag{10}$$

It can be seen from the Eqs. (9) and (10) that the surf-riding motion within the waves is an 1DOF model.

4   Phase Diagram Analysis

Phase analysis is the main tool to study the mechanism of ship's surf-riding. What presents in the phase diagram is a velocity vs. displacement plot. Each curve of the phase diagram is called a phase trajectory, and each phase trajectory corresponds to a set of initial conditions. The following will be specifically analyzed by fishing boat combined with the wave parameters given in Table 2.

Table 2. The calculation of the critical Froude number

Method	Values
The Melnikov method	0.306
Direct method	0.308

4.1   Change the Propeller Speed with the Given Initial Conditions

This section will first calculate the critical Froude number of the fishing boat by Melnikov method. The results are shown in Table 2.

It can be seen from the Table 2 that the results calculated with Melnikov method is almost as good as the results calculated with direct method. Therefore, Table 2 is selected as the reference for calculating the initial state in this section. Next, the calculation results of the maneuvering equation are used for argumentation as followed.

$$T(c; n) - R(c) + X_W(\xi_G) = 0 \tag{11}$$

It can be seen from Fig. 1(a) that the trajectory tends to a certain point slowly, indicating the position and state of surf-riding. To show this, the calculation time is increased to 250 s. As shown in Fig. 1(b), it is easy to find that the phase diagram trajectory is finally fixed at one point with coordinates (−0.922, 7.335). The speed is close to the wave speed, and the displacement have a certain gap from −1.2048 mentioned above.

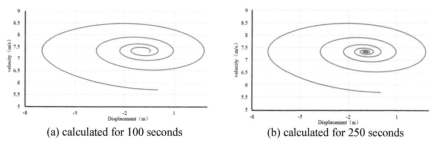

(a) calculated for 100 seconds   (b) calculated for 250 seconds

Fig. 1. Surf-riding phase diagram

In this case it is very difficult to simulate the wave motion exceeding the ship's surge motion. Next, the propeller speed is changed to 6 m/s, the remaining values are unchanged and calculated for 100 s, as shown in Fig. 2. It's easy to catch the direction of the trajectory in the surf-riding phase diagram, that is, the initial point is finally focused on one point. The surging phase diagram will be an infinitely long curve. Therefore, the arrow given in Fig. 2 is the direction of the trajectory. The ship speed increases shapely before 40 s and reaches a stable state. The reason for occurring an oscillation state is that the ship constantly passes through the wave crests and troughs and is constantly subjected to positive wave forces and negative wave forces.

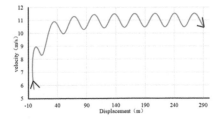

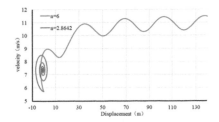

Fig. 2. The surging phase diagram at n = 6

Fig. 3. A comparison of the surf-riding and surging phase diagrams

Next, the Fig. 1(a) and Fig. 2 are now placed in the same phase diagram for comparison, as shown in Fig. 3. The two trajectories start from the same point and finally enter two completely different motion states. The major similarity for these two curves is that the two trajectories' speeds are increasing at first. However, the orange curve is ultimately affected by the wave force, and the ship speed approaches the wave speed, while the blue curve cannot maintain a stable speed affected by the wave exciting force.

4.2 Change the Initial Speed with the Given Initial Conditions

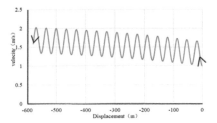

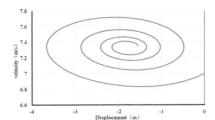

Fig. 4. The ship's surging phase diagram

Fig. 5. The ship's surf-riding phase diagram

Figure 4 shows that the ship is captured by the waves, accelerated by the wave force but does not reach the wave speed and finally becomes a surging motion mode. Figure 5 shows that the propeller thrust cannot be maintained at the current speed and decelerated and is captured by the waves and eventually accelerated to the wave speed. In conclusion, the closer the ship speed is to the wave speed, the easier the ship is surf-riding.

4.3 The Calculation of the Critical Froude Number Using the Phase Analysis Method

Through analysis, it can be found that the propeller speed and the initial speed of the ship are the two important parameters affecting a ship's surf-riding. In this section, the phase analysis method is used to obtain the critical Froude number.

After changing the initial speed, it is obvious that it takes longer to calculate and judge the ship's motion. This is because the ship will perform a surging motion firstly

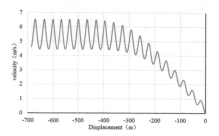

Fig. 6. The ship's surging phase diagram   Fig. 7. The ship's surf-riding phase diagram

when the ship speed accelerates to a speed close to that in still water. At this time, the state of motion will change. Taking the fishing boat as an example, the simulation time is approaching 300 s. The result shows that the ship tends to surf-riding. The surging movement shows a completely periodic change. If the surging of the periodic variation is to be simulated, a longer calculation time is required. The propeller speeds in Fig. 6 and Fig. 7 are 2.7 and 2.9, respectively, with very little difference. However, the ship presents a completely different motion state, and its motion parameters also change greatly.

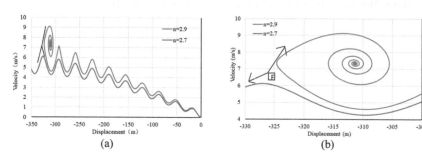

Fig. 8. A comparison of surging phase diagram and surf-riding phase diagram

As can be seen from Fig. 8(a), before the ship moves to the wave-300 m, the form of motion is similar. When the ship speed accelerates to 6 m/s, the two diagrams bifurcate. In the surging phase diagram, the ship speed is still changing alternately between acceleration and deceleration with little change trend. In the phase diagram of surf-riding, the ship speed suddenly increases from 5 m/s to 9 m/s and finally stabilizes at the wave speed. The black line in the figure is equivalent to the asymptotic line of the two trajectories, and the phase diagram of the section from −330 m to −300 m is magnified to compare, as shown in Fig. 8(b).

5   Conclusion

In this paper, the 4DOF maneuvering equation is simplified into a 1DOF maneuvering equation to study the critical conditions for the ship's surf-riding and surging in waves.

According to the phase diagram, it can be found that the critical speed is the intermediate value of the changed phase diagram, which is between 2.7 and 2.9. This is consistent with the critical propeller speed 2.8642 calculated by the Melnikov equation. According to the analysis above, 2.8642 is an approximation, not the critical value, in the phase diagram, and the phase diagram can determine the range of the value. It is noteworthy that if a real threshold is input, the phase diagram should enter the surf-riding at the unstable equilibrium point.

References

1. Belenky, V., Bassler, C.G., Spyrou, K.J.: Development of Second Generation Intact Stability Criteria, pp. 87–121 (2011)
2. Spyrou, KJ.: The nonlinear dynamics of ships in broaching. Marie Curie Fellowships Annal (2002)
3. Umeda, N., Hashimoto, H., Matsuda, A.: Broaching prediction in the light of an enhanced mathematical model with higher-order terms taken into account. J. Mar. Sci. Technol. **7**, 145–155 (2003)
4. Yu, C.C., Hu, Y.H., Zhang, B.J., et al.: The evaluation method for weakness criteria of the surf-riding/broaching. J. Shanghai Maritime Univ. **37**(2), 29–34 (2016)
5. Chu, J.L., Lu, J., Han, Y., et al.: Study on prediction of the critical value of ship's surf-riding based on melnikov method. China Shipbuilding **2015**(A01), 89–96 (2015)
6. Umeda, N.: Nonlinear dynamics of ship capsizing due to broaching in following and quartering. seas. J. Marine Sci. Technol. **4**(1), 16–26 (1999)

Research on Virtual and Real Fusion Maintainability Test Scene Construction Technology

Yi Zhang, Zhexue Ge[✉], and Qiang Li

School of Mechatronics Engineering and Automation, Laboratory of Science and Technology on Integrated Logistics Support, National University of Defense Technology, De Ya Road, 109 Changsha, Hunan, People's Republic of China
gzx@nudt.edu.cn

Abstract. The construction of maintenance test scenes is the premise of accurate assessment of equipment maintenance. In order to reduce the cost and simulate the actual maintenance scene of the product with high fidelity, the construction method of virtual and real fusion maintainability test scene based on partial physical devices is studied in depth. The position and posture of the physical equipment are recognized by binocular vision, and the virtual environment is registered around the physical equipment. Firstly, the ORB (Oriented FAST and Rotated BRIEF) feature extraction of the physical product is carried out and compared, the ICP (iterative closest point) method is then used to perform the matching of physical product features and digital prototype features. Secondly, the virtual maintenance environment is register accurately. Thirdly, the experimental evaluation method of qualitative and quantitative indexes of virtual and real fusion maintainability is formulated. Finally, a case study of a virtual and real fusion maintainability test is carried out with an engine as an example, which verifies the effectiveness and feasibility of the maintenance evaluation based on the virtual and real fusion test scene.

Keywords: Maintainability assessment · ORB feature extraction · Virtual and real fusion · Augmented reality

1 Introduction

Maintainability is important to reflect whether product maintenance is convenient, fast and economical [1]. In order to ensure that the product has high availability and low life cycle cost, the product must have good maintainability, so as to reduce the maintenance requirements for manpower, time and resources [2, 3]. Therefore, during the development process of industrial products, sufficient maintainability tests must be carried out to verify and evaluate their maintainability to ensure that they meet the required maintainability requirements.

© The Author(s) 2022
Z. Qian et al. (Eds.): WCNA 2021, LNEE 942, pp. 576–585, 2022.
https://doi.org/10.1007/978-981-19-2456-9_59

The traditional method of physical maintainability evaluation relies too much on physical prototype, which is expensive and sometimes impractical [4]. The method of virtual maintainability simulation evaluation using digital prototypes is difficult to accurately evaluate the maintenance force characteristics and maintenance time indicators due to the difficulty of accurate human-machine force interaction. However, virtual and real fusion can present the real world and the virtual world at the same time, providing information extensions for real scenes. In the field of maintenance and assembly, the application of virtual and real fusion has made certain progress. Deshpande designed AR-assisted visual features and interactive modes for support-as-assembly (RTA) furniture [5], and developed an application on Microsoft Hololens™ headsets, which enabled users to quickly conceive the spatial relationship of their components and can support assembly tasks that require high spatial knowledge. And it was tested on the users of RTA furniture for the first time. Vicomtech studies the creation method of AR workspace with interaction and visualization mode as the core, and provides more effective support means for the assembly task of hybrid man-machine production line [6]. It can be considered that the virtual and real fusion maintainability test has good accuracy and economy by reducing the hardware scale, which has a huge application prospect. The key issue here is to integrate the physical equipment and the virtual environment according to the actual positional relationship. The three-dimensional pose of the physical equipment must be accurately identified and then the virtual environment is superimposed. The paper focuses on this research and conducts the application of maintainability evaluation.

2 Overall Solution

In the process of a virtual and real fusion maintainability test, a full set of digital prototypes of the product are usually provided as the basic information for the test. The digital prototypes reflect the relationship between the physical product and the surrounding environment. In order to superimpose the virtual maintenance environment model on the periphery of the physical product object and make it sure that it is a part of the maintenance environment, it is necessary to identify the physical product and make the virtual world fully aligned with the physical world. In this paper, the binocular camera is used to obtain the video stream of the real maintenance scene and the characteristics of the video image are extracted on the basis of calibrating the internal parameters of the camera. The transformation matrix is solved for pose estimation. Then, the virtual scene is registered to the real scene through coordinate transformation to complete the construction of virtual and real fusion maintainability test scene. The overall process is shown in Fig. 1.

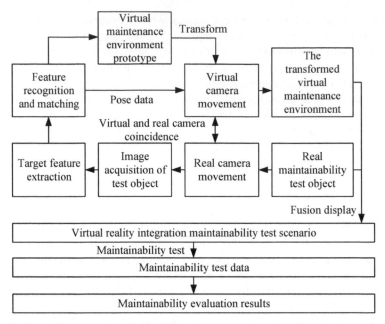

Fig. 1. Overall process of maintainability assessment based on virtual and real fusion.

3   Key Technology Implementation

The key problem to achieve seamless integration of virtual and real maintenance scene is how to accurately identify physical objects and match them with virtual models. In order to construct a realistic scene of virtual and real fusion maintainability test, the main research is based on the ORB feature extraction method, and ICP matching is carried out with the corresponding equipment model in the digital prototype. On this basis, the qualitative and quantitative maintainability index evaluation method based on virtual reality information fusion is formulated.

3.1   Image Feature Extraction of Maintainability Test Object Based on ORB

At present, many local features such as SIFT, SURF, ORB, BRISK, FREAK, etc. are widely used in the fields of image matching and object recognition [7]. Since the object of the maintainability test process is usually a mechanical product, its surface sometimes lacks rich texture features. Considering the stability and rapidity based on feature point extraction and matching, the ORB local feature is selected here. ORB local features use FAST as the feature point detector, and use the improved BRIEF as the feature descriptor, and use the BF pattern matching algorithm for feature descriptor matching.

FAST feature points are not directional, and the directional parameters are determined by obtaining the center of gravity of the feature point neighborhood. The neighborhood moment is:

$$m_{pq} = \sum_{x,y} x^p y^q I(x, y) \tag{1}$$

where $I(x, y)$ is the gray value at point (x, y), $x, y \in [-r, r]$, r is the radius of the circle, p and q are non-negative integers, when p is 1 and q is 0, the value I_x of I in the x direction can be obtained, when p is 0 and q are 1, the value I_y of I in the y direction can be obtained, and the C coordinate of the image center of gravity can be obtained as:

$$C = \left(\frac{m_{10}}{m_{00}}, \frac{m_{01}}{m_{00}} \right) \tag{2}$$

The angle between the feature point and the center of gravity is defined as the direction of the FAST feature point:

$$\theta = \arctan\left(\frac{m_{01}}{m_{10}}\right) = \arctan\left(\frac{\sum\limits_{x,y} yI(x, y)}{\sum\limits_{x,y} xI(x, y)}\right) \tag{3}$$

ORB extracts the BRIEF descriptor according to the direction parameters obtained in the above formula. However, due to environmental factors and the introduction of noise, the direction of feature points will change, and the correlation of random pixel block pairs will be relatively large, thereby reducing the discrimination of the descriptor. ORB adopts a greedy algorithm to find random pixel block pairs with low correlation. Generally, 256 pixel block pairs with the lowest correlation are selected to form a 256-bit feature descriptor. Note two descriptors:

$$K_1 = x_0 x_1 \cdots x_{255}, K_2 = y_0 y_1 \cdots y_{255}$$

3.2 Matching of Physical Equipment Characteristics and Virtual Environment Registration

The ORB feature set is extracted from the real maintainability test object and the virtual maintenance environment model, and the corresponding feature descriptors K_1, K_2 are obtained. The similarity between two ORB feature descriptors is characterized by the sum of the exclusive ORB Hamming distances:

$$D(K_1, K_2) = \sum_{i=0}^{255} x_i \oplus y_i \tag{4}$$

The smaller the $D(K_1, K_2)$, the higher the similarity, and the greater the probability that the two describe the same feature. Conversely, the lower the similarity, the more likely they are not describing the same feature.

Use BF matcher to get all possible matching feature pairs, assuming that the minimum Hamming distance of feature pairs is MIN_DIST. In order to select the best matching pair and improve the operating efficiency, an appropriate threshold is selected and the matching pair smaller than the threshold is selected for the next camera pose estimation. The threshold value cannot be too small, which will affect the final effect, and it is necessary to select the best threshold value through experiments on the image frame.

Given the point k_{1i} in K_1, find the point k_{2i} with the shortest Euclidean distance of k_{1i} from K_2, and take k_{1i} and k_{2i} as the corresponding points to obtain the transformation matrix. Through continuous iteration, the following formula is minimized and the iteration is terminated, and finally the most Optimal transformation matrix is obtained to make them coincide.

$$f(R, T) = \frac{1}{n} \sum_{i=1}^{n} \|k_{1i} - (Rk_{2i} + T)\|^2 \qquad (5)$$

In the formula, R indicates the rotary transform matrix, T indicates the translation transform matrix.

The essence of the ICP algorithm is to calculate the transformation matrix between the feature sets, minimize the registration error between the two through rotation and translation and then achieve the best registration effect. Assuming two feature point sets $K_1 = \{k_{1i} \in R^3, i = 1, 2, \cdots, n\}$ and $K_2 = \{k_{2i} \in R^3, i = 1, 2, \cdots, n\}$, the registration process using the ICP algorithm is introduced below:

(1) Sample set K_1, $K_{10} \subset K_1$, K_{10} represents a subset of set K_1;
(2) Search in set K_2, find the closest point to each point in K_{10}, and get the initial correspondence between K_1 and K_2;
(3) Remove the wrong corresponding point pairs using algorithms or constraints;
(4) Calculate the transformation relationship between the two according to the corresponding relationship in step (2), minimize the value of the objective function and apply the calculated transformation matrix to K_{10} to obtain the changed new K_{10}';
(5) Determine whether the iteration is terminated according to $d = \frac{1}{n} \sum_{i=1}^{n} \|K_{2i} - K_{1i}\|^2$.

If d is greater than the preset threshold, return to step (2) to continue the iteration; if d is less than the preset threshold or reach the set number of iterations, the iteration stops.

By obtaining the transformation matrix through the above steps, the pose transformation relationship between the physical equipment and the virtual maintainability test environment can be obtained, and then virtual registration can be performed to complete the construction of the virtual and real fusion maintainability test environment.

4   Experimental Verification

Take the auxiliary engine room of a ship as a case to carry out the test verification to verify the correctness and applicability of the virtual and real fusion maintainability test evaluation method studied in this paper. The auxiliary engine room is powered by a diesel engine, which is composed of a crank connecting rod mechanism, a gas distribution structure, a fuel system, a lubrication system, a cooling system, a starting system, etc. The engine needs to replace consumable parts such as fuel filter and air filter, and the cylinder and starter motor have a certain failure rate. It needs to be well designed for maintenance to ensure rapid maintenance at the crew level.

In the ship cabin environment, the equipment maintenance process has certain complexity, and other equipment around the equipment and peripheral pipelines and cables are easy to cause insufficient accessibility of the maintenance objects and insufficient operating space. Therefore, in the process of maintainability test of the engine, it is necessary to be able to simulate actual cabin maintenance scenes and maintenance space, and fully consider the impact of various operational obstacles on maintainability, so as to obtain more accurate maintainability test results.

Since the establishment of a 1:1 full-physical maintainability test condition is very costly and has a long cycle, the virtual and real fusion maintainability test evaluation method studied in this paper is adopted, and a small part of the physical equipment and a large number of virtual environments are used to realistically simulate a complete test scene. During the test, the available test conditions include the YN92 physical diesel engine and the complete digital model of the auxiliary engine compartment, as shown in Fig. 2 and Fig. 3. Next, take the repairing and replacing the starting motor as an example to verify.

Fig. 2. YN92 diesel engine.

Fig. 3. Virtual maintenance scene of ship auxiliary engine cabin.

4.1 Verification of the Establishment Method of Virtual and Real Fusion Test Scene

In order to build a realistic virtual and real fusion maintenance scene, it is necessary to consider the impact of multiple factors on the registration accuracy of the virtual environment. The feature extraction method is an important factor affecting the registration accuracy.

Firstly, the feature extraction and recognition of diesel engine are carried out. Different feature extraction methods have different feature extraction results. The feature extraction of the same object (diesel engine) is performed using SIFT, SURF, and ORB methods respectively, and the comparison of the diesel engine feature extraction results of the three methods is shown in Fig. 4.

(a) SIFT   (b) SURF   (c) ORB

Fig. 4. The comparison of the diesel engine feature extraction results of the three methods.

The data results of the three methods for feature extraction are shown in Table 1.

Table 1. Experimental results of different feature extraction methods.

	Physical feature points	Model feature points	Match points	Consume time (ms)
SIFT	502	522	112	62.90
SURF	454	426	168	21.76
ORB	1023	1004	136	13.92

Through experimental analysis and comparison, the feature points detected by SIFT, SURF and ORB are 502, 454 and 1023 respectively under the same experimental conditions. The feature points matched by SIFT, SURF and ORB are 112, 168 and 136 respectively. It can be found that although the number of feature points matched by the three methods is roughly the same, the time required for ORB matching is significantly shorter and the operation efficiency is obviously higher.

Inject the above two algorithms into AR glasses, obtain the three-dimensional visual information of the physical equipment through the binocular lens of the glasses, and then perform feature extraction and match them with the virtual model one by one. The resulting virtual and real fusion ship cabin repair scene is shown in Fig. 5.

Fig. 5. The obtained virtual and real fusion ship engine maintenance scene.

4.2 Maintainability Test Operation and Result Analysis

Next, according to the established virtual and real fusion maintainability test scene of YN92 physical diesel engine, the maintainability operation test of the replacement of the starting motor is carried out. The tester wears AR glasses to carry out maintainability test operation and obtain basic test data.

A total of 5 groups of tests are carried out, and each group of tests is carried out in three scenes of real environment, virtual and real fusion and without surrounding environment respectively, and the comparison of maintenance operations in three scenes is shown in Fig. 6.

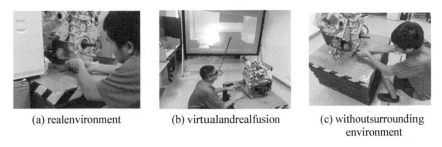

(a) realenvironment   (b) virtualandrealfusion   (c) withoutsurrounding
                                                                                environment

Fig. 6. The comparison of maintenance operations in three scenes

In the virtual and real integration maintenance test, the maintenance personnel can feel the existence of the surrounding cabin equipment through vision. During mainte-nance, in order to avoid collisions with the virtual cabin equipment, the bending angle of the arm will be smaller and the movement range will not be large. The posture of the maintenance personnel should be adjusted accordingly to be closer to the real maintenance situation, so the maintainability evaluation error is smaller.

5 Conclusion

This paper proposes a method of constructing a maintainability test scene based on the fusion of virtual and reality for maintainability evaluation. The ORB feature of the equipment is extracted based on binocular vision, and then the ICP method is used for feature matching and recognition according to the feature extraction results, and the virtual environment is registered to complete the construction of virtual reality fusion maintainability test scene. Experiments show that the use of orb features can effectively extract equipment features, with high speed and high precision. The ICP method can be used to realize the registration of the physical object and the virtual environment, thereby completing the registration of the virtual environment. The maintainability test is carried out and evaluated in the built virtual and real fusion test scene. The results show that the surrounding virtual environment has a certain impact on the maintenance process, and the maintainability verification is closer to the maintenance process in the real maintenance environment.

The virtual and real fusion maintainability test method studied in this paper provides a novel and efficient method for simulating the real maintenance performance of the products under complex maintenance conditions. It can carry out the main test operations on the real object and simulate the spatial characteristics at low cost, so as to make the index evaluation of visibility, accessibility and maintenance time more accurate.

References

1. Fedele, L.: Methodologies and Techniques for Advanced Maintenance. Springer, London (2011)
2. MIL-HDBK-470A. Designing and developing maintainable products and systems. Department of Defense Handbook (1997)
3. Guo, Z., et al.: A hybrid method for evaluation of maintainability towards a design process using virtual reality. Comput. Ind. Eng. **140**(1), 106227 (2020)
4. Slavila, C.A., Decreuse, C., Ferney, M.: Fuzzy approach for maintainability evaluation in the design process. Concurr. Eng. **13**(4), 291–299 (2005)
5. Deshpande A, Kim I. The effects of augmented reality on improving spatial problem solving for object assembly. Adv. Eng. Inform. **38**, 760–775 (2018)
6. Simões, B., Álvarez, H., Segura, A., Barandiaran, I.: Unlocking augmented interactions in short-lived assembly tasks. Adv. Intell. Syst. Comput. **771**(1), 270–279 (2018)
7. Wang, Y., Zhang, S., Bai, X.: Stuten, enhanced realistic assembly system, enhanced reality assembly system, integrated reality assembly system. J. Northwestern Univ. Technol. **37**(01): 143–151 (2019)

Automatic Scoring Model of Subjective Questions Based Text Similarity Fusion Model

Bo Xie and Long Chen[✉]

Zhejiang GongShang University, Hangzhou, China
boxie@mail.zjgsu.edu.cn, 987465580@qq.com

Abstract. AI In this era, scene based translation and intelligent word segmentation are not new technologies. However, there is still no good solution for long and complex Chinese semantic analysis. The subjective question scoring still relies on the teacher's manual marking. However, there are a large number of examinations, and the manual marking work is huge. At present, the labor cost is getting higher and higher, the traditional manual marking method can't meet the demand The demand for automatic marking is increasingly strong in modern society. At present, the automatic marking technology of objective questions has been very mature and widely used. However, by reasons of the complexity and the difficulty of natural language processing technology in Chinese text, there are still many shortcomings in subjective questions marking, such as not considering the impact of semantics, word order and other issues on scoring accuracy. The automatic scoring technology of subjective questions is a complex technology, involving pattern recognition, machine learning, natural language processing and other technologies. Good results have been seen in the calculation method-based deep learning and machine learning. The rapid development of NLP technology has brought a new breakthrough for subjective question scoring. We integrate two deep learning models based on the Siamese Network through bagging to ensure the accuracy of the results, the text similarity matching model based on the birth networks and the score point recognition model based on the named entity recognition method respectively. Combining with the framework of deep learning, we use the simulated manual scoring method to extract and match the score point sequence of students' answers with standard answers. The score recognition model effectively improves the efficiency of model calculation and long text keyword matching. The loss value of the final training score recognition model is about 0.9, and the accuracy is 80.54%. The accuracy of the training text similarity matching model is 86.99%, and the fusion model is single. The scoring time is less than 0.8s, and the accuracy is 83.43%.

Keywords: Subjective question automatic scoring · Text similarity · Siamese network · Named entity recognition · Natural language processing · Machine learning

1 Introduction

The scale of China's online education market is increasing year by year. As a test method for learning effect and knowledge mastery, due to the large number and scale of various

© The Author(s) 2022
Z. Qian et al. (Eds.): WCNA 2021, LNEE 942, pp. 586–599, 2022.
https://doi.org/10.1007/978-981-19-2456-9_60

training examinations, the demand of education and training institutions for automatic marking is increasingly strong, so that manual marking can't meet the demand. At present, there is no formed Chinese marking system applied to the market. Because of the complexity of Chinese text and the differences in semantic level, the development of Chinese subjective intelligent marking system is frequently hindered. By reasons of the complexity and the difficulty of natural language processing technology in Chinese text, most of the automatic marking systems stop at the objective question marking and simple English composition marking. Due to the growth of data and the improvement of computing power, deep learning has made a great breakthrough. The deep learning methods based on neural network have been applied into NLP field. At the same time, information extraction, part of speech tagging, named entity recognition and other research directions have been improved, which greatly improves the accuracy of automatic marking.

With the development of computer and network technology, a lot of subjective marking systems about English have sprouted abroad, such as PEG, IEA, Criterion and so on. However, the domestic research on subjective question marking has only been carried out gradually in the past 20 years. At present, no formed Chinese marking system has been applied to the market. Due to the complexity of Chinese text and the differences in semantic level, the development of Chinese subjective question intelligent marking system is frequently hindered.

Three main technical methods about the automatic marking system are introduced at present: the method based on templates and rules, based on the traditional machine learning method, based on the deep learning method.

(1) Rule based and template-based method: this method relies on artificial features and templates, and the trained model does not have generalization. For example, auto mark system [1] makes multiple scoring templates of correct or wrong answers for each question in advance, matches the candidates' answers with the templates one by one, judges the correctness and gives scores, which is in line with people's way of thinking. Bachman et al. Proposed that [2] generate regular expressions automatically according to the reference answers, and each regular expression matches a score. When the students' answers are consistent with the generated expressions, they get a score. This method is suitable for students with low diversity of answers and low difficulty of questions. Jinshui Wang et al. [3]. introduced professional terms in the field of power system analysis into the dictionary to improve the ability of word segmentation of professional terms. At the same time, they introduced ontology and synonym forest in the field of power system analysis to improve the word similarity calculation ability between common words and professional terms. However, the disadvantage is that it costs huge human resources to build the scoring data set, which makes it impossible to comprehensively evaluate Objective to evaluate the effectiveness and universality of the automatic scoring method. Fang Huang proposed [4] to design a new text translation information automatic scoring system based on XML structure. By setting weights, the valuable information in the answers is extracted, the closeness between candidates' answers and standard answers is analyzed, and the corresponding scores are given.

(2) Based on the traditional machine learning method. In traditional machine learning, we usually need to define features manually, and use regression, classification or

a combination of them to get a score. For example, Sultan et al. [5]. constructed a random forest classifier using text similarity, term weight and other features. Kumar et al. [6]. defined a variety of features including key concept weight, sentence length and word overlap features, and scored them by decision tree, and achieved good results on ASAP dataset. Jie Cao et al. [7]. proposed that after preprocessing the student answer text and the reference answer text, the similarity of the topic probability distribution between the student answer and the reference answer can be calculated through LDA model training, so as to realize the evaluation.

(3) With the rapid increase of big data storage capacity and computing power, deep learning has been successfully applied into the field of image recognition and natural language processing. Shuai Zhang [8] Based on the Siamese Network subjective question automatic scoring technology, at the same time input student answers and reference answers for similarity calculation, so as to estimate the score of student answers, improve the similarity calculation method based on sentence surface features, and improve the accuracy. Yifan Wang et al. [9]. used the extended named entity recognition method to extract some keywords from the candidate answers of subjective questions, and used the improved synonym forest word similarity calculation method to calculate the similarity between the candidate keywords and the target keywords in the standard answers of subjective questions. The method solves the problem of low matching efficiency in similarity calculation of long text words and preferentially extracts keywords for similarity calculation, which effectively improves the performance of similarity calculation of key words in shortening the calculation time compared with the traditional word similarity methods.

Subjective question scoring faces many challenges. How to calculate the similarity between standard answers and students' answers is an important problem in subjective question scoring model. Traditional models only consider the surface features of sentences by using words, words and other indicators to calculate text similarity, so the accuracy is not high. There are some researches on the automatic score of composition by analyzing text coherence in China. Due to the limitation of short text in the answer text of subjective question, accuracy is not effectively improved by simply increasing the coherence of the text. In addition, the method of word similarity calculation based on synonym forest has achieved good results in Chinese text, while applying into long text may lead to the decline of the method performance and accuracy.

In order to solve the mentioned problems, this paper proposes a fusion method based on Siamese Network and named entity recognition. On the basis of general lexical features, Siamese Network model is added to judge the similarity between students' answers and reference answers, so as to score students' answers. Compared with other neural network models, Siamese Network is special in that it inputs two subnets at the same time Network, and these two subnetworks share weight. The characteristics of Siamese Network make it have a good effect in measuring similarity. But the disadvantage is that as a kind of neural network, Siamese Network can only get the scoring results, and can't make a reasonable explanation for the scoring results. The extended named entity recognition method is used to extract some keywords from the candidate answers of the subjective questions, and the improved synonym forest word similarity calculation method is used to calculate the similarity between the candidate keywords and the target keywords in the standard answers of the subjective questions, which improves the performance of the original algorithm and effectively shortens the calculation time.

2   Model Presentation

Neural network can accurately measure the similarity between standard answers and students' answers. To simulate the process of manual scoring and make a reasonable explanation for the results of the model, this paper proposes a text similarity matching model (TSMM) based on Siamese Network, Text similarity matching model and scoring point identification model (SPRM) based on named entity recognition are used to fuse the models. The model is able to score according to the scoring points of user answers and the interpretation in the answers. We adopt a two-pronged strategy: on the one hand, we use deep learning method to extract the scoring points of user answers and highly simulate "manual marking" to realize the judgment of scoring points hit; on the other hand, we use Siamese Network model to compare the standard answers with students' answers. The final subjective score results are obtained through the fusion of dual-strategy model, and the overall route diagram is shown in Fig. 1.

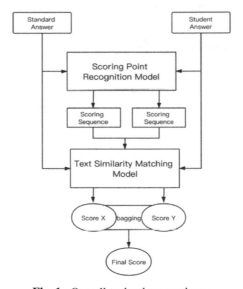

Fig. 1. Overall technology roadmap.

3   Related Technologies

Text similarity calculation is the core of the intelligent evaluation system of subjective questions. The method of text similarity calculation is related to the accuracy and practicability of the whole intelligent evaluation system of subjective questions. The following is the text similarity calculation technology involved in the development of the subjective question automatic evaluation model, including long-term and short-term memory (LSTM), conditional random field (CRF), pre-training model, Siamese Network and other text similarity models.

3.1 Long Short-Term Memory (LSTM)

The normal RNN has no solution to the long-term memory function. For example, trying to predict the last word of "I majored in logistics when I was in University… I will be engaged in logistics after graduation." Recent information shows that the next word may be the name of an industry. However, if we want to narrow the selection range, we need to include the context of "logistics major" and infer the following words from the previous information. Similarly, in terms of score point prediction, whether the user's answer or the standard answer is a long text, the interval between the relevant information and the predicted position It's quite possible. However, RNNs are incapable of solving this problem. As one of the most popular RNNs, long-short term memory network (LSTM) successfully solved the defects of the original recurrent neural network which has been applied into many fields such as speech recognition, picture description, and natural language processing. LSTM is quite suitable for processing and predicting important events with relatively long interval and delay in time series [10].

3.2 Conditional Random Field (CRF)

In order to make our scoring point recognition model perform better, the marking information of adjacent data can be considered when marking data. This is difficult for ordinary classifiers to do, and also a good place for CRF. CRF is the conditional random field, which represents the Markov random field of another group of output random variables y given a group of input random variables X. the attribute of CRF is to assume that the output random variables establish the Markov random field [11].

The CRF is refered as the speculation of the Maximum Entropy Markov model in the labeling problem. The CRF layer can be used to predict the final result of the sequence labeling task, some constraints are added to guarantee that the predicted label is reasonable. During the training process, these constraints can be adapted consequently through CRF layer [12].

- The first word in the sentence is constantly begun with the name "O-" or "B-", rather than "I-".
- Label stands for name entity (person name, organization name, time, etc.). The label "B-L1 I-L2 I-L3 I-…", L1, L2, L3 are supposed to be entity of the same type.
- A tag sequence that starts with "I-label" is usually unreasonable. A logical sequence would start with "B-label".

These constraints will greatly reduce the probability of unreasonable sequence occurrence in label sequence prediction.

3.3 Pretraining

The pretraining model is a deep learning architecture, which has been prepared to perform explicit assignments on a lot of data. This kind of training is relatively hard to implement, and always requires a great deal of resources. Therefore, the large number of parameters it gets make the model implementation results closer to the actual results. The pretraining model learns a context-dependent representation of each member of an input sentence using almost unlimited text, and it implicitly learns general syntactic semantic knowledge. It can migrate knowledge learned from the open domain to downstream tasks to improve low-resource tasks, and is also very helpful for low-resource language processing [13].

The pretraining model has achieved good results in most of NLP tasks, and the BERT model is a language representation model released by Devlin et al. [14] (Google) in October 2018. the BERT swept the optimal results of 11 tasks in the NLP field, which can be considered as the most important breakthrough in NLP field recently. Because of its flexible training mode and outstanding effect, the BERT model has been deeply studied and applied in many tasks of NLP. This paper applies few BERT modules for pretraining tasks.

3.4 Siamese Network

Siamese Network is a kind of neural network architecture which contains two or more identical subnetworks, which sets the same configuration, same parameters and weights [15]. Parameter updating is carried out in two subnets. The structure of Siamese Network is shown in Fig. 2.

Siamese Networks are popular in tasks involving finding similarities or relationships between two comparable things [15]. Examples of how similar the input or output of two signatures are from the same person verify whether they are. Usually, in such a task,

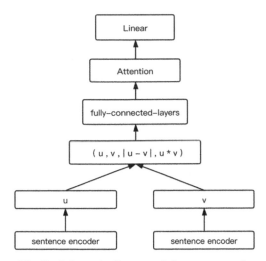

Fig. 2. Schematic diagram of siamese network.

two identical subnetworks are used to process two inputs, and another module will take their output and produce the final output.

The advantages are as follows: 1. Subnet sharing weight means that training needs less parameters, which means that it needs less data and is not easy to over fit. 2. Each subnet essentially produces a representation of its input. It makes sense to use a similar model for the same type of input (for example, matching two images or two paragraphs). Representation vectors with a similar semantics, making them simpler to compare.

4   Model Composition and Fusion

For the sake of scoring user's answers reasonably, this paper proposes an automatic evaluation model of subjective questions, which is composed of text similarity matching model (TSMM) and score point recognition model (SPRM). The TSMM calculates the semantic similarity between the standard answer with the user's answer. The SPRM is used to extract the scores of the answers, which is regard as "manual marking" simulation. Finally, the final subjective score is obtained by the model fusion.

4.1   The Automatic Evaluation Model of Subjective Questions

Input the standard answer text and student answer text into the score recognition model after training respectively, then we can extract the score point sequence of two strings of text, and further match the score points of the two strings of text through the text similarity matching model after training, so as to calculate the score of each score point and accumulate it to get the final score X; at the same time, the standard answer and student answer text are compared Students' answer text is directly input into the text similarity matching model to get the overall similarity, that is, the score Y.

Ensemble learning is a paradigm of machine learning. Training multiple models to solve the same problem and combining them to get better results [16]. One of the most important assumption is that when the weak models are combined correctly, we can get more accurate and more robust models.

Considering that both TSMM and SPRM are homogeneous weak learners, bagging can be used to learn these weak learners independently and in parallel. This method does not operate the model itself, but acts on the sample set. We use the random selection training data, then construct the classifier, and finally combine them. Different from the interdependence and serial operation among classifiers in boosting method, there is no strong dependency between base learners in bagging method, and parallel operation is generated at the same time [16].

We use bagging based method to get the final model fusion result through TSMM and SPRM model, that is, bagging the two scores obtained from the score recognition model and the text similarity matching model to get the final score.

4.2   Scoring Point Recognition Model (SPRM)

Named entity recognition is to identify entities with specific meaning in text. From the perspective of knowledge map, it is to obtain entities and entity attributes from unstructured text [17]. Therefore, we consider using named entity recognition method to extract score points. Bi-LSTM refers to bidirectional LSTM; CRF refers to conditional random field. In SPRM, Bi-LSTM is mainly used to give the probability distribution of the corresponding label of the current word according to the context of a word, which can be regarded as a coding layer. The CRF layer can add some restrictions on the final prediction labels to ensure that the results are valid. These limitations can be learned from the CRF layer's automatic training data set during the training process. The text sequence is processed by Bi-LSTM model, the output result is transferred to CRF layer, and finally the prediction result is output [18].

The part of preprocessing prediction data, that is, sequence labeling has been completed in data preprocessing.

Take a sentence as a unit, record a sentence with n words as:

$$x = (x_1, x_2, \ldots, x_n)$$

x_i represents the ID of the ith word of a sentence in the dictionary, thus obtaining the one-hot vector of each word (dimension is the dictionary size).

Look-up layer is the first layer of the model, each word x_i in a sentence is mapped from a one-hot vector to a low dimensional character embedding using a pretrained or randomly initialized embedding matrix $x_i \in R_d$, d is the dimension of embedding. Set dropout to ease over fitting before entering the next layer [19].

Bidirectional LSTM layer is the second layer of the model that automatically extracts sentence features. The char embedding sequence $(x_1, x_2, \ldots, x_n)$ of each word of a sentence is used as the input of each time step of bidirectional LSTM, and then the hidden state sequence $\left(\overrightarrow{h}_1, \overrightarrow{h}_2, \ldots, \overrightarrow{h}_n \right)$ of forward LSTM output and the hidden state sequence of reverse LSTM $\left(\overleftarrow{h}_1, \overleftarrow{h}_2, \ldots, \overleftarrow{h}_n \right)$ output in each position are spliced according to the position $h_t = \left| \overrightarrow{h}_t; \overleftarrow{h}_t \right| \in R^m$ to obtain a complete hidden state sequence

$$(h_1, h_2, \ldots, h_n) \in R^{n \times m}$$

After dropout is set, a linear layer is connected, and the hidden state vector is mapped from m dimension to k dimension. K is the number of tags in the annotation set, so the automatically extracted sentence features are obtained and recorded as matrix $P = (p_1, p_2, \ldots, p_n) \in R^{n \times m}$. Each dimension p_{ij} of $p_i \in R^k$ can be regarded as the scoring value of the j-th tag. If softmax is used for P, it is equivalent to k-class classification for each position independently. However, it is impossible to make use of the information that has been labeled when labeling each position, so a CRF layer will be connected to label next [19].

CRF layer is the third layer of the model, which is used for sequence annotation at sentence level. The parameter of CRF layer is a matrix A of $(k + 2) \times (k + 2)$, and A_{ij} represents the transfer score from the i-th tag to the j-th tag. When labeling a location, it can use the previously labeled data. The reason for adding 2 is to add a start state to the beginning of the sentence and an end state to the end of the sentence. If we remember a tag sequence $y = (y_1, y_2, \ldots, y_n)$ whose length is equal to the length of the sentence, the score of the model for the tag of Sentence x equal to y is as follows [19]:

$$score(x, y) = \sum_{i=1}^{n} P_{i,y_i} + \sum_{i=1}^{n+1} A_{y-1,y_i}$$

The score of the whole sequence is equal to the sum of the scores of each position, and the score of each position is obtained by combining pi of LSTM output and transfer matrix A of CRF. Then, the normalized probability can be obtained by Softmax:

$$P(y|x) = \frac{\exp(score(x, y))}{\sum_{y} \exp(score(x, y'))}$$

By maximizing the log likelihood function in the model training, the log likelihood of a training sample (x, y_x) is given by the following formula:

$$\log P(y^x|x) = score(x, y^x) - \log\left(\sum_{y} \exp\left(score(x, y')\right)\right)$$

In the process of prediction (decoding), The Viterbi algorithm of dynamic programming is used to solve the optimal path:

$$y^* = \underset{y}{\arg\max} score(x, y')$$

The structure is shown in Fig. 3 SPRM structure diagram [20–22]:

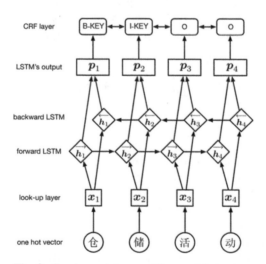

Fig. 3. Scoring point recognition model structure

4.3   Text Similarity Matching Model (TSMM)

The main idea of TSMM is: mapping the input to the target space through a function, and comparing the similarity in the target space using distance. During the training stage,

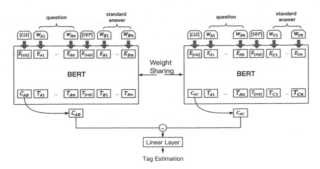

Fig. 4. Text similarity matching model structure.

we minimize the loss function values of a pair of samples from the same category and maximize the loss function values of a pile of samples from different categories. Its feature is that it receives two pieces of text as input instead of one piece of text as input.

It can be summarized as the following three points:

- Input is no longer a single sample, but a pair of samples, no longer give a single sample exact label, and given a pair of sample similarity labels.
- Designed as like as two networks, the network shared weight W, and the distance measurement of output, L1, L2, etc., were carried out in two.
- According to whether the input sample pairs come from the same category or not, a loss function is designed in the form of cross entropy loss.

In the Siamese Network, the loss function is comparative loss, which can effectively deal with the relationship of paired data in the t Siamese Network. The expression of contrastive loss is as follows [23]:

$$L = \frac{1}{2N} \sum_{n=1}^{N} yd^2 + (1 - y)\max(\text{margin} - d, 0)^2$$

The specific purpose of Siamese Network is to measure the similarity of two input texts [24]. In the process of training and testing, the encoder part of the model shares weight, which is also the embodiment of the word "Siamese". The choice of encoder is very wide, traditional CNN, RNN and attention, transformer can be used.

After getting the features u and V, we can directly use the distance formula, such as cosine distance, L1 distance, Euclidean distance, to get the similarity between the two texts. However, a more general approach is to build feature vectors based on u and V to model the matching relationship between them, and then use additional models (MLP, etc.) to learn the general text relational function mapping.

5 Experiment and Results

5.1 Experimental Data

The data of this paper comes from the official logistics industry corpus and professional questions provided by China outsourcing service competition in 2020. The data features are as follows: short answer questions in the field of logistics vocational education are basically noun explanation and concept explanation questions, and the sentence structure is relatively simple; the composition of a piece of data includes serial number, question description, answer, keyword and keyword description, and the data is divided into three parts 600.

For the above 600 pieces of data, we expanded the data according to the score points, and got 5924 pieces of augmented data as the data set for the training of TSMM model. The characteristics of this training set are: it belongs to the field of Logistics Vocational Education, and the data composition includes question number, question, standard answer and user answers with 0 to 10 points.

5.2 Analysis of SPRM Experimental Results

First, we preprocess the existing 600 pieces of data, mainly including sequence annotation, word segmentation, and data cleaning and formatting. For the preprocessed 600 pieces of data, 70% is used as training set, and the remaining 30% is used as test set and verification set.

Table 1. Scoring point recognition model training results

	Accuracy	Precision	Recall
SPRM	80.54%	57.12%	58.75%

Experimental results: the model loss in the training set is reduced from 53.138512 to 0.93004, and the accuracy rate is 80.54%. For SPRM, the processing in each layer is relatively simple compared to the existing work, and there is room for improvement in the future. For instance, the initialization method of word vector embedding we used in the experiment is simple random initialization. Besides, due to the small size of corpus, we can consider the pretraining value on a larger corpus. SPRM may over fit in this case because of the large number of iterations, so it is necessary to draw a verification set for early stopping.

5.3 Analysis of TSMM Experimental Results

For the expanded 5924 data, 70% is used for training set, and the remaining 30% is used for test set and verification set. The loss value of the model is reduced from 174.2736382 to 21.5801761, and the accuracy rate reaches 86.99%. It can be seen that the calculation effect of using twin network to input standard answers and student answers at the same time is higher than that only based on the surface features of sentences.

5.4 Experimental Analysis of the Automatic Evaluation Model for Subjective Questions

After the recognition of the score point sequence by SPRM model, through the word similarity matching calculation based on Synonymy Thesaurus and CNKI, the subjective score can be obtained, which can be used as the comparison between TSMM model and fusion model. This experiment uses real short answer questions of logistics final examination, a total of 10 questions as experimental data. After scoring by SPRM, TSMM and model fusion, the calculated evaluation indexes are as follows lower.

Table 2. The performance of the grading approaches.

	MSE	RMSE	MAE
SPRM	1.96	1.40	1.16
TSMM	0.80	0.89	0.60
Fusion model	0.32	0.57	0.57

Table 2 compares the calculation results of SPRM, TSMM and fusion model under different indexes. Results show that the fusion model has the advantages of MSE, RMSE, MAE is the minimum, which shows that the fusion model has more advantages than the single model of SPRM and TSMM, and the score sequence of SPRM is interpretable to the fusion model.

References

1. Rudner, L., Gagne, P.: An overview of three approaches to scoring written essays by computer. Practical Assessment **151**(3), 501 (2001)
2. Bachman, L.F., Carr, N., Kamei, G., et al.: A reliable approach to automatic assessment of short answer free responses. In: Proceedings of the 19th International Conference on Computational Linguistics - Volume 2. DBLP (2002)
3. Wang, J., Guo, W., Tang, Z.: Automatic scoring method for subjective questions based on domain ontology and dependency parsing. J. Guizhou University (Natural Science) **37**(06), 79–84+124 (2020)
4. Huang, F.: Design of XML structure based automatic scoring system for text translation information. Modern Electron. Tech. **42**(23), 177–181 (2019)
5. Sultan, M.A., Salazar, C., Sumner, T.: Fast and easy short answer grading with high accuracy. In: Proceedings of the 2016 Conference of the North American Chapter of the Association for Computational Linguistics: Human Language Technologies, pp. 1070–1075 (2016)
6. Kumar, Y., Aggarwal, S., Mahata, D., et al.: Get IT scored using auto SAS — an automated system for scoring short answers. In: International Conference on Artificial Intelligence, 2019, vol. 33(01), pp. 9662–9669 (2019)
7. Jie, C.A.O., Mengyao, L.I., Dawei, C.H.E.N.: Automatic scoring algorithm of subjective questions based on LDA topic model. Comput. Programm. Skills Maintenance **04**, 119–121 (2020)
8. Zhang, S.: Automatic scoring technology of subjective questions based on twin neural network. Modern Comput. **2020**(05), 23–25 (2020)
9. Yifan, W.A.N.G., Guoping, L.I.: Automated scoring method for subjective questions based on semantic similarity and named entity recognition. Electron. Measur. Technol. **42**(02), 84–87 (2019)

10. Xie, X., Wu, D., Liu, S., et al.: IoT data analytics using deep learning. arXiv preprint arXiv: 1708.03854 (2017)
11. Yang, E., Ravikumar, P., Allen, G.I., et al.: A general framework for mixed graphical models. arXiv preprint arXiv:1411.0288 (2014)
12. Panchendrarajan, R., Amaresan, A.: Bidirectional LSTM-CRF for named entity recognition. In: Proceedings of the 32nd Pacific Asia Conference on Language, Information and Computation (2018)
13. Tamborrino, A., Pellicano, N., Pannier, B., et al.: Pre-training is (almost) all you need: An application to commonsense reasoning. arXiv preprint arXiv:2004.14074 (2020)
14. Yuanzhi, W.A.N.G., Ziying, C.A.O.: Chinese named entity recognition based on bert-BLSTM-CRF model. J. Anqing Normal Univ. (Natural Sci. Edition) **27**(01), 59–65 (2021)
15. Manocha, P., Badlani, R., Kumar, A., et al.: Content-based representations of audio using siamese neural networks. In: 2018 IEEE International Conference on Acoustics, Speech and Signal Processing (ICASSP) pp. 3136–3140. IEEE (2018)
16. Ganaie, M.A., Hu, M.: Ensemble deep learning: A review. arXiv preprint arXiv:2104.02395 (2021)
17. Adnan, K., Akbar, R.: Limitations of information extraction methods and techniques for heterogeneous unstructured big data. Int. J. Eng. Bus. Manage. **11**, 1847979019890771 (2019)
18. Zhang, M., Geng, G., Chen, J.: Semi-supervised bidirectional long short-term memory and conditional random fields model for named-entity recognition using embeddings from language models representations. Entropy **22**(2), 252 (2020)
19. Ji, B., Liu, R., Li, S., et al.: A hybrid approach for named entity recognition in Chinese electronic medical record[J]. BMC Med. Inform. Decis. Mak. **19**(2), 149–158 (2019)
20. Ma, X.Z., Eduard, H.: End-to-end sequence labeling via bi-directional LSTM-CNNs-CRF. Ann Meet Assoc Comput Linguist (ACL) (2016)
21. Dong, C., Zhang, J., Zong, C., et al.: Character-based LSTM-CRF with radical-level features for Chinese named entity recognition. In: International conference on computer processing of oriental languages, vol. 10102, pp. 221–230. Springer, Cham (2017). Doi: https://doi.org/10.1007/978-3-319-50496-4_20
22. Chen, T., Xu, R.F., He, Y.L., et al.: Improving sentiment analysis via sentence type classification usint BiLSTM-CRF and CNN. In: Experts Systems with Applications, pp. 260–270 (2016)
23. Hadsell, R., Chopra, S., LeCun, Y.: Dimensionality reduction by learning an invariant mapping. In: 2006 IEEE Computer Society Conference on Computer Vision and Pattern Recognition (CVPR' 2006), vol. 2, pp. 1735–1742. IEEE (2006)
24. Neculoiu, P., Versteegh, M., Rotaru, M.: Learning text similarity with siamese recurrent networks. In: Proceedings of the 1st Workshop on Representation Learning for NLP, pp. 148–157 (2016)
25. Aderhold, J., et al.: 2001 J. Cryst. Growth 222 701
26. Dorman, L.I.: Variations of Galactic Cosmic Rays (Moscow: Moscow State University Press), p. 103 (1975)
27. Caplar, R., Kulisic, P.: Proc. Int. Conf. on Nuclear Physics (Munich), vol. 1 (Amsterdam: North-Holland/American Elsevier) p. 517 (1973)

28. Szytula, A., Leciejewicz, J.: 1989 Handbook on the Physics and Chemistry of Rare Earths, vol. 12, ed K A Gschneidner Jr and L Erwin (Amsterdam: Elsevier), p. 133 (1989)
29. Kuhn, T.: Density matrix theory of coherent ultrafast dynamics Theory of Transport Properties of Semiconductor Nanostructures (Electronic Materials vol 4) ed E Schöll (London: Chapman and Hall) chapter 6, pp. 173–214 (1998)

Research on Positioning Technology of Facility Cultivation Grape Based on Transfer Learning of SSD MobileNet

Kaiyuan Han, Minjie Xu, Shuangwei Li, Zhifu Xu, Hongbao Ye, and Shan Hua$^{(\boxtimes)}$

Institute of Agricultural Equipment, Zhejiang Academy of Agricultural Sciences, Hangzhou 310021, China
huashan@zaas.ac.cn

Abstract. There is an urgent need of developing grape picking robot with intelligent recognition function due to the decrease of grape picking workers' population. Acquiring the 3D information of picking coordinate is the key process of constructing intelligent picking equipment. In this paper, based on SSD MobileNet neural network model, transfer learning and central deviation angle method were used to realize the positioning of picking coordinate points of facility cultivation grape by machine vision. After testing 720 fruit labels, 633 stem labels and 603 leaf labels labelled by pretreatment, the general precision was 79.5%, which was close to the inherent accuracy of the original model before transfer learning.

Keywords: Agricultural equipment · Object detection · Automatic picking · Transfer learning

1 Introduction

Grape picking is one of the most important links in grape production, which directly affects the market value of grapes. Picking is time-consuming and laborious, and its labor input accounts for 50% to 70% of the labor input in the entire grape planting process. The aging population of China is increasing, on the other hand the number of agricultural workers is decreasing. The inefficient manual picking will inevitably lead to higher and higher picking costs, and with the prevalence of large-scale and facility viticulture, the previous manual picking operations are difficult to adapt to the needs of market development. Therefore, the development of a grape picking robot with intelligent recognition function has become a hot research issue for scholars at home and abroad. One of the key issues in the development of intelligent recognition picking robots is the recognition and positioning of the target fruit. Zhiyong Xie and others used RGB channel recognition technology to realize the contour recognition of strawberry fruit, with an accuracy rate higher than 85%. Using the characteristic spectrum of apple reflection, Zhaoxiang Liu and others used PSD three-dimensional calibration technology to realize the positioning of the apple fruit, and the maximum deviation was controlled within 13 mm. Traditional optical recognition technology has the advantages of fast recognition

© The Author(s) 2022
Z. Qian et al. (Eds.): WCNA 2021, LNEE 942, pp. 600–608, 2022.
https://doi.org/10.1007/978-981-19-2456-9_61

speed and low structural complexity. However, it has insufficient processing capacity for obscured branches and leaves and overlapping fruits in a complex environment, and is difficult to use in actual production.

In recent years, there have been related researches on target positioning based on deep learning. Grishick proposed R-CNN (Regions with Convolutional Neural Network Features), which is a regional convolutional neural network [1]. The neural network uses a selective search algorithm to select 2000 candidate regions in the input image, and uses the volume of the image of each candidate region, producting neural network for feature extraction and recognition. This method is the first to combine deep learning with object detection algorithms. After that, Fast R-CNN and Faster R-CNN were successively proposed. Fast R-CNN solves the repeated convolution of candidate regions in R-CNN, and adds ROI pooling (Region of interest pooling) to the last layer of the extracted feature network [2], which significantly speeds up the recognition speed. Faster R-CNN builds RPN (Region Proposal Networks) on the basis of Fast R-CNN, which directly generates candidate regions and realizes high-accuracy end-to-end detection [3–5]. Its derivative iterative network model includes SSD (Single Shot Multibox Detector) etc.

Based on the SSD network model, this paper conducts further transfer learning and transformation, and uses the mode of multi-image combined analysis to study the location of grapes cultivated in facilities.

2 Materials and Methods

2.1 Image Acquisition

The image of grapes to be picked was collected as the training set and test set of SSD MobileNet model transfer learning training. The image of the training set would directly affect the microstructure of the model, and then affect the final accuracy [6]. Therefore, when selecting the image, it was necessary to collect representative and wide coverage images, and pay attention to the complexity of the background to avoid over fitting. The model of image acquisition equipment was Sony IMX363 with CMOS resolution of 4032×3024 pixels, using a lens with an equivalent focal length of 28 mm. In order to ensure the robustness of the target network model under various light sources, the light sources were not strictly limited. In the process of image acquisition, the light sources were randomly distributed. 30 clusters of Pujiang grapes with different shapes were selected as the experimental object. The cluster height was distributed between 17.3 cm–31.1 cm. The grapes were hung vertically downward perpendicular to the cross bar of facility cultivation. With the grape stem as the axis center, the lens was 50 cm away from the axis center. An image was taken every $15°$, and a total of 720 color images were taken.

2.2 Image Pretreatment

The image analysis and processing platform was a computer equipped with windows10 operating system, Intel i7-7700 CPU, 8 GB ram, NVIDIA Quadro P620 2 GB VRAM professional graphics card.

The training mode adopted in this paper is supervised learning, that is, it is necessary to input the label and previous frame content into SSD MobileNet model, and use the model to construct the mapping function of grape object detection. Manually mark the collected image with labelimg tool, place the grape fruit string in the rectangular box of the marking tool, and the upper, lower, left and right edges need to coincide with the rectangular box. Mark the position of grape stem. The edge marking is the same as that of fruit string. If the stem is blocked by fruit or leaves, it will not be marked. At the same time, if there are blades, the blades shall also be marked accordingly. A total of 720 fruit string labels, 633 fruit stem labels and 201 leaf labels were marked (Fig. 1).

Fig. 1. Manual marking of fruit.

Before the transfer learning training of the image, it is necessary to preprocess the image to remove some noise that may affect the accuracy, or lift the weight of some low weight training sets to prevent under fitting [7]. Because the number of 201 leaf labels collected was much less than the other two types, and there were large differences among leaves in different viewing angles, this paper oversampled the images with leaves. We transformed each image with a clockwise tilt of 10° and a counterclockwise tilt of 10°, so that the images with leaves were expanded to 603. After amplification, label the new samples manually with labelimg tool. Because there were images captured at various angles with the grape stem as the axis center, the training set samples were no longer subject to geometric preprocessing.

2.3 Transfer Learning

In this article, we used a programming environment with Tensorflow 1.14.0-gpu and CuDNN 7.6.0 to build a new SSD MobileNet.

SSD MobileNet is a neural network model combining MobileNet and SSD algorithm. MobileNet is used for image classification in the front end of the model, and SSD algorithm is placed in the back end to realize object detection [8]. MobileNet belongs to a lightweight convolutional neural network structure with relatively low network complexity. It can obtain better recognition rate on platforms with low computing power, such as mobile processor or embedded chip carried by agricultural machinery. This network contains the depthwise separable convolution [9]. In the conventional convolution calculation process, the total number of parameters is the number of channels plus the size of the convolution cores. A mature neural network model often involves the combination of several dozens of layers of convolution and pooling layers, so the size of parameters is large and will affect its rate. Depth Separable Convolution divides the traditional convolution calculation into two steps. First, Depthwise Convolution is performed, a separate feature map is generated in each channel. Then Pointwise Convolution is implemented by using a 1×1 convolution core. The weighted operation of the individual feature map in the depth direction gives a feature map consistent with the number of traditional convolution processes [10]. Because the number of parameters is significantly reduced during channel-by-channel convolution, this method can significantly reduce the number of parameters, improve the recognition rate, increase the network depth and increase the recognition accuracy in the neural network architecture mode with the same number of parameters.

The MobileNet V1 network structure has 28 layers. The entire network uses only an average pooling layer of $7 \times 7 \times 1024$ size at the end and a SoftMax classifier at the front. A serial combination of multiple convolution layers and deep detachable convolution layers is used at the front, which reduces the computing time required for pooling. This network model also introduces two superparameters: Width Multiplier α and Resolution Multiplier β, Width Multiplier α in the convolution result operation is $Dk \times Dk \times \alpha M \times Df \times Df + \alpha M \times \alpha N \times Df \times Df$, where $\alpha \in (0,1]$, when α is 1, for standard MobileNet model, when α is less than 1, it is a reduction model. Width factor α can make each layer in the network smaller, further accelerate training and recognition speed, but will affect accuracy. Resolution Multiplier β is to reduce the length and width of the input parameter, which can reduce the length and width of the output feature map in equal proportion [11].

The back-end SSD network model is a modification of the VGG16 network. SSD has 11 blocks, converting the sixth and seventh layers of the VGG16 full connection layer to a 3×3 convolution layer, removing the eighth layer of the Dropout layer and the full connection layer, and adding a new convolution layer to increase the number of feature maps. SSD uses a combination of feature maps of multiple resolutions to monitor. For different size targets, small size feature maps have low resolution and can be used for large-scale object detection. For fine texture targets, there is also a corresponding large size feature map to detect. This network is end-to-end, no longer requires candidate areas, and is more efficient than Faster R-CNN.

In transfer learning, the source domain is the built-in classification in the recognition classifier inherent in the MobileNet part [12], while the target domain is the classification set containing fruit strings, fruit stems and leaves. First, the labelled XML identification file needs to be converted to Tensorflow identifiable TFRecord format data. This paper divides 80% of the sample data into training set, 10% into test set and 10% into validation set. When configuring files and pipeline profiles, it is necessary to adjust the parameters of one training sample according to the size of graphics card's video memory. The size used in this paper is 16. We used fixed feature extractor for transfer learning. Solidify network structures such as mature convolution layers at the front end of the model, were used as feature extractors for the process required by the target domain. At last, train classifiers at the end of the network and related parts of the structure for constructing new classifiers [13–15].

2.4 Position Calibration

After getting the network model completed by transfer learning, the network model can identify the contents of the target domain in the image and provide the coordinate points of the rectangular vertex of the recognition box in the image. During the harvesting process, the end executor uses the method of cutting the fruit stem and receiving at the bottom of the fruit string. Therefore, this paper mainly carries out location calibration on the center of the fruit stem and the bottom of the fruit string.

Depth distance acquisition was carried out with a micro laser range finder. The measurement accuracy of the range finder is < 1 mm, the measurement range is 0.03–80.00 m, the spot diameter is less than 0.6 mm under normal working conditions, and it was parallel to the camera on a 360° rotatable electronic pedestal. The camera lens center had a horizontal distance of 2.5 cm from the center of the transmission module of the distance sensor.

When collecting the 3-D coordinate data of the target object, the fruit stem is located by the return value of the object detection. When the picture combination is only fruit strings and blades, it prompts for moving until the fruit stem appears. After the object detection identifies the fruit stem, the target object is placed in the center of the picture by rotating the rotatable support, and the horizontal and vertical rotation angles of the support are recorded at this time. Sweep left and right to get the return value characteristic spectrum of the range sensor. The minimum value x of the characteristic spectrum is determined as the depth distance, then the three-dimensional coordinate of the target point is (x·cosβ·sinα, x·sinβ, x·cosβ·cosα) (Fig. 2).

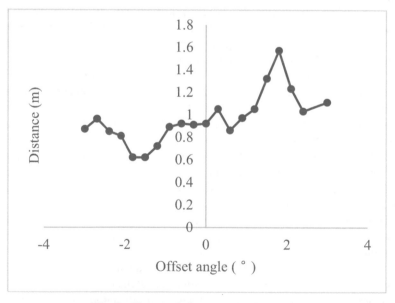

Fig. 2. Characteristic spectrum of distance.

3 Experimental Results and Analysis

3.1 Object Detection Results

The model was migrated using fixed feature extraction, which included 576 training sets, 72 test sets and 72 validation sets of fruit strings; 506 training sets of fruit stems, 63 testing sets and 63 validation sets; 482 training sets of leaf blades, 60 testing sets and 60 validation sets. Setting batch size to 16, initial learning rate to 0.003, maximum training times to 10,000, when iterating training at 5000 times, the recognition accuracy reached a maximum of 82.9%. Where IOU > 0.85, it was determined correct (Table 1).

Table 1. Results of object detection.

Number of iterations	Comprehensive accuracy (%)	Loss
1000	56.6	3.117727
2000	62.8	2.222301
3000	75.7	2.473103
4000	76.8	1.623363
5000	82.9	1.607318
6000	80.1	1.509823
7000	76.3	1.586632
8000	75.1	1.593135
9000	77.6	2.698133
10000	73.2	2.553231

When the batch size was reduced to a smaller scale, loss begins to fluctuate greatly with the increase of the number of iterations, so it is difficult to carry out good normalization conversion, and it is impossible to accurately calculate the mean and variance of all data. At the same time, the recognition accuracy will also decline significantly [16]. As the batch size increased, the number of parameter updates was less and the gradient decreases more accurately. However, because of a too large batch size, the training stops due to the insufficient display memory. At the same time, too large batch size also affects the performance of the random function [17, 18].

3.2 Comprehensive Test Results

Due to the combination of object detection output, comprehensive image analysis, and side-axis ranging data, the final three-dimensional coordinate points need to be determined with the accuracy of the data. 20 strings of fruits were measured with a camera and a ranging sensor installed on a rotatable support. A single target was repeated 10 times, totaling 200 times. The target recognition model identified the stem and the bottom of the string. When the IOU is >0.85, the recognition is correct. When the error between the three-dimensional coordinate points and the actual measurement was less than 1.5 cm, the calculation was correct. Among them, the correct number of target recognition was 159, the accuracy was 79.5%, and the accurate number of three-dimensional coordinate positioning was 159, that is, the error of coordinate calculation of all correctly identified targets was within the allowable range, and the overall accuracy was 79.5%.The aliasing frame rate remains around 20 fps, which achieved good recognition results.

4   Discussion

In Tensorflow platform, SSD MobileNet V1 was used to transfer and learn the characteristics of grape picking samples, and the recognition accuracy was close to the original model. Through the central deviation angle method and the depth data of rangefinder, the picking three-dimensional coordinate information is constructed.

Transfer learning significantly speeds up the efficiency of model construction, and eliminates the process of repeatedly adjusting network structure, optimizing network node parameters, collecting and labeling a large number of sample sets. In the fixed feature extraction process, there is a better generalization ability of the original mature network for feature extraction, which makes the recognition rate and accuracy of the target domain task close to or even exceed the original model. It is very suitable for the model construction of target recognition of grapes and other fruits and vegetables.

In this paper, the three-dimensional coordinate information obtained by the combination of object detection and central deviation angle method is constructed from the orientation of the image receiving end. In the future construction of picking machinery, the coordinate information can be transformed into the final coordinate point required for the positioning of the end effector by re-calibration. When the object detection is completed, the calculation accuracy of three-dimensional coordinate information is close to 100%. The focus of further improving the comprehensive accuracy lies in the further transformation and optimization of the object detection model.

5   Conclusion

According to the subdivision steps of grape picking process, SSD MobileNet V1 network model is used for grape picking sample transfer learning by using fixed feature extraction. Combined with the central deviation angle method, we achieved 79.5% comprehensive accuracy in 200 physical samples, which is close to the inherent accuracy of the original model before transfer learning. It shows that the method in this paper has achieved ideal migration effect in the target domain.

Acknowledgements. This research is funded by the project "Research on new technology of intelligent facility agricultural safety production - research and development of intelligent multi ecological three-dimensional planting in greenhouse and multi-functional electric operation platform" from Key Research and Development Projects in Zhejiang Province (Subject No.: 2019C02066).

References

1. Zeiler, M., Fergus, R.: Visualizing and understanding convolutional networks. In: Fleet, D., Pajdla, T., Schiele, B., Tuytelaars, T. (eds.) ECCV 2014. LNCS, vol. 8689, pp. 818–833. Springer, Cham (2014). https://doi.org/10.1007/978-3-319-10590-1_53
2. Girshick, R.: Fast R-CNN. Computer Ence (2015)
3. Shaoqing, R., Kaiming, H., Girshick, R., Jian, S.: Faster R-CNN: towards real-time object detection with region proposal networks. IEEE Trans. Pattern Anal. Mach. Intell. 1137–1149 (2017)
4. Lanchantin, J., Singh, R., Wang, B., Yanjun, Q.: Deep motif dashboard: visualizing and understanding genomic sequences using deep neural networks. In: PSB 2017: Pacific Symposium on Biocomputing, pp. 254–265 (2017)
5. Li, Y., Huang, H., Xie, Q., Yao, L., Chen, Q.: Research on a surface defect detection algorithm based on MobileNet-SSD. Appl. Sci. **8**(9), 1678 (2018)
6. Weiss, K., Khoshgoftaar, T.M., Wang, D.: A survey of transfer learning. J. Big Data **3**(1), 1–40 (2016). https://doi.org/10.1186/s40537-016-0043-6
7. Zheng, Q., Zhaoning, Z., Shiqing, Z., Hao, Y., Yuxing, P.: Merging-and-evolution networks for mobile vision applications. IEEE Access **6**, 31294–31306 (2018)
8. Ni, Z., Yan, Y., Si, C., Hanzi, W., Chunhua, S.: Multi-label learning based deep transfer neural network for facial attribute classification. Pattern Recogn. **80**, 225–240 (2018)
9. Zhu, J., Liao, S., Yi, D., Lei, Z., Li, S.: Multi-label CNN based pedestrian attribute learning for soft biometrics. In: 2015 International Conference on Biometrics (ICB), pp. 535–540 (2015)
10. Vishal, P., Raghuraman, G., Ruonan, L., Ca, R.: Visual domain adaptation: a survey of recent advances. IEEE Signal Process. Mag. **32**(3), 53–69 (2015)
11. Yuhua, C., Wen, L., Christos, S., Dengxin, D., Luc, G.: Domain adaptive faster R-CNN for object detection in the wild. In: 2018 IEEE/CVF Conference on Computer Vision and Pattern Recognition, pp. 3339–3348 (2018)
12. Bertasius, G., Torresani, L., Shi, J.: Object detection in video with spatiotemporal sampling networks. In: Ferrari, V., Hebert, M., Sminchisescu, C., Weiss, Y. (eds.) ECCV 2018. LNCS, vol. 11216, pp. 342–357. Springer, Cham (2018). https://doi.org/10.1007/978-3-030-01258-8_21
13. Yi, T., Wenbin, Z., Zhi, J., Yuhuan, C., Yang, H., Xia, L.: Weakly supervised salient object detection with spatiotemporal cascade neural networks. IEEE Trans. Circuits Syst. Video Technol. **29**(7), 1973–1984 (2019)
14. Wei, F., Lin, W., Peiming, R.: Tinier-YOLO: a real-time object detection method for constrained environments. IEEE Access **8**, 1935–1944 (2020)
15. Srivastava, N., Hinton, G., Krizhevsky, A., Sutskever, I., Salakhutdinov, R.: Dropout: a simple way to prevent neural networks from overfitting. J. Mach. Learn. Res. **15**, 36–47 (2014)
16. Bach, S., Binder, A., Montavon, G., Klauschen, F., Müller, K.-R., Samek, W.: On pixel-wise explanations for non-linear classifier decisions by layer-wise relevance propagation. PloS One **10**, 21–30 (2015)
17. Mottaghi, R., et al.: The role of context for object detection and semantic segmentation in the wild. In: 2014 IEEE Conference on Computer Vision and Pattern Recognition (CVPR), pp. 891–898 (2014)
18. Long, J., Shelhamer, E., Darrell, T.: Fully convolutional networks for semantic segmentation. In: Proceedings of the IEEE Conference on Computer Vision and Pattern Recognition, pp. 3431–3440 (2015)

Application of Big Data Technology in Equipment System Simulation Experiment

Jiajun Hou, Hongtu Zhan, Jia Jia, and Shu Li[✉]

Department of the Second System Integration of CEC Great Wall ShengFeiFan Information System Co., Ltd., Shenzhen, China
lishu@greatwall.com.cn

Abstract. In order to solve the problem of single utility between system data in simulation experiment, the simulation experiment method and experimental framework of equipment system development are analyzed. This paper constructs the experimental framework of big data technology in equipment system simulation, uses big data analysis technology, analyzes the application process in equipment system simulation experiment, and puts forward the shortcomings and difficulties of applying big data technology in equipment system simulation experiment. By introducing big data technology, it provides a reference basis for weapon equipment system development demonstration.

Keywords: Big data · Simulation experiment · Data application

1 Introduction

In recent years, the Key Laboratory of complex ship system simulation has accumulated a large amount of equipment system demand demonstration data, equipment construction scheme data, equipment performance data, and equipment performance data in the process of using the simulation experiment system for equipment system development to provide support for equipment combat demand demonstration [1], equipment development strategy demonstration, equipment planning plan demonstration and equipment key technology demonstration Force deployment data, equipment combat effectiveness data, battlefield environment data, key technology data and other multi type data [2–4]. Due to the different use characteristics and storage structure of the data of each system in the experimental environment, the utility of each system is single, and the value of the data can not be fully realized [5–7]; Therefore, the author introduces big data technology to find out the relationship in the process of operational demand demonstration, development strategy demonstration, planning plan demonstration and key technology demonstration, mine and give full play to the maximum utility of existing data, and realize the integrated and collaborative demonstration among operational demand, development strategy, equipment construction and key technology [8].

© The Author(s) 2022
Z. Qian et al. (Eds.): WCNA 2021, LNEE 942, pp. 609–614, 2022.
https://doi.org/10.1007/978-981-19-2456-9_62

2 Application Mode of Big Data Technology in Simulation Experiment

A simulation experiment system has been built with the operational experiment database and key technology management platform as the data support and the operational deduction research, operational simulation research, military value analysis method, system evolution simulation method and technology maturity evaluation method as the theoretical support, so as to complete the demonstration from equipment operational requirements to equipment development strategy, and then to equipment construction planning, Until the whole process and multi angle demonstration process of equipment key technology demonstration, so as to realize the construction and development of weapon equipment demonstration system [9]. Equipment system of systems experimental framework is shown in Fig. 1.

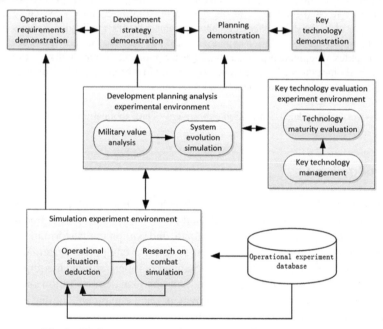

Fig. 1. Equipment system of systems experimental framework

The operational experiment database mainly provides data support for operational deduction research and operational simulation research. Operational deduction research and operational simulation research jointly provide theoretical support for the equipment system of systems confrontation simulation experimental environment to support the demonstration of operational requirements.

Military value analysis method and system evolution simulation method jointly provide theoretical support for the simulation analysis experimental environment of equipment development planning.

The key technology management platform provides data support for the technology maturity evaluation method, and together constitutes the equipment key technology evaluation experimental environment to support the key technology demonstration.

The equipment system of systems confrontation simulation experiment environment, equipment development planning simulation analysis experiment environment, equipment key technology evaluation experiment environment and combat experiment database support each other and cooperate organically, which constitute the experimental framework for the development of equipment system of systems.

3 Application Process of Big Data Technology in Simulation Experiment

In the demonstration process for the development of equipment system, a set of integrated demonstration methods are provided by using the above experimental framework. There is still no actual coordinated demonstration in the data flow, and the systems only achieve logical consistency. When facing the demonstration task, they mostly rely on the experience analysis of arguers, Independently use each system to provide corresponding experimental support.

Introduce big data technology, adopt big data storage and management technology, break through the data barriers between systems, comprehensively analyze the heterogeneous data of multiple systems and scenarios by using big data analysis technologies such as data mining and in-depth learning, identify valuable information from massive data information, analyze and judge the laws of strategic and tactical application, equipment development and construction According to the evolution law of equipment structure and the iteration law of key technologies, starting from the top level of operational requirements, clarify the equipment development strategy, put forward the equipment construction plan, sort out the framework system and development roadmap of key supporting technologies, and provide scientific experimental support for the better and faster development of weapon equipment system. The application process of big data technology is shown in Fig. 2.

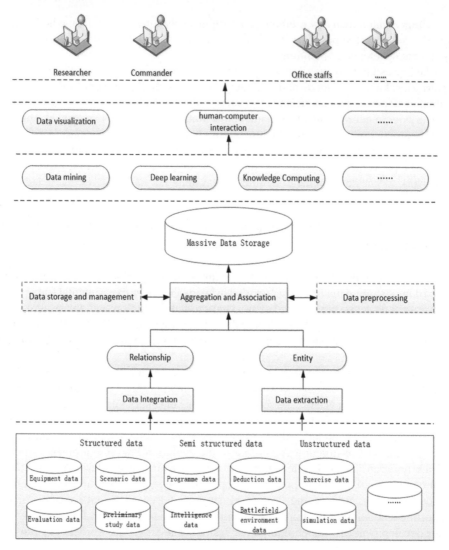

Fig. 2. Application process of big data in equipment system of systems simulation experiment

4   Application Analysis of Big Data Technology

At this stage, the data in the equipment system simulation experiment fails to meet the requirements of big data characteristics, the data volume is not enough, the data acquisition process is slow, the real-time contact with the real equipment state cannot be established, and the battlefield environment data is one-sided and untrue. To make full use of the advantages of big data technology, we must recognize the shortcomings and difficulties of applying big data technology in the current equipment system simulation experiment.

1) Data acquisition channels are blocked.

   At present, major units have established multiple data centers. Due to the poor organizational relationship, the interconnection between data centers has not been solved. Many important data are basically distributed in the hands of business organs, and a complete and easy-to-use data warehouse has not been formed.

2) Poor real-time data.

   In the equipment system simulation experiment, the data sources include military exercises, major subject research, data engineering construction, scheme evaluation, etc. the real-time performance of the data is difficult to be guaranteed, especially the equipment strength statistics, which is updated once a year. Even if the data is obtained, it is the equipment state one year ago, and the research and analysis conclusions can not reflect the latest equipment situation.

3) The doubling of data volume challenges data storage capacity.

   Video, audio, battlefield environment monitoring data and other huge data sources require the use of special database technology and special data storage equipment. The doubling of the amount of data is a great challenge to the data storage capacity.

4) Diverse data types challenge data processing capabilities.

   With the increase of multi-source data storage, data types become more complex, including not only traditional relational data types, but also unprocessed, semi-structured and unstructured data in the form of web pages, video, audio and documents. The diversification of data types challenges the traditional data analysis platform.

5) Data heterogeneity and incompleteness challenge the ability of data management.

   Equipment system of systems simulation experiments involve a wide range of data. The data directly obtained or precipitated by experiments are generally heterogeneous, which is difficult to describe with a simple data structure.

6) Data security challenges organizational management.

   Data is faced with security risks in the process of storage, processing and transmission. For military data, data security is the top priority. In order to achieve big data security and ensure the efficient and rational use of data, it has brought challenges to the current organization and management mode.

5   Conclusion

In the simulation experiment for the development of equipment system, big data technology is introduced to mine and analyze the hidden laws of equipment application, development and evolution according to systematic thinking, so as to provide a scientific basis for the demonstration of system development of weapons and equipment. It can break the traditional modeling and simulation technology based on accurate calculation, and realize the fuzzy The new scientific research paradigm without hypothesis injects new vitality into the equipment system of systems simulation experiment.

References

1. Lin, X., Jia, L., Wu, X.: Application mode and difficulty analysis of big data technology in simulation experiment of equipment system. Ordnance Ind. Autom. **38**(7), 26–29 (2019)

2. Luo, R., Xiao, Y., Wang, L., Sheng, L.: Application of big data in command information system of Naval Battle field. Ship Electron. Eng. (3), 1–5 (2019)
3. Poelmans, J., Dmitry, I.I., Sergei, O.K., et al.: Fuzzy and rough for-mal concept analysis: a view. Int. J. General Syst. **43**(2), 105–134 (2014)
4. Arasu, A., Chaudhuri, S., Chen, Z., et al.: Experiences with using data cleaning technology for Bing services. IEEE Data Eng. Bull. **35**(2), 14–23 (2012)
5. Sun, D.W., Zhang, G., Zheng, W.M.: Big data stream computing: technologies and instances. Ruan Jian XueBao/J. Softw. **25**(4), 839–862 (2014)
6. Philip, R.: Big Data Analytics. TDWI Best Practices Report, TDWI, USA (2011)
7. Hinton, G.E., Osindero, S., Teh, Y.W.: A fast learningalgorithm for deep belief nets. Neural Comput. **18**(7), 1527–1554 (2006)
8. Xu, W.B.J.: Development trend of big data technology in command information system. Command Inf. Syst. Technol. **5**(3), 12–16 (2014)
9. Zhang, Y.: Big data ecosystem application and countermeasures in naval warfare. Natl. Defense Sci. Technol. **36**(3), 101–103 (2015)

A User-Interaction Parallel Networks Structure for Cold-Start Recommendation

Yi Lin[✉]

Beijing National Day School, Beijing, China
lykyyx2021@163.com

Abstract. The goal of the recommendation system is to recommend products to users who may like it. The collaborative filtering recommendation algorithm commonly used in recommendation systems needs to collect explicit/implicit feedback data, and new users do not leave behavioral data on the product, which leads to cold-start problem. This paper proposes a parallel network structure based on user interaction, which extracts features from user interaction information, social media information, and comment information and forms a matrix. The graph neural network is introduced to extract high-level embedded correlation features and the role of parallelism is to reduce computing cost further. Experiments based on standard data sets prove that this method has a certain degree of improvement in NDCG and HR indicators compared to the baseline.

Keywords: Recommendation system · Cold-start problem · Parallel GCN · High-level correlation features

1 Introduction

With the widespread deployment of the Internet and mobile Internet, billions of people have experienced online shopping. In online shopping applications like Amazon, one of the most important intelligent systems is the recommendation system, that is, the system recommends potential products to users or expands users' interests in other areas; recommendation systems are also widely used in social networks to automate the social process of recommending friends or news to users [1].

One kind of recommendation system connect two different areas together, Zero-Shot learning (ZSL) and Cold-Start Recommendation (CSR) use their own Low-rank Linear Auto-Encoder (LLAE) [2]. The important challenge faced by online recommendation systems is the well-known cold start problem: how to provide advice to the new user? The embedded Influential-context Aggregation Unit (ICAU) as their ways to solve the problem for CSR. Their ICAU-based Heterogeneous Relations for Sparse model was presented in the passage to learn the user's behaviour and give appropriate recommendations [3]. In the recommendation system, a MAML-based user preference estimator for movie recommendation. The MeLU model was separated into several layers that could be constantly updated to suit for new users based on its fast-learning speed. When user plug in their basic information, the model will adjust the movies for users to evaluate

© The Author(s) 2022
Z. Qian et al. (Eds.): WCNA 2021, LNEE 942, pp. 615–622, 2022.
https://doi.org/10.1007/978-981-19-2456-9_63

based on their ages and work previously collected by the system, then give the recommended movies based on the ratings the user gives. The feature or advantage of the model could give better results than regular methods, such as PPR and Wide & Deep, when encounter new users or new items [4]. Another approach of meta-learning to deal with CSR questions. This model proposed in the paper has the features of fast-learning speed and offers satisfying results just based on small datasets. Another unique feature of this model is its adaptive learning based on HINs to cope with different tasks easily. The result of the researcher's experiments shows that, in both normal and new conditions, the HIN-based meta-learning model gives better results than regular models used in previous researches [5].

The recommendation complete current condition of the CSR problems and proposes their two separate solutions. The first solution is the framework of investigating the CF approach and machine learning algorithms to improve the performance for CS items. Then the second solution proposed is based on the first solution's general framework. The original timeSVD++ model was presented by researchers to deal with the problem. This model make uses of CCS items with non-CS items' similarity, and make use of different biases predictors to fully demonstrate the ability of the model. The results show that the timeSVD++ based IRCD-ICS model has the best performance of the five tested model [7]. The paper [9] proposed one linear-based model to deal with the CSR problems. To begin their researches, this paper analyzes three popular models that commonly used in solving CSR recommendations, and leads to the result that they are all the special case of the linear content-based model. Based on this results, the researchers gives their own model, the Low-Rank Linear CSR model.

This paper proposes a parallel network structure based on user interaction. The parallel graph neural network structure is used to process a matrix containing user interaction information, social media information and comment information at the same time. The purpose is to form a unified information among the three. The embedded structure fully captures the high-level relevance of the three, and reduces the computational dimension through parallel GNN. Experiments based on standard data sets prove that this method is better than baseline in standard measures and has a certain improvement in efficiency.

The rest of this paper is: the part II gives the general method of cold start of the recommendation system; the part III introduces the parallel network structure based on user interaction; the fourth part is the score results on the dataset; the last part gives the conclusion.

2   Cold-Start Recommendation Structure

In the recommendation system application, there are two types of entities, which we call users and items. The main purpose of the recommendation system is to filter based on the user's preference for a certain item (such as a movie or book), generally using content-based item features or user social data based on collaborative filtering. The general structure of the recommendation system is shown in Fig. 1. In the past ten years, due to the popularization of the Internet, the massive amount of data generated has provided a rapid development opportunity for the recommendation system. The increasing demand for recommendation systems has caused many difficulties and challenges. Methods similar to cluster filtering and enhanced collaborative filtering have been proposed as a rich research field, recommendation system still needs continuous improvement.

Bi-clustering and Fusion [12] is a method that combines clustering and scoring to provide accurate recommendations for social recommendation systems. It tries to construct dense areas of the item-user rating matrix to solve the cold start problem. First, the method determines the popular items and extracts the scores into the item-user rating matrix; next, the role of Bi-clustering is to reduce the sparseness problem, smooth the ratings and aggregate similar users/items to form clusters, so that the items can be recommended to the classified customers Bi-clustering and Fusion. Its advantage is that it improves the accuracy of the recommendation while further reducing the dimension of the item-user rating matrix. In addition, the solution to the cold start problem is to remove the impact of sparseness and cluster users/items for smoothing.

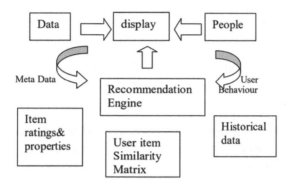

Fig. 1. Recommendation system framework [10].

The starting point for the design of neural networks is that computers learn to a certain extent similar to the way the human brain processes information. For the cold-start problem of the recommendation system, neural network [13] could optimize the similarity scoring process, which especially in the hybrid recommendation system by using neural network to learn user parameters or in the cluster recommendation system to learn voting information, such as Widrow-Hoff and other methods are used to learn user/item information to refine user parameter granularity.

The mathematical description of the cold start problem is as follows [8]: U is the group of users and $\mathcal{P}$ is the group of products. $a_{u'}$ represents whether current user purchased p. Each $u \in \mathcal{U}$ connected with $\mathcal{P}$ and has a timestamp. A small number of U linked to their social media content. $\mathcal{A}$ denote the social media features and each account has a $|\mathcal{A}|$ size vector. The social media account $u \notin \mathcal{U}$ is a new user to the e-commerce platform because it has no record of purchasing on the platform. In order to generate a unique product purchase recommendation ranking for each account from its social media account, due to the heterogeneous problem of social media and product purchase, the information from the social media account cannot be directly useful for product recommendation. Change the user's social account information to feature $\mathbf{V}_{u'}$, where the purpose of u is to make platform recommendations.

Common inputs in collaborative filtering include user set $\mathcal{U} = u_1, u_2, ..., u_n$ and item $\mathcal{J} = v_1, v_2, ..., v_m$. The recommendation level in the system can be represented by a matrix $Y \in R^{m \times n}$ that each item y_{ij} corresponds to the score of i by j. The general CF matrix decomposition is based on the rank $Y \approx UV$ form, where $U \in \mathbb{R}^{m \times k}$ and $V \in \mathbb{R}^{k \times n}$ characterization matrices represent potential factors, and the error is mainly obtained by minimizing reconstruction [11].

3   Parallel Network Structure Based on User Interaction

The latent factor model for users is one of the useful methods of the user recommendation system [6], but the interaction between users is often sparse, that is, there is a cold start problem, which limits the role of the latent factor model. The improved methods include normalized matrix decomposition for more relationship information similar to those embodied on social media, which to establish a standardized user-comment similarity evaluation model, and the use of word2vec to build an embedded model.

Graph representation is a method of describing data structure objects and their relationships in the form of nodes and edges [14, 15]. In recent years, many researchers have used machine learning to achieve graph representation, that is, graphs can be used to represent data structures in complex systems such as social networks for classification, Prediction and clustering operations. The graph neural network based on deep learning has interpretability and good performance. GNN draws on the ability of convolutional neural networks to express multi-scale spatial features, but CNN can only process European data (Fig. 2).

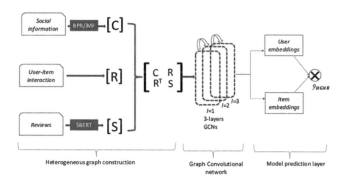

Fig. 2.   User interations and expected social connections [6].

Aiming at the problem of data sparseness caused by cold start, this paper proposes a parallel network cold start recommendation method based on user interaction information, social media information, and comment information, which is shown in Fig. 3. The purpose is to extract the embedded structure between the three types of information at the same time and obtain more information of high-level correlation inference. The purpose of the parallel structure is to compress further sparse data to achieve the purpose of reducing the computational dimension.

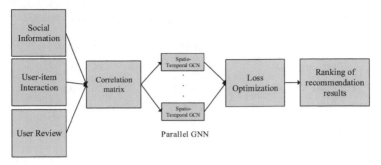

Fig. 3. Parrallel GNN structure based on three information.

In the input part, the user interaction information, social media information and comment information are combined to extract the embedded structure and form an embedded matrix. In the parallel GNN, multiple Spatio-Temporal GCN parallel methods are mainly used to divide the matrix into multiple sub-matrices through the connection structure, where each most sub-matrix is adjusted to achieve parallel compression of sparse data and reduce the amount of calculation. Finally, loss optimization is performed and the recommended ranking result is output.

4 Experimental Results

E-commerce platforms like amazon can provide a large amount of user and product data. Founded in 2004, Yelp is a well-known merchant review website in the United States, covering merchants in restaurants, shopping malls, hotels, tourism, etc. from all over the world. Users use the Yelp website to rate merchants and submit reviews.

This paper selects Yelp's 2014 dataset [16], which has more than 40k business items and 110k text comments from Phoenix and other regions. Yelp Reviews format is divided into two types: JSON and SQL, which contains user/check-in/business/tip/review saved in JSON files with specified ID. Comments for different business categories maybe very different in their contents. Therefore, it is necessary to clean and preprocess the data set to ensure the consistency of the data distribution.

First, we selected 100,000 reviews and converted the JSON format of these reviews into CSV format. From these reviews, we selected Cold-start users, that is, users with less than 5 user-item interactions. The model we used was pre-trained on the adjusted 2014 dataset training set. In order to verify the performance of this network structure, we compared and evaluated the baseline and the method proposed in this article on the above data set, and then selected part of the data for parameter fine-tuning, and the number of iterations in the fine-tuning stage is determined based on experience, and finally tested on the test dataset (Table 1).

Table 1. NDCG/HR score and average improvement of two methods.

	NDCG@10	HR @10
NeuMF structure	0.1285	0.2671
Proposed structure	0.1324	0.2798
Average improvement	3.0%	4.8%

Normalized Discounted Cumulative Gain is the evaluation index of the sorting result to evaluate the accuracy of the sorting, where Gain represents the relevance score of each item in the list, and Cumulative Gain represents the accumulation of the gain of K items. The calculation formula is nDCGp = $DCGp/IDCGp$. Here for $p < 0.05$, the improvement is statistically significant compared to all other methods.

The baseline in this paper uses Neural collaborative filtering [17], which is a collaborative filtering method in recommendation systems. Unlike other algorithms that use neural networks to extract auxiliary features, user and item are still calculated using matrix inner products.

Table 1 shows the NDCG and HR scores obtained under the condition of cold start under Neural collaborative filtering (NeuMF) and the structure proposed in this paper. On the sparse Yelp dataset selected based on the cold start problem, the percentage improvements on the NDCG@10 and HR@10 indicators were 3.0% and 4.8%, respectively. This result shows that the proposed structure obtains better scores than the classical NeuMF method.

5   Conclusion

In online recommendation systems, products are recommended based on a large amount of user information. The cold start problem has always been one of the thorny issues that commercial recommendation platforms need to solve. Commonly used collaborative filtering methods are very unsuccessful for new users who do not have a lot of information. This paper proposes a parallel graph neural network based on user interaction, and extracts the embedded information of the user interaction letter/social media/comment information matrix to obtain high-level correlation. The parallel method further reduces the computational cost. Experiments based on the yelp data set prove that the standard index of this method under cold start conditions has certain advantages compared with NeuMF.

References

1. Park, S.T., Chu, W.: Pairwise preference regression for cold-start recommendation. In: Proceedings of the Third ACM Conference on Recommender Systems, pp. 21–28 (2009)
2. Li, J., Jing, M., Lu, K., et al.: From zero-shot learning to cold-start recommendation. In: Proceedings of the AAAI Conference on Artificial Intelligence, vol. 33, no. 01, pp. 4189–4196 (2019)

3. Hu, L., Jian, S., Cao, L., et al.: HERS: modeling influential contexts with heterogeneous relations for sparse and cold-start recommendation. In: Proceedings of the AAAI Conference on Artificial Intelligence, vol. 33, no. 01, pp. 3830–3837 (2019)
4. Lee, H., Im, J., Jang, S., et al.: MeLU: meta-learned user preference estimator for cold-start recommendation. In: Proceedings of the 25th ACM SIGKDD International Conference on Knowledge Discovery & Data Mining, pp. 1073–1082 (2019)
5. Lu, Y., Fang, Y., Shi, C.: Meta-learning on heterogeneous information networks for cold-start recommendation. In: Proceedings of the 26th ACM SIGKDD International Conference on Knowledge Discovery & Data Mining, pp. 1563–1573 (2020)
6. Liu, S., Ounis, I., Macdonald, C., et al.: A heterogeneous graph neural model for cold-start recommendation. In: Proceedings of the 43rd International ACM SIGIR Conference on Research and Development in Information Retrieval, pp. 2029–2032 (2020)
7. Wei, J., He, J., Chen, K., et al.: Collaborative filtering and deep learning based recommendation system for cold start items. Expert Syst. Appl. **69**, 29–39 (2017)
8. Zhao, W.X., Li, S., He, Y., et al.: Connecting social media to e-commerce: cold-start product recommendation using microblogging information. IEEE Trans. Knowl. Data Eng. **28**(5), 1147–1159 (2015)
9. Sedhain, S., Menon, A., Sanner, S., et al.: Low-rank linear cold-start recommendation from social data. In: Proceedings of the AAAI Conference on Artificial Intelligence, vol. 31, no. 1 (2017)
10. Sharma, L., Gera, A.: A survey of recommendation system: research challenges. Int. J. Eng. Trends Technol. **4**(5), 1989–1992 (2013)
11. Liu, N.N., Meng, X., Liu, C., et al.: Wisdom of the better few: cold start recommendation via representative based rating elicitation. In: Proceedings of the Fifth ACM Conference on Recommender Systems, pp. 37–44 (2011)
12. Zhang, D., Hsu, C.H., Chen, M., et al.: Cold-start recommendation using bi-clustering and fusion for large-scale social recommender systems. IEEE Trans. Emerg. Top. Comput. **2**(2), 239–250 (2013)
13. Bobadilla, J.S., Ortega, F., Hernando, A., et al.: A collaborative filtering approach to mitigate the new user cold start problem. Knowl.-Based Syst. **26**, 225–238 (2012)
14. Qiu, J., Chen, Q., Dong, Y., et al.: GCC: graph contrastive coding for graph neural network pre-training. In: Proceedings of the 26th ACM SIGKDD International Conference on Knowledge Discovery & Data Mining, pp. 1150–1160 (2020)
15. Zhou, J., Cui, G., Zhang, Z., et al.: Graph neural networks: a review of methods and applications. arXiv preprint arXiv:1812.08434 (2018)
16. Asghar, N.: Yelp dataset challenge: review rating prediction. arXiv preprint arXiv:1605.05362 (2016)
17. He, X., Liao, L., Zhang, H., et al.: Neural collaborative filtering. In: Proceedings of the 26th International Conference on World Wide Web, pp. 173–182 (2017)

A Brief Comparison of K-means and Agglomerative Hierarchical Clustering Algorithms on Small Datasets

Hassan I. Abdalla[(⊠)]

College of Technological Innovation, Zayed University, P.O. Box 144534, Abu Dhabi, UAE
hassan.abdalla@zu.ac.ae

Abstract. In this work, the agglomerative hierarchical clustering and K-means clustering algorithms are implemented on small datasets. Considering that the selection of the similarity measure is a vital factor in data clustering, two measures are used in this study - cosine similarity measure and Euclidean distance - along with two evaluation metrics - entropy and purity - to assess the clustering quality. The datasets used in this work are taken from UCI machine learning depository. The experimental results indicate that k-means clustering outperformed hierarchical clustering in terms of entropy and purity using cosine similarity measure. However, hierarchical clustering outperformed k-means clustering using Euclidean distance. It is noted that performance of clustering algorithm is highly dependent on the similarity measure. Moreover, as the number of clusters gets reasonably increased, the clustering algorithms' performance gets higher.

Keywords: Clustering · K-means · Hierarchical clustering · Clustering comparison · Cosine · Euclidean

1 Introduction

Clustering algorithms are a vital techniques of machine learning, and are widely used in almost all scientific application including databases [1, 2], collaborative filtering [3], text classification [4], indexing, etc. The clustering is an automatic process of assembling of data points into similar assembles so that points in the same cluster are highly similar to each other, and maximally dissimilar to points in other assembles. With the constantly-increasing volumes of daily data and information, clustering is being undeniably helpful technique in organizing collections of data for an efficient and effective navigation [1]. However, with the dynamic characteristics of the collected data, the clustering algorithms have to be able to cope and deal with the newly-added data in every second so it would help in discovering knowledge effectively and timely. As one of the most commonly known techniques for the unsupervised learning, clustering comes with the main objective finding the natural clusters among the assigned patterns. It simply groups data points into categories of similar points.

This paper is organized as follows: in Sect. 2, related work is briefly covered. Section 3 covers methodology including clustering algorithms and similarity measures used in

© The Author(s) 2022
Z. Qian et al. (Eds.): WCNA 2021, LNEE 942, pp. 623–632, 2022.
https://doi.org/10.1007/978-981-19-2456-9_64

this work. Section 3 introduces performance evaluation including experimental setup, datasets description, evaluation metrics and results. Discussion is concisely covered in Sect. 4. Finally, conclusions and future work is given in Sect. 5.

2 Related Work

In literature, the Hierarchical clustering is often seen to give solutions of better quality than k-means. However, it is limited due to its complexity in terms of quadratic time. Opposed to hierarchical, K-means has a linear time complexity. It is linear in the number of points to be assigned. However, it is seen to give inferior clusters comparing with hierarchical. Most of earlier works used both algorithms with K-means algorithm (with Euclidean distance) is used more frequently to assemble the given data points. In its nature, K-means is linked with the finding of centroids. The centroids comes from the Euclidean Geometry itself. K-means also enjoys its being scalable and more accurate than hierarchical clustering algorithm chiefly for document clustering [5].

In [5], on the other hand, the experimental results of agglomerative hierarchical and K-means clustering techniques were presented. The results showed that hierarchical is better than k-means in producing clusters of high quality. In [6] authors compared two similarity measures - cosine and fuzzy similarity measures - using the k-means clustering algorithm. The results showed that fuzzy similarity measure is better than cosine similarity in terms of time and clustering solutions quality. In [7], several measures for text clustering were described approaches using affinity propagation. In [8] different clustering algorithms were explained and implemented on text clustering. In [9] some problems that that text clustering have been facing was explained. Some key algorithms, and their merits and des-merits were discussed in details. The feature selection and the similarity measure were the corner stones for proposing an effective clustering algorithm.

3 Methodology

3.1 Term Weighting

The Term Frequency (TFIDF) technique, as the most widely used, of weighting is adapted in this work.

3.2 K-Means Clustering Algorithm

The k-means clustering algorithm is widely used in data mining [1, 4] for its being more efficient than hierarchical clustering algorithm. It is used in our work as follows;

1. The number of clusters is one of these K values [2, 4]. That means K-means is run three times with one different K value each time.
2. The centroids has been chosen at first step randomly.
3. The standard k-means is run by getting all the data points involved in the first loop. The results are saved for next iteration and centroids are modified. Then, the clustering process run over for successive iteration by setting all points of clusters free, and randomly selecting new centroids.
4. Step 3 is iteratively continued till either number of iterations reach 30 iterations or each cluster has been seen in stable state.

3.3 The Hierarchical Clustering (HC)

Initialization: Given a set of points N, the data point matrix between points, initial clusters were initiated by randomly picking head for each cluster [10]. Then, in each loop, for any new data point, the data point cost between the new point and each cluster is calculated. The cluster whose average cost is the lowest would contain the relative point at hand. The step (1) is repeated till all points were clustered. Like K-means, number of clusters is selected to be one of these K values [2, 4]. That means hierarchical clustering is run three times with one different K value each time.

3.4 Similarity Measures

The similarity measures, used in this study, are Cosine and Euclidean [1].

Euclidean Distance (ED). In ED, each document is seen as a point in 2D space based on the term frequency of N terms that would represent the N dimension. ED measures the similarity between each point pair in this space using their coordinate based on the following equation:

$$D_{Euc}(x, y) = \sum \sqrt{x1 - y1)^2 + x2 - y2)^2 + \cdots xn - yn)^2} \tag{1}$$

Cosine Similarity Measure. The Cosine similarity, as one of the most widely-used measure, computes the pairwise similarity between ach document pair using the dot product and the magnitude of both vectors of both documents. It is computed as follows:

$$Sim_{Cos}(x, y) = \frac{\sum_{i=1}^{n} (x * y)}{\sqrt{\sum_{i=1}^{n} x^2} * \sqrt{\sum_{i=1}^{n} y^2}} \tag{2}$$

The union is used to normalize the inner product. Where x and y are the point pair needed to be clustered.

3.5 Experimental Setup

Machine Description. Table 1 displays the machine and environment descriptions used to perform this work.

Table 1. Machine and environment description.

Task	Tool	Specification
Clustering	Language	Python 3, Development Software: Jupyter Notebook
	OS	Windows 8 (64 bit)
	Memory	RAM 4 GB
	CPU	Intel I Core ™ (i5)
	Dataset	Glass & Iris

3.6 Dataset Description

Tables 2, 3 hold the datasets description which is taken literally from UCI (Machine Learning Repository).

Table 2. Iris dataset

Dataset characteristics:	Multivariate	Number of instances:	150	Area	Life
Attribute characteristics:	Real	Number of attributes:	4	Date donated	1988–07-01
Associated tasks:	Classification	Missing values?	No	Number of web hits:	3536252

Table 3. Glass identification dataset

Data set characteristics:	Multivariate	Number of instances:	214
Attribute characteristics:	Real	Number of attributes:	10
Associated tasks:	Classification	Missing Values?	No

3.7 The Clustering Evaluation Criterions

The evaluation metrics used to assess clustering quality are Entropy and Purity.

Purity (also known as Accuracy): It determines how large the intra-cluster is, and how less the inter-cluster is [1]. In other words, it is use to evaluates how much coherent the clustering solution is, and is formulated as follows;

$$\text{Purity} = \frac{1}{N} \sum_{i=1}^{k} max_j |c_i \cap t_j| \tag{3}$$

where N is the number of objects (data points), k is the number of clusters, ci is a cluster in C, and tj is the classification which has the max count for cluster ci.

Entropy. It is used to measure the extent to which a cluster contain single class and not multiple classes. It is formulated as follows:

$$Entropy = \sum\nolimits_{i=1}^{c} ci * log(ci) \tag{4}$$

Unlike purity, the best value of entropy is "0" and the worst value is "1".

4   Results and Discussion

In this section, we provide the obtained results of running both algorithms on both datasets using both measures – Cosine and Euclidean. Three K values for clusters – 2, 4, and 8 – along with using two evaluation metrics.

Table 4. Iris dataset - Cosine

AHC			
Metric/K	2	4	8
Entropy	4.60517	4.60937	**3.70626**
Purity	0.66667	0.66667	0.68
K-means			
Metric/K	2	4	8
Entropy	4.60517	**4.47621**	4.81686
Purity	0.66667	**0.97333**	**0.95333**

Table 5. Iris dataset - Euclidean

AHC			
Metric/K	2	4	8
Entropy	**3.91202**	**3.93659**	**3.82572**
Purity	0.66667	0.68667	0.7
K-means			
Metric/K	2	4	8
Entropy	3.97029	4.68630	4.7789
Purity	0.66667	**0.88667**	**0.97333**

For Iris dataset, k-means with cosine outperformed AHC. However, AHC with Euclidean outperformed k-means. On the other hand, for Glass dataset, AHC with cosine

Table 6. Glass dataset - Cosine

AHC			
Metric/K	2	4	8
Entropy	**4.72739**	**4.60619**	**4.62534**
Purity	0.48131	0.49065	0.53738
K-means			
Metric/K	2	4	8
Entropy	4.96284	4.99857	5.09285
Purity	**0.67757**	**0.71963**	**0.85981**

Table 7. Glass dataset - Euclidean

AHC			
Metric/K	2	4	8
Entropy	0.69315	**4.93907**	**4.85886**
Purity	0.36449	0.62617	0.67290
K-means			
Metric/K	2	4	8
Entropy	**4.68213**	4.98090	5.09710
Purity	**0.51402**	**0.74766**	**0.83178**

and Euclidean outperformed k-means in terms of entropy. In contrast, k-means outweighed AHC in terms of purity for both cosine and Euclidean. If we took this analysis as points for both algorithm, Table would hold these points.

Table 8. K-means and AHC in points

AHC		
Dataset/Measure	Cosine	Euclidean
Iris	**0**	**1**
Glass	1	1
K-means		
Dataset/Measure	Cosine	Euclidean
Iris	1	0
Glass	1	**1**

From Table 8, it can be noted that both algorithms have similar trend performance on both datasets. However, AHC preferred giving smaller entropy than k-mean, when k-means preferred giving higher purity.

In next Tables 9, 10, 11 and 12, Mean and Standard Deviation (STD) of both Entropy and Purity were taken in an average of all K values (2, 4, and 8) of each algorithm with respect to each evaluation metric -Entropy and Purity. Booth Mean and STD are interpreted using the basic values of entropy and purity that are drawn in Tables 4, 5, 6 and 7).

Table 9. Iris dataset - Cosine

AHC		
	Mean	STD
Entropy	4.30693	0.42474
Purity	0.67111	0.00629
K-means		
Metric/K	Mean	STD
Entropy	4.63275	0.14043
Purity	0.86444	0.14009

Table 10. Iris dataset - Euclidean

AHC		
	Mean	STD
Entropy	3.89144	0.04754
Purity	0.68444	0.01370
K-means		
Metric/K	Mean	STD
Entropy	4.47851	0.36135
Purity	0.84222	0.12908

Table 11. Glass dataset - Cosine

AHC		
	Mean	STD
Entropy	4.65297	0.05320
Purity	0.50312	0.02453

(*continued*)

Table 11. (*continued*)

K-means		
Metric/K	Mean	STD
Entropy	5.01809	0.05484
Purity	0.75234	0.07791

Table 12. Glass dataset – Euclidean

AHC		
	Mean	STD
Entropy	3.49703	1.98291
Purity	0.55452	0.13572
K-means		
Metric/K	Mean	STD
Entropy	4.92035	0.17509
Purity	0.69782	0.13443

Mean (Purity) in k-means is always better than AHC. However, Mean (Entropy) in AHC is always better than K-means. This confirms our previous analysis that AHC always produces solutions of lower entropy and K-means always gives solutions of higher purity. However, STD in AHC is better than K-means on both Iris and Glass datasets for both Euclidean and Cosine respectively. On the other hand, K-means is better than AHC on both Iris and Glass datasets for both Cosine and Euclidean respectively. As a rule of thumb, when STD is >=1, that implies a relatively high variation. However, when STD <=1, it is seen low. This means that the distributions with STD higher than 1 are seen of high variance whereas those with STD lower than 1 are seen of low-variance. In General, STD is better when it is kept as much low as possible which means that data has less variations around the mean with different K values for clusters.

5   Conclusions and Future Work

In this paper, we tried to briefly investigate the behavior of hierarchical and k-means clustering algorithms using cosine similarity measure and Euclidean distance along with using two evaluation metrics – Entropy and Purity. In general, AHC produced clustering solution of lower entropy than k-means. In contrast, k-means produced clustering solution of higher purity than AHC. Both algorithms look to have a similar performance trend on both datasets with AHC being slightly superior in terms of clustering solution quality. On the other hand, although we have not discussed the run time, we found from experiments that AHC suffers from the computational complexity comparing with K-means which was faster. However, the hierarchical clustering produced a clustering solutions of slightly high-quality than K-means. As a matter of fact, the performance of both algorithms on both "small" datasets could not be taken as a decisive factor for the report on behavior of both algorithm.

Therefore, the future work is directed towards extending this study significantly by: (1) Proposing new clustering algorithm, (2) including medium-sized and big datasets, (3) investigating more similarity measures [12], (4) considering more evaluation metrics, and finally, (5) studying one more clustering algorithm [13]. The ultimate aim of future work is to draw a valuable comparison study between all algorithms on target datasets so that the best combination of clustering algorithm and the relative similarity measure is captured. Moreover, the effect of using a different incremental number of clusters "K" is investigated.

Acknowledgments. The author would like to thank and appreciate the support received from the Research Office of Zayed University for providing the necessary facilities to accomplish this work. This research has been supported by Research Incentive Fund (RIF) Grant Activity Code: R20056–Zayed University, UAE.

References

1. Amer, A.A.: On K-means clustering-based approach for DDBSs design. J. Big Data **7**(1), 1–31 (2020). https://doi.org/10.1186/s40537-020-00306-9
2. Amer, A., Mohamed, M., Al_Asri, K.: ASGOP: an aggregated similarity-based greedy-oriented approach for relational DDBSs design. Heliyon **6**(1), e03172 (2020)
3. Amer, A., Abdalla, H., Nguyen, L.: Enhancing recommendation systems performance using highly-effective similarity measures. Knowl.-Based Syst. **217**, 106842 (2021)
4. Amer, A.A., Abdalla, H.I.: A set theory based similarity measure for text clustering and classification. J. Big Data **7**(1), 1–43 (2020). https://doi.org/10.1186/s40537-020-00344-3
5. Lee, C., Hung, C., Lee, S.: A comparative study on clustering algorithms. In: 14th ACIS International Conference on Software Engineering, Artificial Intelligence, Networking and Parallel/Distributed Computing, Honolulu, HI, pp. 557–562 (2013)
6. Scheunders, P.: A comparison of clustering algorithms applied to color image quantization. Pattern Recogn. Lett. **18**(11–13), 1379–1384 (1997)
7. Steinbach, M., Karypis, G., Kumar, V.: A comparison of document clustering techniques. In: KDD Workshop on Text Mining, vol. 400, pp. 1–2 (2000)
8. Goyal, M., Agrawal, N., Sarma, M., Kalita, N.: Comparison clustering using cosine and fuzzy set based similarity measures of text documents. arXiv, abs/1505.00168 (2015)
9. Kumar, S., Rana, J., Jain, R.: Text document clustering based on phrase similarity using affinity propagation. Int. J. Comput. Appl. **61**(18), 38–44 (2013)
10. Kamble, R., Sayeeda, M.: Clustering software methods and comparison. Int. J. Comput. Technol. Appl. **5**(6), 1878–1885 (2014)
11. Xu, D., Tian, Y.: A comprehensive survey of clustering algorithms. Ann. Data Sci. **2**(2), 165–193 (2015). https://doi.org/10.1007/s40745-015-0040-1
12. Abdalla, H., Amer, A.: Boolean logic algebra driven similarity measure for text based applications. PeerJ Comput. Sci. **7**, e641 (2021)
13. Abdalla, H., Artoli, A.: Towards an efficient data fragmentation, allocation, and clustering approach in a distributed environment. Information **10**(3), 112 (2019)

Based on Internet of Things Platform Using NB-IoT Communication Low-Power Weather Station System

Zhenxin Wang[1], Zhi Deng[2], Ke Xu[1], Ping Zhang[1], and Tao Liu[1(✉)]

[1] College of Computer and Information, Anhui Polytechnic University, Wuhu, China
liutao@ahpu.edu.cm
[2] School of Computer Science, Northwestern Polytechnical University, Xi'an, China

Abstract. In recent years, meteorological environment has become a topic of concern to people. Various meteorological disasters threaten human life and production. Accurate and timely acquisition of meteorological data has become a prerequisite for dealing with various aspects of production and life, and also laid a foundation for weather prediction. For a long time, meteorological data acquisition system combined with modern information technology has gradually become a hot spot in the field of meteorological monitoring and computer research. The continuous development of NB-IoT technology has brought new elements to the research of meteorological monitoring system. This paper designs a weather station system based on NB-IoT, including data acquisition module, main controller module, NB-IoT wireless communication module, energy capture module, low power consumption scheme, etc.

Keywords: NB-IoT · Meteorological monitoring · Low power consumption scheme · Internet of Things platform

1 Introduction

Due to the changeable climate and environment of the Earth, People's Daily production and life are greatly affected. In order to obtain meteorological data accurately and timely, meteorological stations are established all over the world for meteorological monitoring [1]. In order to develop meteorological monitoring, our country has also made great efforts to build meteorological stations, most of which are centralized and the system used is relatively backward. Because there are many manufacturers of domestic weather stations, their quality is mixed and their technology is uneven, so there is no certain standard for meteorological data. In addition, China has introduced a variety of foreign weather stations for direct application. Due to geographical and human factors [2], these weather stations are really not suitable for China's actual situation.

The rapid development of Internet of Things (IoT) technology has triggered more scholars to explore the application of NB-IoT in industrial and commercial fields. The current wireless data transmission modes, such as WiFi and Bluetooth modes, have a

© The Author(s) 2022
Z. Qian et al. (Eds.): WCNA 2021, LNEE 942, pp. 633–643, 2022.
https://doi.org/10.1007/978-981-19-2456-9_65

series of disadvantages, such as high power consumption and unstable data transmission efficiency [3, 4]. The existence of this phenomenon makes it necessary for the Internet of Things to study a new wireless data transmission technology to solve the above disadvantages [5]. NB-IoT technology is well suited for data transmission in IoT related applications. Driven by operators and device manufacturers, it has developed rapidly, and in a very short period of time [6], pilot projects have been opened in many cities. It can be seen that NB-IoT technology has developed rapidly in a very short period of time, from project landing to pilot in a very short time. The biggest factor is that NB-IoT technology has the advantages of low power consumption, low cost and long distance.

In this paper, STM32L051C8T6 development board is used as the master controller to connect with the weather sensor, and NB-IoT technology is used for wireless communication to optimize the traditional weather station. The main work is as follows:

(1) Design a low-consumption system scheme, so that the system can keep running for a long time, low power demand.
(2) Solar panels and ultracapacitors are used to construct energy capture and storage of the system, and NB-IoT wireless communication module is built based on BC35 series chips, which has the characteristics of low cost and stable communication.
(3) The cloud platform for data upload to the Internet of Things is realized to provide an interface for real-time acquisition of meteorological data, with the goal of building smart weather.

2   Related Work

Literature [7] proposes a multi-functional integrated weather station, which is mainly applied to precision agriculture and urban climate. Compared with the reference station, it is very consistent in most standard weather variables, and has the characteristics of low cost, low maintenance cost and low power demand. Literature [4] proposes portable automatic weather station, which is mainly used to measure glaciers, and includes three important components: The data recorder records wind direction, wind speed, relative humidity, atmospheric pressure, freezing temperature, temperature, solar radiation meteorological elements. The power system consists of a 10 W, 20 × 30 cm solar panel. The tripod is made of carbon fiber and stainless steel, a recyclable material. Literature [8] proposes the ZigBee-based intelligent weather station, which is mainly used to provide data for weather prediction. It is composed of the measurement unit based on the SiLab C8051F020 microcontroller to measure the data of temperature, relative humidity, atmospheric pressure and solar radiation, which is sent to the base station by the XBee module. Then the base station will store the data to the Access database. Literature [9] based on Internet of things technology and automatic meteorological monitoring system, embedded system is mainly used in monitoring of air and weather conditions, to collect the meteorological data such as temperature, relative humidity and atmospheric pressure, and then sends the data to the remote application or database, finally the data, can be in the form of graphics and tables to visualize, Provides remote access and mail alerts. Literature [10] proposes wireless portable meteorological monitoring station, which is mainly used to collect weather data and provide shared data. The meteorological sensor is connected with THE PIC16F887 microcontroller to measure wind speed, wind direction, relative humidity, atmospheric pressure, rainfall, solar radiation, ground and environmental temperature, and the industrial standard Modbus communication protocol is realized. Upload data to the online MYSQL data server for data sharing. Literature [11] is proposed based on NB-IoT communication model and the Internet of things technology of automatic meteorological station, is mainly used in intelligence, wisdom, meteorological city, based on the technology of digital sensors and independent power supply, intelligent sensor run independently and wireless data transmission, data through data platform for data analysis, data interface for networking meteorological information.

In this paper, the NB-IoT wireless data transmission technology is adopted to optimize the weather station and upload the acquired data to the cloud platform for users to monitor the meteorological data in real time [12]. The research results solve the disadvantages of traditional weather stations to a certain extent, and have a certain research significance for the development of weather stations and NB-IoT.

3   Hardware Design

The hardware design of intelligent weather station based on NB-IoT is BME680 sensor, ZPH02 dust sensor, VEML6070 ultraviolet sensor used to collect data, and the main controller module STM32L051C8T6 is used to ensure the stability of data transmission [13], signal control order, and program implementation efficiency. The energy capture module uses 3 W 9 V small solar panel to capture energy, and the NB-IoT wireless communication module uses BC35G chip to transmit data to the Cloud platform of the Internet of Things (Fig. 1).

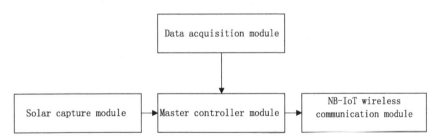

Fig. 1. Overall hardware architecture

The main controller module is the core of the whole hardware. It is connected with the data acquisition module through the serial port to collect meteorological data. The main controller module is connected to the wireless data transmission module to realize data uploading to the cloud platform [7]. As for the main controller module, STM32L051C8T6 development board is selected, which can carry out high-speed data processing under the condition of low power consumption and is equipped with high-speed embedded storage and memory protection unit and rich input and output data interfaces. In order to ensure the stable operation of the hardware part of the system, it is necessary to design the circuit, and the stable voltage required by each device is different.

The design of low-power system is carried out. The main controller of STM32L051C8T6 uses 1.8 V voltage. The working voltage of NB-IoT wireless communication module is 3.3 V; BME680 sensor and VEML6070 sensor in the data acquisition module need 3.3 V working voltage, while ZPH02 sensor needs 5 V power supply; The MICROcontroller uses XC6206P182 ultra-low pressure difference 1.8 V LDO to supply power; The sensor and wireless communication module use the automatic pressure raising chip TPS63070, and the PM2.5 sensor needs 5 V power supply. To sum up, the system circuits need to be designed to allow each module to operate at a normal operating voltage (Fig. 2).

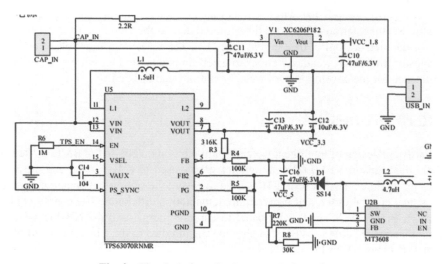

Fig. 2. Circuit design of voltage regulation scheme

The energy capture module uses a 3 W 9 V small solar panel to capture energy, and two 2.8 V 3000 F supercapacitors in series to store energy. The LM2596S stabilized power module can stabilize the output voltage of the supercapacitor. The NB-IoT wireless communication module selects BC35G chip, which has the characteristics of wide coverage, low power consumption, low cost and large connection. It can transmit the data in the data acquisition module to the Cloud platform of the Internet of Things. In the data acquisition module, BME680 sensor was used to detect temperature and humidity, air pressure and smoke resistance, ZPH02 dust sensor was used to collect PM2.5, and VEML6070 ultraviolet sensor was used to detect ultraviolet. The hardware PCB design of the system adopts AD20, which ensures the normal working voltage of the whole system module when the main control board is designed. In addition, the pins of the main controller are directly set out to facilitate the access of the data acquisition module. In order to ensure the small size of the intelligent weather station, the SIM card slot is welded on the back of the PCB board (Fig. 3).

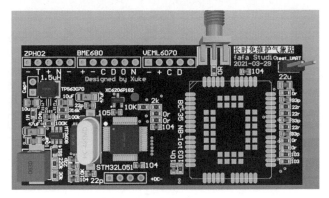

Fig. 3. System PCB

4 Software Design

In the system software design stage, mainly including: data acquisition module design, low power design, NB-IoT wireless communication module design. On the basis of low-power design, the data acquisition module collects temperature, relative humidity, atmospheric pressure, smoke resistance, PM2.5, ultraviolet data and transmits it to the Cloud platform of the Internet of Things through the NB-IoT wireless communication module to realize data storage.

4.1 Data Collection Module

Temperature, relative humidity, atmospheric pressure, smoke resistance collection: The SDA and SCL of BME680 sensor in the data acquisition module communicate with the IIC interface of PB15 and PB13 of the master controller respectively. When PB15 and PB13 are used as the IIC bus interface, the IIC working mode needs to be configured for MCU. Turn on the GPIO Clock using the built-in firmware library function RCC_APB2Periph Clock Cmd() and set PB15 and PB13 to IIC mode with GPIO_Init struct.pin. At the same time, use GPIO_Init struct. Speed to set the transfer Speed to GPIO_SPEED_FREQ_LOW, use gpio_initstruct. Mode to set the open output Mode, and use HAL_GPIO_Init() to initialize the GPIO port. Collect environmental parameters after port configuration (Fig. 4).

Collection of PM2.5 concentration: The COLLECTION of PM2.5 concentration is mainly connected to the PA2 pin of the main controller module through the RX pin of the ZPH02 dust sensor. The pin outputs electrical signals in serial port mode, which is converted into digital signals through the A/D of the main controller, and outputs the CONCENTRATION of PM2.5 after processing. PM2.5 detection procedures are as follows:

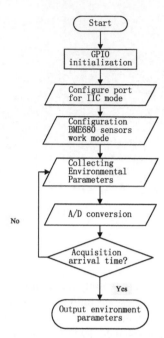

Fig. 4. Flow chart of BME680 sensor subsystem

```
int Get_dust25(void)
{
    if(USART2_RX_BUF[0]==0xff&&USART2_RX_BUF[1]==0x18)
    {
            dust25=(USART2_RX_BUF[3]*100)+USART2_RX_BUF[4];
            return 1;
}
    else
            return 0;
}
```

Ultraviolet parameter collection: In the design of ultraviolet data acquisition program, the VEML6070 ultraviolet sensor itself can directly convert the ultraviolet light sensitivity into digital signal. VEML6070 UV sensor detection procedures are as follows:

```
u16 VEML6070_ReadValue(void)
   {
       u8   value_h=0;
       u8   value_l=0;
       VEML6070_ReadData(VEML6070_ARA);
          value_h = VEML6070_ReadData(VEML6070_READ_VALUE2);
       VEML6070_ReadData(VEML6070_ARA);
          value_l = VEML6070_ReadData(VEML6070_READ_VALUE1);
       veml6070_val = (value_h<<8) + value_l;
       VEML6070_ReadData(VEML6070_ARA);
       VEML6070_WriteCmd(VEML6070_SLAVE_ADDRESS,VEML6070_SET
_VALUE);
       return veml6070_val;
   }
```

4.2 Low Power Solution

The main function module is the program design of the whole main controller module to control other devices, which is mainly reflected in two aspects of system power consumption processing and data processing. The program design of the main controller module mainly realizes the clock setting, the use of serial port initialization and the process of data sending and receiving. After the system is powered on, the clock and peripherals of the system are automatically initialized. After that, the low-power mode of the system exits, and RTC is used for periodic wake up. After wakeup, I/O and peripherals to be used are reconfigured to send data. After the data acquisition module obtains the environmental data from the area to be tested, data transmission is carried out through IIC communication or UART communication.

In the design of low power consumption, wireless data transmission is adopted. The single chip microcomputer turns off the power of The ZIGBEE module, sets all IO except the burning port to analog input mode, and turns off the clock of all peripherals. Then the single chip microcomputer enters the STOP mode and uses RTC to wake up at a certain time. Wake up and reconfigure IO and peripherals to be used and send data.

4.3 NB-IoT Wireless Communication Module

The NB-IoT module connects to the Cloud platform of the Internet of Things. The implementation code is as follows:

```
void NB_SetCDPServer(uint8_t *ncdpIP,uint8_t *port)
{
    memset(cmdSend,0,sizeof(cmdSend));
    strcat(cmdSend,"AT+NCDP=");
    strcat(cmdSend,(const char *)ncdpIP);
    strcat(cmdSend,",");
    strcat(cmdSend,(const char *)port);
    strcat(cmdSend,"\r\n");
    NB_SendCmd((uint8_t*)cmdSend,(uint8_t*)"OK",DefaultTimeout,1);
}
```

The wireless data module needs the CoAP protocol to transmit data to the cloud server, and the BC35G device is designed to register with the route T/R of the Cloud server of the Internet of Things. The CDP server subscribes to the T/D resources of the BC35G device and waits for the BC35G device to send CoAP instructions to it. If the BC35G device receives the +NMGS instruction, it transmits data to the CDP server through the CoAP instruction.

The CDP server serves as the CoAP client and the BC35G serves as the CoAP server. The CDP server sends downlink data to the T/D resource of the BC35G device through THE POST method.

5   Tests and Results

After the design and implementation of the software and hardware of the system, the design of energy capture module, data acquisition module, main controller module and wireless communication module is completed. In order to verify the feasibility and stability of the system in practical application, we need to test the data collection, NB-IoT communication and power consumption of the system.

Comparing the temperature data collected by the sensor with the readings of the traditional thermometer, it is found that the readings are basically the same, and the humidity, pressure, smoke resistance, PM2.5 and ULTRAVIOLET data are basically the same as the data obtained by the traditional weather station (Table 1).

Table 1. Part of the data

Temperature	Relative humidity	Atmospheric pressure	PM2.5	Ultraviolet light	Smoke resistance	Time
24.67 °C	65.21%	100448.0 Pa	10.2 ug	4 uW	1069.0hΩ	20210519 14:58
24.22 °C	65.85%	100342.0 Pa	12.3 ug	2 uW	837.0hΩ	20210519 17:34

(*continued*)

Table 1. (*continued*)

Temperature	Relative humidity	Atmospheric pressure	PM2.5	Ultraviolet light	Smoke resistance	Time
24.79 °C	60.76%	100756.0 Pa	9.8 *ug*	5 *uW*	894.0hΩ	20210520 12:44
25.36 °C	62.27%	100568.0 Pa	10.4 *ug*	3 *uW*	952.0hΩ	20210521 15:27
25.42 °C	62.97%	100543.0 Pa	10.3 *ug*	3 *uW*	992.0hΩ	20210521 15:51

5.1 Collect Data

Before the system data acquisition test, you need to use multimeter on every pin detection circuit of the system, respectively, in order to confirm whether can normal between circuit electricity, and need to check each device in the circuit board welding in normal state, the electricity, note that each sensor, the main controller and wireless communication module if there is a fever more serious phenomenon, To ensure the normal operation of the hardware circuit, the data acquisition function of each sensor is tested.

5.2 Wireless Communication Module Data Transmission Test

As the terminal software is programmed to upload data once every minute (for the convenience of testing, usually once an hour), after testing, the data collected by the data acquisition module can be normally uploaded to the cloud platform within a certain collection time through the wireless communication module after being processed by the primary controller.

5.3 System Power Test

The solar energy capture module is connected to the data acquisition module, and the 3W9V solar panel is used to capture the electricity, and two 2.8 V 3000 F supercapacitors are used to store the electricity, so as to realize the long-term automatic power supply of the system, which has a very long battery life and low maintenance cost.

Through the current test of the whole system, the electricity situation table of the system is obtained (Table 2).

Table 2. System power usage

Current of the system in standby mode	Working state current of the machine
About 5 uA	About 80 mA

6 Conclusion

Through the design and development of hardware and software, the NB-IoT meteorological monitoring station is realized. The hardware is composed of standard weather sensors and interfaces with the STM32L051C8T6 master controller to detect temperature, humidity, air pressure, smoke resistance, PM2.5 and ULTRAVIOLET data in the environment, and upload the data to the cloud platform of the Internet of Things through the NB-IoT wireless transmission module. In the design, solar panels and ultracapacitors are used to build the energy capture module of the system [14], NB-IoT wireless communication module is built based on BC35 series chips, and a low-power system scheme is designed with the characteristics of low cost, low power demand, low maintenance cost and easy to use [15, 18]. Future research directions are as follows:

Solar energy capture method is adopted in this system design, and the volume of ultracapacitors is large. In addition, solar energy capture is easily affected by weather, so a more environmentally friendly power generation method can be adopted in subsequent studies.

Add the NB-IoT wireless data transmission module to the storage system to prevent the failure of data uploading to the cloud platform due to network connection failure.

Acknowledgments. This work was supported by the Industry Collaborative Innovation Fund of Anhui Polytechnic University and Jiujiang District under Grant No. 2021cyxtb4, and the Science Research Project of Anhui Polytechnic University under Grant No. Xjky2020120.

References

1. Kull, D., Riishojgaard, L.P., Eyre, J., Varley, R.A.: The Value of Surface-based Meteorological Observation Data. World Bank 2021-01-01
2. Liu, T., Zhang, D.: Advances in the quality control methods of air temperature data at surface automatic weather stations. IOP Conf. Ser. Earth Environ. Sci. **769**(2), 022060 (2021)
3. Tian, G.P.: Application of wireless communication technology in automatic weather station. Electron. Meas. Technol. **44**(07), 154–158 (2021). (in Chinese)
4. Netto, G.T., Arigony-Neto, J.: Open-source automatic weather station and electronic ablation station for measuring the impacts of climate change on glaciers. HardwareX **5**, e00053 (2019)
5. Bian, Z.Q., Liu, X., Shao, L.J., et al.: Design and implementation of wind tunnel for wind resistance test of meteorological sensor. Electron. Meas. Technol. **43**(21), 15–18 (2020). (in Chinese)
6. Guerrero Osuna, H.A., Luque-Vega, L.F., Carlos-Mancilla, M.A., Ornelas, V.G., Castañeda Miranda, V.H., Carrasco-Navarro, R.: Implementation of a MEIoT weather station with exogenous disturbance input. Sensors **21**(5), 1653 (2021)
7. Dombrowski, O., Hendricks Franssen, H.J., Brogi, C., Bogena, H.R.: Performance of the ATMOS41 all-in-one weather station for weather monitoring. Sensors (Basel Switzerland) **21**(3), 741 (2021)
8. Haefke, M., Mukhopadhyay, S.C., Ewald, H.: A Zigbee based smart sensing platform for monitoring environmental parameters. In: Conference Record IEEE Instrumentation & Measurement Technology Conference, pp. 1–8 (2011)

9. Mabrouki, J., Azrour, M., Dhiba, D., Farhaoui, Y., El Hajjaji, S.: IoT-based data logger for weather monitoring using Arduino-based wireless sensor networks with remote graphical application and alerts. Big Data Min. Anal. **4**(01), 25–32 (2021)
10. Devaraju, J.T., Suhas, K.R., Mohana, H.K., Patil, V.A.: Wireless portable microcontroller based weather monitoring station. Measurement **76**(5), 189–200 (2015)
11. Chen, J., Sun, Y., et al.: Research on application of automatic weather station based on Internet of Things (2017)
12. Wellyantama, P., Soekirno, S.: Temperature, pressure, relative humidity and rainfall sensors early error detection system for automatic weather station (AWS) with artificial neural network (ANN) backpropagation. J. Phys. Conf. Ser. **1816**(1) (2021)
13. Benghanem, M.: Measurement of meteorological data based on wireless data acquisition system monitoring. Appl. Energy **86**(12), 2651–2660 (2009)
14. Zeng, Y., Ji, B.Y.: Design of automatic weather station system with self-checking function. Foreign Electron. Meas. Technolo **39**(10), 88–93 (2020). (in Chinese)
15. Fausto, R.S., Abermann, J., Ahlstrøm, A.P.. Annual surface mass balance records (2009–2019) from an automatic weather station on Mittivakkat Glacier, SE Greenland. Front. Earth Sci. **8**, 251 (2020)
16. Wu, H.Y., Li, Z.H., Li, W.Y., et al.: Characteristics analysis of extremely severe precipitation based on regional automatic weather stations in Guangdong. Meteor. Mon. **46**(6), 801–812 (2020). (in Chinese)

Complex Relative Position Encoding for Improving Joint Extraction of Entities and Relations

Hua Cai[✉], Qing Xu, and Weilin Shen

Algorithm Research Center, UniDT Technology, Shanghai, China
hanscalcai@163.com

Abstract. Relative position encoding (RPE) is important for transformer based pretrained language model to capture sequence ordering of input tokens. Transformer based model can detect entity pairs along with their relation for joint extraction of entities and relations. However, prior works suffer from the redundant entity pairs, or ignore the important inner structure in the process of extracting entities and relations. To address these limitations, in this paper, we first use BERT with complex relative position encoding (cRPE) to encode the input text information, then decompose the joint extraction task into two interrelated subtasks, namely head entity extraction and tail entity relation extraction. Owing to the excellent feature representation and reasonable decomposition strategy, our model can fully capture the semantic interdependence between different steps, as well as reduce noise from irrelevant entity pairs. Experimental results show that the F1 score of our method outperforms previous baseline work, achieving a better result on NYT-multi dataset with F1 score of 0.935.

Keywords: Complex relative position encoding · Pretrained language model · Joint extraction

1 Introduction

Transformer recently has drawn great attention in natural language processing because of its superior capability in capturing long-range dependencies [1]. Extracting entity pairs with relations from unstructured text is an essential step in the construction of automatic knowledge database. Joint extraction of entities and all the possible relations between them at once, which considers the potential interaction between the two subtasks and eliminates the error propagation issue in traditional pipeline method [2, 3]. A typical joint extraction scheme is ETL-Span [4], which transforms information extraction into a sequence labelling problem with multi-part labels. It also proposed a novel decomposition strategy to decompose the task into simpler modules, that is, to decompose the task into several sequence label problems hierarchically. The key point is to distinguish all candidate head entities that may be related to the target relation starting from the beginning of the sentence, and then mark the corresponding tail entity and relation for

© The Author(s) 2022
Z. Qian et al. (Eds.): WCNA 2021, LNEE 942, pp. 644–655, 2022.
https://doi.org/10.1007/978-981-19-2456-9_66

each extracted head entity. This method achieves excellent performance in overlapping entity extraction.

Despite the efficiency of this framework, it is weak for the limited feature representation comparing with other complex models, especially transformer-based encoder BERT [5]. Using BERT to encode sentence extraction features could share feature representation with advanced semantic information. However, the Transformer [6] based network structure is a superposition of self-attention mechanism, which is inherently unable to learn the sequential relations of sentences. The position and order of words in the text are very important features, which will affect the accuracy of information extraction task in which the target is determined by the boundary.

To address the aforementioned limitations, we present our cRPE-Span model, which makes the following contributions:

1. The shared embedding module is improved through BERT, and the complex field relative position encoding is added to represent the relative position information between entities, so that the extractor can consider the semantic and position information of the given entity when marking the tail entity and relation.
2. The hierarchical boundary marker only marks the entity start and end position in a cascade structure and ignores the entity category, which could reduce the task difficulty for one step prediction process, and then alleviate the accumulated error.
3. Our method achieves consistently better performances on three benchmark datasets of entity and relation joint extraction, obtaining a better result on NYT-multi dataset with F1 score of 0.935.

2   Related Works

The entity-relation extraction task has always been widely concerned for its crucial role in information extraction. For most traditional methods ignore the interaction between entity recognition and relationship extraction, researchers have proposed a variety of joint learning methods with end-to-end neural architectures [4, 7–9]. Unfortunately, due to the shared encoder limitation, these methods cannot fully exploit the inter-dependency between entities and relations.

Introducing powerful transformer-based BERT to encode the input information could enhance the capability of modeling the relationship of tokens in a sequence. The core of transformer is self-attention, however, the self-attention has an inherent deficiency that it does not contain sequential order information of input tokens, so that it needs to add positional representations to encode information explicitly. The approaches for positional representations of transformer-based network can fall into two categories. The first one is the absolute position encoding, which inject the positional information to the model by encoding the positions of input tokens from 1 to maximum sequence length. Typically, sinusoidal position encoding in Transformer and learned position encoding in BERT, GPT [10]. However, such absolute positions cannot model the interaction information between any two input tokens explicitly. Therefore, the second relative position encoding (RPE) extends the self-attention mechanism to consider the relative positions or distances between sequential elements. Such as the model NEZHA [11], Transformer-XL [1], T5

[12] and DeBERTa [13]. As such information is not necessary for non-entity tokens, and may introduce noise on the contrary. Different from the relative positions mentioned above, we introduce complex relative position encoding (cPRE) into BERT for entity and relation joint extraction.

3 Method

cRPE-Span joint extraction structure is an end-to-end neural architecture, which jointly extracts entities and overlapping relations. We first add the cRPE to the powerful transformer-based BERT, and then use it to encode the input information for more accurate representation of the relative position information between entities. In the joint extraction structure, we use span-based tagging scheme as well as the reasonable decomposition strategy. In essence, the framework reduces the influence of redundant entity pairs, and captures the correlation between the head entity and the tail entity, thus obtaining better joint extraction performance. Figure 1 shows the framework diagram of our cRPE-Span extraction system.

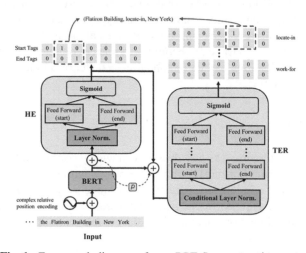

Fig. 1. Framework diagram of our cRPE-Span extraction system

3.1 Shared Feature Representation Module

Intuitively, the distance between entities and other context tokens provides important evidence for entity and relation extraction. So we inject location information for this network structure by adding position encodings to the input token embedding. In Transformer, absolute positional encoding in the form of sine and cosine function is generally used, which can ensure that each position vector is not repeated and there is a relationship between different positions. However, Yan et al. [14] found that the location information of trigonometric function, which is commonly used in Transformer, will lose its relative relationship in the process of forward propagation. Similarly, the embedding vectors of different positions have no obvious constraint relationship in transformer-based BERT. Because the embedding vectors of each position are independently trained in BERT, so they can only model absolute position information, and not model the relative relationship between different positions (such as adjacency and precursor relationship).

In order to make the model capture more accurate relative position relationship, we add the cRPE to the input of BERT except its origin learned position embedding. The continuous function of complex field is adopted to encode the representation of words in different positions. In this paper, the input embedding vector of BERT is the super-position of four embedding features, namely piece-wise word embedding, segmentation embedding, learned position embedding and complex field position embedding.

Relative Position Embedding in Complex Field. Typically, the method to encode the relative position between the token x_i and x_j into vectors $p_{ij}^V, p_{ij}^Q, p_{ij}^K \in \mathbb{R}^{d_z}$ is encoding the positional vectors into the self-attention module, which reformulates the self-attention module as

$$z_i = \sum_{j=1}^{n} \alpha_{ij}(x_j W^V + p_{ij}^V) \tag{1}$$

each weight coefficient α_{ij} is computed using a softmax:

$$\alpha_{ij} = \frac{\exp(e_{ij})}{\sum_{k=1}^{n} \exp(e_{ik})} \tag{2}$$

where e_{ij} is calculated using a scaled dot-product attention:

$$e_{ij} = \frac{\left(x_i W^V + p_{ij}^Q\right)\left(x_i W^V + p_{ij}^Q\right)^T}{\sqrt{d_z}}. \tag{3}$$

Instead of simply adding the word vector and the position vector, we use a function to add the position information modeling the relative position of words. This function is continuously changing with the position. Like complex relative position representations proposed by Wang et al. [15], we first define a function to describe the word in the text with position pos and index j as:

$$f(j, pos) = g_j(pos) \subset \mathbb{R}^D \tag{4}$$

g is a vector-valued function, which satisfies the following two properties:

1. There exists a function $T : \mathbb{C} \times \mathbb{R} \rightarrow \mathbb{R}$ such that for all $pos \geq 0, n \geq 0$, $g_i(pos + n) = T(n, g_i(pos))$. Namely, if we know the word vector representation of a word at a certain position, we can calculate the word vector representation of it at any position. That is to say, linear transformation has nothing to do with position, but only with relative position.

2. There exists $\delta \in \mathbb{R}_+$ such that for all position pos, $g_i(pos) <= \delta$. That is, the norm of the word vector is bounded.

If T is a linear function, then $g_i(pos)$ admits only one solution in vector:

$$r_j e^{i(w_j pos + \theta_j)} \tag{5}$$

it can also be written in the form of components as:

$$\left[r_{j,1}e^{i(w_{j,1}pos+\theta_{j,1})}, r_{j,2}e^{i(w_{j,2}pos+\theta_{j,2})}, \ldots, r_{j,D}e^{i(w_{j,D}pos+\theta_{j,D})} \right] \qquad (6)$$

In this way, we hope to get the word order modeling in this smooth way. Where r_j is the amplitude, θ_j is the initial phase, w_i is the angular frequency. Amplitude is only related to the index of words in the sentence, which represents the meaning of words and corresponds to ordinary word vectors. Phase $w_j pos + \theta_j$ is not only related to the word itself, but also related to the position of the word in the text. It corresponds to the position of a word in the text. When the angular frequency is small, the word vectors of the same word in different positions are almost constant. In this case, the word vector in complex field is not sensitive to position, which is similar to the ordinary word vector without considering position information. When the angular frequency is very large, the complex-valued word vector is very sensitive to the position and will change dramatically with the change of position.

3.2 Joint Extraction of Entities and Relations

The joint entity and relation extraction task is transformed into a sequential pointer marking problem. Firstly, the hierarchical boundary marker is used to mark the start and end positions in a cascade structure, and then the multi span decoding algorithm is used to jointly decode the head entity and tail entity based on the range marker, and the index of the start and end positions is predicted to identify the entity boundary.

Joint Extractor. The extractor consists of a head entity extractor (HE) and a tail entity and relationship extractor (TER). For entity extraction, the HE and TER are decomposed into two sequential marking subtasks. The subtasks are to identify the entity starting and end position by using pointer network [16]. The difference HE and TER is that the TER would predict the relations at the same time. It is worth to note that the entity category information does not involve in this sequential marking process, that is, the model is no need to predict the entity category first, and then predict the relationship according to the category, and only need to predict the relationship according to the entity location information. Therefore, the task difficulty is reduced for the only one step prediction process, as well as the accumulated error is alleviated.

The purpose of HE extractor is to distinguish candidate entities and exclude irrelevant entities. Firstly, the triple library is constructed by training set, and after that the embedding vector sequence h_i is obtained by embedding module. Then, the training data is searched remotely to obtain the prior information representation vector p. Finally, the feature vector $x_i = [h_i; p]$ is obtained by connecting the feature coding vector sequence with the prior information representation vector. $h_{HE}(h_{HE} = \{x_1, \ldots, x_n\})$ is used to represent the vector representation of all the words used for HE extraction. Inputting h_{HE} into HE to extract the head entity, which includes all the head entities in the sentence and the corresponding entity location labels.

Similar to HE extractor, TER also uses basic representation h_i and prior information vector p as input feature. However, the combination of h_i and p is insufficient to detect tail entities and relationships with specific head entities. The key information needed for TER extraction includes: (1) words in tail entities (2) dependent header entities (3) context representing relationships. In a comprehensive way, we combine the head entity and context-related feature representation. That is, given a header entity h, x_i is defined as follows:

$$x_i = [h_i; p; h^h] \tag{7}$$

Here, $h^h = [h_{sh}; h_{eh}]$. h_{sh} and h_{eh} are the index of the beginning and end position of the head entity h, respectively. $[p; h^h]$ is the auxiliary feature vector of tail entity and relation extraction. We will take h_{THE} ($h_{THE} = \{x_1, \ldots, x_n\}$) as the input of hierarchical boundary annotation, and the output is obtained as $\{(h, rel_o, t_o)\}$, which contains all triples in sentence s given header entity h.

In general, for a sentence with m entities, the whole joint decoding task includes two sequence annotation tasks for HE tags and 2 m for TER tags.

Loss Function. In the training process, we aim to share the input sequence among tasks and carry out joint training. So for each training instance, we do not input sentences repeatedly in order to use all the triple information in the sentences, but randomly select a head entity from the labeled head entities as the input of TER extractor. At the same time, two loss functions are used to train the model, one is L_{HE} for HE extraction, and the other is L_{TER} for TER extraction.

$$L = L_{HE} + L_{TER} \tag{8}$$

This optimization function can make the extraction of head entity, tail entity and relationship interact with each other, so that the element in each triplet can be constrained by another element. L_{HE} and L_{TER} can be defined as the sum of negative logarithm probability of real start tag and end tag:

$$L_{HE,TER} = -\frac{1}{n}\sum_{i=1}^{n}\left(logP\left(y_i^{sta} = \hat{y}_i^{sta}\right) + logP\left(y_i^{end} = \hat{y}_i^{end}\right)\right) \tag{9}$$

Here, $\hat{y}_i^{sta}$ and $\hat{y}_i^{end}$ are real tags that represent the beginning and end positions of the i-th word, respectively. n is the length of the sentence. P_i^{sta} and P_i^{end} represent the prediction probabilities of the starting and ending positions of the i-th word as the target entity respectively.

$$P_i^{sta,end} = sigmoid\left(w_{sta,end}x_i + b_{sta,end}\right) \tag{10}$$

$$y_i^{sta,end} = \chi_{\{p_i^{sta,end} > threshold_{sta,end}\}} \tag{11}$$

Here, χ is an indicator function such that $\chi_A = 1$ if and only if A is true.

4 Experiments

4.1 Datasets

We have conducted experiments on three datasets: (1) CoNLL04 was published by Dan et al. [17], we used segmented dataset with 5 relation types defined by Gupta and Adel et al. [18, 19], which contains 910 training data, 243 evaluation data and 288 test data. (2) NYT-multi was published by Zeng et al. [20]. In order to test the overlapping relation extraction in 24 relation types, they selected 5000 sentences from NYT-single as the test set, 5000 sentences as the verification set, and the remaining 56195 sentences as the training set. (3) WebNLG was released by Claire et al. [21] and used for natural language generation task. We used the WebNLG data preprocessed by Zeng et al. [20], including 5019 training data, 500 evaluation data,703 test data and 246 relation types.

4.2   Experimental Evaluation

We follow the evaluation metric in previous work [4, 22]. If and only if the relation type and two corresponding entities of a triple are correct, the triple is labeled as correct. If the head and tail position boundaries are correct, the entity is considered to be correct. We used standard Micro Precision, Recall and F1 scores to evaluate the results.

4.3   Experimental Parameters

We use the mini-batch mechanism to train our model, the batch size is 8, using the weighted moving average Adam to optimize the parameters. The learning rate is set to be 1e−5 and the stacked bidirectional transformer has 12 layers and 768 dimensions of hidden state. We used pretrained BERT base model [Uncased-BERT-Base]. The maximum length of the input sentence in our model is set to be 128. We did not adjust the threshold of the joint extractor, and set the threshold to 0.5 by default. All super parameters are adjusted on the validation set. In each experiment, we use an early stop mechanism to prevent the model from over fitting, and then report the test results of the optimal model on the test set. All our training and test results were performed on 32 GB Tesla V100 GPU.

5   Results and Analyses

5.1   Comparison Models

We mainly compare our model with the following baseline models: (1) Multi-Head [22] and (2) ETL-Span [4]. We reimplement these models on CoNLL04, NYT-multi and WebNLG datasets, marked with * in Table 1 and Table 2.

Table 1. Comparison of model results on CoNLL04 dataset (%)

Model	Prec.	Rec.	F1
Biaffine-attention [23]	–	–	64.4
Relation-Metric [24]	67.9	58.2	62.3
Multi-Head* [22]	70.5	57.8	63.5
ETL-Span* [4]	66.0	68.1	67.1
cRPE-Span	67.1	**68.7**	**67.6**

Table 1 reports the results of our models against other baseline methods on CoNLL04 dataset. Our model achieved a comparable result with F1 score of 67.6%, and with the recall of 68.7%. We found that the result of our model is better than the method based on sequence-by-sequence encoding, such as Biaffine-attention and Multi-Head. This is probably due to the inherent limitation for RNN expansion to generate triples.

In Table 2, It can be seen that our proposed joint extraction based on complex position embedding method, cRPE-Span, significantly outperforms all other methods, especially on NYT-multi dataset with precision, recall and F1 score of 94.6%, 92.5% and 93.6%, respectively.

Table 2. Comparison of model results on NYT-multi and WebNLG datasets (%)

Model	NYT-multi			WebNLG		
	Prec.	Rec.	F1	Prec.	Rec.	F1
Multi-Head* [22]	84.4	79.3	81.7	85.5	79.9	82.6
ETL-Span* [4]	85.9	73.8	79.4	86.8	82.2	84.4
TPLinke$_{LSTM}$ [25]	86.0	82.0	84.0	91.9	81.6	86.4
TPLinke$_{BERT}$ [25]	91.4	92.6	92.0	88.9	84.5	86.7
SPN [26]	92.5	92.2	92.5	–	–	–
cRPE-Span	**94.6**	**92.5**	**93.6**	89.1	**84.8**	**86.9**

Compared with ETL-Span, a joint extraction method based on span scheme, the F1 scores of cRPE-Span on NYT-multi and WebNLG datasets have increased by 17.9% and 2.9%, respectively. In comparison with Multi-Head, the F1 scores of cRPE-Span on NYT-multi and WebNLG datasets increased by 14.6% and 5.2%, respectively. We consider that it is because (1) we decompose the difficult joint extraction task into several more manageable subtasks and handle them in a mutually enhancing way, this suggests that our HE extractor and TER extractor actually work in a mutually enhancing manner; (2) our shared feature extractor based on BERT with cRPE effectively captures the semantic and position information of the dependence of the first entity, while ETL-Span uses LSTM to shared encoding, and it needs to predict the category of entity, and then predict the relationship based on the category, that may cause error propagation issues. Overall, these results demonstrate that our extraction paradigm first extracts the head entity, and then marks the corresponding tail entity, and can better capture the relationship information in the sentence.

5.2 Ablation Study

To demonstrate the effectiveness of each component, we conducted ablation experiments by removing one particular component at a time to understand its impact on the performance. We study the influence of cRPE (complex relative positional encoding) and RSS (remote supervised search) on the WebNLG dataset, as shown in Table 3.

In the table we can find that: (1) when we delete the cRPE, the F1 score drops by 1.4%. This shows that relative position encoding plays a vital role in information extraction, allowing the tail entity extractor to know the position information of a given head entity, so as to filter out irrelevant entities through implicit distance constraints. Secondly, by predicting the entities in the HE extractor, we can explicitly integrate the entity location information into the entity representation, which is also very helpful for subsequent TER mark; (2) after removing the remote supervised search strategy, the F1 score dropped by 0.2%. The above comparison tests once again confirm the effectiveness and rationality of our cRPE and RSS strategy.

Table 3. Comparison of simplified model results (%)

Model	WebNLG		
	P	R	F1
cRPE-Span	**89.1**	84.8	**86.9**
- cRPE	85.8	85.6	85.7
- RSS	85.9	84.9	85.5

5.3 Model Convergence Analysis

In order to analyze the convergence of our model, we conducted further experiments on three test datasets and selected our baseline model RSS-Span for comparison. The RSS-Span model is with the remote supervised search strategy, but without the complex relative positional encoding. To differentiate the test results of baseline and cRPE-Span model, the baseline results are drawn with black hollow circles, and the cRPE-Span results are drawn with blue solid circles, as shown in Fig. 2. The dash lines in the table are benchmark scores which are relatively smaller scores value in the best F1 scores. For the NYT-multi dataset, we select 92.8% of the F1 score between cRPE-Span and the baseline model, which is the smaller of 93.6% and 92.8%. Similarly, for the CoNLL04 and WebNLG datasets, the selected F1 benchmark scores are 66.1% and 85.7%, respectively. That is to say, we analyze the number of training epochs at this time when the benchmark score is reached.

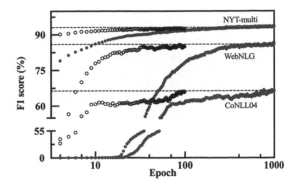

Fig. 2. Comparison results of model convergence

From Fig. 2, we observe that the convergence of cRPE-Span is slightly inferior to that of RSS-Span. After training for about 100 epochs, RSS-Span reaches the F1 benchmark score, while cRPE-Span needs to be iterated to about 1000 epochs. This is because the cRPE-Span position embedding layer is a continuous function in the complex domain to encode the representation of words at different positions, which involves to be learned new parameters including amplitude, angle frequency and the initial phase. The parameters will increase the parameter amount of the embedding vector, and furthermore, it takes longer to train iteratively. In addition, we also observe that the performance stability is better than RSS-Span. The possible reason is that the increase in the number of parameters makes the model have better generalization ability, which further proves the superiority of our embedding method based on the relative position of complex domain.

6 Conclusion

In this paper, we improve a joint extraction method of entities and relationships based on an end-to-end sequence labeling framework with complex relative position encoding. The framework is based on a shared encoding of a pre-trained language model and a novel decomposition strategy. The experimental results show that the functional decomposition of the original task simplifies the learning process and brings a better overall learning effect. Compared with the baseline model, it reaches a better level on the three public datasets. Further analysis proves the ability of our model to handle multi-entity and multi-relation extraction. In the future, we hope to explore similar decomposition strategies in other information extraction tasks, such as event extraction and concept extraction.

Acknowledgement. The work presented in this paper is supported by the International Science and Technology Cooperation Foundation of Shanghai (grant No. 18510732000).

References

1. Dai, Z., Yang, Z., Yang, Y., et al.: Transformer-XL: attentive language models beyond a fixed-length context. In: ACL, pp. 2978–2988 (2019)
2. Lin, Y., Shen, S., Liu, Z., Luan, H., Sun, M.: Neural relation extraction with selective attention over instances. In: ACL, pp. 2124–2133 (2016)
3. Miwa, M., Sætre, R., Miyao, Y., Tsujii, J.: A rich feature vector for protein-protein interaction extraction from multiple corpora. In: EMNLP, pp. 121–130 (2009)
4. Yu, B., Zhan, Z., Shu, X., Liu, T., Wang, Y., et al.: Joint extraction of entities and relations based on a novel decomposition strategy. In: ECAI, pp. 2282–2289 (2019)
5. Devlin, J., Chang, M., Lee, K., Toutanova, K.: BERT: pre-training of deep bidirectional transformers for language understanding. In: NAACL, pp. 4171–4186 (2019)
6. Vaswani, A., Shazeer, N., Parmar, N., Uszkoreit, J., Jones, L., et al.: Attention is all you need. In: NIPS, pp. 6000–6010 (2017)
7. Sun, C., Wu, Y., Lan, M., Sun, S., Wang, W., et al.: Extracting entities and relations with joint minimum risk training. In: Proceedings of the 2018 Conference on Empirical Methods in Natural Language Processing, pp. 2256–2265 (2018)
8. Tan, Z., Zhao, X., Wang, W., Xiao, W.: Jointly extracting multiple triplets with multi-layer translation constraints. In: Proceedings of the AAAI Conference on Artificial Intelligence (2019)
9. Dai, D., Xiao, X., Lyu, Y., et al.: Joint extraction of entities and overlapping relations using position-attentive sequence labeling. In: Proceedings of the AAAI Conference on Artificial Intelligence, vol. 33, pp. 6300–6308 (2019)
10. Radford, A., Narasimhan, K., Salimans, T., Sutskever, I.: Improving language understanding by generative pre-training (2018)
11. Wei, J., Ren, X., Li, X., et al.: NEZHA: neural contextualized representation for Chinese language understanding. arXiv preprint arXiv:1909.00204 (2019)
12. Raffel, C., Shazeer, N., Roberts, A., et al.: Exploring the limits of transfer learning with a unified text-to-text transformer. J. Mach. Learn. Res. **21**, 1–67 (2020)
13. He, P., Liu, X., Gao, J., Chen, W.: DeBERTa: decoding-enhanced BERT with disentangled attention. In: Proceedings of ICLR (2021)
14. Yan, H., Deng, B., Li, X., Qiu, X.: TENER: adapting trans-former encoder for named entity recognition. arXiv preprint arXiv:1911.04474 (2019)
15. Wang, B., Zhao, D., Lioma, C., Li, Q., Zhang, P., Simonsen, J.G.: Encoding word order in complex embeddings. In: ICLR (2020)
16. Li, X., Feng, J., Meng, Y., et al.: A unified MRC framework for named entity recognition. In: ACL, pp. 5849–5859 (2020)
17. Roth, D., Yih, W.: A linear programming formulation for global inference in natural language tasks. In: Proccedings of CoNLL (2004)
18. Gupta, P., Schütze, H., Andrassy, B.: Table filling multi-task recurrent neural network for joint entity and relation extraction. In: COLING, pp. 2537–2547 (2016)
19. Adel, H., Schütze, H.: Global normalization of convolutional neural networks for joint entity and relation classification. In: EMNLP, pp. 1723–1729 (2017)
20. Zeng, X., Zeng, D., He, S., Liu, K., Zhao, J.: Extracting relational facts by an end-to-end neural model with copy mechanism. In: Proceedings of the 56th Annual Meeting of the Association for Computational Linguistics, vol. 1, pp. 506–514 (2018)
21. Gardent, C., Shimorina, A., Narayan, S., Perez-Beltrachini, L.: Creating training corpora for NLG micro-planning. In: 55th Annual Meeting of the Association for Computational Linguistics, pp. 179–188 (2017)

22. Bekoulis, G., Deleu, J., Demeester, T., Develder, C.: Joint entity recognition and relation extraction as a multi-head selection problem. Expert Syst. Appl. **114**, 34–45 (2018)
23. Nguyen, D.Q., Verspoor, K.: End-to-end neural relation extraction using deep biaffine attention. In: Azzopardi, L., Stein, B., Fuhr, N., Mayr, P., Hauff, C., Hiemstra, D. (eds.) ECIR 2019. LNCS, vol. 11437, pp. 729–738. Springer, Cham (2019). https://doi.org/10.1007/978-3-030-15712-8_47
24. Tran, T., Kavuluru, R.: Neural metric learning for fast end-to-end relation extraction. arXiv preprint arXiv:1905.07458 (2019)
25. Wang, Y., Yu, B., Zhang, Y., Liu, T., Zhu, H., Sun, L.: TPLinker: single-stage joint extraction of entities and relations through token pair linking. In: COLING, pp. 1572–1582 (2020)
26. Sui, D., Chen, Y., Liu, K., Zhao, J., Zeng, X., Liu, S.: Joint entity and relation extraction with set prediction networks. arXiv preprint arXiv:2011.01675 (2020)

CTran_DA: Combine CNN with Transformer to Detect Anomalies in Transmission Equipment Images

Honghui Zhou[1], Ruyi Qin[1], Jian Wu[2], Ying Qian[2(✉)], and Xiaoming Ju[2]

[1] Ningbo Power Supply Company of State Grid,
Zhejiang Electric Power Co., Ltd., Ningbo, China
[2] Zhejiang Jierui Electric Power Technology Co., Ltd., Ningbo, China
51205901057@stu.ecnu.edu.cn, yqian@cs.ecnu.edu.cn,
xmju@sei.ecnu.edu.cn

Abstract. With the development of the State Grid, the power lines, equipment and transmission scale are expanding. In order to ensure the stability and safety of electricity, it is necessary to patrol and inspect the power towers and other equipment. With the help of deep learning, neural networks can be used to learn the features in patrol image. In this paper, feature learning model named CNN Transformer Detect Anomalies (CTran_DA) is proposed to detect anomalies in patrol images. CTran_DA uses CNN to learn local features in the image, and uses Transformer to learn global features. This paper innovatively combines the advantages of CNN and Transformer to learn the local details as well as the global feature associations in images. By comparing experiments on out self-constructed dataset, the model outperforms state-of-the-art baselines. Moreover, the Floating Point Operations (FLOPs) and parameters of the model in this paper are smaller than other algorithms. In general, CTran_DA is an efficient and lightweight model to detect anomalies in images.

Keywords: Deep learning · Convolution neural network · Transformer · Feature learning · Lightweight

1 Introduction

With the rapid development and construction of the State Grid, all kinds of circuit equipment and power transmission equipment are constantly on the rise. As the power line equipment are in the outdoor, and by the natural environment and human factors, the pole tower will appear interface rust, collapse, wear and other phenomena. In order to ensure the proper transportation of electricity, frequent patrol inspections of outdoor power towers and other equipment are required. Determining whether there are any anomalies in power equipment by analyzing patrol photos is a very problematic issue.

Deep learning of images in performing analysis is currently a popular topic in the field of artificial intelligence. The method of machine learning not only can significantly

Z. Qian et al. (Eds.): WCNA 2021, LNEE 942, pp. 656–666, 2022.
https://doi.org/10.1007/978-981-19-2456-9_67

improve the efficiency of detection also reduces the cost. Due to the specificity of patrol images, the vast majority of images captured are fault-free and only a few have anomalies. Most researchers base on improving the quality of raw data acquired by image acquisition terminals to obtain the transmission equipment patrol images needed for intelligent analysis. Thus, many framing correction techniques based on angle perception and research end devices have emerged. Researchers are devoted to realizing real-time detection of some abnormal feature quantities and fast filtering of low-quality repetitive images. However, the limited computational resources of the terminal equipment limit the research methods for analysis of transmission equipment inspection images. Thus, an effective, fast and low-power method for image detection is essential for circuit device inspection.

This paper focuses on feature learning analysis of power tower transmission equipment detection images, which is essentially the problem of detecting anomalies in the images on the dataset. The model proposed in this paper named CNN Transformer Detect Anomalies (CTran_DA) which combines the advantages of Convolution Neural Network (CNN) [1] and Transformer [2]. We use CNN to learn local features in the image, and Transformer to learn global features. According to data characteristics, we construct three datasets from the data set of total patrol photos samples. Compared with traditional computer vision classification methods, CTran-DA achieve the best performance in our dataset. CTran_DA is also much smaller than other algorithms or models in terms of the number of parameters. Finally, various experimental results prove that the model proposed in this paper is not only efficient in detecting anomalies in images but also lightweight.

2 Related Work

In recent years, Convolutional Neural Networks (CNNs) has achieved breakthrough results in various fields related to pattern recognition [3]. Especially in the field of image processing, CNNs can reduce the number of parameters of artificial neural networks, which motivates researchers to use large CNNs to solve complex tasks. One of the biggest points of CNNs is that they can learn local features in images very well and work very well with image details, and only a small number of samples are needed to learn a well-designed model.

The basic functionality of CNNs can be divided into four key sections: the input layer, the convolutional layer, the pooling layer and the fully connected layer. The convolutional layer, as the core layer in CNNs, can significantly reduce the complexity of the model by optimizing its output, which can be achieved by setting three hyperparameters: kernel size, stride and padding. Through the inspiration of CNNs, more and more effective models such as AlexNet [4], VGG [5], GoogleNet [6] and ResNet [7] have emerged accordingly. All these models have achieved excellent results in the field of computer vision and are constantly being improved.

Transformer was first applied in the field of natural language processing and was a deep neural network mainly based on a self-attentive mechanism [2]. Many recent NLP scenarios have applied the Transformer structure and have achieved excellent results in various NLP tasks [2, 8, 9]. Inspired by the significant success of the transformer architecture in the field of Natural Language Processing (NLP), researchers have recently applied transformer to computer vision (CV) task [10]. Alexey Dosovitskiy et al. [11] have proposed vision transformer (ViT) model, which applies a pure transformer directly to sequences of image patches [10]. Wenhai Wang et al. [12] proposed the Pyramid Vision Transformer (PVT) model based on the fact that ViT consumes a lot of computational resources and the computational parameters are too large. PVT not only can effectively filter some redundant information in ViT model to achieve the lightweight of the model, it also achieves better results in various tasks of CV. Microsoft Asia Research used the structural design concept of CNN to reconstruct a new transformer structure named Swin Transfomer [13]. The current borrowing of better models from various fields and then transferring learning [14] to other tasks all provide a new way of thinking for researchers in the current field [15].

3   Method

3.1   Overall Architecture

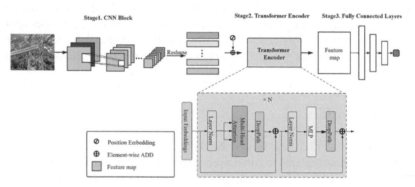

Fig. 1. Overall architecture of the proposed CTran_DA.

Our goal is to fully learn the features and the relationships between features in an image. An overview of CTran_DA is depicted in Fig. 1. Our model consists of three stages as CNN block, Transformer Encoder and Fully Connected Layers. The output of each stage is the input of the next stage, and the final result is obtained by the output of the fully connected layers.

3.2   CNN Block

In the first stage, given an input image with the size of $H \times W \times 3$. Then, we use CNN to learn local features and details of the images. The CNN block contains convolution layers (Conv), batch normalization layers (BN), activation layers (LeakyReLu [16]) and max pooling layers. The process of CNN block is shown in Fig. 2.

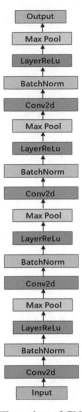

Fig. 2. Flow char of CNN block.

Convolution Layer. The convolutional layer is a feature extraction of the input data, and its process of processing images is just like the human brain recognizes images. It first perceives each feature in the image locally, and then performs a comprehensive operation to get the global information [3]. This convolution operation can be expressed as:

$$P^i_{out} = f(P * W) + b. \tag{1}$$

where $P \in \mathbb{R}^{h \times h}$ denotes the image input, W and b are the parameter matrix and bias of the convolution kernel respectively. P^i_{out} denotes the convolution output of the ith layer.

Batch Normalization Layer. The BN layer is to first find the mean and variance of each batch data, then subtract the mean and divide the variance by the data, and finally add two parameters [17]. BN layer has the following three roles: 1. speed up convergence. 2. prevent gradient exploding and gradient vanishing. 3. prevent overfitting. The result of the convolution, P^i_{out}, as the input to the BN layer can be expressed as:

$$B^i_{out} = BN\left(P^i_{out}\right). \tag{2}$$

Activate Layer. One of the important roles of the activation function is to incorporate nonlinear factors, to map features to high-dimensional nonlinear intervals for interpretation, and to solve problems that cannot be solved by linear models. In nonlinear activation layer, we use LeakyReLu [16] as the activation function and the formula is as followed:

$$LeakyReLU(x) = \begin{cases} x, & if\ x \geq 0 \\ \alpha x, & otherwise \end{cases}. \tag{3}$$

Max Pooling Layer. The pooling layer, also known as the downsampling layer, reduces the resolution of the features to reduce the number of parameters in the model and the complexity of the computation, enhancing the robustness of the model.

After the CNN module, we get the feature map of the local features of the image. Each feature map reshaped to an m-dimensional vector, and then combine them into n*m-dimensional embeddings based on the number of channels n to be used as the input of Transformer encoder.

3.3 Transformer Encoder

Transformer was first used in the field of neural language processing on machine translation tasks [2]. Our encoder contains Layer Normalization (LN) Layer, multi-head attention layer, Dropout layer and MLP block.

Layer Normalization. LN and BN work similarly. Since the length of each piece of data may be different when processing natural language, LN is used to process input embeddings.

Multi-Head Attention. Multiheaded attention is a mechanism that can be used to improve the performance of the self-attention layer. In self-attention layer, the input vector is first transformed into three different vectors: the query vector q, the key vector k and the value vector v. These vectors are packed into different matrices Q, K and V. The attention function of the input vectors is the calculated as followed:

Step 1: Compute scores between query matrix Q and key matrix K with: $S = Q \cdot K^T$
Step 2: Normalize the fraction of gradient stability with: $S_n = S/\sqrt{d_k}$
Step 3: Convert scores to probabilities using softmax function $P = softmax(S_n)$.
Step 4: Obtain the weighted value matrix with Attention $= V \cdot P$.

This whole process can be unified into a formula such as:

$$Attention(Q, K, V) = softmax\left(\frac{(Q \cdot K^T)}{\sqrt{d_k}}\right) \cdot V \tag{4}$$

However, self-attention is not sensitive to position information, and there is no position information in the calculation of the attention score. To solve this problem, the same dimensional position encoding is added to the original input embedding, and the position encoding is given by the following equation:

$$PE(pos, 2i) = sin\left(\frac{pos}{10000^{\frac{2i}{d_{model}}}}\right) \tag{5}$$

$$PE(pos, 2i + 1) = cos\left(\frac{pos}{10000^{\frac{2i}{d_{model}}}}\right) \tag{6}$$

where pos represents the position of the word in the sentence and i denotes the current dimension of the positional encoding. d_{model} is the dimension initially defined by our model.

On the multi-headed attention mechanism, we are given an input vector and the number of heads h. The input vectors are then converted into three different groups of vectors: the query group, the key group and the value group. In each group, the dimensions for a group are equally divided according to h heads. So, the total attention then consists of the combination of the attention of multiple heads with the following equation:

$$MultiHead\left(Q', K', V'\right) = Concat(head_1, \ldots, head_h)W^O \tag{7}$$

where $head_i = Attention(Q_i, K_i, V_i)$ and $W^O \in \mathbb{R}^{d_{model} \times d_{model}}$ is a linear projection matrix.

DropPath. DropPath is a regularization strategy that randomly deactivates the multi-branch structure in a deep learning model [18].

MLP. MLP a traditional neural network that is designed to solve the nonlinear problem that cannot be solved by a single layer perceptron. In addition to the input and output layers, it can have multiple hidden layers in between.

3.4 Model Optimization

In this model, due to the specificity of patrol photos, cb_loss [19] is selected as the method to process the data set in this paper, and then Focal Loss [20] is selected as the loss function.

4 Experiment

In the experiments, the learning rate is 0.001 and batch size equals 64. Our experiments were done on Pytorch 1.6 and GeForce RTX 3080.

4.1 Datasets

We obtained a total sample of 1,886 by manually screening the patrol photos, of which 270 were positive samples. In order to solve the problem of imbalanced sample distribution, we used two different methods to construct two new datasets. Firstly, we filtered and then removed some images with less obvious features from the negative samples to get a small dataset which we named SMALL [21]. In this dataset, the negative sample was removed to only 282 images, and the positive sample was 270 images to reach a balanced sample. Secondly, we replicate the 270 negative samples of the original data 6 times to reach 1620. This results in a balanced set of 1616 positive samples, which is called LARGE [21]. The original dataset is named MIDDLE [21]. The specific data set is shown in Table 1. In our experiments we divide the datasets into training set, validation set and test set in the ratio of 8:1:1, respectively. We train the model on the training set, tune the parameters by the validation set, and finally test the model on the test set [21].

Table 1. Summary of the datasets.

	Samples	Positive	Negative
SMALL [21]	522	270	282
MIDDLE [21]	1,886	270	1,616
LARGE [21]	3,236	1,620	1,616

4.2 Result

In the field of computer vision, many methods used for image classification have achieved excellent results. Therefore, we choose many of these models and modify the final output layer to serve as a reference comparison object for our experiments. Due to the specificity of the image and the specificity of the task, we are required to detect whether the positive sample from photos.

The residual network solves the degradation problem of deep neural network well, and achieves great results on image tasks such as ImageNet and CIFAR-10. The residual network also converges faster with the same number of layers. [7] VGG [5] is a very classical network structure, which adjusts the model effect by constructing different layers of CNN. Therefore, VGG11 and VGG13 are selected as the reference objects for comparison. MLP-mixer [22] builds a pure MLP architecture and communicates in two different dimensions. ViT [11] is a network model that takes a pure Transformer, which applies a pure transformer directly to sequences of image patches. PVT [12] introduces the pyramid structure into Transformer on the basis of ViT, which not only achieves good results but also greatly reduces the number of model parameters. The Swin Transformer is a hierarchical Transformer structure built by learning the hierarchical structure of CNN. In ViT, PVT and Swin Transformer, we set the same parameters, the attention heads to 12 and the depth of transformer blocks to 6.

Table 2. Comparison results of proposed model and other methods on three different datasets.

	SMALL			MIDDLE			LARGE		
	AUC	Recall	ACC	AUC	Recall	ACC	AUC	Recall	ACC
ResNet [21]	0.856	0.926	0.732	0.658	0.963	0.259	0.963	0.981	0.917
VGG11 [21]	0.815	0.963	0.696	0.666	0.889	0.434	0.890	0.957	0.809
VGG13 [21]	0.685	0.926	0.536	0.671	0.889	0.455	0.832	0.975	0.710
Mlp-mixer	0.613	0.963	0.554	0.646	0.852	0.497	0.680	0.944	0.565
ViT	0.566	0.961	0.518	0.556	0.926	0.275	0.668	0.988	0.556
PVT	0.510	0.926	0.554	0.540	0.519	0.582	0.640	0.963	0.546
Swin Transformer	0.605	0.963	0.536	0.535	0.926	0.233	0.674	0.675	0.540
CTrans1	0.815	0.963	0.696	0.766	0.926	0.566	0.910	0.994	0.867
CTrans3	0.833	0.926	0.732	0.798	0.852	0.640	0.931	0.994	0.830
CTrans5	0.890	0.889	0.750	0.497	0.963	0.185	0.898	0.975	0.781

We build our model based on the number of layers of transformer blocks in our model. We set the number of layers of the Transformer Encoder to 1, 3 and 5, and name them CTran-1, Ctran-3 and CTran-5 respectively. We compare our model with above methods on three metrics: Recall scores, Area Under ROC Curve (AUC) and ACC scores. The results of compared with above methods are shown in Table 2.

Table 3. Font sizes of headings.

	Params(M)	FLOPs
ResNet	21.29	3.68
VGG11	128.77	7.63
VGG13	128.96	11.34
MLP-mixer	18.59	1.0
ViT	43.27	8.48
PVT	2.84	0.41
Swin Transformer	18.19	2.27
CTrans-1	5.3	1.21
CTrans-3	5.8	1.22
CTrans-5	6.3	1.21

The experimental results on the three different data sets demonstrate that the method of obtaining the total number of balanced samples by replication achieves the best results. For SMALL dataset, a small sample balanced dataset, it is also slightly higher than the original dataset in all three metrics. After comparing with the traditional convolutional approach, our method achieves the best results on all three datasets. This shows that using only convolution for learning representation misses the global information of the image. After comparing with the latest Transformer-based model it was seen that both the pure Transformer model ViT and the simplified ViT did not achieve great results. When patching images, it is easy to lose details in complex images when using only the transformer to learn them. In particular, the task of processing for details is difficult to identify accurately. Table 3 shows the number of parameters and the amount of computation for each model. It can be seen that our model achieves better results on each dataset while using fewer parameters and consuming less FLOPs.

5 Conclusion

On the problem of abnormal detection for patrol photos, this paper proposes a novel scheme based on the features of pictures that are learned simultaneously by local and global features. In this paper, a new model CTran-DA is proposed which can effectively learn the feature details and global structure of the images. Secondly, it is a lightweight model with a lighter model structure than the current mainstream image classification models. The results from three different datasets show that our proposed model is also very effective and lightweight enough. This model can also provide a new idea for other researchers to follow and is very suitable for some restricted terminal devices. It provides a new solution for tasks that are highly complex and require light weight.

Acknowledgments. The work is supported by State Grid Zhejiang Electric Power Co., Ltd., science and technology project (5211nb200139), the key technology and terminal development of lightweight image elastic sensing and recognition based on AI chip.

References

1. Kalchbrenner, N., Grefenstette, E., Blunsom, P.: A convolutional neural network for modelling sentences. arXiv preprint arXiv:1404.2188 (2014)
2. Vaswani, A., et al.: Attention is all you need. In: Advances in Neural Information Processing Systems 2017, pp. 5998–6008 (2017)
3. Albawi, S., Mohammed, T.A., Al-Zawi, S.: Understanding of a convolutional neural network. In: 2017 International Conference on Engineering and Technology (ICET) 2017, pp. 1–6. IEEE (2017)
4. Krizhevsky, A., Sutskever, I., Hinton, G.E.: ImageNet classification with deep convolutional neural networks. In: Advances in Neural Information Processing Systems 2012, vol. 25, pp. 1097–1105 (2012)
5. Simonyan, K., Zisserman, A.: Very deep convolutional networks for large-scale image recognition. arXiv preprint arXiv:1409.1556 (2014)
6. Szegedy, C., et al.: Going deeper with convolutions. In: Proceedings of the IEEE Conference on Computer Vision and Pattern Recognition 2015, pp. 1–9 (2015)

7. He, K., Zhang, X., Ren, S., Sun, J.: Deep residual learning for image recognition. In: Proceedings of the IEEE Conference on Computer Vision and Pattern Recognition 2016, pp. 770–778 (2016)
8. Devlin, J., Chang, M.-W., Lee, K., Toutanova, K.: BERT: pre-training of deep bidirectional transformers for language understanding. arXiv preprint arXiv:1810.04805 (2018)
9. Brown, T.B., et al.: Language models are few-shot learners. arXiv preprint arXiv:2005.14165 (2020)
10. Han, K., et al.: A survey on visual transformer. arXiv preprint arXiv:2012.12556 (2020)
11. Dosovitskiy, A., et al.: An image is worth 16x16 words: transformers for image recognition at scale. arXiv preprint arXiv:2010.11929 (2020)
12. Wang, W., et al.: Pyramid vision transformer: a versatile backbone for dense prediction without convolutions. arXiv preprint arXiv:2102.12122 (2021)
13. Liu, Z., et al.: Swin transformer: hierarchical vision transformer using shifted windows. arXiv preprint arXiv:2103.14030 (2021)
14. Torrey, L., Shavlik, J.: Transfer learning. In: Handbook of Research on Machine Learning Applications and Trends: Algorithms, Methods, and Techniques, pp. 242–264. IGI Global (2010)
15. Pan, S.J., Yang, Q.: A survey on transfer learning. IEEE Trans. Knowl. Data Eng. **22**, 1345–1359 (2009)
16. Maas, A.L., Hannun, A.Y., Ng, A.Y.: Rectifier nonlinearities improve neural network acoustic models. In: Proceedings of the ICML 2013, vol. 30, p. 3. Citeseer (2013)
17. Ioffe, S., Szegedy, C.: Batch normalization: accelerating deep network training by reducing internal covariate shift. In: International Conference on Machine Learning 2015, pp. 448–456. PMLR (2015)
18. Larsson, G., Maire, M., Shakhnarovich, G., FractalNet: ultra-deep neural networks without residuals. arXiv preprint arXiv:1605.07648 (2016)
19. Cui, Y., Jia, M., Lin, T.-Y., Song, Y., Belongie, S.: Class-balanced loss based on effective number of samples. In: Proceedings of the IEEE/CVF Conference on Computer Vision and Pattern Recognition 2019, pp. 9268–9277 (2019)
20. Lin, T.-Y., Goyal, P., Girshick, R., He, K., Dollár, P.: Focal loss for dense object detection. In: Proceedings of the IEEE International Conference on Computer Vision 2017, pp. 2980–2988 (2017)
21. Chen, J., Luo, W., Hao, Y., Xu, H., Wu, J., Ju, X.: Using convolution neural networks to build a LightWeight anomalies detection model. In: 2021 IEEE 4th International Conference on Automation, Electronics and Electrical Engineering (AUTEEE) 2021, pp. 157–160. IEEE (2021)
22. Tolstikhin, I., et al.: MLP-mixer: an all-MLP architecture for vision. arXiv preprint arXiv:2105.01601 (2021)

Orchard Energy Management to Improve Fruit Quality Based on the Internet of Things

Pingchuan Zhang[✉], Sijie Wang, Xiaowen Li, Zhao Chen, Xu Chen, Yanjun Hu, Hangsen Zhang, Jianming Zhang, Mingjing Li, Zhenzhen Huang, Yan Li, Liutong Li, Xiaoman Xu, Yiwen Yang, Huaping Song, Huanhuan Huo, Yiran Shi, Xueqian Hu, Yabin Wu, Chenguang Wang, Feilong Chen, Bo Yang, Bo Zhang, and Yusen Zhang

School of Information Engineering, Henan Institute of Science and Technology, Xinxiang 453003, China
362764053@qq.com

Abstract. The crop growth is an energy conversion process, and energy management has an important impact on the quality and yield of crop products. As IoT (the Internet of Things) is widely used in agriculture, for example, orchard IoT is often used to realize water-saving irrigation, this paper innovatively proposes a scheme to improve fruit quality by using IoT to realize orchard energy management. The designed Internet of things, in addition to the usual orchard environmental parameters and water-saving irrigation, can further adjust the temperature difference between day and night according to the local temperature, that is, by spraying low-temperature water mist at 16 °C to reduce the ambient temperature of the orchard at night, creating an environment conducive to the conversion of carbohydrate into sugar. The experiment in peach orchard shows that the orchard energy management method based on Internet of Things works effectively, which can reduce the peach orchard temperature to 20° at night in summer, which is beneficial to improve the peach fruit sweetness.

Keywords: Energy management · Orchard IoT · Day and night temperature difference · Fruit quality

1 Introduction

The Internet of Things (IoT) is the fourth information revolution after computers, the Internet, and mobile communication technologies. Since 1999, the Massachusetts Institute of Technology introduced the concept to major countries in the world such as the United States. Planet) ", the European Union proposed the" Internet of Things Action Plan "in 2009, China proposed," Perceive China "and made the Internet of Things one of the strategic emerging industries [1–4].

In agriculture, various sensing terminals have been used to comprehensively sense collection facilities, Environmental information of production processes such as field planting, breeding, etc. to gradually achieve the optimal control and intelligent management of agricultural production processes [5].

© The Author(s) 2022
Z. Qian et al. (Eds.): WCNA 2021, LNEE 942, pp. 667–674, 2022.
https://doi.org/10.1007/978-981-19-2456-9_68

For example, the Orchard Internet of Things is mainly used to collect the related data such as soil or air temperature, humidity, light and the weather condition in the orchard environment, and can carry out independent irrigation, integrated water and fertilizer management, and insect forecasting, which improves the orchard Information level, management efficiency and fruit yield [6–12].

However, China as the biggest fruit production of the world, Chinese fruits also have problems such as low sugar content [13]. As for the sugar content of fruits, according to the literature [14–17], the crop growth is an energy conversion process, and energy management has an important impact on the quality and yield of crop products. The level and variation of ambient temperature have a crucial influence on the sweetness and quality of crops such as fruits, during fruit growth, carbohydrates are produced during the day by photosynthesis. Under the same conditions as water and fertilizer, high temperatures can enhance photosynthesis to produce more carbohydrates; these carbohydrates are converted into sugars at night. Temperature is the main factor affecting sugar conversion, which is the temperature difference between day and night. The greater the temperature difference between day and night, the more favorable the sugar conversion is, and the sweetness of the fruit is higher.

Spray cooling technology has been widely used in industrial and urban areas to reduce environmental temperature or dust pollution, in the agricultural field has also been used to cool the breeding environment or orchard to prevent frost [18].

To sum up, with the wide application of the Internet of Things in the field of agriculture, how to use the Internet of Things to regulate the environmental temperature of the orchard to achieve energy management of the fruit growth environment and create an environment conducive to the improvement of fruit quality has become a topic worth exploring.

2 The Orchard IoT for Temperature Difference Regulation

2.1 The Cooling Principle of Spraying Water Mist in Orchard

The cooling principle of artificial fog space environment is the double flow of air fog and the principle of evaporation and heat absorption [19, 20]. The sprayer diffuses the fog particles with a diameter of 1–10 μ to the cooling area, evaporates continuously in the diffusion process, and absorbs a lot of heat energy in the area. Scientific statistics of a kilogram of water to stimulate the floating state of artificial fog, the effect is equal to the dissolution of seven kilograms of ice, generally up to 6 °C–10 °C cooling effect, extreme cases can be reduced by 14 °C. Per gram of water can be for outdoor air cooling, the spray cooling efficiency is very high, in theory, the spray cooling is the amount of energy needed to overcome the surface tension of the water increases, the energy needed to 1 m^3 of water into the cube, 10 μ needed by its surface tension, and the latent heat of evaporation is as high as 2.2 billion joules, its theory can effect comparing is as high as 50000, And air conditioning is limited by the law of thermodynamics, 30 °C cooling 5 °C theoretical maximum energy efficiency ratio is about 60.

2.2 Principle and Process of Temperature Difference Regulation in Orchard

Photosynthesis and respiration occur simultaneously in cells of green plants such as fruit trees. During the day, Photosynthesis is the main process because of the light intensity and the temperature is high. During the photosynthesis process, the chloroplast in the cell synthesizes solar energy, CO_2, H_2O, and other organic matter, stores energy and releases O_2. At night, the light intensity is small, and the respiration is stronger than photosynthesis. Cell mitochondria decompose organic matter produced by photosynthesis and releases energy and oxygen. Respiratory effects include aerobic and anaerobic respiration.

In the summer of temperate plains, temperatures are high during the day and fruits accumulate nutrients. At night, the ambient temperature drops, however, in general, declines less and the decline rate is slower. Therefore, the mist cooling method can be used to accelerate the reduction of the ambient temperature. In summer, the sun enters the sunset point relatively late. In order to make full use of the photosynthesis of fruit trees after the sunset, under non-rainfall conditions, it is generally chosen to spray the water misting in the orchard at 8:00 pm every day. According to the wind direction collected by the wind direction sensor, the data center transmits the command to the sprayer node through LoRa, adjusts the direction of the sprayer nozzle, and sprays water mist.

2.3 Orchard IoT for Temperature Difference Regulation

The proposed orchard IoT scheme is shown in Fig. 1. The basic functions including collection of orchard environmental information, soil temperature, soil pH, soil humidity, carbon dioxide CO_2 concentration, air temperature and humidity, light intensity, wind speed and direction, rainfall, etc.; monitoring fruit tree pest by hyperspectral sensors; remote monitoring achieved on a computer or smartphone devices [6–8].

According to the three-layer basic architecture of the Internet of Things: the sensing layer, the transmission layer, and the application layer. The sensing layer contains 4 types of sensor nodes and 2 types of actuator nodes. The sensor node mainly implements the orchard information collection. Actuator node 1 completes automatic orchard irrigation. Actuator 2 reduces the ambient temperature of the orchard at night by spraying the mist and increases the temperature difference between day and night in the summer. Water mist is conducive to fruit expansion after the fruit enters the expansion stage [11]. The

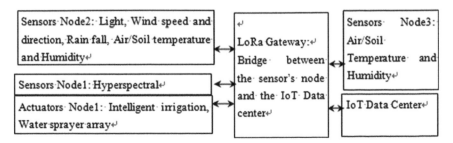

Fig. 1. Internet of orchard things system scheme

basic composition of a sensor node is: a sensor, an ARM microcontroller, LoRa module; the basic composition of an actuator node is: a relay, an ARM microcontroller, LoRa module.

The ARM microcontroller is a low power, high-performance embedded system as the node control core. It is an MCU based on the STM32 F401 series ARM® Cortex ™ -M4. It has a 12-bit ADC and a 16-bit/32-bit timer. FPU floating-point unit, communication peripheral interface (USART, SPI, I2C, I2S) and audio PLL. The operating frequency reaches 84 MHz, 105 DMIPS/285 Core-Mark, the flash ROM capacity is up to 256 kB, the SRAM capacity is 64 kB, and the chip's operating voltage ranges from 1.7 to 3.6 V.

In order to reduce costs, each node is provided with several related sensors. In order to control the day and night temperature difference of the orchard, sensor node 2 collects four orchard meteorological parameters such as air temperature, humidity, CO_2 and light intensity, and actuator node 1 executes relevant commands sent by the data center.

Sensor node 2 selects OSA-F7, which can measure four parameters: air temperature, relative humidity, CO_2 concentration, and illumination. The measurement range and accuracy of the four parameters are air temperature -30–70 ± 0.2 °C; relative humidity 0–100% RH $\pm$ 3% RH; carbon dioxide concentration 0–10000 ppm (optional 2000, 5000 ppm) $\pm$ 20 ppm; light intensity 0–200k lx (optional 2k, 20k lx and other ranges) $\pm 3\%$.

3 System Software

Based on the functions analysis of the orchard IoT, the system program includes 6 subroutines: parameter collection, irrigation, spraying mist, insect analysis, data server and mobile clients. The display can ensure the normal operation of the orchard's data access, data storage, and visual display programs; the interactive platform uses the B/S (Brower/Server) mode. Mist spraying operation procedure flow is shown in Fig. 2.

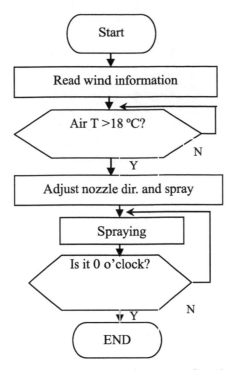

Fig. 2. Orchard mist operation process flow chart

4   System Experiment and Results Analysis

The experiment was conducted on July 20, 2018 in a Peach Orchard, an area of 1hm2, and an Internet of orchard Things. There is a water well in the orchard with a depth of 30 m. The weather: sunny, temperature 37 °C–28 °C, south wind 3–4 level. The water temperature of well is 16 °C. Mist spraying machine parameters: electric high-pressure remote sprayer, rated flow: 30–40 L/min; adjustable working pressure: 10–40 MPa; horizontal range: up to 100M. The sensor node 2 is shown in Fig. 3, and the pressure spray equipment is shown in Fig. 4. There are five sprayers, one at each corner of the orchard and the center. The temperature data is shown in Table 1.

Fig. 3. Sensor node 2.

Fig. 4. The pressure spray equipment.

Table 1. Collected data of temperature

Time	20:00	20:30	21:00	21:30	22:00	22:20	22:40	23:00	23:20	23:40	0:00
Temp.1	33.5	31	28.5	25.5	21.5	19.5	19	18.5	18	18	17.5
Temp.2	33	32.5	31	29	27.5	26	24	23	22.5	20	18
Temp.3	33	30.5	28	26	21	19	19	18.5	18	18	17.5
Temp.4	33	32	32.5	31	31	31	30	29.5	29.5	29	28.5

It can be seen from Table 1 that during the misting operation of the orchard, the ambient temperature in the orchard is reduced by a maximum of 10 °C compared with the temperature outside the orchard, and the cooling effect is obvious. Compared with the maximum temperature of 37 °C during the day, the temperature difference between day and night reaches 20 °C.

5 Conclusion

Regulating the ambient temperature of orchards is the key way to realize the energy management of orchards. At present, the research on energy management of orchards by reducing the night temperature of orchards has not attracted enough attention from scholars at home and abroad. The main reason is that the low temperature media with low cost is not easy to be found.

Compared with the simulation of fluent or CFD [19, 20], it's quite different that this paper has done a beneficial trial to implement orchard energy management based on Internet of Things to improve fruit quality. The Internet of orchard Things was designed and implemented. The experiments show that:

(1) The ambient temperature of the orchard at night can be effectively reduced by spray well water mist of perennial constant temperature at 16 °C, and the maximum temperature reduction can reach 10 °C so that the day-night temperature difference of the orchard on that day can reach 20 °C.
(2) Spray cooling system equipment is cheap, simple installation, and at the same time increases the air humidity, and can improve the yield and quality of peaches.

Acknowledgments. This work was supported by the Science and Technology Department of Henan Province under Grant 212102310553; 182102110301, and Henan Institute of Science and Technology: Innovation Project 2021CX58. Ministry of Education Industry-University Cooperation Collaborative Education Projects (Bai Ke Rong Chuang 201602011006, HuaQing YuanJian 201801082039, NANJING YunKai 201902183002, WUHAN MaiSiWei 202101346001.

References

1. Linnhoff-Popien, C.: Internet of things. Digitale Welt **3**, 58 (2019)
2. Baldini, G., Botterman, M., Neisse, R., Tallacchini, M.: Ethical design in the Internet of Things. Sci. Eng. Ethics **24**(3), 905–925 (2016). https://doi.org/10.1007/s11948-016-9754-5
3. Hammoudi, S., Aliouat, Z., Harous, S.: Challenges and research directions for Internet of Things. Telecommun. Syst. **67**(2), 367–385 (2017). https://doi.org/10.1007/s11235-017-0343-y
4. Navarro, E., Costa, N., Pereira, A.: A systematic review of IoT solutions for smart farming. Sensors **20**, 4231–4259 (2020)
5. Ramli Muhammad, R., Daely Philip, T., Kim Dong, S., Lee, J.M.: IoT-based adaptive network mechanism for reliable smart farm system. Comput. Electron. Agric. **170**, 1884–2022 (2020)

6. Khanna, A., Kaur, S.: Evolution of Internet of Things (IoT) and its significant impact in the field of precision agriculture. Comput. Electron. Agric. **157**, 218–231 (2019)
7. Wenxing, Z., Zhijing, W., Deli, L.: Design of orchard environmental intelligent monitoring system based on internet of agricultural things. Jiangsu Agric. Sci. **45**, 391–394 (2016)
8. Zhengyu, D.: Research on vineyard information acquisition and intelligent irrigation system design based on Internet of Things. J. Agric. Mechanization Res. **45**, 391–394 (2016)
9. Yajun, W.: Agricultural engineering application of Internet of Things technology in agricultural planting. Agric. Eng. **11**, 37–39 (2018)
10. Yue, Z., Hui, D., Jun, Z.: Study on strawberry moisture content monitoring system based on Internet of Things. Inf. Syst. Eng. **7**, 64–66 (2017)
11. Zhilong, Z.: Fruit Culture Science. Agricultural Science and Technology Press, Beijing (2012)
12. He, J., Wei, J., Chen, K., Tang, Z., Zhou, Y.: Multitier fog computing ith large-scale IoT data analytics for smart cities. IEEE Internet Things J. **5**, 677–686 (2018)
13. Lili, P., Weibing, J., Jian, H.: Factors affecting night respiration of early-maturing peach leaf coloring differently. Jiangsu J. Agric. Sci. **29**, 1131–1135 (2013)
14. Yi, S., Wenwen, Y., Kai, X.: Effect of temperature stress on photosynthesis in Myrica rubra leaves. Chin. Agric. Sci. Bull. **25**, 161–166 (2009)
15. Catherine, C.: Estimating Daily Primary Production and Nighttime Respiration in Estuaries by an In Situ Carbon Method. University of Rhode Island, Kingston (2015)
16. Dimitra, L.: Effect of High Night Temperature on Cotton Respiration, ATP Content and Carbohydrate Accumulation. University of Arkansas, Fayetteville (2008)
17. Pessarakli, M.: Handbook of Photo Synthesis, 3rd edn. CRC Press, Florida (2016)
18. Yongxin, C.: Design and Experimental Research of Cooling Fan System for Horticultural Plants. Jiangsu University, Zhenjiang (2019)
19. Heng, Z., Xiaoyun, L., Yungang, B., Hongbo, L.: Water saving irrigation fluent simulation of spray cooling system under grape trellis, **6**, 67–70 (2018)
20. Shengnan, T., Xiaochan, W., Zhimin, B., Zhao, L.: CFD simulation of spray cooling system in Greenhouse. Jiangsu J. Agric. Sci. **29**, 283–287 (2013)

Research on the Relationship Between User Attribute Context and Micro Implicit Interaction Behavior Based on Mobile Intelligent Terminal

Wei Wang[⊠], Xiaoli Zheng, Yuxuan Du, and Chuang Zhang

School of Information and Electrical Engineering, Hebei University of Engineering, Handan 056038, China
wangwei83@hebeu.edu.cn

Abstract. User browsing behavior is an important kind of implicit feedback data reflecting users' interests and preferences in the field of recommendation system. How to make full use of user browsing behavior data and combined with other context information to improve recommendation efficiency has become a research hotspot. This paper analyzes the user micro network implicit feedback behavior of mobile intelligent terminal, and studies the influence of user attribute context on user micro network implicit feedback behavior by using binary and multiple regression analysis. The results show that the user's age attribute, regional attribute and occupation attribute are a kind of very important context information.

Keywords: Recommended system · Mobile intelligent terminal · Implicit feedback behavior · User attribute

1 Introduction

The analysis of users' network behavior characteristics is the design basis of many Internet products. Through in-depth analysis of user behavior, personalized recommendation can bring users a better application experience. In the field of market driven software engineering, user behavior analysis also provides new ideas and improvement directions for application development to meet the requirements of the new situation.

User network behavior can be divided into explicit feedback behavior and implicit feedback behavior. At present, a relatively stable and unified view has been formed on the definition, characteristics, differences and types of the two types of behavior. The display feedback behavior data can accurately express the user's intention, but it interferes with the user's normal interaction process in the network, increases the cognitive burden and reduces the user experience, so it is difficult to obtain the data. On the contrary, for the implicit feedback behavior data of users, it is much less difficult to obtain, and the information abundance is large. Therefore, although such information has low accuracy,

© The Author(s) 2022
Z. Qian et al. (Eds.): WCNA 2021, LNEE 942, pp. 675–684, 2022.
https://doi.org/10.1007/978-981-19-2456-9_69

large data noise, large context sensitivity, this research field is still getting more and more attention.

The research on recommendation methods based on user implicit feedback behavior has made some progress in recent years. Such research relies on user browsing, attention, purchase, transaction and other key intention behaviors to complete commodity recommendation, without fully considering the context of implicit feedback behavior. At the same time, some recommendation systems also explore the direct application of context information, especially time and location context, to recommendation systems, and have made some progress. In addition, by mining the interaction data of network applications in different context, collecting user network activity logs and questionnaires, some research results have been accumulated in understanding user network behavior, and some of them have been applied to the field of software design and human-computer interaction, However, such achievements have not been well extended to the field of personalized recommendation. In this work, we take context implicit feedback behavior personalized recommendation as a whole to supplement the previous research work.

Users' implicit feedback network behavior is easily affected by the context of time, environment, user attributes, application content, interactive terminal, personality and emotional state. Especially for mobile intelligent terminals, the context sensitivity of implicit feedback network behavior is more prominent due to the scattered use time period, changeable environment, diverse crowd attributes and different device terminals. When using the implicit feedback behavior of mobile intelligent terminals for content recommendation, the recommendation results also show a certain sensitivity to the context. Therefore, it is more necessary to discuss the impact of context differences on the implicit feedback behavior applied to personalized recommendation.

2 Related Work

With the rapid development of social networks and e-commerce, the number of Internet users has greatly increased, and the demand for personalized recommendation services is also increasing. Accurately and effectively deal with the massive multi-source heterogeneous data generated by users browsing the mobile Internet is the focus and difficulty of the current research.

The original personalized recommendation service is mainly for PC based users. The relevant research is mainly divided into the following four aspects: Research on a certain application scenario, research on a certain class or technology, research on evaluation methods of recommendation system, and research on a certain kind of common problems in the recommendation system.

The study of user network behavior was initially applied in the field of information retrieval, which significantly improves the performance of information filtering compared to other feedback, and quickly filters from massive information sets, providing the retrieval set [1] with the highest correlation with their interest preferences. By comparing the results of user browsing time preference analysis with user explicit ratings, Morita [2] found the fact that users spend more energy and much longer time reading the preferring tidings on newspaper than regular tidings, representing user browsing time is a available information showing the user's interest preferences. Konstan [3] applied

Usenet News with browsing time-based collaborative filtering methods in 1997. Moreover, Oard and Kim validated the behavior when browsing a website like bookmarking, printing and saving could show user interest preferences and could be used to compensate for insufficient explicit feedback score data. While the Internet develop rapidly, the increase of the number of users, the data overload problem get significant. And the stability of the recommending results accuracy of relying merely on the behaviors which are called explicit feedback decreases, and the significance and requirement of the behaviors which are called implicit feedback, for example, exploring the website behavior in personal recommending models increase. When lots of scientists invest in implicit recommendation study, there are also ordinary solutions in manufacturing. Moreover, the behavior estimation from user website exploring in the recommending system with implicit cues is the most significant one of its core. In the Oard and Kim [4] and Kelly [5] opinions, who research on the website exploring behaviors, there are three groups about the user browsing behavior. They are saving behaviors [6], operational behaviors, and repetitive behaviors. a) the first behavior type- save: it includes download behavior, collection, printing, subscribe to, and bookmarks adding or deleting; b) the second behavior type- operation: it includes mouse clicking, searching information, browsing time on one web page, scroll bar dragging, page size adjusting, and copy data behavior; c) the third behavior type - repeat: it includes accessing a website or web page repeatedly, purchasing goods repeatedly, click on a item repeatedly.

Anyway, insufficient researches about the behaviors of website exploring that indicates user's favorite. While users change their interaction devices in particular, their website exploring behaviors in the mobile network environment may be different. Therefore, carrying on studies about micro network behaviors with implicit attributes is essential.

By analyzing these behavioral data, we can obtain the behavioral habits of mobile users, which are helpful to enhance the servicing character and users' enjoyment. Depending on the users' website or web page exploring behaviors in mobile condition, paper [7] studied personal recommending method. In addition, group recommending method and the mining algorithm of uncertain attribute were also considered. The results were good. Relevant research focused on the direction of recommendation system. The website exploring behaviors in mobile condition is not deeply studied. Literature [8] combines users' website exploring behaviors including mobile location data to analyze the influence from scenes and studied the users' website exploring behaviors in the dimension of space and time. Not only it concerned users' web page exploring behavior, but also it pays attention to the users' mobile behavior. The researches about implicit behaviors are hot [9–11]. According to the statement above, this paper has finished the following work: 1) investigation of user micro network implicit feedback behavior for mobile intelligent terminal. 2) The influence of user attribute context on the implicit feedback behavior of user micro network.

3 Problem Description and Correlation Analysis

3.1 Problem Description

Users' network implicit behavior contains their preference information, but it is generally not clearly expressed, so it is difficult to correctly judge their preferences. Researchers have done more work in this regard. At present, there are many researches on macro network implicit behavior, such as behavior sequence analysis or item recommendation based on browsing, adding shopping cart, shopping and so on. For the implicit feedback behavior of user micro network, there are few relevant studies and conclusions due to the problems of small data scale, few data categories and low data dimension. This paper intends to analyze the implicit feedback behavior of users in micro networks, focusing on the relationship between user attribute context and micro network implicit behavior.

3.2 Users' Micro Implicit Behavior

Acquiring approach of users' micro implicit behavior includes two ways. The first one is direct acquiring way, which is conducted by running some software in background. The other is indirect way, generally speaking, which is acquired by questionnaire. In direct acquisition, there are some problems such as sparse data, few categories and low dimensions, which is not conducive to subsequent analysis and deterministic conclusions. This paper analyzes the micro implicit feedback behavior by using the data obtained indirectly. Based on the questionnaire in literature [12], some survey contents (Q4–Q15) are extracted from the questionnaire, in addition, matched to users' micro implicit behavior, which is demonstrated as below in Table 1.

Table 1. Micro implicit behavior.

Raw data (users' behavior)	Description	Corresponding behavior (micro implicit behavior)
Which app store do you use? (Q4)	Discrete, type: 10, Category mutual exclusion	Category selection of application market (IFB1)
How frequently do you visit the app store to look for apps? (Q5)	Discrete, type: 9, Category mutual exclusion	Access frequency of application market (IFB2)
On average, how many apps do you download a month? (Q6)	Discrete, type: 6, Category mutual exclusion	Number of monthly attention to items (IFB3)
When do you look for apps? (Q7)	Discrete, type: 6, Categories are not mutually exclusive	Query frequency of item (IFB4)
How do you find apps? (Q8)	Discrete, type: 9, Categories are not mutually exclusive	Query method for item (IFB5)
What do you consider when choosing apps to download? (Q9)	Discrete, type: 13, Categories are not mutually exclusive	Detail level of item browsing (IFB6)
Why do you download an app? (Q10)	Discrete, type: 15, Categories are not mutually exclusive	Focus on item (purchase possibility) (IFB7)
Why do you spend money on an app? (Q11)	Discrete, type: 12, Categories are not mutually exclusive	Purchase behavior of item (IFB8)
Why do you rate apps? (Q13)	Discrete, type: 7, Categories are not mutually exclusive	Evaluation behavior of item (IFB9)

(*continued*)

Table 1. (*continued*)

Raw data (users' behavior)	Description	Corresponding behavior (micro implicit behavior)
What makes you stop using an app? (Q14)	Discrete, type: 15, Categories are not mutually exclusive	Cancel attention to item (IFB10)
Which type of apps do you download? (Q15)	Discrete, type: 23, Categories are not mutually exclusive	Category focus behavior on item (IFB11)

For the sake of easing the correlation analysis about influenced factors and users' implicit behavior, according to the questionnaire data in literature [12], this paper divides users' micro implicit feedback behavior into two categories: 1) mutually exclusive type, and 2) non-mutually exclusive type. In Table 1, IFB1-IFB3 are commonly clustered into the mutually exclusive type, which means every behavior exists once. For example, there are selecting only one application market category, determining a certain frequency of access and attention to element. IFB4-IFB11 belongs to non-mutually exclusive type. Ever person could select multiple behavior. For example, the frequency of inquiring items, while the person is discouraged or bored, or desires to accomplish a duty, etc.

3.3 User Attribute Context

To study the relationship between user attribute context and micro network implicit behavior, it is necessary to determine the content of user attributes. Based on the questionnaire in literature [12], the determined user attributes are shown in Table 2.

Table 2. User attributes.

User attributes	Data description
Age (Q17)	Discrete*, type: 8, Category mutual exclusion
Marital Status (Q18)	Discrete, type: 7, Category mutual exclusion
Nationality (Q19)	Discrete, type: 16, Category mutual exclusion
Country of Residence (Q20)	Discrete, type: 16, Category mutual exclusion
First Language (Q21)	Discrete, type: 11, Category mutual exclusion
Ethnicity (Q22)	Discrete, type: 7, Category mutual exclusion
Highest Level of Education (Q23)	Discrete, type: 8, Category mutual exclusion
Years of Education (Q24)	Discrete*, type: 7, Category mutual exclusion
Disability (Q25)	Discrete, type: 3, Category mutual exclusion
Current Employment Status (Q26)	Discrete, type: 9, Category mutual exclusion
Occupation (Q27)	Discrete, type: 25, Category mutual exclusion

*Indicates that the original data is a continuous quantity.

3.4   Correlation Analysis Between User Attribute Context and Implicit Behavior

This paper researches on the relations of users' characteristics and micro implicit behaviors. That is, in the view of users' characteristics, impact on users' micro implicit behaviors is discussed. In addition, big impact factors are chosen. As statements earlier about users' micro implicit behavior and users' characteristics data, this paper selects IFB1-IFB11 as the dependent variable and user attributes Q17-Q27 as the independent variable, and uses logistic regression to complete the correlation analysis between users' characteristics background and implicit behaviors.

Multiple Logistic Regression Analysis. Through the observation of dataset, the type of IFB1, IFB2 and IFB3 is multi-classified micro implicit behavior, in which IFB1 is a disordered variable and IFB2 and IFB3 are ordered ones. Multiple logistic regression analyzing method is used to study the impact on micro implicit behaviors from users' attributes.

Binary Logistic Regression Analysis. Based on the observation of the data, IFB4-IFB11 is consist of multiple subsets. Moreover, this type of behaviors is described as binary. Therefore, binary logistic regression analyzing method to study impact on micro implicit behavior from users' attributes is used in this paper.

4   Results and Discussion

4.1   User Attributes and Influencing Factors of IFBn

According to the significance index of model fitting, shown in Table 3, the fitting models of IFB1 and IFB3 are statistically significant and pass the test. The Pearson Chi-square significance of IFB1 model is 1. The model fitting status, as described in the column, to initial data passes the test. However, its pseudo r square value is flat, and the fitting degree is not actually distinguished.

In accord with the significance of likelihood ratio test in Table 4, for the micro implicit behavior IFB1, there exists results as below: eight user attribute influencing factors such as age, marital status, current country of residence, first language, years of education, physical barrier, current employment status and occupation all contribute significantly to model configurations, which is the crucial component effecting IFB1.

Table 3. Fitting information and forecast percentage (IFB1-IFB3).

	Model fitting significance	Significance of goodness of fit (Pearson)	Pseudo R-square (Cox Snell)	Forecast correct percentage
IFB1	.000	1.000	.523	43.9%
IFB2		.000	.000	29.1%
IFB3	.000	.000	.289	50.7%

Table 4. Likelihood ratio test significance (IFB1-IFB3).

	Q17	Q18	Q19	Q20	Q21	Q22	Q23	Q24	Q25	Q26	Q27
IFB1	0	0	1	0	0	0.987	0.225	0	0	0	0.001
IFB2		0		0				0		0	
IFB3	0	0.139	0.288	0.003	0.618	0.752	0.019	0.782	0.347	0.076	0.003

In agreement with the exhaustive test dataset of model factors in Table 5, for the type of IFB4, the fitting mode of these micro implicit behaviors is commonly essential. Meanwhile, goodness of fit test and prediction correct percentage information show that, considering the IFB4 subgroup, the model fitting goodness of IFB4-1, IFB4-3 and IFB4-6 behavior subset is higher and the fitting model is better.

Table 5. Model sparsity test, goodness of fit and prediction percentage (IFB4).

	Omnibus test of model coefficients	Hosmer lemeshow test	Forecast correct percentage
IFB4-1	.000	.856	68.9%
IFB4-2	.000	.490	65.7%
IFB4-3	.000	.752	68.5%
IFB4-4	.000	.571	67.0%
IFB4-5	.000	.108	61.1%
IFB4-6	.000	1.000	98.3%

Table 6. Variable significance (IFB4).

	Q17	Q18	Q19	Q20	Q21	Q22	Q23	Q24	Q25	Q26	Q27
IFB4-1	.000	.005	.375	.000	.193	.094	.138	.784	.999	.000	.376
IFB4-2	.000	.567	.000	.857	.725	.028	.000	.058	.151	.154	.000
IFB4-3	.000	.094	.198	.000	.406	.318	.114	.018	.203	.774	.112
IFB4-4	.000	.688	.763	.000	.399	.004	.306	.036	.507	.001	.431
IFB4-5	.127	.324	.942	.000	.720	.034	.001	.492	.997	.002	.218
IFB4-6	.656	.408	.028	.975	.966	.298	.083	.404	.798	.013	.091

According to the significance index of each variable in Table 6 (in which the gray shadow part commonly shows the significance index >0.05), the micro implicit feedback behavior of item query frequency (IFB4) as a whole, age, current country of residence and current employment status are the main influencing factors of user attributes. Specifically, for the behavior subset IFB4-1 of micro implicit feedback behavior IFB4, four user attribute influencing factors such as age, marital status, current country of residence and current employment status contribute significantly to the model configurations and are the important factors impacting IFB4-1. Given the type of IFB4-3 behavior subset of micro implicit feedback behavior IFB4, age, current country of residence and years of education are the main factors affecting IFB4-3. For the behavior subset IFB4-6 of micro implicit feedback behavior IFB4, two user attribute influencing factors, nationality and current employment status, contribute significantly to the model configurations and are the important factors impacting IFB4-6. Analysis about user attributes and influencing factors of IFBn (n = 5–11) is similar as above.

4.2 Influence Ranking of User Attributes

Through the above analysis of user attribute influencing factors that make a significant contribution to user micro implicit feedback behavior IFBn, the ranking of influencing factors is obtained, as shown in Table 7. It can be seen that the user's age attribute has a great impact on the micro implicit feedback behavior. The user attributes such as the current country of residence, the first language and the current employment status also affect the user behavior to a certain extent.

Table 7. User attribute impact.

User attribute	Influence ranking	Number of times as the main influencing factor of IFBn
Age (Q17)	1	20
Country of Residence (Q20)	2	10
First Language (Q21)	3	9
Current Employment Status (Q26)	4	8
Years of Education (Q24)	5	7
Ethnicity (Q22)	6	6
Occupation (Q27)	6	6
Marital Status (Q18)	6	6
Highest Level of Education (Q23)	7	5
Disability (Q25)	7	5
Nationality (Q19)	8	3

5 Conclusion

This paper analyzes the user micro network implicit feedback behavior of mobile intelligent terminal, and studies the influence of user attribute context on the user micro network implicit feedback behavior. The results reveal that users' age attributes, regional attributes and professional attributes will have an impact on users' behavior. The outcomes above establish a groundwork for future researches around users' micro implicit behavior data in recommendation area.

Acknowledgments. This work was supported by The National Natural Science Foundation of China (No. 61802107); Science and technology research project of Hebei University (No. ZD2020171); Jiangsu Planned Projects for Postdoctoral Research Funds (No. 1601085C).

References

1. Seo, Y.W., Zhang, B.T.: Learning user's preferences by analyzing web browsing behaviors. In: 4th International Conference on Autonomous Agents, pp. 381–387. ACM Press, New York (2000)
2. Morita, M., Shinoda, Y.: Information filtering based on user behavior analysis and best match text retrieva. In: 17th Annual International ACM-SIGIR Conference on Research and Development in Information Retrieval, pp. 272–281. Springer-Verlag, Berlin (1994)
3. Konstan, J.A., Miller, B.N., Maltz, D., et al.: GroupLens: applying collaborative filtering to Usenet news. Commun. ACM **40**(3), 77–87 (1997)
4. Oard, D.W., Kim, J.: Implicit feedback for recommender systems. In: AAAI Workshop on Recommender Systems, p. 83. AAAI Press, Palo Alto (1998)
5. Kelly, D., Teevan, J.: Implicit feedback for inferring user preference: a bibliography. In: ACM SIGIR Forum, pp. 18--28. ACM Press, New York (2003)
6. Yin, C., Deng, W.: Extracting user interests based on analysis of user behaviors. Comput. Technol. Dev. **5**, 37–39 (2008)
7. Ding, Z.: Research on Mining and Recommendation Algorithm based on Mobile User Behaviors. University of Electronic Science and Technology of China (2017)
8. Lv, Q.J.: Analysis and Application of User Mobility based on Cellular Data Network Traffic. Beijing University of Posts and Telecommunications (2017)
9. Bian, T.Y.: User Behavior Analysis and Purchase Prediction based on Implicit Feedback Data. Nanjing University of Posts and Telecommunications (2020)
10. Wang, Zh.Y.: Research on BPR Algorithm based on Commodity Content and User Behavior Feedback. Donghua University (2021)
11. Xiao, Zh.B., Yang, L.W., Jiang, W., et al.: Deep multi-interest network for click-through rate prediction. In: 29th ACM International Conference on Information & Knowledge Management, pp. 2265--2268. ACM Press, New York (2020)
12. Soo, L., Peter, J.B.: Investigating country differences in mobile app user behavior and challenges for software engineering. IEEE Trans. Softw. Eng. **41**(1), 40–64 (2015)

Traffic Sign Detection Based on Improved YOLOv3 in Foggy Environment

Luxi Ma, Qinmu Wu[✉], Yu Zhan, Bohai Liu, and Xianpeng Wang

School of Electrical Engineering, Guizhou University, Guiyang 550025, China
wqm-watlei@163.com

Abstract. Aiming at the problem of poor detection accuracy and inaccurate positioning of traffic signs under foggy conditions, this paper proposes an improved YOLOv3 detection algorithm. Firstly, a data set of Chinese traffic signs in a foggy environment is constructed; The dark channel a priori algorithm based on guided filtering is used to process the image with fog, which overcomes the problem of image quality degradation caused by fog. Mosaic data enhancement is performed on the annotated data set image, which speeds up the convergence speed of the network. Increased the feature scale of YOLOv3 algorithm. The loss function of the network is optimized, CIOU is used as the positioning loss, and the positioning accuracy is improved. At the same time, the method of transfer learning is used to overcome the problem of insufficient samples. The enhanced yolov3 algorithm proposed in this paper has higher detection accuracy and shorter detection time than the standard yolov3 algorithm and SSD algorithm.

Keywords: Traffic sign detection YOLOv3 · Improved YOLOv3 model · Foggy environment · Transfer training

1 Introduction

Traffic sign detection and recognition is one of the research hotspots of environment perception in the three major modules of unmanned driving [1]. Traffic sign recognition plays an important role in unmanned driving. However, in foggy weather, there are some problems in traffic sign detection, such as small target, unclear target and so on. The designed algorithm needs to take into account the characteristics of high precision and real-time. At the same time, it is necessary to ensure that the training image data is sufficient so that the neural network model can learn the characteristics of traffic signs in different complex environments [2].

Yu fuses the dark channel prior algorithm with MSR to defog, and uses the Faster R-CNN two-stage target detection algorithm to detect traffic signs in foggy environments. Compared with the first stage target detection algorithm, the detection speed is slower and the calculation amount is larger [3]. Xu uses image enhancement to defog, and proposes an improved convolutional neural network design to recognize traffic signs. The method of image enhancement is not to remove the fog, but to sharpen the image. This method can only be used for traffic sign detection under light and medium fog,

© The Author(s) 2022
Z. Qian et al. (Eds.): WCNA 2021, LNEE 942, pp. 685–695, 2022.
https://doi.org/10.1007/978-981-19-2456-9_70

and the effect is not ideal under dense fog. Chen and others first used the deep learning algorithm IRCNN to remove the haze, and then proposed a multi-channel convolutional neural network (Multi-channel CNN) model to identify the image after the haze removal [4]. However, the defogging method based on deep learning requires a large number of images in the data set and the speed is relatively slow. Moreover, none of the above methods has constructed a traffic sign data set in a foggy environment.

2   Image Defogging Preprocessing

2.1   Data Set Construction

In the research of traffic sign detection and recognition, researchers mostly use the American traffic sign data set (LISA) and other algorithms for performance testing. However, most of the above data set samples are collected under good lighting conditions, and no domestic researcher has constructed and published a rich comprehensive for the identification of China. The traffic sign data set of China in the foggy environment. For the traffic sign detection of YOLOv3 in the foggy environment, this article must have the Chinese traffic sign data set in the foggy environment [5].

Based on this, for traffic sign detection in a foggy environment, on the one hand, some clear traffic sign pictures are downloaded from the Internet, and on the other hand, it is collected on the spot by taking pictures in heavy fog. The images are divided into training set and test set according to the ratio of 8:2, a total of 3415 images, including 2390 training set and 1025 test set. Use LabelImg software to label each image. The label information includes the category attribute of the traffic sign, the illumination of the image, the upper left and lower right coordinates of the sign border (in pixels), and the information is saved in xml format. The data is divided into 3 categories: indication signs, prohibition signs, and warning signs.

2.2   Dehazing Algorithm

The existing defogging algorithms are mainly divided into three categories: One is the defogging algorithm based on image enhancement. The second is a defogging algorithm based on image restoration. Three defogging algorithms based on deep learning [6].

This paper compares several algorithms. The dehazing effect is shown in Fig. 4; the best effect is the DehazeNet algorithm. Its disadvantage is that it takes a long time and the average running time is 1.14 s. Therefore, in combination with traffic sign detection scenarios, this paper uses dark channel based on guided filtering. Empirical algorithm for image restoration [7]. The dark channel a priori principle believes that in most non-sky local areas, one of the three RGB color channels of each image has a very low gray value, almost tending to zero. According to the above principles, the dark channel map can be obtained first, and then the atmospheric light value and transmittance can be estimated by using the dark channel map, and the transmission function is refined by the guided filter, and the transmittance value is optimized. Finally, the result obtained is substituted into the atmospheric scattering The model can get the restored image. The steps of the algorithm are shown in Fig. 1 (Fig. 2):

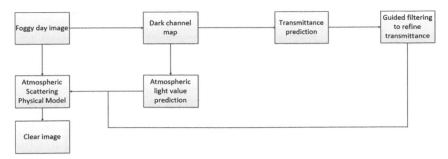

Fig. 1. Flow chart of dark channel restoration algorithm

a. Original image   b. Retinex   c. Dark channel prior   d. DahazeNet

Fig. 2. Comparison of dehazing effects

3 YOLOv3 Algorithm and Improvement

This article chooses YOLOv3 model to complete this research because YOLOv3 has made improvements in category prediction, bounding box prediction, multi-scale fusion prediction, and feature extraction [8]; YOLOv3's mAP can be comparable to RetinaNet, but the speed is increased by about 4 times. At the same time, there have been significant improvements in detecting small objects. Therefore, it is ideal to apply to the detection and recognition of traffic signs in complex environments [9].

3.1 YOLOv3 Detection Network

As shown by the dotted line in Fig. 3, in order to improve the accuracy of the algorithm for small target detection, YOLOv3 uses 5 downsampling of the input image and predicts the target in the last 3 downsampling. It can output 3 features of different scales, respectively Output 1, 2, 3 for prediction. The rule of side length is 13:26:52, and the depth is 255. The up-sample and fusion method of FPN (feature pyramid networks) is adopted; the advantage of choosing up-sample in the network: the expression effect is determined by the network level, and the effect becomes better as the network level deepens, so that you can directly use the deeper object characteristics to perform the object predict [10].

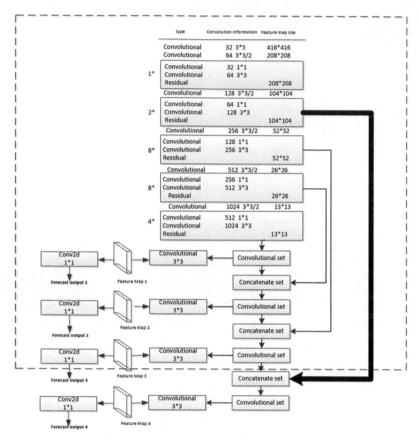

Fig. 3. Improved multi-scale prediction structure

3.2 YOLOv3 Network Optimization

Improved Multi-scale Prediction YOLOv3 Model. YOLOv3 only uses three-scale features, and the shallow information used is not sufficient [11]. Aiming at the problems that the detection and classification of traffic sign targets in complex environments are affected by different environments and the target is small, an improved YOLOv3 deep neural network was designed and proposed, and the fourth feature scale was added: 104×104; as shown in Fig. 6 As shown. The thick line in Fig. 3 shows an improved multi-scale network structure.

The specific method is: in the YOLOv3 network, after the feature layer with a detection scale of 13×13 is up-sampled twice, the original feature scale of 52×52 can be increased to 104×104. If you want to make full use of deep features and For shallow features, the 109th layer and the 11th layer of the feature extraction network should be feature fused through the route layer. The remaining feature fusion is: the 85th and 97th layers outputted after 2 times upsampling. The feature maps of the 85th and 61st layers, and the 97th and 36th layers are respectively merged through the route layer. As shown in Table 1, each feature layer is specific.

Table 1. YOLOv3 feature map

Feature layer	Feature map size	Number of preset bounding boxes
Feature layer 1	13 × 13	13 × 13 × 3
Feature layer 2	26 × 26	26 × 26 × 3
Feature layer 3	52 × 52	52 × 52 × 3
Feature layer 4	104 × 104	104 × 104 × 3

Mosaic Image Enhancement. Traditional data enhancement methods only enrich the number of data set by changing the characteristics of the image [12]. Mosaic image enhancement is a process in which a new image is obtained by combining 4 random images to train the network, which increases the diversity of data and the number of targets provide a more complex and effective training background. At the same time, the original image annotation information still exists. As shown in Fig. 4. This can further improve the accuracy and recall rate. At the same time, because multiple images are input to the network at the same time, the batch size of the input network is increased in disguise. Inputting an image stitched by four images is equivalent to inputting four original images (batch size = 4) in parallel, which reduces the need for training. The performance requirements of the equipment. Effectively improve the efficiency of statistical mean and variance of the BN (Batch Normalization) layer.

Fig. 4. Effect diagram of mosaic image enhancement algorithm

Loss Function. YOLOv3 loss is divided into three parts: positioning loss Lloc (o, c), confidence loss Lconf (o, c), classification loss Lcla (o, c) three parts, as shown in formula 1:

$$L(o, c, O, C, l, g)$$

$$= \lambda_1 L_{conf}(o, c) + \lambda_2 L_{cla}(O, C) + \lambda_3 L_{loc}(l, g) \tag{1}$$

Among them, $\lambda1$, $\lambda2$, and $\lambda3$ are balance coefficients.

Intersection-to-Union Ratio (IOU) When performing bounding box regression prediction, when two bounding boxes (target bounding boxes) do not intersect, according to the definition of IOU, the IOU is zero at this time, and the propagation loss cannot be calculated at this time. In order to solve this defect, this paper introduces the CIOU loss function for the regression prediction of the bounding box. An excellent regression positioning loss should consider three geometric parameters: overlap area, center point distance and aspect ratio. The calculation formula is shown in formula (2):

$$\text{CIoU} = \text{IoU} - \left(\frac{P^2(\text{b},\text{b}^{gt})}{c^2} + av \right) \tag{2}$$

$$L_{CIoU} = 1 - CIoU \tag{3}$$

Among them, α is the weight function, and v is used to measure the similarity of the aspect ratio, and the definition is shown in formula (4) (5).

$$v = \frac{4}{\pi^2}(\arctan \frac{w^{gt}}{h^{gt}} - \arctan \frac{w}{h})^2 \tag{4}$$

$$a = \frac{v}{(1 - IoU) + v} \tag{5}$$

When the CIOU does not overlap with the target box, it can still provide the moving direction for the bounding box. The distance between the two target frames can be minimized directly, and the convergence is much faster. After adding aspect ratio considerations, it can further quickly converge and improve performance.

Retraining Based on Transfer Learning. In the experiment, the idea of middle-level migration in migration learning is adopted. The training of the network model requires a large number of traffic signs. However, the database selected in this experiment only has 3,415 images. The lack of image data will make the network model under-fitting and ultimately reduce the detection accuracy. This article first initializes the pre-trained model (trained on the coco data set on the YOLO official website), Then use this model to retrain the system in this article. The training time is greatly reduced, and the probability of model divergence and fitting process is also reduced. There are a large amount of weight information and feature data in the pre-trained training model [13]. Weight information, these feature information can usually be shared by different tasks, transfer learning is to avoid relearning this knowledge by transferring specific and common feature data and information, and achieve rapid learning.

4   Evaluation of Training Results

4.1   Experimental Environment and Data

See Tables 2 and 3.

Table 2. Experimental environment configuration

Equipment name	Device Information
CPU	Intel(R) Xeon(R) CPU E5–2620
GPU	Tesla P4
Operating system	Windows 7 64 bit
CUDA version	10.0
CUDNN version	7.6.5
TensorFlow version	2.0.0
Python version	3.7.9

Table 3. Configuration file parameters

Parameter name parameter value	Parameter name parameter value
Width	416
Height	416
Batch size	8
Learning rate	0.001
Epochs	200

4.2   Evaluation Indicators

The evaluation indicators are the mean Average Precision (mAP) of all traffic sign types in a complex environment and the time required for each picture t = 1/N, in ms. First, you need to understand the confusion matrix, as shown in Table 4 [14]:

Table 4. Confusion matrix

Confusion matrix		Prediction	
		Positive (P)	Negative (N)
Actual	True(T)	TP	FN
	False(F)	FP	TN

Calculate precision and recall:

$$precision = \frac{TP}{TP + FP} \tag{6}$$

$$recall = \frac{TP}{TP + FN} \tag{7}$$

In the formula: TP, FN, FP, TN respectively represent the negative sample that is incorrectly detected, the positive sample that is correctly detected, the positive sample that is incorrectly detected, and the negative sample that is correctly detected.

mAP: The calculation of mAP is divided into two steps. The first step is to calculate the average precision AP (Average Precision) of each category, and the second step is to average the average precision, which is defined as follows:

$$AP_i = \sum_{k=1}^{N} P(k)\Delta r(k) \tag{8}$$

$$mAP = \frac{1}{m} \sum_{i=1}^{m} AP_i \tag{9}$$

where: m is the number of categories. The evaluation index uses mAP and the time required to detect a picture. The mAP value is directly proportional to the detection effect, and the detection time is inversely proportional to the detection speed.

4.3 Improved YOLOv3 Algorithm Test

In order to compare the detection effect of the improved network, the collected Chinese traffic sign detection data set were used to train and test the improved YOLOv3 network model and SSD model. The precision/recall curves of the three categories are shown in Fig. 5. It can be seen that the accuracy and recall rate of the improved network are better than the YOLOv3 model. Among them: the SSD model has the lowest accuracy rate; the average accuracy of the three categories of improved networks are 85.82%, 80.56% and 80.12%, which are higher than the detection results of YOLOv3. In terms of real-time performance, based on an image of 416 × 416, the standard YOLOv3 and the enhanced YOLOv3 methods in this article require 31.4 ms and 34.2 ms to detect an image, respectively, which meets the real-time requirements (Table 5).

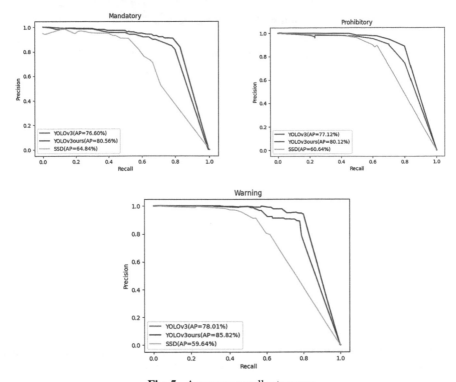

Fig. 5. Accuracy-recall rate curve

Table 5. Comparison of AP value, mAP and running time of the three categories

	Warning signs	Instruction signs	Prohibition signs	mAP	Operation hours
SSD	59.64	64.84	60.64	74.79	34.7 ms
YOLOv3	78.01	76.60	77.12	76.59	31.4 ms
Improved YOLOv3	85.82	80.56	80.12	82.73	34.2 ms

4.4 Experiment to Improve the Detection Ability Under Foggy Conditions

The experiment is divided into 3 groups; as shown in Table 6; the training set and test set of the first group of experiments are all original pictures, so as to compare with the following models. The second set of training sets are the images in the foggy environment after image restoration based on the dark channel algorithm of guided filtering. The test set remains unchanged. The training set and test set of the third group use pictures after image restoration processing.

Table 6. Data set classification

	Training set image restoration	Test set image restoration
First group	Unused	Unused
Second group	Use	Unused
The third group	Use	Use

Table 7. Comparison of AP value and mAP value

	Warning sign	Prohibitory sign	Prohibition sign	mAP
First group	82.92	80.46	80.21	80.69
Second group	82.77	80.35	80.16	80.56
The third group	84.67	81.28	84.36	83.41

It can be seen from Table 7 above that the AP and mAP values of the first group are slightly better than those of the second group, but there is not much difference overall. Compared with the first two groups, the mAP value of the third group is about 2.5% higher, so we can conclude that the detection effect after dehazing based on image restoration on both the training set and the test set is the best.

5   Conclusion

This paper constructs a traffic sign target detection training data set in foggy environments. The dark channel prior algorithm based on guided filtering is used to add image restoration steps to enhance the detection ability under bad foggy weather. Based on the YOLOv3 network, in order to solve the problems of insufficient data set and small amount of data, a Mosaic image enhancement training method is proposed, which improves the training efficiency and model accuracy. Aiming at the poor detection effect of YOLOv3 in complex environments, an improved YOLOv3 algorithm with increased feature scale is proposed. Aiming at the problems of small and fuzzy targets in foggy conditions and inaccurate positioning, the loss function of the target detector is redesigned using the CIOU loss function to further improve its detection accuracy of traffic signs in foggy conditions. In view of the fact that there are not many samples and the accuracy is not high, transfer learning training is adopted. The detection effect has been greatly improved.

Acknowledgments. This study was funded by National Natural Science Foundation of China (grant number 51867006, 51867007) and Natural Science and Technology Foundation of Guizhou province of China (grant number [2018]5781, [2018]1029).

References

1. Wu, X.: Research on traffic sign recognition algorithm based on deep learning. Beijing Architecture University (2020)
2. Chen, F., Liu, Y., Li, S.: Overview of traffic sign detection and recognition methods in complex environments. Computer Engineering and Applications 1–11, 22 June 2021
3. Tiantian, D., Haixiao, C., Xi, K., Deyou, W.: Research on multi-target recognition method in traffic scene in complex weather. Inf. Commun. **11**, 72–74 (2020)
4. Mogelmose, A., Trivedi, M.M., Moeslund, T.B.: Vision-based traffic sign detection and analysis for intelligent driver assistance systems: perspectives and survey. IEEE Trans. Intell. Transp. Syst. **13**(4), 1484 (2012)
5. Stallkamp, J., Schlipsing, M., Salmen, J., et al.: Man vs. computer: benchmarking machine learning algorithms for traffic sign recognition. Neural Netw. **32**, 323 (2012)
6. He, K., Sun, J., Tang, X.: Single image haze removal using dark channel prior. IEEE Trans. Pattern Anal. Mach. Intell. **33**, 2341–2353 (2011)
7. Redmon, J., Farhadi, A.: YOLOv3 an incremental improvement. Computer Vision and Pattern Recognition (2018)
8. Wu, Z.: Research on detection, recognition and tracking of ships based on deep learning in the dynamic background. Three Gorges University (2019)
9. Du, X.: Research on traffic sign recognition based on improved YOLOv3 network in natural environment. Dalian Maritime University (2020)
10. Gudigar, A., Chokkadi, S., Raghavendra, U., Rajendra Acharya, U.: Local texture patterns for traffic sign recognition using higher order spectra. Pattern Recogn. Lett. **94**, 202–210 (2017)
11. Gu, S., Ding, L., Yang, Y.: A new deep learning method based on AlexNet model and SSD model for tennis ball recognition. In: IEEE 10th International Workshop on Intelligence and Applications (IWCIA) (2018)
12. Lin, T.-Y., Dollár, P., Girshick, R., He, K., Hariharan, B., Belongie, S.: Feature pyramid networks for object detection. In: Computer Vision and Pattern Recognition (2017)
13. Bharat Singh, L.S.D.: An analysis of scale invariance in object detection-SNIP. In: 2018 IEEE Conference on Computer Vision and Pattern Recognition (CVPR 2018) (2018)
14. Zhang, Y., Li, G., Wang, L., Zong, H., Zhao, J.: A method for principal components selection based on stochastic matrix. In: 2017 13th International Conference on Natural Computation, Fuzzy Systems and Knowledge Discovery (ICNC-FSKD) (2017)

Development of Deep Learning Algorithms, Frameworks and Hardwares

Jinbao Ji, Zongxiang Hu$^{(\boxtimes)}$, Weiqi Zhang, and Sen Yang

Beijing Key Lab of Earthquake Engineering and Structural Retrofit, Beijing University of Technology, Beijing 100124, China
1536306845@qq.com

Abstract. As the core algorithm of artificial intelligence, deep learning has brought new breakthroughs and opportunities to all walks of life. This paper summarizes the principles of deep learning algorithms such as Autoencoder (AE), Boltzmann Machine (BM), Deep Belief Network (DBM), Convolutional Neural Network (CNN), Recurrent Neural Network (RNN) and Recursive Neural Network (RNN). The characteristics and differences of deep learning frameworks such as Tensorflow, Caffe, Theano and PyTorch are compared and analyzed. Finally, the application and performance of hardware platforms such as CPU and GPU in deep learning acceleration are introduced. In this paper, the development and application of deep learning algorithm, framework and hardware technology can provide reference and basis for the selection of deep learning technology.

Keywords: Artificial intelligence · Deep learning · Neural network · Deep learning framework · Hardware platforms

1 Introduction

The development of deep learning experienced three upsurges: from 1940s to 1960s, the idea of artificial neural network was born in the field of control; from 1980s to 1990s, neural networks were interpreted as connectionism; After entering the 21st century, it was revived in the name of deep learning [1]. The concept of deep learning originates from the research of deep neural network, which is also the core branch of machine learning field. For example, multi-layer perceptron is a simple network learning structure. Generally speaking, deep learning is to realize complex nonlinear mapping by stacking and feature extraction of multi-layer artificial networks. In essence, compared with traditional artificial neural networks, deep learning does not add more complex logical structures, but significantly improves the feature extraction and nonlinear approximation capabilities of the model only by adding hidden layers. Since Hinton formally proposed the concept of "deep learning" [2] in 2006, it immediately triggered a research upsurge in the academic world and the investment of the industry, and many excellent deep learning algorithms began to emerge. For example, during the Visual Recognition Contest (ILSVRC) from 2010 to 2017, CNN demonstrated its powerful image processing capability and confirmed its leading position in the field of computer vision image [3]. In

Z. Qian et al. (Eds.): WCNA 2021, LNEE 942, pp. 696–710, 2022.
https://doi.org/10.1007/978-981-19-2456-9_71

2016, the intelligent Go program AlphaGo [4] developed by Google defeated the world Go champion Lee Sedol by an absolute advantage. The success of AlphaGo marked the arrival of the era of artificial intelligence with deep learning as the core.

After years of development, the rise of deep learning has led to the creation of common programming frameworks such as Tensorflow, Caffe, Theano, MXNet, PyTorch and Keras, It also promotes the rapid development of AI hardware acceleration platforms and dedicated chips, including GPU, CPU, FPGA and ASIC. This paper focuses on the current research hotspots and mainstream deep learning algorithms in the field of artificial intelligence. The basic principles and applications of Autoencoder (AE), Boltzmann Machine (BM), Deep Belief Network (DBM), Convolutional Neural Network (CNN), Recurrent Neural Network (RNN) and Recursive Neural Network (RNN) are summarized. The performance characteristics and differences of deep learning framework, AI hardware acceleration platform and dedicated chip are compared and analyzed.

2 Deep Learning Algorithms

2.1 Auto-Encoder (AE)

As a special multi-layer perceptron, Auto-encoder (AE) is mainly composed of encoder and decoder [5]. As shown in Fig. 1, the basic Auto-encoder can be regarded as a three-layer neural network, from input 'x' to 'a' is the process of encoding, and from 'a' to 'y' is the process of decoding. The learning of auto-encoder is a process to reduce the error between output 'y' and input signal 'x'. The output expectation of Auto-encoder is the input, so it is generally regarded as an unsupervised learning algorithm, mainly used for data dimension reduction or feature extraction. In the training process of neural network, Auto-encoder is often used to determine the initialization parameters of the network. The principle is that if the encoded data can be restored accurately after decoding, the weight of the hidden layer is considered to be able to store the data information better.

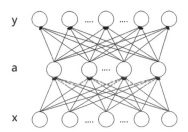

Fig. 1. Auto-encoder (AE)

The approximation ability of Auto-encoder for input and output is not the stronger the better, especially when the output of Auto-encoder is exactly equal to the input, the process only realizes the replication of the original data, and does not extract the inherent characteristics of the input information. Therefore, in order to enable the Auto-encoder to learn the key features, usually impose some constraints on the Auto-encoder. As a result, a variety of improved Auto-encoder emerged, such as: Sparse Auto-encoder

(SAE) makes neurons inactive in most cases by adding penalty items, and the number of nodes in the hidden layer is less than that in the input layer, so as to represent the input data with fewer characteristic parameters [6]. Stack Autoencoders (SAE) make it possible to extract deeper data features by stacking multiple autoencoders in series to deepen the layers of the network [7]; The Denoising Autoencoder (DAE) improves the robustness by adding noise interference during training [8]. Contraction Autoencoder (CAE) can learn mapping relations with stronger contraction by adding regular terms [9]. In addition, Deep Autoencoder (DAE), Stacked Denoised Autoencoder (SDAE), Sparse Stacked Autoencoder (SSAE), etc. [10–12].

2.2 Boltzmann Machine

Boltzmann Machine (BM) is a generative random neural network proposed by Hinton [13]. Traditional BM does not have the concept of layers, and its neurons are in a fully connected state, which is divided into visible unit and hidden unit. These two parts are binary variables, and the state can only be 0 or 1. Due to the complexity of the fully connected structure of BM, the variant of BM - Restricted Boltzmann machine is widely used at present (Fig. 2).

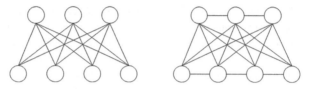

Fig. 2. BM (left) and RBM (right)

Restricted Boltzmann Machine (RBM) was first proposed by Smolensky [14] and has been widely used in data dimension reduction, feature extraction, classification and collaborative filtering. RBM is a shallow network similar to BM in structure, the difference is that RBM cancels the connection between layers and the neurons between layers do not affect each other, thus simplifying the model.

2.3 Deep Boltzmann Machine and Deep Belief Network

Deep Boltzmann Machine (DBM) is a model composed of multiple Restricted Boltzmann Machine, and the network layers are bidirectional connections [15]. Compared with RBM, DBM can learn higher-order features from unlabeled data and has better robustness, so it is suitable for target recognition and speech recognition.

Deep Belief Network (DBN) is also a deep neural network composed of multiple RBM, which differs from DBM in that only the network layer at the output part of RBM is bidirectional propagation [16]. Different from general neural models, DBM aims at establishing joint distribution between data and expected output, to make the network generate the expected output as much as possible, so as to extract and restore data features more abstractly. DBN is a practical deep learning algorithm, and its excellent scalability and compatibility have been proved in the application of feature recognition, data classification, speech recognition and image processing. For example, the combination of DBN and Multi-layer Perceptron (MLP) has good performance in facial expression recognition [17]. The combination of DBN and Support Vector Machine (SVM) has excellent performance in text classification [18].

2.4 Convolutional Neural Network

Convolutional Neural Network (CNN) was originally a deep learning algorithm derived from the discovery of 'Receptive Field' [19], which has excellent ability in image feature extraction. With the successful application of Lenet-5 model in the field of handwritten number recognition, scholars from all walks of life began to study the application of CNN in the fields of speech and image. In 2012, The AlexNet model proposed by Krizhevsky beats many excellent neural network models in the Image Net Image classification competition, which also pushed the application research of CNN to a climax [20] (Fig. 3).

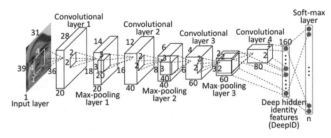

Fig. 3. Convolutional neural network [21]

Convolutional neural network is mainly composed of input layer, convolutional layer, excitation layer, pooling layer, full connection layer and output layer, among which the convolutional layer and pooling layer are the core structure of CNN. Different from other deep learning algorithms, CNN mainly uses convolution kernel (filter) for convolution calculation, and uses pooling layer to reduce inter-layer connections to further extract features. It obtains high-level features through repeated extraction and compression of features, and then uses the output for classification and regression.

Weight sharing mechanism and local perception field are two major features of CNN. They have similar functions with pooling layer and can reduce the risk of overfitting by reducing inter-layer connections and network parameters. Weight sharing means that a filter will be used multiple times, it will slide across the feature surface and do multiple convolution computations [22]. Local perception field is inspired by the process of human observing the outside world, which is from the local to the whole. Therefore, a single filter does not need to perceive the whole, but only needs to extract local features and summarize them at a higher level.

In recent years, CNN has gradually emerged in various industries, such as Alphago, speech recognition, natural language processing, image generation and face recognition, etc. [23–26]. At the same time, many improved CNN models were born, such as VGG, ResNet, GoogLeNet and MobileNet.

VGG. In 2014, Simonyan and Zisserman [27] proposed the VGGmodel, it won the first prize in positioning task and the second prize in classification task in the ImageNet Challenge. In order to improve the fitting ability, the network layer of VGG is increased to 19 layers, and the convolution kernel with small receptive field (3×3) is used to replace the large one (5×5 or 7×7), thus increasing the nonlinear expression ability of the network.

ResNet. VGG proved that the deep network structure can effectively improve the fitting ability of the model, but the deeper network tends to cause gradient dispersion, which makes the network unable to converge. In 2015, Kaiming [28] proposed ResNet, which effectively alleviated the problem of neural network degradation, and won the first prize of classification, positioning, detection and segmentation tasks with absolute superiority in ILSVRC and COCO competitions. To solve the problem of gradient disappearance, Kaiming introduces a Residual Block structure in the network, which enables the model use Shortcut to implement Identity Mapping.

GoogLeNet. To solve the problem of too many parameters in large-scale network model, Google proposed Inception V1 [29] network architecture in 2014 and constructed GoogLeNet, which won the first prize in the ImageNet Challenge classification and detection task in the same year. Inception V1 abandons the full connection layer and changes the convolutional layer to a sparse network structure, that results in a significant reduction of the network parameters. In 2015, Google proposed Batch Normalization operation and improved the original GoogLeNet based on this technology, obtained a better model—Inception V2 [30]. In the same year, Inception V3 [31] is also born. Its core idea is to decompose the convolution kernel into smaller convolution, such as splitting 7×7 into 1×7 and 7×1, to further reduce network parameters. In 2016, Google launched Inception V4 by combining Inception and ResNet, which has been improved in training speed and performance [32]. When the number of filters is too large (More than 1000), the training of Inception V4 will become unstable, but it can be alleviated by adding an Activate Scaling factor.

MobileNet. In recent years, in order to promote the combination of neural network model and mobile devices, neural network model began to develop towards the direction of lightweight. In 2017, Google designs MobileNet V1 by Depthwise Convolution [33] and allows users to change the network width and input resolution, thus achieving a tradeoff between latency and accuracy. In 2018, Google introduced The Inverted Residuals and Linear Bottlenecks on the basis of MobileNet V1, and put forward MobileNet V2 [34]. In 2019, Google proposed MobileNet V3 by combining Depthwise Convolution, Inverted Residuals and Linear Bottlenecks [35]. It is proved that MobileNet has excellent performance in multi-objective tasks, such as classification, target detection and semantic segmentation.

2.5 Recurrent Neural Network

Recurrent neural network (RNN) is a kind of deep learning model that is good at dealing with time series. RNN expands neurons at each layer in time dimension, realizes forward transmission of data in the network through sequential input of information, and stores information in 'long-term memory unit' to establish sequential relations between data.

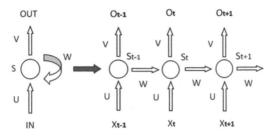

Fig. 4. Convolutional neural network

As shown in Fig. 4, RNN reduces the computation of the network by sharing parameters (W, U, V). RNN mainly uses Back Propagation Through Time algorithm [36] to update the parameters of each node. Its forward Propagation can be expressed as:

$$S_t = \sigma(w * S_{t-1} + X_t * U). \tag{1}$$

$$Q_t = soft\max(V * S_t) \tag{2}$$

Although RNN can consider the correlation between information, traditional RNN is usually difficult to achieve long-term preservation of information. Due to the excitation function and multiplication, when RNN has a large number of network layers or a long time sequence of data, sometimes the gradient will grow or decay exponentially with iteration, resulting in gradient disappearance and gradient explosion [37].

LSTM. In order to solve the shortcomings of traditional RNN, Hochreiter [38] proposed LSTM. LSTM introduces three types of gated units in RNN to realize information extraction, abandoned and long-term storage, which not only improves the problems of gradient disappearance and excessive gradient, but also improves the long-term storage capacity of RNN for information. Each memory cell in the LSTM contains one cell and three gates. A basic structure is shown in the Fig. 5: In the three types of gating units, input gate is used to control the proportion of the current input data X(t) into the network; Forget gate is used to control the extent to which the long-term memory unit abandons information when passing through each neuron. Output gate is used to control the output of the current neuron and the input to the next neuron.

Three types of gate control units are shown:

$$i_t = \sigma(w_{ii}x_t + w_{ih}h_{t-1}) \tag{3}$$

$$f_t = \sigma(w_{fi}x_t + w_{fh}h_{t-1}) \tag{4}$$

$$O_t = \sigma(w_{Oi}x_t + w_{Oh}h_{t-1}) \tag{5}$$

The calculation of Cell is shown:

$$g_t = \tanh(g_{gi}x_t + w_{gi}h_{t-1})$$

(6)

The calculation of long-term memory unit C and hidden layer output h are as follows:

$$C_t = f_t C_{t-1} + i_t g_t$$

(7)

$$h_t = o_t \tanh(c_t)$$

(8)

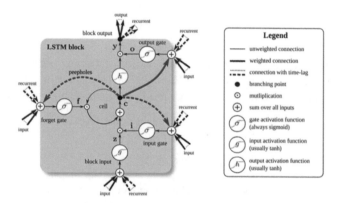

Fig. 5. LSTM memory cell [39]

LSTM has many excellent variants, of which the more successful improvement is the bi-directional LSTM. Bi-directional LSTM realizes the simultaneous utilization of past and future information through two-way propagation of data in the time dimension [40]. In some problems, its prediction performance is better than one-way LSTM. Greff [39] discussed the performance of 8 variants based on Vanilla LSTM, and conducted experimental comparisons in the three fields of TIMIT speech recognition, handwritten character recognition and polyphonic music modeling. The results showed that the performance of 8 variants did not significantly improve; Forgetting gate and output gate are the two most important parts of LSTM model, and the combination of these two gate units can not only simplify the LSTM structure, but also will not reduce the performance.

GRU. As a simplified model of LSTM, GRU only uses two gating units to save and forget information, including update gate for input and forget, and reset gate for output [41]. GRU replaces forget gate and Input gate with Update gate compared with LSTM, simplifying structure and reducing computation without reducing performance. At present, there is no final conclusion to show the performance of LSTM and GRU, but a large number of practices have proved that the performance of the two network models is often similar in general problems [42].

2.6   Recursive Neural Network

Recursive neural network is a deep learning model with tree-like hierarchical structure, its information will be collected layer by layer from the end of the branch, and finally reach root end, that is, to establish the connection between information from the spatial dimension. Compared with recurrent neural network, recursive neural network can map words and sentences expressing different semantics into a vector space, and use the distance between statements to determine semantics [43], rather than just considering word order relations. Recursive neural networks have powerful natural language processing capabilities, but constructing such tree-structured networks requires manual annotation of sentences or words as parsing trees, which is relatively expensive (Fig. 6).

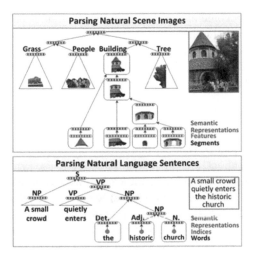

Fig. 6. Syntax parse tree and natural scene parse tree [44]

3   Deep Learning Framework

In the early stage of the development of deep learning, in order to simplify the process of model building and avoid repeated work, some researchers or institutions packaged codes that could realize basic functions into frameworks for the public to use. Currently, commonly used deep learning frameworks include Tensorflow, Caffe, Theano, MXNet, PyTorch, Keras, etc.

3.1   Tensorflow

Tensorflow is an open source framework for machine learning and deep learning developed by Google. It uses the form of a Data Flow Graph to build models and provides TF. Gradients for quickly calculating gradients. Tensorflow is highly flexible and portable, it supports multiple language interfaces such as Python and C++. It can not only be deployed on servers with multiple cpus and gpus, but also run on mobile phones [48]. Therefore, Tensorflow is widely used in many fields such as voice and image. Although it is not superior to other frameworks in terms of running speed and memory consumption, it is relatively complete in terms of theory, functions, tutorials and peripheral services, which is suitable for most deep learning beginners.

3.2 Caffe

Caffe is an open source framework for deep learning, and is maintained by Berkeley Vision Center (BVLC). Caffe can flexibly modify and design new network layers according to different requirements, and is very suitable for modeling deep convolutional neural networks [49]. Caffe has demonstrated excellent image processing skills in ImageNet competitions and has become one of the most popular frameworks in computer vision. Caffe's models are usually implemented in text form, which is easy to learn. In addition, Caffe can use GPU for training acceleration through Nvidia's CUDA architecture and cuDNN accelerators. However, Caffe is not flexible enough to modify or add the network layer, and is not good at dealing with language modeling problems.

3.3 Theano

Theano is an efficient and convenient mathematical compiler developed by the Polytechnic Institute of Montreal, it is the first architecture to use symbolic tensor diagrams to build network models. Theano is a framework developed based on Python that relies on the Numpy toolkit, and is well suited for large-scale deep learning algorithm design and modeling, especially for language modeling problems [50]. Theano's disadvantages are also obvious, it is slow to run both as a toolkit import and during its compilation, and the framework is currently out of development, so it is not recommended as a research tool.

3.4 MXNet

MXNet is a deep learning framework used and maintained by Amazon officially. It has a flexible and efficient programming mode, supporting both imperative and symbolic compilation methods [51], and can perfectly combine the two methods to provide users with a more comfortable programming environment. MXNet has many advantages. It not only supports distributed training of multiple CPU/GPU, but also can realize true portability of micro-devices from servers and workstations to smart phones. In addition, MXNet supports JavaScript, Python, Matlab, C++ and other languages, which can meet the needs of different users. However, MXNet is not widely used by the community due to the difficulty of getting started and the incomplete tutorials.

3.5 PyTorch

Facebook introduced the Torch framework early on, but it struggled to meet market demand due to its lack of support for the Python interface. Instead, Facebook built

Pytorch, a deep learning framework specifically designed for Python programming and GPU acceleration [52, 53]. Pytorch uses a dynamic data flow diagram to build the model, giving users the flexibility to modify the diagram. Pytorch is highly efficient at encapsulating code and runs faster than frameworks such as TensorFlow and Keras, and providing users with a more user-friendly programming environment than other frameworks.

3.6 Keras

Keras is a neural network library derived from Theano. The framework is mainly developed based on Python language and has a complete function chain in the construction, debugging, verification and application of deep learning algorithms. Keras architecture is designed for object-oriented programming, which encapsulates many functions in a modular manner, simplifying the process of building complex models. Meanwhile, Keras is compatible with Tensorflow and Theano's deep learning software package, which supports most of the major algorithms including convolution and cyclic neural networks (Table 1).

Table 1. Deep learning framework

Framework	Caffe	Theano	TensorFlow	MXNet	PyTorch	Keras
Language	C++/cuda/Python	Python/C++/cuda	C++/Python	C++/cuda/Python	Python	Python/R
Hardware support	CPU/GPU	CPU/GPU	CPU/GPU/Mobile	CPU/GPU/Mobile	CPU/GPU	CPU/GPU/Mobile
Speed	Fast	Medium	Medium	Fast	Very fast	Medium
Flexibility	Low	Very high	High	High	High	Medium
Maintain	BVLC	Epdm	Google	Amazon	Facebook	Fchollet

4 Hardware Platform and Dedicated Chip

4.1 CPU

CPU is one of the core parts of the computer, usually composed of control parts, logic parts and registers, its main function is to read, execute computer instructions and process data. As a general-purpose chip, CPU is originally designed to be compatible with all kinds of data processing and computation, and it is not a special processor for neural network training and acceleration. There are a lot of matrix and vector calculations in the training process of deep network, and the computing efficiency is not high by using CPU, and upgrading CPU to improve performance is not cost-effective. Therefore, CPU is generally only suitable for small-scale network training.

4.2 GPU

In 1999, NVIDIA launched GeForce-256 as its first commercial GPU, and began working on developing high-performance GPU technology in the early 2000s. In 2004, gpus evolved to the point where they could carry early neural network computing. In 2006, Kumar Chellapilla [54] successfully used GPU to accelerate CNN, which was the earliest known attempt to use GPU for deep learning.

GPU is a microprocessor specially used for processing image calculation. Different from the generality of CPU, GPU focuses on the calculation of complex matrix and geometric problems, especially good at processing image problems [55]. In the face of complex deep learning model, GPU can greatly increase the training speed. For example, Coates [56] used GPU for training acceleration in the target detection system, which increased its running speed by nearly 90 times. Currently, companies such as Nvidia and Qualcomm have advanced capabilities in developing GPU hardware and acceleration technologies, and support multiple programming languages and frameworks. For example, Pytorch can use the GPU to help model training through CUDA and cuDNN that developed by Nvidia, which can significantly reduce network training time.

4.3 ASIC

ASIC is a professional chip with extremely high flexibility. Its performance can be customized according to actual problems to meet different computing power requirements. Therefore, when dealing with deep learning problems, its performance and power consumption are far higher than CPU, GPU and other general chips. For example, TPU [57], launched by Google in 2015, is a very representative integrated circuit chip. It has been proved that its execution speed and efficiency are dozens of times higher than CPU and GPU. It has been applied and promoted in Google's search map, browser and translation software. In recent years, Google has continuously released the second and third generation of TPU and TPU Pod [58], which not only greatly improves chip performance, but also extends its application to the broader field of artificial intelligence. In addition, the Cambrian series chips [59] proposed by The Chinese Academy of Sciences also have great advantages in improving the running speed of neural networks. ASIC has broader development prospects and application value, but due to long development cycle, high investment risk and high technical requirements, only a few companies have the development ability at present.

4.4 FPGA

FPGA, also known as field programmable gate array, is a variable circuit derived from custom integrated circuit (ASIC) technology. FPGA directly operates through gate circuit, which not only has high speed and flexibility, but also enables users to meet different needs by changing the wiring between internal gate circuits [60]. FPGA generally have lower performance than ASIC, but their development cycle is shorter, risk is lower, and cost is also relatively lower. When processing specific tasks, the efficiency can be further improved through parallel computing. Although FPGA has many advantages and can better adapt to rapidly developing deep learning algorithms, it is not recommended for individual users or small companies to use due to their high cost and difficulty (Table 2).

Table 2. Deep learning hardware technology comparison [61]

Hardware	Performance	Flexibility	Power consumption	Enterprise
CPU	Low	High	Low	Intel
GPU	Medium	High	Medium	Nvidia/Qualcomm
ASIC	High	Low	High	Google
FPGA	High	Medium	Low	Xilinx/Altera

5 Conclusion

Around the current popular research fields in artificial intelligence, this paper summarizes the basic principles and application scenarios of current mainstream deep learning algorithms, introduces and compares common deep learning programming frameworks, hardware acceleration platforms and dedicated chips. Obviously, deep learning algorithms are in a stage of rapid development, and also promote the rise of its surrounding industries. However, problems such as single model type and insufficient algorithm performance also limit the development of some industries, so how to innovate and improve new algorithms is still the focus of future research. In addition, the intelligence of deep learning algorithm also brings a lot of convenience to our daily life, but its application is not widely at present. That mean how to promote and utilize deep learning more efficiently is still a long way to go.

Acknowledgements. The research work of this paper is supported by the National Nature Science Foundation of China (51978015 & 51578024).

References

1. Goodfellow, I., Bengio, Y., Courville, A.: Deep Learning. The MIT Press, Cambridge (2016)
2. Hinton, G.E., Salakhutdinov, R.R.: Reducing the dimensionality of data with neural networks. Science **313**(5786), 504–507 (2006)
3. Krizhevsky, A., Sutskever, I., Hinton, G.E.: ImageNet classification with deep convolutional neural networks. Commun. ACM **60**(6), 84–90 (2017)
4. Fu, M.C.: AlphaGo and Monte Carlo tree search: the simulation optimization perspective. In: Winter Simulation Conference Proceedings, vol. 26, pp. 659–670 (2016)
5. Hinton, G.E., Zemel, R.S.: Autoencoders, minimum description length and Helmholtz free energy. Adv. Neural Inf. Process. Syst. **6**, 3–10 (1993)
6. Le, Q.V., Ngiam, J., Coates, A., et al.: On optimization methods for deep learning. In: International Conference on Machine Learning. DBLP (2011)
7. Scholkopf, B., Platt, J., Hofmann, T.: Greedy layer-wise training of deep networks. Adv. Neural Inf. Process. Syst. **19**, 153–160 (2007)
8. Vincent, P., Larochelle, H., Bengio, et al.: Extracting and composing robust features with denoising autoencoders. In: International Conference on Machine Learning. ACM (2008)

9. Rifai S, Vincent P, Muller X, et al.: Contractive auto-encoders: explicit invariance during feature extraction. In: ICML, vol. 6, pp. 26–46 (2011)
10. Ma, X., Wang, H., Jie, G.: Spectral-spatial classification of hyperspectral image based on deep auto-encoder. IEEE J. Sel. Top. Appl. Earth Obs. Remote Sens. 9(9), 1–13 (2016)
11. Vincent, P., Larochelle, H., Lajoie, I., et al.: Stacked denoising autoencoders: learning useful representations in a deep network with a local denoising criterion. J. Mach. Learn. Res. 11(12), 3371–3408 (2010)
12. Jiang, X., Zhang, Y., Zhang, W., et al.: A novel sparse auto-encoder for deep unsupervised learning. In: Sixth International Conference on Advanced Computational Intelligence, vol. 26, pp. 256–261. IEEE (2013)
13. Hinton, G.E.: Learning and relearning in Boltzmann machines. Parallel Distrib. Process.: Explor. Microstruct. Cogn. 1, 2 (1986)
14. Smolensky, P.: Restricted Boltzmann machine. Stellenbosch Stellenbosch Univ. 16(04), 142–167 (2014)
15. Salakhutdinov, R., Hinton, G.E.: Deep Boltzmann Machines. J. Mach. Learn. Res. 5(2), 1967–2006 (2009)
16. Hinton, G.E., Osindero, S., Teh, Y.W.: A fast learning algorithm for deep belief nets. Neural Comput. 18(7), 1527–1554 (2014)
17. Shi, X.G., Zhang, S.Q., Zhao, X.M.: Face expression recognition based on deep belief network and multi-layer perceptron. J. Small Micro Comput. Syst. 36(07) (2015). (in Chinese)
18. Tao, L.: A novel text classification approach based on deep belief network. In: Neural Information Processing Theory & Algorithms-international Conference. DBLP (2010)
19. Hubel, D.H., Wiesel, T.N.: Receptive fields, binocular interaction and functional architecture in the cat's visual cortex. J. Physiol. 160(1), 106–154 (1962)
20. Krizhevsky, A., Sutskever, I., Hinton, G.: ImageNet classification with deep convolutional neural networks. Adv. Neural. Inf. Process. Syst. 25(2), 22–27 (2012)
21. Yi, S., Wang, X., Tang, X.: Deep learning face representation from predicting 10,000 classes. In: IEEE Conference on Computer Vision & Pattern Recognition. IEEE (2014)
22. Lecun, Y., Boser, B., Denker, J., et al.: Backpropagation applied to handwritten zip code recognition. Neural Comput. 1(4), 541–551 (2014)
23. Silver, D., Huang, A., Maddison, C.J., et al.: Mastering the game of Go with deep neural networks and tree search. Nature 529(7587), 484–489 (2016)
24. Abdel-Hamid, O., Mohamed, A.-R., et al.: Convolutional neural networks for speech recognition. IEEE/ACM Trans. Audio Speech Lang. Process. (TASLP) 22(10), 1533–1545 (2014)
25. Donahue, J., Hendricks, L.A., Rohrbach, M., et al.: Long-term recurrent convolutional networks for visual recognition and description. In: 2015 IEEE Conference on Computer Vision and Pattern Recognition (CVPR), pp. 677–691. IEEE (2017)
26. Schroff, F., Kalenichenko, D., Philbin, J.: FaceNet: a unified embedding for face recognition and clustering, pp. 815–823 IEEE (2015)
27. Simonyan, K., Zisserman, A.: Very deep convolutional networks for large-scale image recognition. Comput. Sci. (2014)
28. He, K., Zhang, X., Ren, S., et al.: Deep residual learning for image recognition. In: 2016 IEEE Conference on Computer Vision and Pattern Recognition (CVPR) (2016)
29. Szegedy, C., Liu, W., Jia, Y., et al.: Going deeper with convolutions. IEEE Computer Society (2014)
30. Ioffe, S., Szegedy, C.: Batch normalization: accelerating deep network training by reducing internal covariate shift. In: Bach, F., Blei, D. (eds.) Proceedings of Machine Learning Research, vol. 37, pp. 448–456 (2015)
31. Szegedy, C., Vanhoucke, V., Ioffe, S., et al.: Rethinking the inception architecture for computer vision. In: Proceedings of IEEE, pp. 2818–2826. IEEE (2016)

32. Szegedy, C., Ioffe, S., Vanhoucke, V., et al.: Inception-v4, inception-ResNet and the impact of residual connections on learning (2016)
33. Howard, A.G., Zhu, M., Chen, B., et al.: MobileNets: efficient convolutional neural networks for mobile vision applications (2017)
34. Sandler, M., Howard, A., Zhu, M., et al.: MobileNetV2: inverted residuals and linear bottlenecks. In: 2018 IEEE/CVF Conference on Computer Vision and Pattern Recognition (CVPR). IEEE (2018)
35. Howard, A., Sandler, M., Chu, G., et al.: Searching for MobileNetV3 (2019)
36. Werbos, P.J.: Backpropagation through time: what it does and how to do it. Proc. IEEE **78**(10), 1550–1560 (1990)
37. Bengio, Y.: Learning long-term dependencies with gradient descent is difficult. IEEE Trans. Neural Netw. **5**(2), 157–166 (1994)
38. Hochreiter, S., Schmidhuber, J.: Long short-term memory. Neural Comput. **9**(8), 1735–1780 (1997)
39. Greff, K., Srivastava, R.K., Koutník, J., et al.: LSTM: a search space odyssey. IEEE Trans. Neural Netw. Learn. Syst. **28**(10), 2222–2232 (2016)
40. Graves, A., Schmidhuber, J.: Framewise phoneme classification with bidirectional LSTM and other neural network architectures. Neural Netw. **18**(5–6), 602–610 (2005)
41. Cho, K., Merrienboer, B.V., Gulcehre, C., et al.: Learning phrase representations using RNN encoder-decoder for statistical machine translation. Comput. Sci. **22**(10), 21–33 (2014)
42. Chung, J., Gulcehre, C., Cho, K.H., et al.: Empirical evaluation of gated recurrent neural networks on sequence modeling. Eprint Arxiv, vol. 32, no. 18, pp. 119–132 (2014)
43. Ying, X., Le, L., Zhou, Y., et al.: Deep learning for natural language processing. Handb. Stat. **56**(20), 221–231 (2018)
44. Socher, R., Lin, C.Y., Ng, A.Y., et al.: Parsing natural scenes and natural language with recursive neural networks. In: Proceedings of the 28th International Conference on Machine Learning, ICML 2011, Bellevue, Washington, USA, June 28–July 2 (2011)
45. Abadi, M.: TensorFlow: learning functions at scale. In: ACM SIGPLAN International Conference on Functional Programming, pp. 1–12. ACM (2016)
46. Jia, Y., Shelhamer, E., Donahue, J., et al.: Caffe: convolutional architecture for fast feature embedding, pp. 144–156. ACM (2014)
47. Al-Rfou, R., Alain, G., et al.: Theano: a Python framework for fast computation of mathematical expressions, vol. 122, no. 05, pp. 1022–1034 (2016)
48. Chen, T., Li, M., Li, Y., et al.: MXNet: a flexible and efficient machine learning library for heterogeneous distributed systems. Statistics (2015)
49. Ketkar, N.: Introduction to PyTorch (2017)
50. Sen, S., Sawant, K.: Face mask detection for covid_19 pandemic using pytorch in deep learning. In: IOP Conference Series: Materials Science and Engineering, vol. 1070, no. 1 (2021)
51. Chellapilla, K., Puri, S., Simard, P.: High performance convolutional neural networks for document processing. In: Tenth International Workshop on Frontiers in Handwriting Recognition (2006)
52. Shenyan: Radio and Television Information, no. 10, pp. 64–68 (2017)
53. Coates, A., Baumstarck, P., Le, Q., et al.: Scalable learning for object detection with GPU hardware. In: 2009 IEEE/RSJ International Conference on Intelligent Robots and Systems, IROS 2009. IEEE (2009)
54. David, K.: Google TPU boosts machine learning. Microprocess. Rep. **31**(5), 18–21 (2017)
55. Kumar, S., Bitorff, V., Chen, D., et al.: Scale MLPerf-0.6 models on Google TPU-v3 Pods, vol. 56, no. 12, pp. 81–89 (2019)
56. Editorial Department of the Journal: Cambrian released the first cloud artificial intelligence chip in China. Henan Sci. Technol. **647**(14), 0–9 (2018)

57. Wei, J., Lin, J.: Deep learning algorithm, hardware technology and its application in future military. Electron. Packag. (12) (2019)
58. Zhang, W.: Deep neural network hardware benchmark testing status and development trend. Inf. Commun. Technol. Policy (012), 74–78 (2019)

Implementation and Application of Embedded Real-Time Database for New Power Intelligent Terminal

Yingjie Shi[1]([✉]), Xiang Wang[1], Wei Wang[1,2], Huayun Zhang[1,2], and Shusong Jiang[1]

[1] China Realtime Database Co. Ltd., Building 6, No. 19, Integrity Avenue, Jiangning District, Nanjing, China
shiyingjie@sgepri.sgcc.com.cn
[2] Nari Group Corporation/State Grid Electric Power Research Institute, No. 19, Integrity Avenue, Jiangning District, Nanjing 211106, China

Abstract. An implementation method of embedded real-time database is proposed. The lightweight high matching of power model is realized through tree structure. The resource consumption of real-time database in embedded device environment is reduced by means of separated storage and non independent process deployment. The efficient access of measuring point data is realized through internal mapping rules and improved breadth first search algorithm. Experiments show that the embedded real-time database realized by this method has good performance and low energy consumption, and is suitable for intelligent terminal equipment in new power system.

Keywords: New power system · Intelligent terminal · Embedded · Real time database · Tree model

1 Introduction

With the in-depth development of the "double carbon" action, the State Grid Corporation of China is accelerating the construction of a new power system with new energy as the main body [1]. While large-scale access of new energy, new equipment and multiple loads, it poses new challenges to the data carrying capacity, real-time and security of the existing intelligent terminal equipment of the power system.

At present, the real-time data storage and processing of power system intelligent terminal mainly rely on embedded real-time database. Most of the existing embedded real-time database cores adopt open-source general products, which lack consideration of power model, especially the new power system intelligent terminal model, and there are great security risks, which affect the stability and security of power system.

In this paper, an implementation method of embedded real-time database for new power intelligent terminal is proposed, which takes dynamic connection library as the carrier, tree structure model as the modeling basis, separated storage as the data basis, memory mapping rules and improved breadth first search algorithm as the logical basis, and constructs a new power intelligent terminal environment with low energy consumption, high timeliness, high security and professional embedded real-time database.

© The Author(s) 2022
Z. Qian et al. (Eds.): WCNA 2021, LNEE 942, pp. 711–720, 2022.
https://doi.org/10.1007/978-981-19-2456-9_72

2 Background and Related Work

2.1 Characteristic Analysis of New Power Intelligent Terminal

One of the main technical features of the new generation power system in the energy transformation is the multi energy complementarity between the power system and other energy systems [2], and one of its key cores is digitization. At the same time, for the power industry, the power intelligent terminal equipment is progressing day by day under the promotion of the policy of "new digital infrastructure". The application scenario type and number of new power intelligent terminals represented by intelligent distribution terminals, intelligent vehicle charging piles and intelligent electricity meters [3] continue to grow. The integration of different types of terminals is imperative, and gradually presents the technical characteristics of "digitization", "intelligence" and "integration". The continuous upgrading of embedded technology, 5g network and other hardware and network technologies will further accelerate the integration process of power, energy and Internet of things.

In terms of digitization, under the new power system, in addition to the metering function of traditional electric energy meters, smart meters also have two-way multi rate metering function, user end control function, two-way data communication function of multiple data transmission modes [4], etc. The real-time data that needs to be stored and processed at the same time will increase exponentially. In the future, the measurement data acquisition frequency of smart meters will be further improved. Taking the power consumption information acquisition system as an example, the current data acquisition frequency of smart terminals has been increased from 1/360 to 1/15, and the amount of data has increased by 24 times. In terms of intelligence, the smart grid puts forward higher requirements for user side metering devices. On the one hand, it should be able to comprehensively monitor the real-time load of users and monitor the real-time load, voltage, current, power factor, harmonic and other grid parameters of each power terminal to ensure power supply; On the other hand, it is necessary to control the electric equipment, and select the appropriate time to automatically operate or stop according to the real-time electricity price of the system and the wishes of users, so as to realize the functions of peak shifting and valley filling. In terms of integration, due to the insepa-rable relationship among power terminals, 5g terminals and Internet of things terminals [5, 6], these infrastructure terminals can usually be integrated. For example, after the integration of power and Internet of things, an industrial Internet of things suitable for power grid, namely power Internet of things, will be formed, which will produce various types of intelligent integration terminal requirements.

Therefore, under the new power system, the power intelligent terminal needs to process a wider range of data, faster frequency and stronger timeliness requirements.

2.2 Relevant Research Work

The research on embedded real-time database abroad started earlier, among which the representative ones are Berkeley DB and SQLite. However, the research shows that their performance in real-time applications is poor [7]. At this stage, the domestic research on embedded real-time database mainly relies on open source database and focuses on application research. Among them, a real-time database implementation method for micro grid intelligent terminal [8] adopts MySQL database, which maps the data tables, fields and records constituting the real-time database to the memory of the intelligent terminal through file mapping to form a database entity. The disadvantage is that data access and submission need complex lexical and syntax analysis, and the CPU resource overhead is huge. The cross platform lightweight database packaging method and system based on mobile terminal [9] realizes the database operation on HTML page (IOS and Android), and solves the problem of repeated development of database operation functions on HTML page based on different mobile intelligent terminal platforms. The disadvantage is that the database adopts open source SQLite products, and the system security is not guaranteed. Design and implementation of embedded real-time database based on ARM platform [10] transplanted the traditional real-time database on ARM platform and realized the basic storage function. The disadvantage is that it needs to call a special interface and is lack of friendliness to the application of power equipment. At the same time, domestic researchers also try to use the embedded real-time operating system to solve the problem of real-time data storage of embedded devices, such as VxWorks, QNX, ucLinux and RTEMS. Since the embedded real-time system essentially belongs to the category of operating system, it is qualitatively different from the embedded real-time database. To sum up, the existing embedded real-time database in China is mainly a general relational database. There are many problems in the embedded equipment of power system, such as high system resource consumption, weak matching with the model of power intelligent terminal equipment, and unable to guarantee security.

3 Design and Implementation of Embedded Real-Time Database

3.1 Design Framework

The overall deployment of the embedded real-time database for the new power intelligent terminal described in this paper is shown in Fig. 1. It is divided into four layers from the outside to the inside, marked with serial numbers ①–④. The outermost layer is layer ①, which represents the entity of the new power system intelligent terminal equipment. It is composed of microprocessor, register, digital I/O interface and other units, which is used to carry the embedded operating system. Layer ② is the embedded container, usually the embedded container represented by docker, which is deployed in the embedded operating system to carry different embedded applications. Layer ③ is embedded application, usually data access application and embedded data center application, which are used to collect and store real-time data. Layer ④ is the embedded real-time database, which is embedded in the embedded application in the form of dynamic link library, coupled with the application through the database interface, does not occupy independent process handles, saves system resources to a great extent, and supports embedded and container deployment.

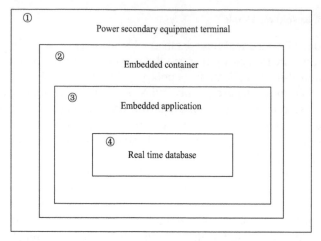

Fig. 1. Deployment diagram of embedded real-time database

The overall system structure of embedded real-time database for new power intelligent terminal is shown in Fig. 2. From bottom to top, the real-time database includes storage layer, model layer and application layer. The storage layer is used to store specific measurement type data, including storage interface, lightweight cache, data compression, data storage, resource optimization and other modules. The model layer is the object model management module, which is used to build and store the device model and associate it with the measuring points, including model interface, model algorithm and model storage modules. The application layer is used for data query and analysis, and provides application capabilities such as model construction and data access through the interface.

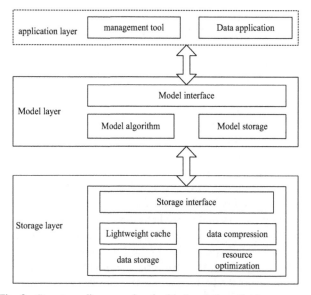

Fig. 2. Structure diagram of embedded real-time database system

3.2   Tree Structure Model Design

The traditional relational data model uses two-dimensional tables to represent the relationship between entities. In data modeling, it is necessary to split the data objects, store their respective information in the corresponding table fields, and connect each table when necessary. This model design generally has storage redundancy in power intelligent terminal. Due to the large amount of correlation calculation required for multi table connection, it needs to consume a lot of CPU system resources, which is easy to affect the performance and stability of embedded applications. According to the technical characteristics of the new power system intelligent terminal and combined with the design of the power equipment IOT terminal model, the object model management module in this paper realizes the organization and management function of the power intelligent terminal model by using the tree structure. As shown in Fig. 3, the tree structure includes leaf nodes and non leaf nodes, in which the non leaf nodes are used as the index of the tree. The leaf node records the measurement point ID when it is created and is associated with the measurement point ID of the database storage layer.

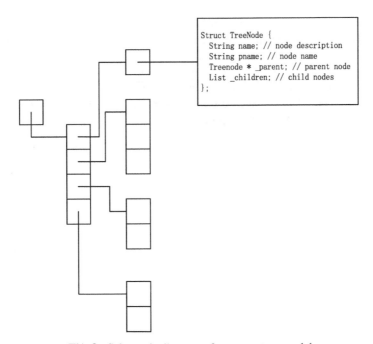

Fig. 3. Schematic diagram of tree structure model

In terms of model storage, this paper uses the improved document structure (i-json) storage device model to store the model in a document as a unit, supports array and document nesting, and the information to be split in the ordinary relational model is represented by a document. Based on the JSON (JavaScript object notation) structure, i-json optimizes and adds the complete path, node type and node attribute information of nodes, and supports nested structures and arrays. The specific structure definition is shown in the Table 1.

Table 1. I-json structure diagram.

No.	Attribute name	Attribute type	Sub attribute name	Sub attribute type
1	Dynamic	int	Name	string[]
2	Dynamic	int	Type	int
3	Static	int	Name	string[]
4	Node	int	Path	string
5	Node	int	Type	int
6	Node	int	Archive	int[]

The object model equipment attributes include dynamic attributes and static attributes. The dynamic attributes are used to describe the collected measurement type data of the equipment, including but not limited to three-phase current, three-phase voltage, active power, reactive power, etc. Static attributes are used to describe the file type data of equipment, including but not limited to serial number, attribute name, type, unit, collection cycle, etc. The specific equipment attributes are different according to the functions of intelligent terminal equipment.

3.3   Separate Storage Design

In order to reduce storage redundancy, this paper adopts a separate storage design, which separates the power IOT terminal model storage process from the collected data storage process, and separates the traditional measurement point model from the measurement data. The dynamic attribute management of power intelligent terminal is realized by hash algorithm, and the association relationship between equipment dynamic attributes and equipment measurement data is established and maintained by measuring point mapping rules.

The measurement point model and data compression storage of the storage layer are associated through the hash algorithm. The hash function adopts the executable link format function elfhash (extensible and linking format, ELF), takes the absolute length of the string as the input, and combines the decimal values of the characters through coding conversion to ensure that the generated measurement point ID positions can be evenly distributed, At the same time, it is convenient to locate the location according to the point nåme, and has high query performance. The model data association process is shown in Fig. 4.

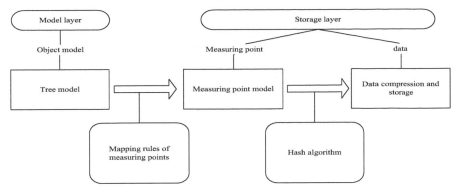

Fig. 4. Schematic diagram of model data association mode

In addition, the tree model node of the model layer is associated with the measuring point model through the measuring point mapping rules, which is mainly combined into the full path equipment attribute according to the model path and node name, and is associated with the measuring point name in the measuring point model through this attribute. Generally, the full path equipment attribute combines the model path and node name through the path symbol "/", and the measuring point name in the measuring point model is defined according to the combined equipment attribute. Since the path and node name can be used to describe the unique equipment attribute, the combined string can also define the unique measuring point name, so as to ensure the uniqueness of the measuring point.

3.4   Heads Improved Breadth First Search Algorithm

Considering that after the introduction of the tree structure, the access to the measured point data needs to be searched and located through the tree model, in order to improve the query performance and reduce the CPU resource consumption of the embedded system, the real-time database adopts the improved breadth first search (e-bfs) algorithm. First, access the starting vertex v, then start from V, access each unreachable adjacent vertex W1, W2, W3… Wn of V in turn, and then access all unreachable adjacent vertices of W1, W2,…, WI in turn. Then, start from these accessed vertices and perform pruning optimization by comparing the initials of adjacent node names with query conditions. Then access all their adjacent vertices that have not been accessed, and so on until all vertices on the way have been accessed. The specific implementation steps are shown in Fig. 5.

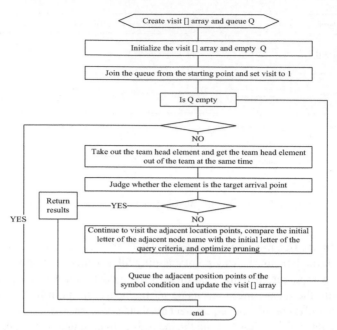

Fig. 5. Flow chart of e-bfs search algorithm

4 Performance Test

The test selected power secondary equipment terminal embedded ARM development board, processor armv7 processor Rev 2 (v7l), memory 240 MB and external memory 216 MB. Simulate the real-time data acquisition and storage connected to 100 power devices, with an average of 40 dynamic attributes for each device, and conduct data submission according to the second frequency. Compare and analyze the CPU resource utilization of the embedded system during the operation of the embedded real-time database (hs-ertdb) and SQLite database described in this paper. The test results are shown in Fig. 6.

The experimental results show that in the process of data submission, the CPU resources of SQLite database fluctuate greatly and have low stability. The minimum utilization rate is 20%, the maximum is 80%, and the average utilization rate is about 45%. The hs-ertdb database CPU utilization realized in this paper has small fluctuation range and high stability. The average utilization rate is about 15%, and the CPU energy consumption in the same scenario is reduced by 30%.

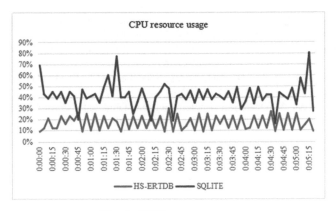

Fig. 6. CPU resource usage

5   Conclusion

In this paper, an embedded real-time database implementation method is proposed for the new power system intelligent terminal equipment, a lightweight power model construction scheme based on tree structure is proposed, a new power terminal model data separation storage mode is constructed, the model search algorithm is optimized, and the lightweight embedded real-time database is realized. Experiments show that the embedded real-time database realized by this method has good performance and low energy consumption, and is suitable for intelligent terminal equipment in new power system.

Acknowledgments. This paper is supported by research on lightweight database for power secondary edge equipment (524623200007).

References

1. Xin, B.: Accelerating the construction of a new power system to help achieve the "double carbon" goal. Economic daily, July 23 (2021)
2. Zhou, X., Chen, S., Lu, Z., et al.: Technical characteristics of China's new generation power system in energy transformation. Chin. J. Electr. Eng. **7**, 1893–1904 (2018)
3. Zhao, Y.: Design and implementation of intelligent terminal monitoring system for distribution network, Zhengzhou (2019)
4. Peng, Z., Chao, L., Yupeng, Z., et al.: Research on security control mechanism of power intelligent terminal equipment under new digital infrastructure. J. Huadian Technol. **43**, 66–70 (2021)
5. Zhu, B., Ye, S., Chen, M., et al.: Research on security protection of mobile operation terminals in power companies. Electromech. Inf. 86--87 (2019)
6. Yi, W., Qixin, C., Ning, Z., et al.: Integration of 5G communication and ubiquitous power Internet of things: application analysis and research prospect. Power Grid Technol. **043**, 1575–1585 (2019)

7. Kang, W., Sang, H.S., Stankovic, J.A.: Design, implementation, and evaluation of a QoS-aware real-time embedded database. IEEE Trans. Comput. **61**, 45–59 (2011)
8. Lin, C., Wang, Q., Zhang, P., et al.: A real-time database implementation method for micro grid intelligent terminal. P. cn102495891a (2012)
9. Zhou, K., Wang, B.: Cross platform lightweight database packaging method and system based on mobile terminal. P. cn106775719a (2016)
10. Li, H., Zhu, T., Xu, X.: Design and implementation of embedded real-time database based on ARM platform. Internet Things Technol. 75–77 (2014)

Sentiment Analysis-Based Method to Prevent Cyber Bullying

Giuseppe Ciaburro[1(✉)], Gino Iannace[1], and Virginia Puyana-Romero[2]

[1] Department of Architecture and Industrial Design, Università degli Studi della Campania
Luigi Vanvitelli, Borgo San Lorenzo, 81031 Aversa, CE, Italy
{giuseppe.ciaburro,gino.iannace}@unicampania.it
[2] Department of Sound and Acoustic Engineering, Universidad de Las Américas,
Quito EC170125, Ecuador
virginia.puyana@udla.edu.ec

Abstract. Cyberbullying is spreading in social networks frequented by young people. Its rapid spread is due to a series of specific preconditions due to the nature of the context within which the cyberbully finds himself operating. Anonymity, the absence of space-time limits, and the lack of responsibility of the individual are the strengths on which the actions of bullies are based. Automatically identifying acts of cyberbullying and social networks can help in setting up support policies for victims. In this study a method based on sentiment analysis is proposed with the use of recurrent neural networks for the prevention of cyberbullying acts in social networks.

Keywords: Sentiment analysis · Cyberbullying · Recurrent neural networks · Deep learning

1 Introduction

The recent explosion of violence involving groups of young people requires a serious discussion: One of the fundamental contexts for the development of such manifestations of violence is the school, both as an institution responsible for the training and transmission of knowledge, and as a relational space between young people and adults [1]. In the evolutionary process of the young person, school life represents an important stage in his social experience, experimenting with different ways of interacting: The young person learns the rules of behavior and strengthens their cognitive, emotional, and social skills. The school, therefore, can become the theater of both prosocial behaviors and aggressive behaviors, occasional or repeated, which have a profound impact on the development of the individuals involved in various capacities [2]. In fact, peer abuse occurs mainly between classmates or schoolmates, or between people who, voluntarily or not, share time, environment, and experiences [3]. People are hurt when they feel rejected, threatened, offended. Young victims, adolescents, and pre-adolescents, who are often ashamed to talk about it with someone, for fear of a negative judgment or for fear of receiving further confirmation of their being weak from the other. Bullying has

© The Author(s) 2022
Z. Qian et al. (Eds.): WCNA 2021, LNEE 942, pp. 721–735, 2022.
https://doi.org/10.1007/978-981-19-2456-9_73

long been under observation, while cyberbullying is a new and perhaps more hidden form, because it is less striking. It's a subtle manifestation of bullying itself, but no less important. Its diffusion is due to the massive use of information technology which has allowed the creation of new meeting spaces [4].

Bullying is a specific form of violence which, unlike the normal quarrels that exist between children, destined to lead to small jokes, acquires persecutory traits. The bully attacks the intended victim with physical and psychological acts, to subdue it until it is annihilated, often inducing the most fragile victims to extreme gestures, or in any case opening wounds destined to remain for life. Most adolescents have experienced bullying, one in three of these cases occurs in the school setting [5].

The term cyberbullying means those acts of bullying and stalking, prevarication carried out through electronic means such as e-mails, chats, blogs, mobile phones, websites, or any other form of communication attributable to the web [6]. Although it comes in a different form, online bullying is also bullying. Circulating unpleasant photos or sending emails containing offensive material can hurt much more than a punch or a kick, even if it does not involve violence or other forms of physics coercion. In online communities, cyberbullying can also be group-based, and girls are usually victims more frequently than boys, often with messages that contain sexual allusion. Usually the heckler acts anonymously, but sometimes he doesn't bother at all about hiding her identity. In this period of pandemic due to the spread of the Covid-19 contagion, with the adoption by many states of prolonged lockdown periods, this form of abuse has taken on even greater weight [7].

Social networks are means through which it is possible to communicate, share information and always stay in contact with people near and far. There are many, which differ from each other in various characteristic aspects aimed at satisfying the needs of some or many, but the purpose remains the same for all: to put the bet on the connection between individuals at the center, making it easier and more accessible. Among these, some of the best known and used are Facebook, Instagram, Twitter, and LinkedIn. Social networks are not limited only to instant messaging such as chats, but allow you to create your own profile, manage your social network and share files of all kinds that persist over time. Electronic bullying mostly occurs through social networks. This is because the web, with the ability to create and share millions of contents, has introduced a large amount of personal data and information into cyberspace [8]. The information ranges from personal data, tastes, favorite activities, places visited. This is because almost all social networks have rather soft personal data access policies, which allow their advertisers, and not just them, to collect thousands of data about their users. In many cases, in fact, it is sufficient to enter your name and surname in a search engine or in a social network, to know the opinions of a person, his romantic and working relationships, his daily activities [9]. The result is the social media paradox: if on the one hand we can more easily modify and shape our virtual identity, it is also true that, following the traces left by the different virtual identities, it is easier for others to reconstruct their real identity. This is because, the insertion of their data, their comments, their photo in a social network builds a historical memory of their activity and personality that does not disappear even when the subject wants it. The Data Protection Act, while helping to prevent the misuse of personal data, does not offer sufficient protection. It is therefore necessary to identify new methodologies capable of detecting possible cases of cyberbullying to intervene promptly and reduce the damage caused by these acts on the psychology of young people [10].

The term Sentiment analysis indicates the set of techniques and procedures suitable for the study and analysis of textual information, to detect evaluations, opinions, attitudes, and emotions relating to a certain entity [11]. This type of analysis has evident and important applications in the political, social, and economic fields. For example, a company may be interested in knowing consumer opinions about its products. But also, potential buyers of a particular product or service will be interested in knowing the opinion and experience of someone who has already purchased or used the product [12]. Even a public figure might be interested in what people think of him. Let's imagine a political figure, who wants to know what people think of his work, to monitor and control the consent for his next eventual re-election. Of course, there are already tools for the detection of consensus and opinions (surveys and statistical surveys); but through Opinion Mining techniques it is possible to obtain significantly lower detection costs and, in many cases, greater informative authenticity. Indeed, people are not obliged to express opinions, on the contrary, they flow freely without any coercion [13].

In recent years, the use of techniques based on Deep Learning for the extraction of sentiment from sources available on the net has become widespread. Deep learning is a branch of machine learning based on algorithms for modeling high level abstractions on data. It is part of a family of targeted techniques learning methods to represent data [14–18]. Recurrent neural networks (RNN) are a family of neural networks in which there are some feedback connections, such as loop within the network structure [19]. The presence of loop allows to analyze time sequences. In fact, it is possible to perform the so-called unfolding of the structure to obtain a feedforward version of the network of arbitrary length which depends on a sequence of inputs. What distinguishes the RNN from a feedforward is therefore the sharing of a state (weights and bias) between the elements of the sequence. So, what is stored within the network represents a pattern that binds the elements temporally of the series that RNN analyzes [20].

In this work, we will first introduce the general concepts underlying sentiment analysis, and then move on to the analysis of the architecture of algorithms based on recurrent neural networks. Subsequently, a practical case of classification of the polarity of the messages extracted from the WhatsApp chat will be analyzed for the identification of possible acts of cyberbullying. The rest of the chapter is structured as follows: Sect. 2 presents the methodology used to extract knowledge from the data. Section 3 describes the analyzed data and the results obtained with these methodologies, discussing them appropriately. Finally, in Sect. 4 the conclusions are reported.

2 Methodology

2.1 Sentiment Analysis Basic Concepts

The problem of text categorization is to assign labels to texts written in natural language. Text classification is a problem addressed in Information Retrieval since 1960. The applications are innumerable: searching for content related to a theme, organizing, and indexing web pages or other documents, other anti-spam, determining the language of a text, rationalization of pre-established archives. In the 1990s, the development of statistical techniques in artificial intelligence led to a paradigm shift in this area as well. In fact, before this period the problem was mostly solved, in practical applications, through what is called knowledge engineering: the construction by experts of a set of empirical rules, based on keywords or regular expressions and combined through Boolean operators, which classified the text [21].

To date, however, the most widespread techniques are those that exploit what is made available by modern machine learning [22]: an algorithm is provided with a series of examples of texts classified by experts, and this returns a mathematical model capable of classifying new texts. Most academic efforts also tend to focus on this technique. The advantages are first and foremost in effectiveness: accuracy is much higher than that obtained through rules-based approaches and is for some problems comparable to that of a human classifier. Furthermore, it is usually much easier and faster for an expert to categorize sample texts than to define, together with a computer scientist, the rules necessary for the categorization: for this there are also economic advantages in terms of the expert's working time. Furthermore, any refinements or updates of the classifier can be carried out systematically, through new sets of examples.

Recently, new text analysis tools are catching attention, not so much related to the extraction of specific characteristics of the text, but to some status of its author. This definition includes those inquiries by their nature aimed at the subject, such as the analysis of the writer's opinions and his feelings towards the object of the text. These two objectives, partly overlapping, are known in the literature as Opinion Mining and Sentiment Analysis, respectively. A third problem, in some ways similar and derivative, is the detection of the agreement, or the measure of the degree of agreement between two authors.

In recent years, the development of the Web has offered numerous possibilities for applying these techniques [23]. In fact, the large amount of textual content containing personal opinions of the authors has allowed several research ideas. Ordering these documents for the opinions they express offers several practical possibilities: For example, we could search for the keywords that are most present in negative reviews of a product, before buying it or to improve its sales strategy. Or, we may automatically have a concise assessment of a blog or comment author's opinion. Furthermore, on a larger scale, it is possible to hypothesize search engines for reviews, which find, classify, and present textual content present on the web that give opinions on a certain object searched for [11].

All these objectives therefore presuppose the identification of subjective contents expressed in a text. The problem is often broken down into two distinct sub-problems:

- the existence or not of these subjective contents, that is, to distinguish objective texts from subjective texts
- identify the polarity of the sentiment present in subjective texts (positive, neutral, or negative) (Fig. 1).

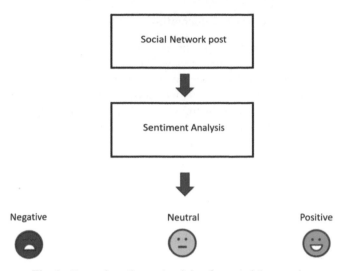

Fig. 1. Extraction of users' opinion from social networks.

An objective text is the opposite of a subjective text, and one with a negative feeling is the opposite of one with a positive feeling; having to distinguish several topics, however, one does not have that one is the opposite of the other. Furthermore, the polarity of sentiment can be framed, contrary to the topic, as a regression problem. For example, we can establish a scale in which -10 corresponds to a negative feeling while 10 to a positive one. Although it is useful to note this difference with respect to other textual classification problems, this does not mean that a regression-based approach is the best. On the contrary, the problem becomes more solvable by framing it as a multiclass problem: negative, neutral, positive. These classes typically have a specific vocabulary, different from contiguous classes. It is also important to note that the neutral class (to which we can associate the value 0) does not express the same concept as the absence of subjectivity [13].

The analysis of textual data, within the new Big Data discipline, represents one of the most important horizons, in terms of volume and relevance of the information obtainable, and is, in fact, one of those fields in which researchers and companies are currently concentrating its efforts. This interest stems from the fact that while systems and methods are available to analyze non-textual data, the same cannot be said for textual data. Obviously, this delay is understandable, the tools were first developed to analyze the data already available historically, that is, the data that are in a structured and numerical form. Furthermore, the value of textual data has acquired real importance only in recent years, thanks to the widespread use of smartphones and the massive entry of social networks into everyday life [12]. The goal today lies precisely in being able to interpret and extract useful information for your activities from this huge amount of data, generated every day. In general, all industries can benefit from text data analysis. In any case, speaking of textual analysis we do not mean the simple identification of keywords and their frequency, but instead we mean a much more in-depth activity and the results of which can be much more precise and useful.

2.2 Extracting Social Networks Information

Social Networks are certainly the most important phenomenon of the contemporary era from a technological and social point of view. We can say that the most popular social networks such as Twitter and Facebook have revolutionized the way in which a very large and heterogeneous part of all of us interacts, communicates, works, learns, and spreads news or, more simply, fills the time for a break or one moving, perhaps by train or bus. Social Networks are virtual platforms that allow us to create, publish and share user-generated content. It is this last feature that allows us to distinguish social media and Content Communities from Social Networks, that is, platforms where users can share specific content with other members of the community.

For a virtual platform to be correctly called a Social Network, three conditions must be met:

- there must be specific users of the platform in question
- these must be linked together
- there must be the possibility of interactive communication between the users themselves.

So, to give an example, Wikipedia is a social media, in fact users are not connected to each other, YouTube is a Content Community, users are connected to each other, but external people can also access the contents, while Twitter and Facebook are Social Networks, in fact, the latter satisfy the three previous conditions. The most interesting aspect of Social Networks and social media is their ability, in addition to the possibility of creating completely new and totally digital relational networks, to create content, and it is this last characteristic that makes the platforms so interesting. Moreover, we must always keep in mind, even if it is not that difficult, the importance that these tools are having on social evolution and daily behavior. Consider that by now about 59% of the world population is active on Social Networks or Media and that some events, political or custom, can generate large volumes of interesting data in a few hours.

In recent years, several researchers have used sentiment analysis to extract the opinion of users from social networks. West et al. [24] proposed a random field Markov-based model for text sentiment analysis. Wang et al. [25] applied data mining to detect depressed users who frequent social networks. They first adopted a sentiment analysis method that uses man-made vocabulary and rules to calculate each blog's inclination to depression. Next, they developed a depression detection model based on the proposed method and 10 characteristics of depressed users derived from psychological research. Zhou et al. [26] studied customer reviews after a purchase to manage loyalty. Satisfaction, trust, and promotion efforts were adopted as the input of the model and the consumer's buyback intention as the output. Five sportswear brands were analyzed by extracting the opinion of the merchants from the reviews to determine the intention to buy back products by consumers. In addition, the relationship between the initial purchase intention and the consumers' intention to buy back was compared to guide the marketing strategy and brand segmentation. Contratres et al. [27] proposed a recommendation process that includes sentiment analysis on textual data extracted from Facebook and Twitter. Recommendation systems are widely used in e-commerce to increase sales by matching product offerings and consumer preferences. For new users there is no information to make adequate recommendations. To address this criticality, the texts published by the user in social networks were used as a source of information. However, the valence of emotion in a text must be considered in the recommendation so that no product is recommended based on a negative opinion.

Wang et al. [28] tried to extract sentiment from images posted on the Internet based on both image characteristics and contextual information from social networks. The authors demonstrated that neither visual characteristics nor textual characteristics are in themselves sufficient for accurate labeling of feelings. Then, they leveraged both information by developing sentiment prediction scenarios with supervised and unsupervised methodologies. Kharlamov et al. [29] proposed a text analysis method that exploits a lexical mask and an efficient clustering mechanism. The authors demonstrate that cluster analysis of data from an n-dimensional vector space using the single linkage method can be considered a discrete random process. Sequences of minimum distances define the trajectories of this process. Vu et al. [30] developed a lexicon-based method using sentiment dictionaries with a heuristic data preprocessing mode: This methodology has sur-passed more advanced lexicon-based methods. Automated opinion extraction using online reviews is not only useful for customers to seek advice, but also necessary for businesses to understand their customers and improve their services.

Liu et al. [31] proposed a deep multilingual hierarchical model that exploits the regional convolutional neural network and the bi-directional LSTM network. The model obtains the temporal relationship of the different sentences in the comments through the regional CNN and obtains the local characteristics of the specific aspects in the sentence and the distance dependence in the entire comment through the hierarchical attention network. In addition, the model improves the gate mechanism-based word vector representation to make the model completely language independent. Li et al. [32] used public opinion texts on some specific events on social networking platforms and combined textual information with sentiment time series to get a multi-document sentiment prediction. Considering the interrelated characteristics of different social user identities and time series, the authors implemented a time + user dual attention mechanism model to analyze and predict textual public opinion information. Hung et al. [33] have applied methods based on machine learning to analyze the data collected by Twitter. Using tweets sourced exclusively from the United States and written in English during the 1-month period from March 20 to April 19, 2020, the study looked at discussions related to COVID-19. Social network and sentiment analyze were also conducted to determine the social network of dominant topics and whether the tweets expressed positive, neutral, or negative feelings. A geographical analysis of the tweets was also conducted.

2.3   Recurrent Neural Network

In the case of problems with interacting dynamics, the intrinsic unidirectional structure of the feedforward networks is highly limiting. However, it is possible to start from it and create networks in which the results of the computation of one unit influence the computational process of the other. The algorithms based on this new network structure converge in new ways compared to the classic models [19]. A recurrent neural network (RNN) is based on the artificial neural networks model but differs from this for the presence of two-way connections. In feed-forward networks the connections propagate the signals only and exclusively in the direction of the next layer. In recurrent networks this communication can also take place from one layer to the previous one or connections between neurons of the same layer as well as between a neuron and itself [20]. This change in the architecture of the neural network affects the decision-making process: The decision made in an instant affects the decision that will take in the next instant.

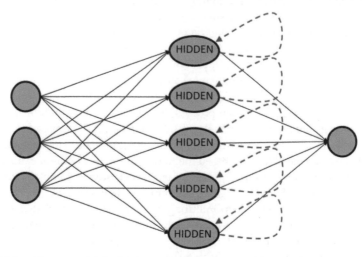

Fig. 2. RNN architecture with indications of bidirectional flows between layers - unfolding of a recurring network.

In recurrent neural network, the present and recent past contribute to determining the response of the system, a common feature in the decision-making process of human beings. The differences compared to feed-forward networks are reflected in the feed-back circuit connected to past decisions: The output of a layer is added to the input of a previous layer, characterizing its processing. This feature gives recurrent networks a memory for the purpose of using information already present in the sequence itself to perform tasks precluded to traditional feed-forward networks. The information in memory is used with content-based access, and not by location as is the case with a computer's memory. The information collected in the memory is processed in the next layer and, therefore, sent back to its origin, in modified form. This information can circulate several times gradually decreasing: In the case of information crucial for the system, the network can keep it without attenuation during several cycles, until the learning process considers it influential. Figure 2 shows an RNN architecture with indications of bi-directional flows between layers.

The RNN architecture shown in Fig. 2 requires that the weights of the hidden layer be regulated based on the information provided by the neurons from the input layer and by the processing obtained from the neurons of the hidden layer that have been activated. It is therefore a variant of the architecture of an artificial neural network (ANN), characterized by a different arrangement of the data flow: In the RNN the connections between the neurons combine in a cycle and propagate in the successive layers to learn sequences.

In the network shown in Fig. 3, the so-called unfolding of the structure is performed to obtain a feedforward version of the network of arbitrary length which depends on a sequence of inputs. The weights and biases of a layer are shared, and each output depends on the processing by the network of all inputs. The number of layers of the unfolded network essentially depends on the length of the sequence to be analyzed.

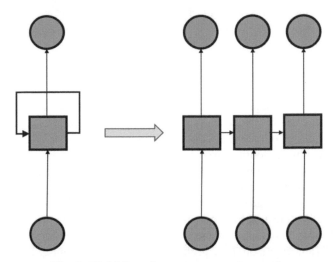

Fig. 3. Unfolding of a recurrent neural network.

What distinguishes the RNN from a feedforward is therefore the sharing of weights and bias between the elements of the sequence. The information stored within the network represents a pattern that temporally binds the elements of the series that the RNN analyzes. In Fig. 2 each input of the hidden layer is connected to the output, but it is possible to mask part of the inputs or part of the outputs to obtain different combinations. For example, it is possible to use a many-to-one RNN to classify a sequence of data with a single output, or to use a one-to-many RNN to label the set of subjects present from an image, as shown in Fig. 4.

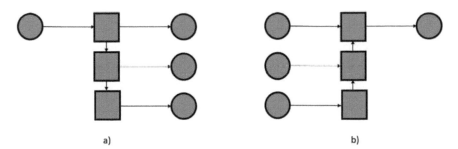

a)   b)

Fig. 4. a) One-to-many RNN architecture; b) Many-to-one RNN architecture.

During the input processing phase, the RNNs keep track of information on the history of all the elements of the past in the sequence in their hidden layers, that is, previous instants of time. Considering the output of the hidden layers at different times of the sequence as the output of different neurons of a deep multi-layer neural network, it becomes easy to apply backward propagation to train the network. However, although the RNNs are powerful dynamic systems, the training phase is often problematic because the gradient obtained with backward propagation either increases or decreases at any discrete time, so after many instants of time it can either become too large or become not very appreciable.

3 Data Processing, Results, and Discussion

WhatsApp is a free messaging application used to keep in touch with friends. Its free of charge and ease of use have made it the most popular instant messaging application. Creating groups is one of the main ways to exploit the potential of WhatsApp, in which dialogue can be a useful tool for exchanging information and concentrating users on a certain topic. These features have made this application very popular among students who use it by creating groups by classes, by topics or by sports groups. To begin, the WhatsApp chats of different school groups were extracted, creating datasets in.csv format. The messages were then cleaned by removing special symbols and various characters and emoticons. These symbols and characters can lead to a wrong classification. To avoid this, special symbols and emoticons have been replaced by their meaning. The next operation involved the labeling of each message by dividing it into the following classes: positive, and negative. To ensure sufficient generalization capacity for the algorithm, about 1000 messages were collected, taking care to distribute them as evenly among the two classes.

Before processing the data, it is necessary to carry out an appropriate subdivision of the data [34]. This procedure is necessary to avoid an excessive fit of the model on the data provided as input. The purpose of a classification model is to allow the correct classification of an occurrence never seen before by the model. To be sure that the model can do this, it is necessary that the performance evaluation is carried out on data that has never been subjected to the model so far [35]. The original data with the labeled examples were then partitioned into two distinct sets, training, and test sets, respectively. The classification model will then be trained using the training data, while its performance will be evaluated using the test set. The proportion of confidential data for training and testing was set at 70% for the training phase and the remaining 30% for the testing phase. This subdivision was made randomly. The accuracy of the classifier is then evaluated based on the accuracy achieved by the classifier itself on the test data [36, 37].

A preliminary step in any computational processing of the text is its tokenization. Tokenizing a text means dividing the sequences of characters into minimal units of analysis called tokens. The minimum units can be words, punctuation, dates, numbers, abbreviations, etc. Tokens can also be structurally complex entities, but they are nonetheless assumed as a base unit for subsequent processing levels. Depending on the type of language and writing system, tokenization can be an extremely complex task. In languages where word boundaries are not explicitly marked in writing, tokenization is also called word segmentation [38].

Another preliminary operation to be performed concerns the removal of the so-called stopwords. Stopwords are common words in a text that do not relate to a specific topic. Articles, propositions, conjunctions, or adjectives are typical examples of stopwords. These words can be found in any text regardless of the subject matter. They are called stopwords because they are eliminated in the search processes of a search engine, this is because they consume a lot of computational resources and do not add any semantic value to the text [39].

The last preliminary operation concerns stemming, a term used to name the linguistic process that aims to eliminate the morphological variations of a word, bringing it to its basic form [40].

Table 1. Sentiment analysis algorithm based on RNN.

Input: WhatsApp Messages
Output: Polarity of the Message (Positive, Negative)
Import the libraries
Load the data (csv format: Two columns: WhatsApp Message, Classification)
Data splitting (70% for training, 30% for testing)
Data Preprocessing
Tokenization
Stopwords removing
Stemming
Model building
Model compile
Model fit
Evaluate model performance

In summary, in the preliminary phase, the lexical analysis of the messages is carried out, in which the tokens are extracted, that is, all the sets of characters delimited by a separator. Then the stopwords are removed, that is all those words that are very frequent but whose informative content is not relevant. Usually they are articles, conjunctions, prepositions, pronouns and are listed in the appropriate stoplists, which obviously vary depending on the language considered. After removing the stopwords, we move on to the stemming phase, in which the words are grouped into their respective linguistic roots, thus eliminating the morphological variations. The next step is related to the composition of terms and the formation of groups of words. In fact, some terms, if grouped, improve the expressiveness of the associated concept or in some cases express a different concept from the individual words that compose it. Table 1 show the algorithm used in this work.

For the setting of the classification model of messages extracted from WhatsApp chats, we used the sequential model of the Keras library. Keras is an open-source neural network library written in Python. It can run on different backend frameworks. Designed to allow rapid experimentation with deep neural networks, it focuses on being intuitive, modular, and extensible [41].

Five-layer classes were imported: Sequential, Embedding, SimpleRNN, Dense, and Activation. The Sequential class is used to define a linear stack of network layers that make up a model. The Embedding layer is used to transform positive integers into dense vectors of fixed size. This level can only be used as the first level in a model. The SimpleRNN level is used to add a fully connected RNN. The Dense class is used to instantiate a Dense layer, which is the fully connected base feedforward layer. The activation level is used to add an activation function to the level sequence. A sigmoid activation function is used, which produces a sigmoidal curve. This is a characteristic curve characterized by its S shape. This is the earliest and most often used activation function.

In the compile procedure we have set the loss, the optimizer, and the evaluation metric. As loss function, we have used the binary_crossentropy loss function, especially suited for binary classification problem. This loss function computes the cross-entropy loss between true labels and predicted labels. As optimizer the RMSProp optimizer was used, and finally for the performance evaluation the accuracy metric was used. This RMSProp optimization algorithm maintains a moving average of the square of the gradients and divides the gradient by the root of this average. The accuracy returns the percentage of predictions correct with a test dataset. Equivalent to the ratio of the number of correct estimates to the total number of input samples. It works well if there are a similar number of examples belonging to each class.

After training the model on the training data, we tried to evaluate the model's performance on a never-before-seen dataset. The model returned approximately 85% accuracy showing clearly that an RNN-based model is capable of correctly classifying the polarity of a message.

4 Conclusion

Cyberbullying is becoming a real social problem and given the young age of the people involved it requires a lot of attention from adults. Young people are now making massive and sometimes excessive use of telematic communication channels. These channels do not have an appropriate control of the contents of the conversations due to the constraints imposed by the respect of privacy. But given the weight assumed by such conversations in the lives of children, it is necessary to think of methodologies that can guarantee vigilance without compromising the freedom of children to have spaces for socialization. Automatic identification of cyberbullying acts on social networks can help set up support policies for victims. In this study, a method based on sentiment analysis was proposed with the use of recurrent neural networks for the identification of the polarity of the message contents of the popular WhatsApp messaging app. The results showed that this methodology can represent a tool for monitoring the contents of conversations between young people.

References

1. Rigby, K.: Bullying in schools: and what to do about it. Aust Council for Ed Research (2007)
2. Iannace, G., Ciaburro, G., Maffei, L.: Effects of shared noise control activities in two primary schools. In: INTER-NOISE and NOISE-CON Congress and Conference Proceedings, vol. 2010, no. 8, pp. 3412–3418. Institute of Noise Control Engineering (June 2010)

3. Smith, P.K., Brain, P.: Bullying in schools: lessons from two decades of research. Aggress. Behav.: Off. J. Int. Soc. Res. Aggress. **26**(1), 1–9 (2000)
4. Juvonen, J., Graham, S.: Bullying in schools: the power of bullies and the plight of victims. Annu. Rev. Psychol. **65**, 159–185 (2014)
5. Olweus, D.: Bullying at School: What We Know and What We Can Do (Understanding Children's Worlds). Blackwell Publishing, Oxford (1993)
6. Menesini, E., Salmivalli, C.: Bullying in schools: the state of knowledge and effective interventions. Psychol. Health Med. **22**(sup1), 240–253 (2017)
7. Elmer, T., Mepham, K., Stadtfeld, C.: Students under lockdown: comparisons of students' social networks and mental health before and during the COVID-19 crisis in Switzerland. PLoS ONE **15**(7), e0236337 (2020)
8. Peng, S., Zhou, Y., Cao, L., Yu, S., Niu, J., Jia, W.: Influence analysis in social networks: a survey. J. Netw. Comput. Appl. **106**, 17–32 (2018)
9. Smith, E.B., Brands, R.A., Brashears, M.E., Kleinbaum, A.M.: Social networks and cognition. Ann. Rev. Sociol. **46**, 159–174 (2020)
10. Kelly, M.E., et al.: The impact of social activities, social networks, social support and social relationships on the cognitive functioning of healthy older adults: a systematic review. Syst. Rev. **6**(1), 1–18 (2017)
11. Feldman, R.: Techniques and applications for sentiment analysis. Commun. ACM **56**(4), 82–89 (2013)
12. Medhat, W., Hassan, A., Korashy, H.: Sentiment analysis algorithms and applications: a survey. Ain Shams Eng. J. **5**(4), 1093–1113 (2014)
13. Agarwal, A., Xie, B., Vovsha, I., Rambow, O., Passonneau, R.J.: Sentiment analysis of Twitter data. In: Proceedings of the Workshop on Language in Social Media (LSM 2011), pp. 30–38 (June 2011)
14. LeCun, Y., Bengio, Y., Hinton, G.: Deep learning. Nature **521**(7553), 436–444 (2015)
15. Ciaburro, G.: Sound event detection in underground parking garage using convolutional neural network. Big Data Cogn. Comput. **4**(3), 20 (2020)
16. Goodfellow, I., Bengio, Y., Courville, A.: Deep Learning. MIT Press, Cambridge (2016)
17. Ciaburro, G., Iannace, G.: Improving smart cities safety using sound events detection based on deep neural network algorithms. In: Informatics, vol. 7, no. 3, p. 23. Multidisciplinary Digital Publishing Institute (September 2020)
18. Schmidhuber, J.: Deep learning in neural networks: an overview. Neural Netw. **61**, 85–117 (2015)
19. Pascanu, R., Mikolov, T., Bengio, Y.: On the difficulty of training recurrent neural networks. In: International Conference on Machine Learning, pp. 1310–1318. PMLR (May 2013)
20. Yin, C., Zhu, Y., Fei, J., He, X.: A deep learning approach for intrusion detection using recurrent neural networks. IEEE Access **5**, 21954–21961 (2017)
21. Yang, L., Li, Y., Wang, J., Sherratt, R.S.: Sentiment analysis for E-commerce product reviews in Chinese based on sentiment lexicon and deep learning. IEEE Access **8**, 23522–23530 (2020)
22. Yadav, A., Vishwakarma, D.K.: Sentiment analysis using deep learning architectures: a review. Artif. Intell. Rev. **53**(6), 4335–4385 (2019). https://doi.org/10.1007/s10462-019-09794-5
23. Ke, P., Ji, H., Liu, S., Zhu, X., Huang, M.: Sentilare: linguistic knowledge enhanced language representation for sentiment analysis. In: Proceedings of the 2020 Conference on Empirical Methods in Natural Language Processing (EMNLP), pp. 6975–6988 (November 2020)
24. West, R., Paskov, H.S., Leskovec, J., Potts, C.: Exploiting social network structure for person-to-person sentiment analysis. Trans. Assoc. Comput. Linguist. **2**, 297–310 (2014)

25. Wang, X., Zhang, C., Ji, Y., Sun, L., Wu, L., Bao, Z.: A depression detection model based on sentiment analysis in micro-blog social network. In: Li, J., Cao, L., Wang, C., Tan, K.C., Liu, B., Pei, J., Tseng, V.S. (eds.) PAKDD 2013. LNCS (LNAI), vol. 7867, pp. 201–213. Springer, Heidelberg (2013). https://doi.org/10.1007/978-3-642-40319-4_18

26. Zhou, Q., Xu, Z., Yen, N.Y.: User sentiment analysis based on social network information and its application in consumer reconstruction intention. Comput. Hum. Behav. **100**, 177–183 (2019)

27. Contratres, F.G., Alves-Souza, S.N., Filgueiras, L.V.L., DeSouza, L.S.: Sentiment analysis of social network data for cold-start relief in recommender systems. In: Rocha, Á., Adeli, H., Reis, L.P., Costanzo, S. (eds.) WorldCIST'18 2018. AISC, vol. 746, pp. 122–132. Springer, Cham (2018). https://doi.org/10.1007/978-3-319-77712-2_12

28. Wang, Y., Li, B.: Sentiment analysis for social media images. In: 2015 IEEE International Conference on Data Mining Workshop (ICDMW), pp. 1584–1591. IEEE (November 2015)

29. Kharlamov, A.A., Orekhov, A.V., Bodrunova, S.S., Lyudkevich, N.S.: Social network sentiment analysis and message clustering. In: El Yacoubi, S., Bagnoli, F., Pacini, G. (eds.) INSCI 2019. LNCS, vol. 11938, pp. 18–31. Springer, Cham (2019). https://doi.org/10.1007/978-3-030-34770-3_2

30. Vu, L., Le, T.: A lexicon-based method for sentiment analysis using social network data. In: Proceedings of the International Conference on Information and Knowledge Engineering (IKE), pp. 10–16. The Steering Committee of The World Congress in Computer Science, Computer Engineering and Applied Computing (World-Comp) (2017)

31. Liu, G., Huang, X., Liu, X., Yang, A.: A novel aspect-based sentiment analysis network model based on multilingual hierarchy in online social network. Comput. J. **63**(3), 410–424 (2020)

32. Li, L., Wu, Y., Zhang, Y., Zhao, T.: Time+ user dual attention based sentiment prediction for multiple social network texts with time series. IEEE Access **7**, 17644–17653 (2019)

33. Hung, M., et al.: Social network analysis of COVID-19 sentiments: application of artificial intelligence. J. Med. Internet Res. **22**(8), e22590 (2020)

34. Ciaburro, G., Puyana-Romero, V., Iannace, G., Jaramillo-Cevallos, W.A.: Characterization and modeling of corn stalk fibers tied with clay using support vector regression algorithms. J. Nat. Fibers 1–16 (2021)

35. Puyana Romero, V., Maffei, L., Brambilla, G., Ciaburro, G.: Acoustic, visual and spatial indicators for the description of the soundscape of waterfront areas with and without road traffic flow. Int. J. Environ. Res. Public Health **13**(9), 934 (2016)

36. Iannace, G., Ciaburro, G.: Modelling sound absorption properties for recycled polyethylene terephthalate-based material using Gaussian regression. Build. Acoust. **28**(2), 185–196 (2021)

37. Ciaburro, G., Iannace, G., Ali, M., Alabdulkarem, A., Nuhait, A.: An Artificial neural network approach to modelling absorbent asphalts acoustic properties. J. King Saud Univ.-Eng. Sci. **33**(4), 213–220 (2021)

38. Kaplan, R.M.: A method for tokenizing text. Inq. Words Constraints Contexts **55**, 79 (2005)

39. Ghag, K.V., Shah, K.: Comparative analysis of effect of stopwords removal on sentiment classification. In: 2015 International Conference on Computer, Communication and Control (IC4), pp. 1–6. IEEE (September 2015)

40. Jivani, A.G.: A comparative study of stemming algorithms. Int. J. Comp. Tech. Appl. **2**(6), 1930–1938 (2011)

41. Manaswi, N.K.: Understanding and working with Keras. In: Deep Learning with Applications Using Python, pp. 31–43. Apress, Berkeley (2018)

Signal Processing

The Research of Adaptive Modulation Technology in OFDM System

Xiuyan Zhang[✉] and Guobin Tao

School of Electric and Automatic Engineering,
Changshu Institute of Technology, Changshu, China
xyzhang_113@163.com

Abstract. Orthogonal frequency division multiplexing (OFDM) as a special multi-carrier transmission technology has good resistance to narrow-band interference and frequency selective fading ability. Compared with traditional modulation techniques, adaptive modulation can enhance bandwidth efficiency and system capacity. Therefore, applying adaptive modulation in OFDM systems can take full advantage of spectrum resources, and it is suitable for the high-speed and reliable mobile communication systems in the future. The purpose of this paper is to improve traditional OFDM adaptive algorithms (Hughes-Hartogs, Chow) to realize bits allocation, power allocation better. In this paper, simulation results demonstrated that the improved Levin-Campello algorithm lowers algorithm's complexity greatly and owns better flexibility, at the same time, it guarantees good the bit error rate (BER) performance and can be applied to speech communication (fixed rate) and data communication (variable rate) in wireless communication systems.

Keywords: OFDM · Adaptive modulation · Bit allocation · Power allocation

1 Introduction

With the high speed data in wireless mobile communication business and the rapid development of multimedia services. The research is importance how to effectively use of spectrum resources to provide high-speed and reliable communication service. In this paper, the improved better Levin-Campello algorithm is researched for ensuring BER, better bit and power allocation by the comparing of two traditional adaptive modulation algorithm.

2 The Principle of OFDM System and the Realization of Adaptive Modulation [1]

The multicarrier transmission way is adopted by OFDM [2] technology after the high speed serial data is decomposed into several parallel data at low speed, then the width of each data element is widened, so that the influence of intersymbol interference can

© The Author(s) 2022
Z. Qian et al. (Eds.): WCNA 2021, LNEE 942, pp. 739–748, 2022.
https://doi.org/10.1007/978-981-19-2456-9_74

reduced. By Orthogonal function sequence is used as subcarrier, so the carrier spacing is reached the minimum, and the band utilization rate of the system is fully enhance. By making fully use channel state information (CSI) in adaptive modulation OFDM system, Low order modulation method is adopted in the smaller decline amplitude subcarrier, and high order modulation method is adopted in the larger decline amplitude subcarrier. and the corresponding power is distributed, so the efficiency of data transmission is greatly improved.

The adaptive modulation block diagram of OFDM system [3] is shown in Fig. 1.

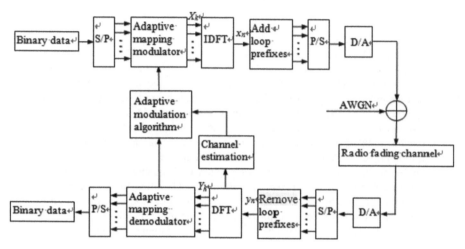

Fig. 1. The adaptive modulation block diagram of OFDM system

3 The Adaptive Modulation Algorithm of the Traditional Raditional OFDM System [4]

3.1 Hughes - Hartogs Algorithm

Optimization criterion of Hughes - Hartogs algorithm [5] is the minimum total power of the system in a condition of the guarantee target BER and data rate.

The algorithm is a kind of algorithm based on the channel gain, the basic idea is the bits of each channel number are set to zero, then all bit will be distributed are assigned to the corresponding sub-channels. Every time allocation, firstly, the channel increasing the minimum power will be found when adding a bit, then the number of bits of sub-channels will increased one, then the process is repeated, until all bits allocated are reached the requirements of a given target bit, finally, the required power of each channel are calculated.

① The initialization process

For all n = 1, 2,... N, make $C_n = 0$. Calculate $\Delta P(n) = P(C_{n+1}) - P(C_n)$.

② The iterative process of bit allocation

The minimum value of $\Delta P(n)(n = 1, 2, \ldots, N)$ is searched, and is recorded label the subcarrier for $\hat{n} = \arg\min$, then increasing power of the subcarrier are recalculated once again:

$$\Delta P(\hat{n})P(C_{\hat{n}+1}) - P(C_{\hat{n}}) \tag{1}$$

③ Repeat step ②, until the R bit allocation are completed.

$\{C_1, C_2, \cdots, C_n\}$ are calculated by the above steps is the last bit allocation scheme. Each bit of information is distributed by searching and sorting in Hughes - Hartogs algorithm, when the total bits number of the carrier and emission is larger, then the complexity of the algorithm is very high.

3.2 Chow Algorithm

Chow algorithm [6–8] is the adaptive bit allocation algorithm of subprime power minimization similar water flooding algorithm, this algorithm is suitable for large transmission capacity ASDL system, the performance is lower than the Hughes - Hartogs algorithm, but it has faster convergence speed, and bit allocation of Chow algorithm is based on the channel capacity of each channel. Its optimization criterion is the system's performance allowance is maked the largest on the premise of maintaining the target bit error rate. Bits are gradually allocated by the iteration process in this algorithm, and at same time the allowance system are gradually sete increased, until all the bits are allocated to complete. A maximum number of iterations is d for keeping the convergence rate of the algorithm. This algorithm has the following three steps to complete:

① Determine the threshold margin for achieving the optimal performance of the system;

② Determine the modulation way of each sub-carrier;

③ Adjust the power of each subcarrier.

4 Levin-Campello Algorithm

Drawbacks of the Hughes - Hartogs algorithm are high complexity, slow convergence speed and unsuitability real-time systems. Chow algorithm based on maximum data rate standard can not meet the sending power minimum requirements of the many systems. In view of the above two algorithms existing problems, and then the improved Campello algorithm based on Chow algorithm and Hughes - Hartogs algorithm is appeared, the improved Campello algorithm with the advantages of the two algorithms can achieve the minimizing sending power.

Levin – Campello [9, 10] algorithm is divided into three step implementation, the specific steps are as follows:

Step 1: Bit and power are initialized allocation according to Chow algorithm ideas, specific implementation process of this step is as follows:

① Calculate SNR of all sub-channels;

② Bit allocation of sub-channels according to the formula:

$$b'_i = \log_2\left(1 + \frac{SNR_i}{gap}\right) \tag{2}$$

where gap is coordinate parameters, it is the function of Coding scheme, the target ber and noise margin.

③ b_i' must be rounded for the integer bit allocation of communication system

$$b_i' = round\left(b_i'\right) \tag{3}$$

④ Because of the modulation mode is usually adopt even, so b_i has a value of 0, 1, 2, 4, 6, 8. Allocation energy of each subcarrier b_i bit is calculated by using the following formula:

$$e_i(b_i) = \frac{2^{b_i} - 1}{GNR_i} \tag{4}$$

where, $GNR_i = SNR_i/gap$.

Step 2: Adjust bit and power allocation according to the Hughes - Hartogs algorithm.

Firstly, an energy increment table must be built, table contained increase energy of average increase a bit in each channel on the original basis, For I sub-channels, originally allocated b-x bit is increased to x bits, and the energy increment is:

$$\Delta e_i(b)_x = e_i(b) - e_i(b - x) \tag{5}$$

Power increment of average every bit is $\Delta e_i(b) = \Delta e_i(b)_x/x$, because each subcarrier is only allocated 8 bits in the system, so bits increment from 8 bits to are set to a very high value, so it is avoided that the subcarrier distribution system is distributed any greater than 8 bits.

The specific implementation steps of the steps are as follows:

① m_i is the maximum number of adjusted bits for each channel, m is the biggest adjustment step length, then the actual change length should satisfy $M_i = min[m_i, m]$. The power increment is $\Delta e_i(b)_{M_i}$ by changing M_i, every bit power increment is:

$$\Delta e_i(b) = \Delta e_i(b)_{M_i}/M_i \tag{6}$$

② The largest or smallest element of energy table is drawn, and its bit is adjusted according to the corresponding adjustment step length M_i of sub-channels, so a new $\Delta e_i(b)$ is got, and new energy increment table is obtained.

③ If the purpose of the distribution don't reach, return step 2, or quit.

Detailed algorithm process is:

Firstly, initial bit numbers for each channel are summed: $B' = sum(b_i)$, then for the following operations:

while $B' \neq B$

　if $B' > B$

　　$n = \text{argmax } \Delta e_i(b)$

　　$b_n = b_n - M_n$

　　$B' = B' - M_n$

　else

　　$n = \text{argmax } \Delta e_i(b)$

　　$b_n = b_n + M_n$

　　$B' = B' + M_n$

End

Step 3: Optimize the last 1 bit.

Through step 1 and step 2, the last one bit may be assigned to subcarrier with the bit number greater than 2 and an even number of bits, so bits of the subcarrier number is odd number greater than 2, if the number of allocation bits of subcarrier is greater than 2, then the subcarrier is allocated an even number bits of less than or equal to 8, so a last bit need to specially treat.

Campello algorithm using RTLB (Resolve The Last Bit) algorithm. RTLB algorithm implementation steps are as follows:

① Check each subchannel, if there is the number bits due to the last 1 bit allocation isn't be supported. If it does not have this kind of channel, distribution is terminated; If the channel r exist, the next step $\Delta e_r(b(r))$ and $\Delta e_r(b(r) + 1)$ are calculated.

② Search subcarrier given 1 bit or 2 bits, subcarrier with most energy reduction by decreasing 1 bit is denoted by i, the energy increment $\Delta e_i(b(i))$ is obtained, calculate the following formula:

$$E1 = \Delta e_r(b(r) + 1) - \Delta e_i(b(i)) \tag{7}$$

③ Collect subcarrier allocated 0 bit or 1 bit, subcarrier with minimum energy increase by increasing 1 bit is denoted by j, the energy increment $\Delta e_j(b(j) + 1)$ is obtained, calculate following formula:

$$E2 = \Delta e_j(b(j) + 1) - \Delta e_r(b(r)) \tag{8}$$

④ Compare E1 and E2, if E1 is less than E2, the subcarrier i reduce a bit, subcarrier increase a bit at the same time; If the E2 is less than E1, the subcarrier j increase a bit, at the same time the subcarrier reduce a bit. At the same time, the corresponding energy allocation is adjusted, the algorithm is over.

5   Levin-Campello Algorithm Simulation and Performance Analysis

In order to verify the correctness of the theory analysis, the Levin - Campello algorithm, Hughes-Hartogs algorithm and Chow algorithm are simulated by using MATLAB, simulations are conducted in the case of the same parameters mentioned earlier, the simulation parameters [11, 12] are shown in Table 1.

Table 1. System simulation parameters

The subcarrier number N of OFDM	32
Cyclic prefix CP	16
The biggest sign bit number	8
Transmitting antenna number	1
Receiving antenna number	1
Fading channel type	Rayleigh

The subchannel gain simulation results, the bit allocation simulation results and the power allocation simulation results of three algorithm are shown in Fig. 2, 3 and 4. The BER simulation of Levin-Campello is shown in Fig. 5.

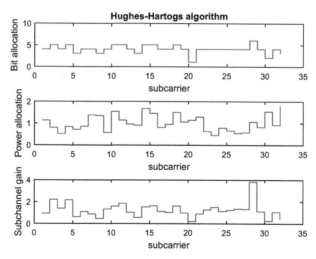

Fig. 2. The simulation results of Hughes - Hartogs algorithm

It can be seen from the simulation results of Fig. 2, Fig. 3 and Fig. 4 that bit allotment of each subcarrier are determined by algorithm according to the subcarrier channel gain, distribution of bit is more in the good channel conditions, Otherwise, distribution of bit is less or no in the poor channel conditions. Hughes - Hartogs algorithm can achieve the ideal performance, in every time for bit allocation, the additional power needed to ensure the transmission bit is minimal. Sorting and searching computation is very big, and complexity is high, and practicability is poor. Rate allocation of Chow algorithm is according to the capacity of each channel, large allowance system is needed, it don't conform to the actual demand. But complexity of Levin - Campello algorithm is not only greatly reduced, but also BER performance is good, it can be seen from Fig. 5 that the BER of system is significantly dropped, until almost don't make a mistake when the SNR is greater than 102, this is the biggest advantage of the algorithm.

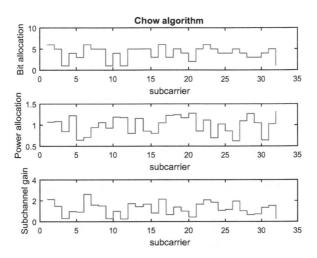

Fig. 3. The simulation results of Chow algorithm

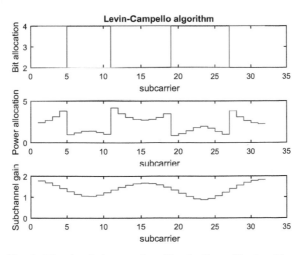

Fig. 4. The simulation results of Levin-Campello algorithm

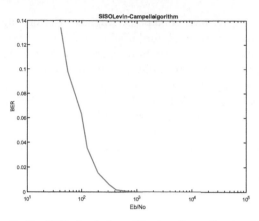

Fig. 5. The BER simulation of Levin - Campello algorithm

Table 2. Data simulation results data of three algorithm

Subcarrier	Hughes-Hartogs algorithm			Chow algorithm			Levin-Campello algorithm		
	Bit	Power	Gain	Bit	Power	Gain	Bit	Power	Gain
1	5	0.9374	1.5364	6	1.0376	2.3487	0	0	0.3362
2	3	0.7323	0.4452	6	0.7435	2.7746	0	0	0.4531
3	2	1.5777	0.1642	6	1.1514	2.2296	4	22.4173	0.5791
4	4	0.7363	0.8710	6	1.4632	1.9779	4	15.2963	0.7011
5	5	0.9220	1.5491	5	1.3796	1.4288	4	11.4332	0.8109
6	4	1.3471	0.6439	5	0.8020	1.8740	6	38.7352	0.9029
7	3	0.7347	0.4445	2	0.7829	0.5900	6	33.3724	0.9727
8	5	1.1318	1.3982	1	1.1034	0.2869	6	30.5020	1.0174
9	4	0.5992	0.9655	6	0.8999	2.5220	6	29.4578	1.0353
10	5	1.3746	1.2687	5	0.9006	1.7684	6	30.0069	1.0258
11	4	0.4819	1.0765	6	1.1381	2.2426	6	32.2414	0.9896
12	3	0.7366	0.4439	6	0.8356	2.6173	6	36.6048	0.9288
13	2	2.3071	0.1357	0	0	0.1038	6	44.0587	0.8466
14	6	0.9352	3.0727	4	1.0284	1.1512	4	13.4449	0.7478
15	3	0.5237	0.5265	2	0.9865	0.5256	4	18.4149	0.6389
16	5	0.9105	1.5589	4	0.8359	1.2769	4	26.8468	0.5292
17	4	0.4798	1.0790	4	0.7797	1.3221	0	0	0.4317
18	4	0.6523	0.9451	4	1.519	0.9494	0	0	0.3654

(continued)

Table 2. (*continued*)

Subcarrier	Hughes-Hartogs algorithm			Chow algorithm			Levin-Campello algorithm		
	Bit	Power	Gain	Bit	Power	Gain	Bit	Power	Gain
19	4	1.3765	0.6370	2	1.0170	0.5177	0	0	0.3480
20	5	0.7818	1.6824	4	1.5109	0.9497	0	0	0.3782
21	4	0.6980	0.8945	5	1.1593	1.5587	0	0	0.4340
22	5	0.8246	1.6381	4	1.1926	1.0690	4	30.9130	0.4931
23	4	1.0864	0.7170	3	0.7889	0.8979	4	25.6226	0.5417
24	5	0.5600	1.9878	0	0	0.1429	4	23.0080	0.5716
25	2	0.9118	0.2159	3	0.7466	0.9229	4	22.4430	0.5788
26	3	1.8894	0.2772	4	0.9954	1.1701	4	23.8501	0.5614
27	5	0.7683	1.6971	6	1.2356	2.1523	4	27.7966	0.5201
28	2	0.6752	0.2509	4	0.7497	1.3482	4	35.9475	0.4573
29	5	1.8648	1.0893	5	1.1362	1.5745	0	0	0.3791
30	5	0.8038	1.6592	2	1.3370	0.4515	0	0	0.2980
31	5	1.3302	1.2897	4	1.5187	0.9473	0	0	0.2418
32	3	1.3361	0.3296	4	1.2318	1.0518	0	0	0.2536

Data simulation results data of three algorithm is shown in Table 2, it can be seen from Table 2 that an obvious characteristic with Levin - Campello algorithm compared to Hughes - Hartogs and Chow algorithm is that subchannels of channel gain under a certain limit (Here is about 0.5) will be discarded, so the quality of the communication is improved.

6 Conclusion

Traditional algorithm of Hughes – Hartogs and Chow algorithm based on the adaptive modulation rule are researched in this paper, aim at the shortcomings of high computation complexity of Hughes – Hartogs and low power efficiency of Chow algorithm, Campello algorithm firstly initializes bit and power allocation on according to Chow algorithm ideas, and then bit and power allocation are adjusted according to Hughes - Hartogs algorithm, so the algorithm complexity is greatly reduced. It is found by the above analysis that Campello algorithm has low computation complexity and high power efficiency, at the same time, it has greater flexibility on condition of the BER, it is suitable for voice communication of the fixed rate in wireless communication system, and it can also be applied to variable speed data communication, so it conforms to the actual requirements of wireless communication system, and it is a kind of better adaptive modulation algorithm.

References

1. Liu, X., Yang, F., Ruan, D., Cheng, G.: Current situation and development of adaptive modulation technology in wireless communication. Microprocessors **38**(03), 34–37(2017)
2. Zhang, H.: The Basic Principle and Key Technology of Orthogonal Frequency Division Multiplexing. National Defence Industry Publishing, Beijing (2006)
3. Li, Z., Wang, W., Zhou, W.: The adaptive transmission technology based on OFDM. Radio Commun. Technol. **13**(87), 34–35 (2014)
4. Wang, D.: The Research of Adaptive Bit Power Allocation Algorithm. Nanjing Information Engineering University, Nanjing, vol. 5 (2009)
5. Huang, X., Wang, G., Ma, Y., Zhang, C., Jiang, H.: An efficient OFDM bit power allocation algorithm. J. Harbin Inst. Technol. **42**(9), 1379–1382 (2010)
6. Ren, G., Zhang, H.: Adaptive orthogonal frequency division multiplexing throughput maximization power allocation algorithm. J. Xi'an Jiaotong Univ. **02**(11), 122–125 (2014)
7. Li, L., Yanjun, H.: The performance analysis of bit allocation optimization algorithm of OFDM system. Inf. Technol. **10**, 66–69 (2007)
8. Stuber, G.L., Barry, J., Mclaughlin, S., Li, G., Pratt, T.: Broadband MIMO-OFDM wireless communications. Proc. IEEE **92**(2), 271–294 (2004)
9. Li, X., Xin, X.: Analysis of adaptive bit allocation algorithm in OFDM system. Radio Commun. Technol. **35**(4), 1379–1382 (2009)
10. Wang, L., Chen, S., Li, Y.: Adaptive efficient power allocation in MIMO-OFDM system. Commun. Technol. **44**(04), 74–76 (2011)
11. Loh, A., Siu, W.: Improved fast polynomial transform algorithm for cyclic convolutions. Circuits Syst. Signal Process. **05**(12), 125–129 (2005)
12. Salo, J., El-Sallabi, H.M., Vainikainen, P.: Impact of double-Rayleigh fading on system performance. Proc of ISWPC **13**(12), 06–09 (2011)

A Fully-Nested Encoder-Decoder Framework for Anomaly Detection

Yansheng Gong[1] and Wenfeng Jing[2(✉)]

[1] First China Railway First Survey and Design Institute Group Co., Ltd., Xi'an 710043, China
[2] Xi'an Jiaotong University, Xi'an 710049, China
wfjing@xjtu.edu.cn

Abstract. Anomaly detection is an important branch of computer vision. At present, a variety of deep learning models are applied to anomaly detection. However, the lack of abnormal samples makes supervised learning difficult to implement. In this paper, we mainly study abnormal detection tasks based on unsupervised learning and propose a Fully-Nested Encoder-decoder Framework. The main part of the proposed generating model consists of a generator and a discriminator, which are adversarially trained based on normal data samples. In order to improve the image reconstruction capability of the generator, we design a Fully-Nested Residual Encoder-decoder Network, which is used to encode and decode the images. In addition, we add residual structure into both encoder and decoder, which reduces the risk of overfitting and enhances the feature expression ability. In the test phase, a distance measurement model is used to determine whether the test sample is abnormal. The experimental results on the CIFAR-10 dataset demonstrate the excellent performance of our method. Compared with the existing models, our method achieves the state-of-the-art result.

Keywords: Anomaly detection · Unsupervised learning · Encoder-decoder · Distance measurement

1 Introduction

Anomaly detection is becoming more and more important in visual tasks. In industrial production, it can greatly improve production efficiency to detect the faults of various parts of machines by means of anomaly detection. Over the years, scholars have done a lot of preliminary works [1–6] to explore the development direction of the field of anomaly detection. The development of CNN offers new ideas for image anomaly detection. From the proposal of LeNet [7] structure, to AlexNet [8], to VGG [9] and Inception series [10–12], the performance of CNN is getting better and better. In the tasks of anomaly detection, the methods of supervised learning based on CNNs have been widely used to detect anomalies. However, in some engineering areas, the lack of anomaly samples hinders the development of supervised anomaly detection methods. Due to the lack of abnormal samples, traditional methods such as object detection, semantic segmentation and image classification are difficult to carry out model training. Therefore, anomaly detection methods based on normal samples need to be proposed urgently.

© The Author(s) 2022
Z. Qian et al. (Eds.): WCNA 2021, LNEE 942, pp. 749–759, 2022.
https://doi.org/10.1007/978-981-19-2456-9_75

The development of GAN in recent years has provided new ideas for the research of anomaly detection methods based on normal samples. As an unsupervised image method, GAN was proposed by Ian Goodfellow et al. [13] in 2014. Subsequently, methods such as LAPGAN, CGAN, InfoGAN, and CycleGAN [14–17] have gradually enhanced the performance of GAN. AnoGAN [18] applied GAN to the field of image anomaly detection, and realized image anomaly detection without abnormal samples. This method only uses normal samples to train DCGAN [19], and introduces an image distance measurement model to judge whether the samples are abnormal. After that, the proposal of Efficient-GAN [20], ALAD [21] and f-AnoGAN [22] further improved the performance of the GAN-based anomaly detection models.

On the basis of the GAN as the backbone network method, Akcay et al. proposed the GANomaly [23], which trains the autoencoder by adversarial mechanism and carries out image reconstruction operation. Skip-GANomaly [24] adds the skip connections between the encoding part and the decoding part of the generator on the basis of GANomaly to reduce information loss and enhance model performance. However, in some small target anomaly detection tasks, such as bird in CIFAR-10 dataset [25], the performance of f-AnoGAN, Skip-GANomaly and GANomaly are not satisfactory. Moreover, the current encoder-decoder networks lack stability and robustness in the training process.

In the paper, we mainly study abnormal detection tasks based on unsupervised learning and propose a Fully-Nested Encoder-decoder Framework. The main body of the anomaly detection method consists of a generating model and a distance measurement model. The generating model includes a generator and a discriminator, which detects data anomalies by a distance measurement model. In the generating model, we design a Fully-Residual Encoder-decoder Network as the generator. Taking into account the needs of different datasets for different network depths, the generator uses encoding-decoding networks of different depths to nest, which enhances the selectivity of different datasets for the best-depth encoding-decoding network. Then, we choose the discriminant network in DCGAN as the discriminator of the model. The experiments of our method on CIFAR-10 dataset demonstrate its excellent performance.

2 Proposed Method

This paper proposes a Fully-Nested Encoder-decoder Framework for anomaly detection. As shown in Fig. 1, the main body of the anomaly detection method consists of two parts, generating model and distance measurement model. Generating model is generated by learning the distribution of the normal data to reconstruct the normal samples. In the process of training generator, the model uses a classification network as discriminator to train with the adversarial mechanism. Furthermore, we introduce the distance measurement model. The distance measurement model is a distance calculation method. In the test phase, the distance between the reconstructed image and the real image is used to determine whether the test sample is abnormal.

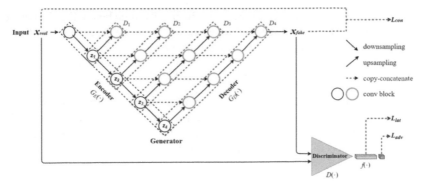

Fig. 1. Pipeline of our proposed framework for anomaly detection

2.1 Generating Model

The generating model reconstructs the image by learning the distribution of normal samples. Choosing a high-performance encoder-decoder network is very important for image reconstruction. The composition of encoder and decoder directly affects the effect of reconstructed image.

In the generating model, generator is a fully nested residual network, which can be divided into encoding part and decoding part, as shown in Fig. 2. The network can be regarded as multiple encoding and decoding networks with different scales nested. The encoder is a shared branch. The decoder decodes the deep semantic feature maps of four different scales generated by the encoder, and produces four parallel decoding branches. The generating model uses a classification network as discriminator and is trained based on the adversarial mechanism. In the whole network structure, Batch Normalization [26] and ReLU activation functions [27] are used.

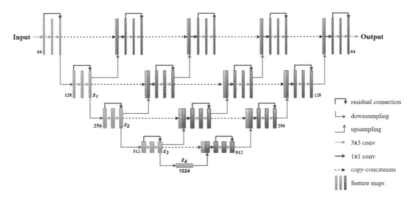

Fig. 2. The architecture of our proposed generator

The encoder is the shared part, as shown in the black dotted box in Fig. 1, represented as G_E, which is used to read in the input image x_{real} to generate the deep semantic feature map $z = (z_1, z_2, z_3, z_4)$, its specific expression is shown in Formula (1),

$$z = G_E(x_{real}) \tag{1}$$

The decoder network decodes (z_1, z_2, z_3, z_4), and produces four parallel branches: D_1, D_2, D_3 and D_4, which are expressed as G_D, as shown in the red dotted box in Fig. 1. Moreover, the internal decoding branches uses dense skip connections to connect to adjacent external decoding branches for feature fusion. Skip connections enhance the transfer of detailed information between different branches, greatly reducing information loss. The final layer of the outermost decoding branch outputs the reconstructed image x_{fake} of the generator, its specific expression is shown in Formula (2),

$$x_{fake} = G_D(z) \tag{2}$$

We add residual structure into both encoder and decoder to improve the feature expression ability and reduce the risk of overfitting. Through back propagation, the model can independently select the suitable depth network for different datasets through the nested model of four scales.

We add a classification network after the generator as the discriminator of the model, which is the classification network of DCGAN model, denoted by $D(\cdot)$. For the input image, the discriminator network identifies whether it is normal sample x_{real} or the image x_{fake} reconstructed by the generator.

The dataset is divided into the training set D_{train} and the test set D_{test}. The training set D_{train} is only composed of normal samples, and the test set D_{test} is composed of normal samples and abnormal samples. At the training phase, the model only uses normal samples to train the generator and discriminator. At the test phase, the distance between the given test images and their reconstructed images generated by the generator are calculated to determine whether they are abnormal.

2.2 Distance Measurement Model

In the test phase, we calculate the anomaly score of the test image to measure whether it is abnormal. Given test set D_{test} and input x_{test}, the anomaly score is defined as $A(x_{test})$. We use two kinds of distances to measure the difference between x_{test} and x_{fake}. First, calculate L_1 distance directly for x_{test} and x_{fake}, represented as $R(x_{test})$, which describes the detailed difference between the reconstructed image and the input image. Secondly, calculate L_2 distance directly for $f(x_{fake})$ and $f(x_{test})$, which describes the difference in semantic feature, is denoted by $L(x_{test})$. The formulas for $A(x_{test})$, $R(x_{test})$, and $L(x_{test})$ are as follows,

$$A(x_{test}) = \lambda R(x_{test}) + (1 - \lambda)L(x_{test}) \tag{3}$$

$$R(x_{test}) = ||x_{test} - x_{fake}||_1 \tag{4}$$

$$L(x_{test}) = ||f(x_{test}) - f(x_{fake})||_2 \tag{5}$$

where λ is the weight to balance the two distances $R(x_{test})$ and $L(x_{test})$. In the proposed model, λ is set to 0.9.

In order to better measure whether the input image is abnormal, it is necessary to normalize the anomaly score of each image in the test set D_{test} calculated according to Formula (3). Suppose set $A = \{A_i : A(x_{test,i}), x_{test} \in D_{test}\}$ is the set of anomaly scores of all images in the test set D_{test}. The model maps the set of anomaly scores A to the interval [0, 1] by Formula (6).

$$A'(x_{test}) = \frac{A(x_{test}) - min(A)}{max(A) - min(A)} \tag{6}$$

We set a threshold for $A'(x_{test})$. Samples with anomaly score greater than the threshold are judged to be abnormal, else normal.

2.3 Training Strategy

The loss function of the model consists of three kinds of loss functions, which are Adversarial Loss, Contextual Loss, and Latent Loss.

In order to maximize the reconstruction ability of the model during the training phase and ensure that the generator reconstructs the normal image x_{real} as realistically as possible, the discriminator should classify the normal image x_{real} and the reconstructed image x_{fake} generated by the generator as much as possible. Use cross entropy to define the Adversarial Loss, the specific expression is shown in Formula (7).

$$L_{adv} = \log(D(x_{real})) + \log(1 - D(x_{fake})) \tag{7}$$

In order to make the reconstructed image generated by the generator obey the data distribution of normal image as much as possible and make the reconstructed image x_{fake} conform to the context image, the model defines the reconstruction loss by calculating the SmoothL1 Loss [28] of the normal image and the reconstructed image, as shown in Formula (8):

$$L_{con} = S_{L1}(x_{real} - x_{fake}) \tag{8}$$

where S_{L1} represents the SmoothL1 Loss function.

$$S_{L1} = \begin{cases} 0.5x^2 & |x| < 1 \\ |x| - 0.5 & |x| \geq 1 \end{cases} \tag{9}$$

In order to pay more attention to the differences between the reconstructed image x_{fake} generated by the generator and the normal image x_{real} in the latent space, the model uses the last convolution layer of discriminator to extract the bottleneck features $f(x_{real})$ and $f(x_{fake})$, and takes the SmoothL1 loss between the two bottleneck features as the Latent Loss. The specific expression is shown in Formula (10).

$$L_{lat} = S_{L1}(f(x_{real}) - f(x_{fake})) \tag{10}$$

In the training phase, the model adopts the adversarial mechanism for training. First, fix the parameters of generator, and optimize the discriminator by maximizing the Adversarial Loss $\mathcal{L}_{adv}$. The objective function is

$$\mathcal{L}_{D-Net} = \max_{D} \mathcal{L}_{adv} \tag{11}$$

Then, fix the parameters of discriminator, and optimize the generator by the objective function:

$$\mathcal{L}_{G-Net} = \min_{G}(w_{adv}\mathcal{L}_{adv} + w_{con}\mathcal{L}_{con} + w_{lat}\mathcal{L}_{lat}) \tag{12}$$

where w_{adv}, w_{con} and w_{lat} are the weight parameters of $\mathcal{L}_{adv}$, $\mathcal{L}_{con}$ and $\mathcal{L}_{lat}$.

3 Experiments

All experiments in this paper are implemented using the Pytorch1.1.0 framework with an Intel Xeon E5-2664 v4 Gold and NVIDIA Tesla P100 GPU.

3.1 Dataset

To evaluate the proposed anomaly detection model, this paper conducted experiments on the CIFAR-10 [25] dataset.

The CIFAR-10 dataset consists of 60,000 color images, and the size of each image is 32×32. There are 10 classes of images in the CIFAR-10 dataset, each with 6000 images. When implementing anomaly detection experiments on the CIFAR-10 dataset, we regarded one class of them as abnormal class, and the other 9 classes as normal class. Specifically, we use 45000 normal images from the other 9 normal classes as normal samples for model training, and the remaining 9000 normal images in the other 9 normal classes and 6000 abnormal images in the abnormal class as test samples for model testing.

3.2 Implementation Details

Model Parameters Setting. The model is set to be trained for 15 epochs and optimized by Adam [29] with the initial learning rate $lr = 0.0002$, with a lambda decay, and momentums $\beta_1 = 0.5$, $\beta_2 = 0.999$. The weighting parameters of loss function are set to $w_{adv} = 1$, $w_{con} = 5$, $w_{lat} = 1$. The weighting parameter λ of the distance metric is empirically chosen as 0.9.

Metrics. In this paper, AUROC and AUPRC are used to assess the performance of our method. Concretely, AUROC is the area under the ROC curve (Receiver Operating Characteristic curve), which is the function plotted by the TPR (true positive rates) and FPR (false positive rates) with varying threshold values. AUPRC is the area under the PR curve (Precision Recall curve), which is the function plotted by the Precision and Recall with varying threshold values.

Results and Discussion. To demonstrate the performance of our method, we compare our method with Skip-GANomaly, GANomaly and f-AnoGAN on the CIFAR-10 dataset. The parameter settings of Skip-GANomaly and GANomaly are consistent with our experimental parameter settings in this paper, and the parameters of f-AnoGAN are the same as the settings in [22].

Table 1 and Fig. 3 show the experimental results of the CIFAR-10 dataset under the AUROC indicator, and Table 2 and Fig. 4 show the experimental results of the CIFAR-10 dataset under the AUPRC indicator. It is apparent from Table 1, Fig. 3, Table 2 and Fig. 4 that the proposed method is significantly better than the other methods in each anomaly classes of the CIFAR-10 dataset, achieving the optimal accuracy under both AUROC and AUPRC indicators. Moreover, the proposed method achieves the best performance among the three class of objects: airplane, frog, and ship, with almost 100% accuracy for anomaly detection. In addition, for the most challenging abnormal classes bird and horse in the CIFAR-10 dataset, the optimal AUROC of the other methods are 0.658 and 0.672, and the optimal AUPRC are 0.558 and 0.501, respectively. Significantly, the AUROC of abnormal classes bird and horse for the proposed method are 0.876 and 0.866, with accuracy increases of 21.8% and 19.4%, and the AUPRC are 0.818 and 0.775, with accuracy increases of 26.0% and 27.4%.

Figure 5 shows the histogram of anomaly scores of Skip-GANomaly and the proposed model on the CIFAR-10 dataset when bird class is considered as abnormal image. This can be seen that compared with Skip-GANomaly, our method can better distinguish between the normal and the abnormal, and achieves a good anomaly detection effect. Taking bird class as abnormal class, Fig. 6 illustrates the reconstruction effect of our method on objects of CIRAR-10 dataset in the test phase.

In conclusion, the anomaly detection performance of the method proposed in this paper on the CIFAR-10 dataset is better than the previous related methods.

Table 1. AUROC results for CIFAR-10 dataset

AUROC	Automobile	Bird	Deer	Cat	Frog	Airplane	Ship	Dog	Truck	Horse	Avg
f-AnoGAN	0.729	0.378	0.356	0.479	0.427	0.532	0.474	0.523	0.695	0.611	0.531
GANomaly	0.689	0.559	0.751	0.634	0.926	0.967	0.926	0.719	0.717	0.637	0.749
Skip-GANomaly	0.872	0.658	0.931	0.751	0.969	0.994	0.975	0.752	0.868	0.672	0.851
Our method	**0.943**	**0.876**	**0.978**	**0.873**	**0.994**	**0.999**	**0.993**	**0.838**	**0.911**	**0.866**	**0.931**

Table 2. AUPRC results for CIFAR-10 dataset

AUPRC	Automobile	Bird	Deer	Cat	Frog	Airplane	Ship	Dog	Truck	Horse	Avg
GANomaly	0.516	0.492	0.666	0.525	0.853	0.929	0.821	0.604	0.525	0.501	0.643
Skip-GANomaly	0.770	0.558	0.911	0.635	0.961	0.997	0.943	0.606	0.803	0.494	0.768
Our method	**0.912**	**0.818**	**0.963**	**0.825**	**0.993**	**0.999**	**0.998**	**0.707**	**0.836**	**0.775**	**0.883**

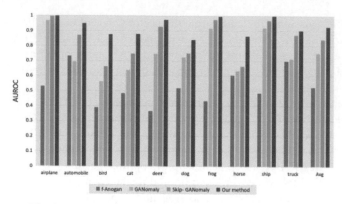

Fig. 3. Histogram of AUROC results for CIFAR-10 dataset

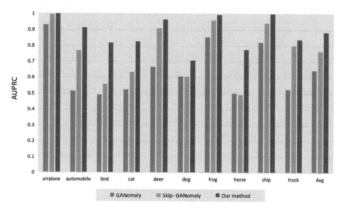

Fig. 4. Histogram of AUPRC results for CIFAR-10 dataset

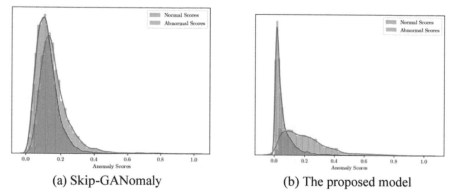

(a) Skip-GANomaly   (b) The proposed model

Fig. 5. Histograms of anomaly scores for the test data when bird is used as abnormal class.

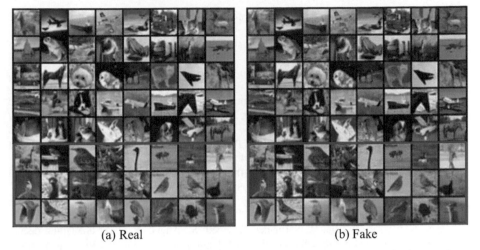

| (a) Real | (b) Fake |

Fig. 6. The reconstruction effect of our method on objects of CIRAR-10 dataset in the test phase.

4 Conclusion

In this paper, we introduce a Fully-Nested Encoder-decoder Framework for general anomaly detection within an adversarial training scheme. The generator in the proposed model is composed of a novel full-residual encoder-decoder network, which can independently select suitable depth networks for different datasets through four-scale nested models. The residual structure is added to the generator to reduce the risk of overfitting and improve the feature expression ability. We have conducted multiple comparative experiments on the CIFAR-10 dataset. And the experimental results show that the performance of the proposed method in this paper has greatly improved compared with previous related work.

Acknowledgement. This research is supported by Major Special Project (18-A02) of China Railway Construction Corporation in 2018 and Science and Technology Program (201809164CX5J6C6, 2019421315KYPT004JC006) of Xi'an.

Conflicts of Interest. The authors declare that there are no competing interests regarding the publication of this paper.

References

1. Hodge, V., Austin, J.: A survey of outlier detection methodologies. Artif. Intell. Rev. **22**(2), 85–126 (2004)
2. Niu, Z., Shi, S., Sun, J., He, X.: A survey of outlier detection methodologies and their applications. In: Deng, H., Miao, D., Lei, J., Wang, F.L. (eds.) AICI 2011. LNCS (LNAI), vol. 7002, pp. 380–387. Springer, Heidelberg (2011). https://doi.org/10.1007/978-3-642-23881-9_50

3. Chandola, V., Banerjee, A., Kumar, V.: Anomaly detection. ACM Comput. Surv. **41**(3), 1–58 (2009)
4. Ahmed, M., Naser Mahmood, A., Hu, J.: A survey of network anomaly detection techniques. J. Netw. Comput. Appl. **60**, 19–31 (2016)
5. Ma, J., Dai, Y., Hirota, K.: A survey of video-based crowd anomaly detection in dense scenes. J. Adv. Comput. Intell. Intell. Inform. **21**(2), 235–246 (2017)
6. Kwon, D., Kim, H., Kim, J., Suh, S.C., Kim, I., Kim, K.J.: A survey of deep learning-based network anomaly detection. Clust. Comput. **22**(1), 949–961 (2017). https://doi.org/10.1007/s10586-017-1117-8
7. Lecun, Y., Bottou, L., Bengio, Y., Haffner, P.: Gradient-based learning applied to document recognition. Proc. IEEE **86**(11), 2278–2324 (1998)
8. Krizhevsky, A., Sutskever, I., Hinton, G.: ImageNet classification with deep convolutional neural networks. Adv. Neural Inf. Process. Syst. **25**(2), 1097–1105 (2012)
9. Simonyan, K., Zisserman, A.: Very deep convolutional networks for large-scale image recognition. arXiv:1409.1556 (2014)
10. Szegedy, C., Liu, W., Jia, Y., Sermanet, P., Reed, S., Anguelov, D., et al.: Going deeper with convolutions. In: Proceedings of the IEEE Conference on Computer Vision and Pattern Recognition, pp. 1–9 (2015)
11. Szegedy, C., Vanhoucke, V., Ioffe, S., Shlens, J., Wojna, Z.: Rethinking the Inception architecture for computer vision. In: Proceedings of the IEEE Conference on Computer Vision and Pattern Recognition, pp. 2818–2826 (2016)
12. Szegedy, C., Ioffe, S., Vanhoucke, V., Alemi, A.: Inception-v4, inception-ResNet and the impact of residual connections on learning. In: Proceedings of 31th AAAI Conference on Artificial Intelligence, pp. 4278–4284 (2017)
13. Goodfellow, I.J., Pouget-Abadie, J., Mirza, M., Xu, B., Warde-Farley, D., Ozair, S., et al.: Generative adversarial networks. arXiv: 1406.2661 (2014)
14. Denton, E., Chintala, S., Szlam, A., Fergus, R.: Deep generative image models using a Laplacian pyramid of adversarial networks. In: Proceedings of 28th International Conference on Neural Information Processing Systems, pp. 1486–1494 (2015)
15. Mirza, M., Osindero, S.: Conditional generative adversarial nets. arXiv: 1411.1784 (2014)
16. Chen, X., Duan, Y., Houthooft, R., Schulman, J., Sutskever, I., Abbeel, P.: InfoGAN: interpretable representation learning by information maximizing generative adversarial nets. In: Proceedings of 30th International Conference on Neural Information Processing Systems, pp. 2180–2188 (2016)
17. Zhu, J., Park, T., Isola, P., Efros, A.A.: Unpaired image-to-image translation using cycle-consistent adversarial networks. In: Proceedings of the IEEE International Conference on Computer Vision, pp. 2242–2251 (2017)
18. Schlegl, T., Seeböck, P., Waldstein, S.M., Schmidt-Erfurth, U., Langs, G.: Unsupervised anomaly detection with generative adversarial networks to guide marker discovery. In: Niethammer, M., et al. (eds.) IPMI 2017. LNCS, vol. 10265, pp. 146–157. Springer, Cham (2017). https://doi.org/10.1007/978-3-319-59050-9_12
19. Radford, A., Metz, L., Chintala, S.: Unsupervised representation learning with deep convolutional generative adversarial networks. arXiv: 1511.06434 (2015)
20. Zenati, H., Foo, C.S., Lecouat, B., Manek, G., Chandrasekhar, V.R.: Efficient GAN-based anomaly detection. arXiv: 1802.06222 (2018)
21. Zenati, H., Romain, M., Foo, C.S., Lecouat, B., Chandrasekhar, V.R.: Adversarially learned anomaly detection. In: Proceedings of 2018 IEEE International Conference on Data Mining, pp. 727–736 (2018)
22. Schlegl, T., Seeböck, P., Waldstein, S., Langs, G., Schmidt-Erfurth, U.: f-AnoGAN: fast unsupervised anomaly detection with generative adversarial networks. Med. Image Anal. **54**, 30–44 (2019)

23. Akcay, S., Atapour-Abarghouei, A., Breckon, T.P.: GANomaly: semi-supervised anomaly detection via adversarial training. In: Jawahar, C.V., Li, H., Mori, G., Schindler, K. (eds.) ACCV 2018. LNCS, vol. 11363, pp. 622–637. Springer, Cham (2019). https://doi.org/10. 1007/978-3-030-20893-6_39

24. Akçay, S., Atapour-Abarghouei, A., Breckon, T.P.: Skip-GANomaly: skip connected and adversarially trained encoder-decoder anomaly detection. In: Proceedings of 2019 International Joint Conference on Neural Networks, pp. 1–8 (2019)

25. Krizhevsky, A.: Learning multiple layers of features from tiny images. Tech Report (2009)

26. Ioffe, S., Szegedy, C.: Batch normalization: accelerating deep network training by reducing internal covariate shift. In: Proceedings of International Conference on Machine Learning, pp. 448–456 (2015)

27. Glorot, X., Bordes, A., Bengio, Y.: Deep sparse rectifier neural networks. In: Proceedings of 14th International Conference on Artificial Intelligence and Statistics, pp. 315–323 (2011)

28. Girshick, R.: Fast R-CNN. In: Proceedings of the IEEE International Conference on Computer Vision, pp. 1440–1448 (2015)

29. Kingma, D.P., Ba, J.: Adam: a method for stochastic optimization. In: Proceedings of the International Conference on Learning Representations (2015)

The Method for Micro Expression Recognition Based on Improved Light-Weight CNN

Li Luo, Jianjun He$^{(\boxtimes)}$, and Huapeng Cai

School of Computer and Network Security (Oxford Brookes College),
Chengdu University of Technology, Chengdu 610059, China
66690059@qq.com

Abstract. In view of the particularity of micro expression, there are some problems, such as resource waste or parameter redundancy in micro expression training and recognition by using large convolutional neural network model alone. Therefore, a method of using lightweight model to recognize micro expression is proposed, which aims to reduce the size of model space and the number of parameters, and improve the accuracy at the same time. This method uses mini-Xception as the framework and Non-Local Net and SeNet as parallel auxiliary feature extractors to enhance feature extraction. Finally, the simulation experiments are carried out on the two public data sets of fer2013 and CK+. After a certain training cycle, the accuracy can reach 74.5% and 97.8% respectively, which slightly exceeds the commonly used classical models. It is proved that the improved lightweight model has higher accuracy, lower parameters and model size than the large convolution network model.

Keywords: Facial expression recognition · Deep learning · Convolutional network · Attention mechanism · SeNet · Non-local net · Xception

1 Introduction

Since this century, with the rapid development of deep learning [1], image recognition technology [2] has also ushered in a golden age, and various improved convolutional neural network models [3] have continuously refreshed the highest accuracy rate in history. Expression recognition includes the recognition of static images and dynamic images. Static image recognition is a recognition technology for a single picture, while dynamic image recognition is a recognition method based on video sequences. But for now, most researches still focus on the recognition of static images.

The development of facial expression recognition can be divided into three stages: from the previous manual design of feature extractors (LBP [4], LBP-TOP [5]) for recognition, and then to shallow learning (SVM [6], Adaboost [7]) Recognition, and now it is based on deep learning [8]. Each stage of development is changing its limitations and making up for deficiencies. For example, traditional hand-designed feature extractors need to rely on manually-designed feature extractors to a certain extent. Its generalization, robustness, and accuracy are slightly insufficient. Shallow learning overcomes the

© The Author(s) 2022
Z. Qian et al. (Eds.): WCNA 2021, LNEE 942, pp. 760–768, 2022.
https://doi.org/10.1007/978-981-19-2456-9_76

shortcomings of requiring excessive manual intervention, but it is accurate There are still shortcomings in terms of rate. Therefore, in this respect, with the development of computer hardware, facial expression recognition based on deep learning has gradually overcome the lack of accuracy of shallow learning.

2   LWCNN

2.1   Related Work

Nowadays, deep learning is a relatively mature field, but in order to improve the accuracy of image recognition, researchers have also begun to improve the neural network of deep learning from other aspects. For example, the activation function [9] is improved, the attention mechanism is added to the neural network [10], and the self-encoding layer [11] is added, all of which have made significant progress. This improved idea has not only made progress in image classification, but also further improved the recognition rate in facial expression recognition. Other problems that have arisen are that the formed network structure superposition leads to more and more bloated convolutional networks. Redundant parameters and complex calculations make computer resources wasted. To solve these problems, many scholars are trying find method to overcome it such as in previous studies, the literature [12] summarizes the characteristics of the past lightweight convolutional networks, which are mainly divided into three categories: lightweight convolution structure, lightweight convolution module, and lightweight convolution operation. A recent literature [13] proposed a lightweight model method based on the attention mechanism combined with a convolutional neural network. This document combines the first two features of the lightweight model together, but there are multiple computational branches in the network model. Road, this will increase the calculation cost.

   Therefore, the improvement of this paper is to cut off the calculation channels of the branches of the neural network model, retain the main calculation channels, reduce the size of the convolution kernel, and add the currently used detachable attention model as a feature auxiliary extractor to assist the main calculation channel for learning.

2.2   Improved LWCNN

The lightweight model in this paper continues to use the attention mechanism combined with the convolutional neural network method, but it strengthens the parallel extraction and fusion of features, increases the Non-Local attention mechanism (Non-Local Net) [14], and reduces the parameter amount of the main calculation channel. To put it simply, the model includes a main calculation channel and an attention mechanism calculation branch. The function of the attention mechanism calculation branch on the main calculation channel is to merge the information extracted by auxiliary features while retaining the original main channel feature information. Similar to the idea of residual structure. As shown in Fig. 1.

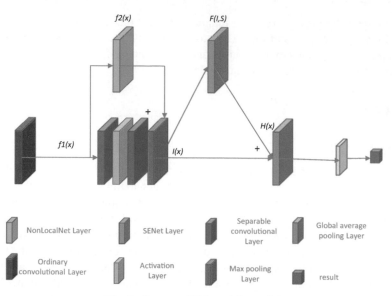

Fig. 1. Improved lightweight model.

The SeNet [15] structure is used near the output of the bottom layer, and the Non-Local Net structure is used near the input of the high layer. Through the use of Non-Local Net in the input layer to establish feature connections between the relevant features of different regions of the image, SeNet is used to merge the features of different channels before the output layer, and finally the predicted value is calculated.

The relevant calculation formula is:

$$H(x) = F_{scale}[I(x), S] + I(x) \tag{1}$$

Among them, $H(x)$ represents the network mapping after the summation, S represents the feature weight value of different channels, F_{scale} represents the weighted calculation, and $I(x)$ represents the input of the previous layer, which can be expressed as:

$$I(x) = f_1(x) + f_2(x) \tag{2}$$

$I(x)$ represents the total network mapping after summation, $f_1(x)$ represents the mapping calculated by ordinary convolution on the main road, and $f_2(x)$ represents the mapping calculated by the Non-Local Net mechanism.

The backbone calculation channel of the model uses the Xception [16] model, but the size of the convolution kernel is optimized and the amount of parameters is reduced. The hierarchy of the entire model is shown in Fig. 2.

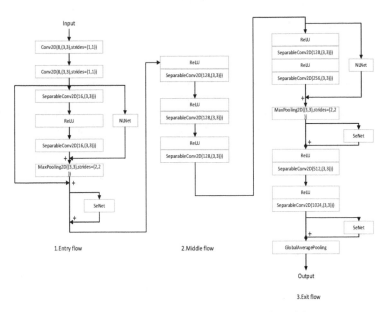

Fig. 2. Hierarchy diagram of improved model.

In Fig. 2, the model is divided into three modules. The first module is Entry flow. In its module, two ordinary convolution operations are first performed on the image, and then the output feature of the second convolution operation is copied as a residual connection and used after the MaxPooling layer is completed. Add, and then copy one to NL-Net to establish the feature correlation of the image and add it before the MaxPooling layer. Following the main channel are two separable convolutional layers with an activation layer in between. This series of operations are repeated 4 times, and the size of the convolution kernel of the separable convolution layer changes from (16, 3 * 3) to (32, 3 * 3), (64, 3 * 3), (128, 3 * 3). After processing, the feature fusion of different dimensional channels is performed through SeNet to adjust the feature value of the output channel, and finally enter the second module Middle flow.

In the second module, the activation layer is above the separable convolutional layer. This is set up according to the research of the paper, repeat the calculation 8 times and enter the third module Exit flow.

In the third module, before entering the activation layer, an input channel is copied to NL-Net for parallel processing, and then the activation layer is followed by a separable convolutional layer with a number of convolution kernels of 128 and the next separable convolution. The number of convolution kernels of the layer becomes 256.

Before the MaxPooling layer, add the output channel characteristics of the NL-Net operation and the output channel characteristics of the separable convolutional layer with the number of convolution kernels of 256, and enter the MaxPooling layer for processing, and then merge the characteristics of different dimensional channels through SeNet. Adjust the characteristic value. Finally, the final result is output through the remaining operations.

3 Experiment

3.1 Configuration

Hardware environment: CPU is AMD 5800X. The graphics card is NVIDIA RTX3060, and the memory is DDR4 3200 MHz 32 GB.

Software environment: operating system is Window10, programming software is PyCharm, python version is 3.6, keras version is 2.2.4, tensorflow version is 1.13.1.

Model parameters: the batch size is set to 64, the period is 200, the photo size of Fer2013 and CK+ is unified to 48 * 48, the initial learning rate is 0.0025, and the learning decline factor is 0.1. The loss function uses the multi-class log loss function, the activation function uses the ReLU function uniformly, and the data enhancement uses the ImageDataGenerator that comes with keras.

3.2 DataSet

At present, the FER-2013 data set contains a total of 27809 training samples, 3589 verification samples and 3859 test samples. The resolution of each sample image is 48 * 48. It contains seven categories of expressions: angry, disgusted, fearful, happy, sad, surprised and neutral. Due to the incorrect labels in this data set, some images do not even have faces, and there are still faces that are occluded. Therefore, the current recognition accuracy of human eyes is only 65% (±5%). However, because Fer2013 is more complete than the current expression data set, and is also in line with daily life scenarios, so this experiment chose FER-2013.As shown in the Table 1, this is one of the various expressions of the enlarged jpg picture of 48 * 48 pixels.

Table 1. The example of Fer2013 expression.

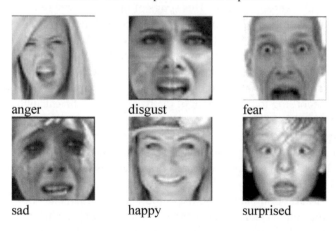

anger   disgust   fear

sad   happy   surprised

The CK+ data set is an extension of the CK data set. It is a data set specifically used for facial expression recognition research. It includes 138 participants, 593 picture sequences, and each picture sequence has an image in the last frame. Tags, including common emoticons, the number of which is consistent with the FER-2013 data set, examples are shown in Table 2.

Table 2. The example of CK+ expression.

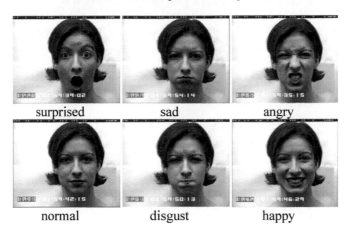

| surprised | sad | angry |

| normal | disgust | happy |

3.3 Result

The accuracy of the experimental results is shown in Fig. 3 and 4.

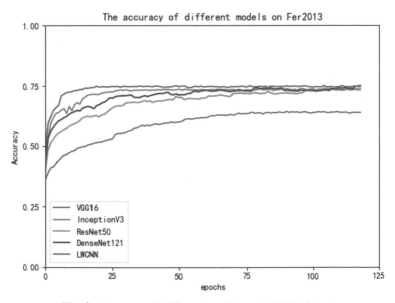

Fig. 3. Accuracy of different models on Fer2013 dataset

It can be seen from Fig. 3 that in the experiment of the Fer2013 data set, VGG16 [17] is the network model with the lowest recognition rate, with the highest accuracy rate of about 64%. The LWCNN and the other three models have similar or even higher accuracy in a certain period. In the last 30 cycles of the experimental data, the average accuracy of LWCNN was 74.5%, the average accuracy of InceptionV3 [18] was 73.3%, the average accuracy of ResNet50 [19] was 73.8%, and the average accuracy of DenseNet121 [20] was 74.1%. It can be seen that the LWCNN model has a higher accuracy rate on the Fer2013 data set than other classic models.

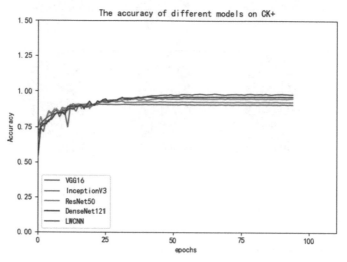

Fig. 4. Accuracy of different models on CK+ dataset

It can be seen from Fig. 4 that in the experiment of the CK+ data set, since the image training data is better than that of Fer2013, each model gradually tends to be flat and stable starting from the 50th cycle. The average accuracy of the last 30 cycles of each model is approximately: 90.4% of VGG16, 92.2% of InceptionV3, 94.6% of ResNet50, 95.9% of DenseNet121, and 97.8% of LWCNN. It can be seen that the accuracy of the LWCNN model on CK+ also exceeds the classic model.

Finally, by comparing the size and parameter amount of each model appearing in the experiment, it can be clearly seen that the improved model not only significantly reduces the size and parameter amount of the model, but also has a certain improvement in the recognition rate. as shown in Table 3.

Table 3. Comparison table of each model

Model	Size	Accuracy on Fer2013	Accuracy on CK+	Params
VGG16	528 MB	0.64	0.904	138,357,544
InceptionV3	92 MB	0.733	0.922	23,851,784
ResNet50	98 MB	0.738	0.946	25,636,712
DenseNet121	33 MB	0.741	0.959	8,062,504
LWCNN	10 MB	0.745	0.978	1,303,223

4 Conclusion

This paper mainly follows the design of the predecessors on the lightweight model, but retains the main calculation channel of the convolutional neural network, and there is no other redundant parallel calculation branch. Focus on optimizing the neural network model combined with the attention mechanism, and add the attention mechanism as a component to the main neural network model. This part draws on the idea of residual structure.

However, the current lightweight model only integrates two of the three design ideas. How to integrate the third design idea into the model requires some time of research and learning. The future will also focus on research in this direction.

References

1. Han, X.H., Xu, P., Han, S.: Theoretical overview of deep learning. Comput. Era. **06**, 107–110 (2016)
2. Zheng, Y., Li, G., Li, Y.: Survey of application of deep learning in image recognition. Comput. Eng. Appl. **55**(12), 20–36 (2019)
3. Li, B., Liu, K., Gu, J., Jiang, W.: Review of the researches on convolutional neural networks. Comput. Era **4**, 8–12+17 (2021)
4. Li, L., Yu, W.: Research on the face recognition based on improved LBP algorithm. J. Mod. Comput. **30**(17), 68–71 (2015)
5. Lu, G., Yang, C., Yang, W., Yan, J., Li, H.: Micro-expression recognition based on LBP-TOP features. J. Nanjing Univ. Posts Telecommun. (Nat. Sci. Ed.) **37**(06), 1–7 (2017)
6. Yan, Q.: Survey of Support Vector Machine Algorithms. China Information Technology and Application Academic Forum, Chengdu, Sichuan, China (2008)
7. Yi, H., Song, X., Jiang, B., Wang, D.: Selection of Training Samples for SVM Based on AdaBoost Approach. In: National Virtual Instrument Conference, Guilin, Guangxi, China (2009)
8. Lu, J.H., Zhang, S.M., Zhao, J.L.: Static face image ex pression recognition method based on deep learning. Appl. Res. Comput. **37**(4), 967–972 (2020)
9. Zhang, H., Zhang, Q., Yu, J.: Overview of the development of activation function and its nature analysis. J. Xihua Univ. (Nat. Sci. Ed.) **06**, 1–10 (2021)
10. Zhu, Z., Rao, Y., Wu, Y., Qi, J., Zhang, Y.: Research progress of attention mechanism in deep learning. J. Chin. Inf. Process. **33**(06), 1–11 (2019)

11. Yuan, F.-N., Zhang, L., Shi, J.-T., Xia, X., Li, G.: Theories and applications of auto-encoder neural networks: a literature survey. Chin. J. Comput. **42**(01), 203 230 (2019)
12. Ma, J., Zhang, Y., Ma, Z., Mao, K.: Research progress in lightweight neural network convolution design. Comput. Sci. Explor. 1–21 (2021)
13. Li, H., Li, J., Li, W.: A visual model based on attention mechanism and convolutional neural network. Acta Metrol. **42**(07), 840–845 (2021)
14. Wang, X., Girshick, R., Gupta, A., He, K.: Non-local Neural Networks. arXiv: 1711.07971 (2017)
15. Hu, J., Shen, L., Samuel, A., Gang, S., Wu, E.: Squeeze-and-excitation networks. IEEE Trans. Pattern Anal. Mach. Intell. **42**(8), 1 (2020)
16. Chollet, F.: Xception: deep learning with depth-wise separable convolutions. In: IEEE Conference on Computer Vision and Pattern Recognition (CVPR). IEEE (2017)
17. Simonyan, K., Zisserman, A.: Very deep convolutional networks for large-scale image recognition. Comput. Sci. (2014)
18. Szegedy, C., Vanhoucke, V., Ioffe, S., et al.: Rethinking the inception architecture for computer vision. In: 2016 IEEE Conference on Computer Vision and Pattern Recognition (CVPR), pp. 2818–2826 (2016)
19. He, K., Zhang, X., Ren, S., et al.: Deep residual learning for image recognition. In: IEEE Conference on Computer Vision and Pattern Recognition (2016)
20. Huang, G., Liu, Z., Laurens, V., et al.: Densely connected convolutional networks. IEEE Computer Society (2016)

Unsupervised MRI Images Denoising
via Decoupled Expression

Jiangang Zhang, Xiang Pan$^{(\boxtimes)}$, and Tianxu Lv

School of Artificial Intelligence and Computer Science, Jiangnan University, Wuxi, China
panzaochen@aliyun.com

Abstract. Magnetic Resonance Imaging (MRI) is widely adopted in medical diagnosis. Due to the spatial coding scheme, MRI image is degraded by various noise. Recently, massive methods have been applied to the MRI image denoising. However, they lack the consideration of artifacts in MRI images. In this paper, we propose an unsupervised MRI image denoising method called UEGAN based on decoupled expression. We decouple the content and noise in a noisy image using content encoders and noise encoders. We employ a noising branch to push the noise decoder only extract the noise. The cycle-consistency loss ensures that the content of the denoised results match the original images. To acquire visually realistic generations, we add an adversarial loss on denoised results. Image quality penalty helps to retain rich image details. We perform experiments on unpaired MRI images from Brainweb datesets, and achieve superior performances compared to several popular denoising approaches.

Keywords: Unsupervised · MRI image denoising · GAN · Decouple expression

1 Introduction

MRI image can provide various kinds of detailed information with respect to physical health. However, external errors, inappropriate spatial encoding, body motion etc. may jointly result in the undesirable effects of MRI and the harmful noise. Clean MRI images could increase the accuracy of computer vision assignments [1, 2], like semantic segmentation [3] and object detection [4]. In the past, *a wide variety of* denoising methods have been proposed such as filtering methods [5, 6], transform domain method [7]. Nevertheless, these methods are restricted to numerous objective factors such as undesirable texture changes caused by violation of assumptions and heavy computational overhead. Recently, deep learning methods have made great progress in the field of image denoising. These means helps to acquire the impressive effects in MRI image denosing. Due to the scarcity of medical images, researchers need to use unpaired data during training. Generative adversarial network (GAN) [8] have been found to be more competitive in image generation tasks [9, 10]. One of the solution might be directly using some unsupervised methods (DualGAN [11], CycleGAN [12]) to find the mappings between clear and noised image domains. However, these general methods often encode some irrelevant

© The Author(s) 2022
Z. Qian et al. (Eds.): WCNA 2021, LNEE 942, pp. 769–777, 2022.
https://doi.org/10.1007/978-981-19-2456-9_77

characteristics such as texture features rather than noise attributes into the generators, and thus will not produce high-quality denoised images.

Under the guidance of aforementioned theories, we present a MRI image denoising method called UEGAN which uses GAN based on decoupled expression to generate visually realistic denoised images. More specifically, we decouple the content and noise from noised images to accurately encode noise attributes into the denoising model. As shown in Fig. 1, the content encoders encode content information and the noise encoder encode noise attributes from unpaired clear and noised MRI images. However, this type of structure can't guarantee that the noise encoder encodes noise attributes only - it may encode content information as well. So we employ the nosing branch to limit the noise encoder to encode the content attributes of n. The denosing generator G_{clear} and the noising generator G_{noised} take corresponding content information on condition of noise attributes to generate denoised MRI images and noised MRI images. Based on CycleGAN [12], we apply the adversarial loss and the cycle-consistency loss as the regularizers to help the generator generate a MRI image which closes to the original image. In order to further reduce the undesirable banding artifacts introduced by G_{noised} and G_{clear}, we apply the image quality penalty into this structure. We conduct experiments on Brainweb MRI datasets, and obtain qualitative and quantitative results that are competitive with several conventional methods and a deep learning method.

2   Related Work

Since the proposed model structure makes most use of the popular denoising network and the latest technology of image disentangled representation, in this part, we briefly review the generative adversarial network, single image denoising and disentangled representation.

2.1   Generative Adversarial Network

Generative adversarial network [8] is brought forward to train generative models. Radford et al. [13] propose GANs of CNN version called DCGANs. Arjovsky et al. [14] introduce a novel loss called wasserstein into GAN at train time. Zhang et al. [15] propose Self-Attention GAN which applies attention mechanism to the field of image creation.

2.2   Disentangled Representation

Recently, there is a rapid development in learning disentangled representations, namely decoupled expression. Tran et al. [16] unravel posture and identity components for face recognition, which called DRGAN. Liu et al. [17] present an identity extraction and elimination autoencoder to disentangle identity from other characteristics. Xu et al. propose FaceShapeGene [18] which correctly disentangles the shape features of different semantic facial parts.

2.3 Single Image Denosing

Image noise has caused serious damages to image quality. There are many deep learning methods that focus on image denoising tasks. Jain et al. [19] firstly introduce Convolutional neural networks (CNN) which has a small receptive field into image denoising. Chen et al. [20] joint Euclidean and perceptual loss functions to find more edge information. According to deep image prior (DIP), present by Ulyanov et al. [21], abundant prior knowledge for image denosing already exist in the pre-train convolutional neural network.

3 Proposed Method

Inspired by GAN, single image denosing, decoupled expression, we proposed a MRI image Unsupervised denoising method called UEGAN which has well designed loss functions based on decoupled expression. This structure combines the advantages of the above three classic models and is made up of four parts: 1) content encoders E_N^{cont} for noisy image domain and E_C^{cont} for clear image domain; 2) noise encoder E^{noise}; 3) noised and clear image generator G_{noised} and G_{clear}; 4) noised and clear image discriminators D_N and D_C. Given a train sample $n \in N$ in the noised image domain and $c \in C$ in the clear image domain, the content encoders E_N^{cont} and E_C^{cont} acquire content information from corresponding samples and E^{noise} extract the noise attributes from N. Then $E^{noise}(n)$ and $E_C^{cont}(c)$ are feed into the G_{noised} to generate a noised image c^n, meanwhile, $E^{noise}(n)$ and $E_N^{cont}(n)$ are feed into the G_{clear} to generate a clear image n^c. The discriminators D_{noise} and D_{clear} differentiate the real from generated examples. The final structure is shown in Fig. 1.

3.1 Decoupling Noise and Content

It is not easy to decouple content information from a noised image because the ground truth image is not available in the unpaired setting. since the clear image c is not affected by noise, the content encoder $E_C^{cont}(c)$ is equivalent to encoding the content characteristics only. We share the weights of the last layer which existing in the $E_N^{cont}(n)$ and $E_C^{cont}(c)$ respectively to encode as much content information from noised image domain as possible.

Meanwhile, the noise encoder should only encode noise attributes. So We feed the outputs of $E^{noise}(n)$ and $E_C^{cont}(c)$ into the G_{noised} to generate c^n. Since c^n is a noised version of c, c^n does not contain any content information of n in the whole process. This nosing branch further limits the noise encoder to encode the content information of n.

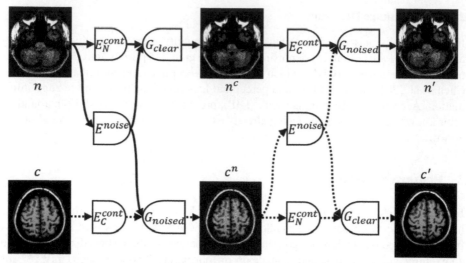

Fig. 1. The architecture of our network. *The denoising branch (bottom noising branch) is represented by full line (dotted line).* E_N^{cont} and E_C^{cont} are content encoders for noised and clear images. E^{noise} is a noise encoder. G_{noised} and G_{clear} are noised image and clear image generators. GAN losses are added to differentiate c^n from noised images, and n^c from clear images. Cycle-consistency loss is employed to n and n', c and c'. IE loss is applied to n and n^c.

3.2 Adversarial Loss

In order to acquire a cleaner output, we introduce the adversarial loss function into the content domain and the noise domain. For the clear image domain, we define the adversarial loss as L_{D_C}:

$$L_{D_C} = \mathbb{E}_{c \sim p(c)}[\log D_C(c)] + \mathbb{E}_{n \sim p(n)}[\log(1 - D_C(G_{clear}(E_N^{cont}(n), z)))]. \quad (1)$$

where $z = E^{noise}(n)$ and D_C devotes to maximize the objective function to differentiate denoised images from real clear images. In contrast, G_{clear} tries to minimize the objective function to make denoised images look similar to real samples in clear image domain. For the clear image domain, we define the loss as L_{D_N}:

$$L_{D_N} = \mathbb{E}_{n \sim p(n)}[\log D_N(n)] + \mathbb{E}_{c \sim p(c)}[\log(1 - D_N(G_{noise}(E_C^{cont}(c), z)))]. \quad (2)$$

3.3 Image Quality Penalty

We have observed that the denoised images n^c usually contains unpleasant banding artifacts in the experiment. So we introduce the Image information entropy (IE) [22] which is utilized to compute the amount of information in an image to reduce the banding artifacts. And IE loss is employed to guide the generator to produce MRI images with less noise. The loss is defined as:

$$L_{IE}(G_{clear}(z)) = \sum_{i=0, \; p(i) \neq 0}^{d} p(i) log \frac{1}{p(i)}. \quad (3)$$

where d is the range of image intensity and $p(i)$, $i = 0, 1, 2, ..., $ d is the probability distribution of the intensity of the output $G_{clear}(x)$.

3.4 Cycle-Consistency Loss

G_{clear} should have the ability to generate visually realistic and clear images after the minmax game. However, without the guidance of pairwise supervision, the denoised image n^c may rarely retains the content information of the original noised sample n. Therefore, we introduce the cycle-consistency loss to ensure that the denoised image n^c can be renoised to construct the original noised image and c^n can be translated back to the original clear image domain. The loss preserves more content information of corresponding original samples. In more detail, we define the forward translation as:

$$n^c = G_{clear}(E_N^{cont}(n),\ E^{noise}(n)),$$

$$c^n = G_{noised}(E_C^{cont}(c),\ E^{noise}(n)). \tag{4}$$

And the backward translation as:

$$n' = G_{noised}(E_C^{cont}(c^n),\ E^{noise}(n^c)),$$

$$c' = G_{clear}(E_N^{cont}(n^c),\ E^{noise}(n^c)). \tag{5}$$

We perform the loss on both domains as follows:

$$L_{cc} = \mathbb{E}_{c\sim p(c)}\big[\|c - c'\|_1\big] + \mathbb{E}_{n\sim p(n)}\big[\|n - n'\|_1\big]. \tag{6}$$

Meanwhile, we carefully balance the weights among the aforementioned losses to prevent n^c from staying too close to n.

The total objective function is a combination of all the losses from (1) to (6) with respective weights:

$$L = \lambda_{adv}L_{adv} + \lambda_{IE}L_{IE} + \lambda_{cc}L_{cc}. \tag{7}$$

3.5 Testing

In the process of testing, the noising branch is removed. Provided a test image a, E_N^{cont} and E^{noise} extract the content information and noise attributes. Then G_{clear} takes the outputs and generates the denoised image A:

$$A = G_{clear}(E_N^{cont}(a),\ E^{noise}(a)). \tag{8}$$

4   Experiments and Analysis

We compare the MRI image denoising performance between our work with non-local means (NLM) [23] and a deep learning method DIP. To analyze the performance of denoising methods quantitatively, peak signal to noise ratio (PSNR), structural similarity index (SSIM) are employed. We evaluate the proposed model on Brainweb MRI datasets. The unpaired train set with 150 MRI images consists of the following two parts:

1) Samples from the noise image domain consist of seventy-five slices, whose slice thickness is 1 mm, and additional gaussian noise standard deviation sigma is 25.
2) Samples (no additional gaussian noise) from the clear image domain consist of seventy-five slices, whose slice thickness is 1 mm.

4.1   Implementation Details

We train our network UEGAN using Pytorch 1.4.0 package on a computer with Intel i9 9300k CPU, NVIDIA RTX 2080Ti GPU, 32 Gb memory and windows10 OS with Brainweb MRI datasets. The UEGAN is optimized using the gradient-based Adam-optimizer whose hyper-parameter is set as $\beta1 = 0.5$, $\beta2 = 0.999$, Nepoch = 100000, and the learning rate of all generators is 2e−4, the learning rate of all discriminators is 1e−4. We utilize 208×176 original size with batch size of 4 for training. We experimentally set hyper-parameters: $\lambda_{adv} = 1$, $\lambda_{cc} = 10$, $\lambda_{IE} = 10$.

4.2   Experimental Results

In this section, we compare our method with NLM and DIP, and the denosing performance is shown in Fig. 2. For NLM, the denoising results is blurry and a great quantity of local details are missing. However, our visual results have the sharper texture and more structure details.

For DIP, it produces artifacts and cannot recover meaningful MRI image information. On the contrary, our model UEGAN obtains more distinct results and less noise especially on local regions.

The UEGAN achieves the best visual performance in denosing and image information recovering.

4.3   Quantitative Analysis

Two quantitative analysis strategies PSNR and SSIM are adopted to assess the effects of a traditional image denoising method NLM, a deep learning method DIP and our work UEGAN. The denoisong results of our work shows superior performance to other algorithms on above two quantitative evaluation indexes as shown in Table 1 and Table 2.

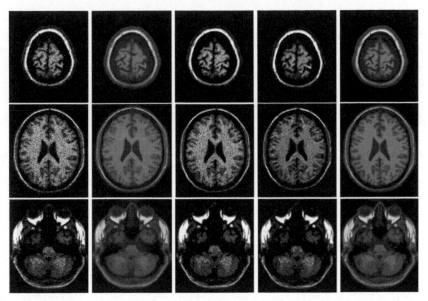

Fig. 2. Visual denoising results in three selected MRI slices. Column: noised image, NLM, DIP, the proposed method UEGAN, noise-free image in order from left to right.

Table 1. PSNR comparison

Methods	Slice 1	Slice 2	Slice 3	Average
NLM	22.4307	23.5221	22.7302	22.8943
DIP	27.5301	27.7642	26.8247	27.3730
UEGAN	28.2248	27.1062	28.1143	27.8151

Table 2. SSIM comparison

Methods	Slice 1	Slice 2	Slice 3	Average
NLM	0.6133	0.5036	0.5725	0.5631
DIP	0.5810	0.7738	0.7285	0.6944
UEGAN	0.7526	0.7310	0.7069	0.7302

5 Conclusion

In this paper, we concentrate on generating high-quality denoised MRI images with a deep-learning method which called UEGAN based on decoupled expression. We utilize the noise encoder and the content encoder to decouple the content information and noise attributes in a noisy MRI image. In order to obtain rich content characteristics from the original image, we add the adversarial loss and the cycle-consistency loss. We add the nosing branch into model so as to limit the noise encoder to encoding noise attributes as much as possible. The IE loss helps to remove the banding artifacts which consisting in the outputs of generator. After competing with several popular methods, both visual effects and quantitative results show that our work is extremely promising.

References

1. He, K., Gkioxari, G., Dollár, P., Girshick, R.: Mask R-CNN. http://arxiv.org/abs/1703.06870 (2017)
2. Huang, G., Liu, Z., van der Maaten, L., Weinberger, K.Q.: Densely connected convolutional networks. http://arxiv.org/abs/1608.06993 (2016)
3. Jégou, S., Drozdzal, M., Vázquez, D., Romero, A., Bengio, Y.: The one hundred layers tiramisu: fully convolutional DenseNets for semantic segmentation. In: CVPR Workshops, pp. 1175–1183. IEEE Computer Society (2017)
4. Kupyn, O., Budzan, V., Mykhailych, M., Mishkin, D., Matas, J.: DeblurGAN: blind motion deblurring using conditional adversarial networks. In: CVPR, pp. 8183–8192. IEEE Computer Society (2018)
5. Ma, J., Plonka, G.: Combined curvelet shrinkage and nonlinear anisotropic diffusion. IEEE Trans. Image Process. **16**, 2198–2206 (2007)
6. Starck, J.-L., Candes, E.J., Donoho, D.L.: The curvelet transform for image denoising. IEEE Trans. Image Process. **11**, 670–684 (2002)
7. Sijbers, J., den Dekker, A.J., Van Audekerke, J., Verhoye, M., Van Dyck, D.: Estimation of the noise in magnitude MRI images. Magn. Reson. Imaging **16**, 87–90 (1998)
8. Goodfellow, I.J., et al.: Generative adversarial networks. http://arxiv.org/abs/1406.2661 (2014)
9. Denton, E.L., Chintala, S., Szlam, A., Fergus, R.: Deep generative image models using a laplacian pyramid of adversarial networks. In: Cortes, C., Lawrence, N.D., Lee, D.D., Sugiyama, M., Garnett, R. (eds.) NIPS, pp. 1486–1494 (2015)
10. Van den Oord, A., Kalchbrenner, N., Kavukcuoglu, K.: Pixel recurrent neural networks. CoRR. abs/1601.06759 (2016)
11. Yi, Z., Zhang, H. (Richard), Tan, P., Gong, M.: DualGAN: unsupervised dual learning for image-to-image translation. In: ICCV, pp. 2868–2876. IEEE Computer Society (2017)
12. Zhu, J.-Y., Park, T., Isola, P., Efros, A.A.: Unpaired image-to-image translation using cycle-consistent adversarial networks. In: ICCV, pp. 2242–2251. IEEE Computer Society (2017)
13. Radford, A., Metz, L., Chintala, S.: Unsupervised representation learning with deep convolutional generative adversarial networks. http://arxiv.org/abs/1511.06434 (2015)
14. Arjovsky, M., Chintala, S., Bottou, L.: Wasserstein GAN. http://arxiv.org/abs/1701.07875 (2017)
15. Zhang, H., Goodfellow, I.J., Metaxas, D.N., Odena, A.: Self-attention generative adversarial networks. CoRR. abs/1805.08318 (2018)
16. Tran, L., Yin, X., Liu, X.: Disentangled representation learning GAN for pose-invariant face recognition. In: CVPR, pp. 1283–1292. IEEE Computer Society (2017)
17. Liu, Y., Wei, F., Shao, J., Sheng, L., Yan, J., Wang, X.: Exploring disentangled feature representation beyond face identification. In: CVPR, pp. 2080–2089. IEEE Computer Society (2018)
18. Xu, S.-Z., Huang, H.-Z., Hu, S.-M., Liu, W.: FaceShapeGene: a disentangled shape representation for flexible face image editing. CoRR. abs/1905.01920 (2019)

19. Jain, V., Seung, H.S.: Natural image denoising with convolutional networks. In: Koller, D., Schuurmans, D., Bengio, Y., and Bottou, L. (eds.) NIPS, pp. 769–776. Curran Associates, Inc. (2008)
20. Chen, X., Zhan, S., Ji, D., Xu, L., Wu, C., Li, X.: Image denoising via deep network based on edge enhancement. J. Ambient. Intell. Humaniz. Comput. **149**, 1–11 (2018). https://doi.org/10.1007/s12652-018-1036-4
21. Ulyanov, D., Vedaldi, A., Lempitsky, V.: Deep image prior. Int. J. Comput. Vis. **128**(7), 1867–1888 (2020). https://doi.org/10.1007/s11263-020-01303-4
22. Tsai, D.-Y., Lee, Y., Matsuyama, E.: Information entropy measure for evaluation of image quality. J. Digit. Imaging **21**, 338–347 (2008)
23. Manjón, J.V., Carbonell-Caballero, J., Lull, J.J., García-Martí, G., Martí-Bonmatí, L., Robles, M.: MRI denoising using non-local means. Med. Image Anal. **12**, 514–523 (2008)

A Lightweight Verification Scheme Based on Dynamic Convolution

Lihe Tang[1,2]([✉]), Weidong Yang[1,2], Qiang Gao[1,2], Rui Xu[1,2], and Rongzhi Ye[1,2]

[1] NARI Group Corporation/State Grid Electric Power Research Institute, Nanjing 211106, China
453927489@qq.com
[2] NARI Information Communication Science and Technology Co. Ltd., Nanjing 210003, China

Abstract. Since Electricity Grid Engineering involves a large number of personnel in the construction process, face recognition algorithms can be used to solve the personnel management problem. The recognition devices used in Electricity Grid Engineering are often mobile, embedded, and other lightweight devices with limited hardware performance. Although a large number of existing face recognition algorithms based on deep convolutional neural networks have high recognition accuracy, they are difficult to run in mobile devices or offline environments due to high computational complexity. In order to maintain the accuracy of face recognition while reducing the complexity of face recognition networks, a lightweight face recognition network based on Dynamic Convolution is proposed. Based on MobileNetV2, this paper introduces the Dynamic Convolution operation. It proposes a Dynamic Inverted Residuals Block, which enables the lightweight neural network to combine the feature extraction and learning ability of large neural networks to improve the recognition accuracy of the model. The experiments prove that the proposed model maintains high recognition accuracy while ensuring lightweight.

Keywords: Dynamic Convolution · Lightweight face recognition network · Electricity Grid Engineering · Recognition accuracy

1 Introduction

The construction span of Electricity Grid Engineering is large, and the construction cycle is long. The handover and acceptance of engineering construction materials cover the whole construction cycle, and there are many handover points and many units involved in the handover of materials. These factors bring certain risks for material storage and confirmation of material handover personnel. There are phenomena that material handover responsibilities are difficult to clarify and non-handover personnel take over the handover.

Z. Qian et al. (Eds.): WCNA 2021, LNEE 942, pp. 778–787, 2022.
https://doi.org/10.1007/978-981-19-2456-9_78

With the continuous promotion of power grid information reform and the increasing information security requirements, it is necessary to informatize the engineering aspects of the power grid and improve the artificial intelligence management capability of Electricity Grid Engineering. Through the automatic authentication of engineering personnel's identity, the material handover and responsibility implementation are transformed from a loose and sloppy management mode to a centralized and lean management mode, thus forming a sound and centralized, lean and efficient management system. The efficient and reliable face verification algorithm can not only improve Electricity Grid Engineering's management services but also effectively improve the information protection and information security of Electricity Grid Engineering personnel.

Currently, high-precision face verification models are mostly built based on deep convolutional neural networks that require high computational resources. These models are trained using large amounts of data, and the models are complex and have a very large number of parameters that require a large amount of computational resources. Therefore, these models are difficult to run in mobile and embedded devices, which are mostly seen in Electricity Grid Engineering scenarios. Therefore, lightweight neural networks with low memory consumption and low computational resource consumption have become a trend in current research.

Non-lightweight face verification networks have higher verification accuracy but are more computationally intensive, such as DeepFace [1], FaceNet [2], etc. This paper proposes a lightweight face verification network based on Dynamic Convolution using the lightweight neural network MobileNetV2 [3] as the baseline network to address the above problems. By learning multiple sets of convolution kernels within a single convolution operation, the feature extraction capability of the lightweight network is improved, making the lightweight neural network also achieve good face verification accuracy. At the same time, the network only enhances the baseline network MobileNetV2 with a very limited amount of computing power and meets the demand for real-time verification recognition.

2 Dynamic Convolution-Based Face Verification Network

2.1 Dynamic Convolution

Dynamic Convolution is a network substructure [4], which can be very easily embedded into other existing network structures. The core idea is to give a layer of convolution the ability to learn multiple groups of convolution kernels so that a single convolution operation has a stronger feature extraction and representation capability. At the same time, an attention mechanism [5] is introduced to learn the weights of the parameters of each group of convolutional kernels through the network so that the effective convolutional kernel parameters have high weights. The remaining parameters have low weights, prompting the model to adaptively capture the high-weight convolutional kernel parameters according to the input, improving the performance of existing convolutional neural networks, especially lightweight neural networks. By introducing Dynamic Convolution operation into the operation of the lightweight neural network, the lightweight network can extract and learn face features more efficiently. The overall structure of Dynamic Convolution is shown in Fig. 1.

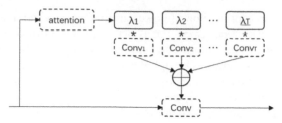

Fig. 1. The overall structure of Dynamic Convolution

The Squeeze operation is performed on the input channels in the first step. That is, feature compression is performed on the input layer to turn each two-dimensional feature channel into a real number with a global perceptual field. The resulting output features are the same as the number of input feature channels. The Squeeze operation used is global average pooling:

$$F_s(u_k) = \frac{1}{W \times H} \sum_{i=1}^{W} \sum_{j=1}^{H} u_k(i,j) \tag{1}$$

where u_k is the input feature, k is the number of channels, W and H are the width and height of an input channel feature, and F_s is the result of the Squeeze operation, which is a vector of length equal to k.

In the second step, the Excitation operation is performed on the result of the Squeeze. This operation outputs the corresponding weights of each set of convolution kernel parameters, which enables the network to adaptively select the appropriate convolution kernel for convolution according to the input features:

$$F_e(F_s, W) = \sigma(W_2 \delta(W_1 F_s)) \tag{2}$$

where W_1 and W_2 are the parameters of the fully connected layer, the dimension of W_1 is $k/r * k$, r is the scaling factor in reducing the output dimension to reduce the operational complexity of the attention mechanism, $r = 0.25$ is used in this paper. The dimension of W_2 is $T * k/r$ to obtain a vector of length T. T is the number of groups of convolution kernel parameters, δ is the nonlinear activation function ReLU [6], and σ is the softmax function. The output weight vector F_e is normalized to be in the interval [0, 1] and summed to 1 using the softmax function, and the length of F_e is T.

In the actual training of the network, in order to ensure that all groups of convolutional kernel parameters can participate in the training at the beginning of the training and avoid falling into local optimal points at the beginning of the training, the softmax used is the temperature-controlled softmax:

$$F_{e,t} = \frac{exp(F_{e,t}/\tau)}{\sum_j exp(F_{e,j}/\tau)} \tag{3}$$

where τ is the temperature parameter. It is set to a larger value at the beginning of the training and decreases until it becomes 1 as the training progresses.

In the third step, according to the weight F_e of each group of convolution kernel parameters obtained from the Excitation operation, each group of convolution kernel parameters is weighted to obtain the real convolution kernel parameters for the convolution operation:

$$W = \sum_{t=1}^{T} F_{e,t} W^t, b = \sum_{t=1}^{T} F_{e,t} b^t, s.t. 0 \le F_{e,t} \le 1, \sum_{t=1}^{T} F_{e,t} = 1 \quad (4)$$

where W^t and b^t are the t-th set of convolutional kernel parameters and $F_{e,t}$ is the tth value of the attention weight, which corresponds to the probability of using the t-th set of convolutional kernel parameters. The adaptive convolutional kernel parameters were obtained by weighting and summing each set of parameters by multiplication. The weights obtained using softmax contain a probabilistic sense, ensuring the scale stability of the obtained convolution kernel parameters. The application of the attention mechanism allows the network to automatically transform the parameters used for convolution in response to the input, greatly increasing the feature extraction and learning capability of the network.

The application of the attention mechanism allows the network to automatically transform the parameters used for convolution in response to the input, greatly increasing the feature extraction and learning capability of the network.

$$v_k = W u_k + b \quad (5)$$

where u_k is the convolutional input feature and v_k is the output feature of Dynamic Convolution. After completing the Dynamic Convolution, the features can be normalized using the common Batch Normalization layer [7] and nonlinear activation operations can be performed using nonlinear activation functions such as ReLU, PReLU [8], etc.

2.2 Bottleneck Layer Structure Design

In order to solve the degradation problem of deep neural networks and accelerate the collection of the network, MobileNetV2 introduces the Inverted Residuals Block bottleneck layer structure [3], as shown in Fig. The traditional residual structure [9] is like an hourglass with narrow middle and fat ends. Using only a small number of convolutional kernels to extract features will lead to poor feature extraction. The number of convolutional kernels in each layer of the lightweight feature extraction network is limited. Using the traditional residual structure will lead to the network not extracting enough information, resulting in a poor network. Therefore, in this paper, we use an inverted residual structure, which is like a spindle with a large middle and small ends. The feature data are first up-dimensioned by 1 * 1 Conv. The convolution operation is performed to extract the feature data, and finally down-dimensioned again by 1 * 1 Conv, which ensures the feature extraction effect and controls the parameters and computation of the network to a certain extent.

It can be seen that the backbone network part of the Inverted Residuals Block is divided into three main blocks. The first block has a similar network structure to the third block, consisting of 1×1 Conv, BN, and ReLU6. Among them, 1×1 Conv is the convolutional layer with a convolutional kernel size of 1, which is mainly used to change the number of channels of the features. BN is the Batch Normalization layer, which normalizes the features after the convolutional layer computation. reLU6 is the activation function, which gives this neuron a layered nonlinear mapping learning capability. Note that the third block of the network structure does not contain an activation function. The second network structure consists of 3 × 3 DwiseConv, BN, and ReLU6 [10], where 3 × 3 DwiseConv refers to the Depthwise Convolution with a convolutional kernel size of 3 [11] (Fig. 2).

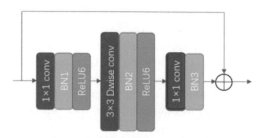

Fig. 2. The Inverted Residuals Block

Inverted Residuals Block is an important component of MobileNetV2. Using a large number of Inverted Residuals Blocks, the input information can flow sufficiently within the network so that the network has enough parameters to understand the input information and record the information characteristics. For this structure, we empirically replace the 1 * 1 convolution in the third block of the network structure with the Dynamic Convolution layer. On the one hand, such a structural replacement can already be sufficient to improve the face verification performance of MobileNetV2. On the other hand, although the increase in the number of operations of Dynamic Convolution is very limited, the increase in the number of parameters is considerable. Replacing only the last 1 * 1 convolutional layer in the Inverted Residuals Block with a Dynamic Convolution layer can also effectively prevent the size of the network model from increasing so much that it can be used in grid-side devices. The modified Inverted Residuals Block will be called Dynamic Inverted Residuals Block.

2.3   Network Architecture Design

The size of the input image used in this paper is 112 * 112. Based on MobileNetV2, the Inverted Residuals Block used in this paper is replaced with the Dynamic Inverted Residuals Block with Dynamic Convolution as described above. As shown in the Table, the network structure mainly consists of four parts. The first part obtains a feature map of size 56 * 56 with rich face feature information by a normal convolution with a kernel size of 3, step size of 2, padding of 1, and output channel number of 64. The second part consists of six Dynamic Inverted Residuals blocks in different configurations. The third part contains 3 convolution operations. First, the number of feature channels is expanded by 1 × 1 convolution, and the 7 × 7 feature map with 512 channels is output. Then, a 7 × 7 convolution layer is used to obtain 512 1 × 1 features. Finally, the feature transform is performed by a 1 × 1 convolution, and after flattening, a 512-dimensional face feature vector is obtained. The fourth part, which is a fully connected layer, implements the face classification at training time.

Table 1. Network structure

Input	op	E	C	d	r	s
$112^2 \times 3$	conv2d	0	64	N	1	2
$56^2 \times 64$	block	2	64	N	2	2
$28^2 \times 64$	block	4	128	N	3	2
$14^2 \times 128$	block	4	128	N	4	1
$14^2 \times 128$	block	4	128	N	3	2
$7^2 \times 128$	block	2	256	N	2	1
$7^2 \times 256$	block	2	256	N	1	1
$7^2 \times 256$	conv, 1×1	0	512	Y	1	1
$7^2 \times 512$	gconv, 7×7	0	512	N	1	1
$1^2 \times 512$	conv, 1×1	0	512	Y	1	1
512	fc	0	–	Y	1	-

In Table 1, op indicates the operation, e is the channel expansion factor, c is the number of output channels (number of dimensions), d indicates whether dropout is used, r indicates the number of repetitions of the block, and s is the step size (only the first repetition module has a step size of s, the rest of the repetition modules have a step size of 1).

3 Analysis of Experimental Results

3.1 Data Set and Experimental Setup

The public dataset CASIA-WebFace [12] contains 494,414 images of 10,575 individuals. In this paper, we use CASIA-WebFace as a training dataset and use the face verification database LFW [13] to check the improvement of the algorithm under different conditions. The dataset has 13233 face images containing 5749 people, containing various types of conditions such as different poses, lighting changes, and background changes. There is no overlap between the training data and the test data.

The input face image size of the model is 112 * 112. For this reason, the data needs to be processed before the face recognition network is trained. The face detection algorithm is used to derive the coordinates of face regions and key points. Based on these coordinates, the face is aligned for correction, and finally, the aligned face image is scaled to 112 * 112. The data augmentation method used contains image mirroring, panning, brightness, color, contrast, sharpness adjustment, etc. The face image is normalized before training by subtracting 127.5 from the pixels and then dividing by 128 to obtain the normalized training data finally.

The experimental hardware platform is Ubuntu 18.04 operating system and Intel Corel NVIDIA Tesla V100 graphics card. The experiments in this paper are based on PyTorch deep learning framework [14] for algorithm model training.

In this paper, all experiments are trained using a stochastic gradient descent optimizer [15]. In order to speed up the convergence and reduce the oscillation in the process of model convergence, the Momentum factor is added to the experimental training process in this paper. Its value is set to 0.9, the weight decay is set to 5e−4, the initial learning rate is set to 0.01, and the learning rate is multiplied by 0.1 at epochs of 40, 50, and 60, and the model is trained for a total of 70 epochs.

In this paper, the loss function used in the training process is the Adacos [16] adaptive scale loss function. Compared with the loss functions used for face recognition, such as CosFace [17] and ArcFace [18], Adacos does not rely on manual adjustment of the hyperparameters of the loss function to achieve good optimization results.

3.2 Analysis of Experimental Results

The comparison between the lightweight face recognition algorithm model based on Dynamic Inverted Residuals Block and the baseline network MobileNetV2 on the LFW validation set is shown in Table 2.

Table 2. The comparison on the LFW validation set

	Recognition rate	Number of model parameters	MAdds	Time/image
MobileNetV2	98.58%	3.50M	292.6M	30.57 ms
MobileNetV2 (Dynamic)	99.28%	7.54M	305.3M	34.97 ms

As can be seen from the Table, the model with the introduction of Dynamic Convolution increases from 292.6M to 305.3M in terms of computing volume, which is only a 4.34% improvement, while the accuracy of face recognition increases from 98.58% to 99.28%, with a significant 50.7% decrease in error rate. This result is not easy for such performance improvement in a long-tail task like face recognition. The number of model parameters and the forward transmission time are kept at the same order of magnitude as the baseline network, ensuring the possibility of applying the network model to all types of end devices on the grid.

In order to fully verify the performance of this algorithm model, an experimental comparison with the current mainstream algorithms in the field of face recognition was conducted, as shown in Table 3.

Table 3. The comparison with other algorithms

Method	Training set size	Accuracy
LMobileNetE [18]	3.8M	99.50%
Light CNN [19]	4M	99.33%
MobileID [20]	0.5M	97.32%
ShuffleNet [21]	0.5M	98.70%
Ours	0.5M	99.28%

LMobileNetE and Light CNN have higher recognition accuracy. Still, their training datasets are 4M and 3.8M. The number of model parameters are 12.8M and 26.7M (one order of magnitude higher than the model in this paper), which are significantly higher than the algorithms in this paper. It is significantly more difficult to migrate them to mobile platforms. Although the model size of MobileID and ShuffleNet is smaller, the performance is weak, failing to reach 99%, and the recognition accuracy is insufficient to meet the standard used by Electricity Grid Engineering. The algorithm model proposed in this paper achieves a good trade-off in recognition accuracy, operation volume, and model size by introducing Dynamic Convolution, which makes it meet both the accuracy requirements of recognition and can be efficiently applied on mobile devices.

4 Conclusion

In this paper, we propose a lightweight face recognition network based on Dynamic Convolution to address the common people management problem in Electricity Grid Engineering. The Dynamic Convolution operation not only gives richer feature extraction and learning capability to individual convolution, but also makes the convolution operation self-adaptive, so that it can automatically construct different convolution kernel parameters for different inputs for convolution. It has been proven that the lightweight face recognition network based on Dynamic Convolution proposed in this paper achieves a good balance of operational efficiency and recognition accuracy.

Acknowledgements. This work is supported by the State Grid Corporation Science and Technology Project Funded "Key technology and product design research and development of power grid data pocket book" (1400-202040410A-0-0-00).

References

1. Taigman, Y., Yang, M., Ranzato, M.A., Wolf, L.: DeepFace: closing the gap to human-level performance in face verification. In: Proceedings of the IEEE Conference on Computer Vision and Pattern Recognition, pp. 1701–1708 (2014)

2. Schroff, F., Kalenichenko, D., Philbin, J.: FaceNet: a unified embedding for face recognition and clustering. In: Proceedings of the IEEE Conference on Computer Vision and Pattern Recognition, pp. 815–823 (2015)
3. Sandler, M., Howard, A., Zhu, M., Zhmoginov, A., Chen, L.C.: MobileNetV2: inverted residuals and linear bottlenecks. In: Proceedings of the IEEE Conference on Computer Vision and Pattern Recognition, pp. 4510–4520 (2018)
4. Chen, Y., Dai, X., Liu, M., Chen, D., Yuan, L., Liu, Z.: Dynamic convolution: attention over convolution kernels. In: Proceedings of the IEEE/CVF Conference on Computer Vision and Pattern Recognition, pp. 11030–11039 (2020)
5. Hu, J., Shen, L., Sun, G.: Squeeze-and-excitation networks. In: Proceedings of the IEEE Conference on Computer Vision and Pattern Recognition, pp. 7132–7141 (2018)
6. Agarap, A.F.: Deep learning using rectified linear units (relu). arXiv preprint arXiv:1803. 08375 (2018)
7. Ioffe, S., Szegedy, C.: Batch normalization: Accelerating deep network training by reducing internal covariate shift. In: International Conference on Machine Learning, pp. 448–456. PMLR (June 2015)
8. He, K., Zhang, X., Ren, S., Sun, J.: Delving deep into rectifiers: surpassing human-level performance on ImageNet classification. In: Proceedings of the IEEE International Conference on Computer Vision, pp. 1026–1034 (2015)
9. He, K., Zhang, X., Ren, S., Sun, J.: Deep residual learning for image recognition. In: Proceedings of the IEEE Conference on Computer Vision and Pattern Recognition, pp. 770–778 (2016)
10. Sheng, T., Feng, C., Zhuo, S., Zhang, X., Shen, L., Aleksic, M.: A quantization-friendly separable convolution for MobileNets. In: 2018 1st Workshop on Energy Efficient Machine Learning and Cognitive Computing for Embedded Applications (EMC2), pp. 14–18. IEEE (March 2018)
11. Howard, A.G., et al.: MobileNets: Efficient convolutional neural networks for mobile vision applications. arXiv preprint arXiv:1704.04861 (2017)
12. Yi, D., Lei, Z., Liao, S., Li, S.Z.: Learning face representation from scratch. arXiv preprint arXiv:1411.7923 (2014)
13. Huang, G.B., Mattar, M., Berg, T., Learned-Miller, E.: Labeled faces in the wild: A database for studying face recognition in unconstrained environments. In: Workshop on Faces in 'Real-Life' Images: Detection, Alignment, and Recognition (October 2008)
14. Paszke, A., et al.: PyTorch: An imperative style, high-performance deep learning library. arXiv preprint arXiv:1912.01703 (2019)
15. Bottou, L.: Large-scale machine learning with stochastic gradient descent. In: Lechevallier, Y., Saporta, G. (eds.) Proceedings of COMPSTAT 2010, pp. 177–186. Physica-Verlag HD (2010). https://doi.org/10.1007/978-3-7908-2604-3_16
16. Zhang, X., Zhao, R., Qiao, Y., Wang, X., Li, H.: AdaCos: adaptively scaling cosine logits for effectively learning deep face representations. In: Proceedings of the IEEE/CVF Conference on Computer Vision and Pattern Recognition, pp. 10823–10832 (2019)
17. Wang, H., et al.: CosFace: large margin cosine loss for deep face recognition. In: Proceedings of the IEEE Conference on Computer Vision and Pattern Recognition, pp. 5265–5274 (2018)
18. Deng, J., Guo, J., Xue, N., Zafeiriou, S.: ArcFace: additive angular margin loss for deep face recognition. In: Proceedings of the IEEE/CVF Conference on Computer Vision and Pattern Recognition, pp. 4690–4699 (2019)
19. Wu, X., He, R., Sun, Z., Tan, T.: A light CNN for deep face representation with noisy labels. IEEE Trans. Inf. Forensics Secur. **13**(11), 2884–2896 (2018)

20. Luo, P., Zhu, Z., Liu, Z., Wang, X., Tang, X.: Face model compression by distilling knowledge from neurons. In: Proceedings of the AAAI Conference on Artificial Intelligence, vol. 30, no. 1 (March 2016)
21. Zhang, X., Zhou, X., Lin, M., Sun, J.: ShuffleNet: an extremely efficient convolutional neural network for mobile devices. In: Proceedings of the IEEE Conference on Computer Vision and Pattern Recognition, pp. 6848–6856 (2018)

Analysis of Purchasing Power Data of Department Store Members and Design of Effective Management Model

Bo Li[1], Henry L. Jiang[2], Hanhuan Yan[3], Yishan Qi[4(✉)], and Zhiwang Gan[5]

[1] School of Data Science and Intelligent Media, Communication University of China, Beijing, People's Republic of China
[2] Heider College of Business, Creighton University, Omaha, NE, U.S.A.
[3] Civil Aviation Management Institute of China, Beijing, People's Republic of China
[4] Beijing Polytechnic, Beijing, People's Republic of China
qiyishan001@qq.com
[5] Beijing Langxin Investment Consulting Co. Ltd., Beijing, People's Republic of China

Abstract. This paper focuses on the consumption situation and discounting strategies of members in large department stores. On this basis, reasonable strategies and suggestions for discounting activities in department stores are proposed. It needs to determine the consumption habits of members, customer value, life cycle, discount effect and other information. The mathematical model was established to calculate the activation rate of non-active members in the life cycle of members, that is, the possibility of transforming from inactive members to active members. Based on the actual sales data, the relationship model between the activation rate and shopping mall promotion was determined. Generally speaking, the higher the commodity price is, the higher the profit will be. IA regression model of activation rate and promotion activities is developed. The appraisal index of market promotion activities is established in terms of both discounts and integral. Lasso regression is used for variable screening, and the correlation between activation rate and the above indicators is studied.

Keywords: IA regression model · Life cycle · The activation rate · Lasso regression · Calculating statistical indicators

1 Instructions

1.1 Model Assumptions

In the era of big data, the general analysis of the basic information of members, to make a correct assessment of their consumer behavior can help managers make the right marketing decisions. In the retail industry, the purchasing power of members reflects the consumption level and consumption level. Understand the purchasing power of consumers, to do more accurate member marketing programs and improve sales. This paper studies the method and model of shopping mall members' purchasing power evaluation in big data environment.

© The Author(s) 2022
Z. Qian et al. (Eds.): WCNA 2021, LNEE 942, pp. 788–796, 2022.
https://doi.org/10.1007/978-981-19-2456-9_79

In the retail industry, the value of membership is reflected in the consistent generation of stable sales and profits for retail operators, as well as the provision of data to support the retail operators' strategy development. The retail industry will adopt various methods to attract more people to become members and to increase the loyalty of members as much as possible. At present, the development of e-commerce has led to continuous loss of mall members, which brings serious losses to retail operators. At this point, operators need to implement targeted marketing strategies to strengthen good relationship with their members. For example, merchants take a series of promotional activities for members to maintain their loyalty. Some people think the cost of maintaining old members is too high. In fact, the cost of developing new members is much higher than the cost of taking certain measures to maintain existing members. Effective ways for the brick-and-mortar retail industry include improving the member portrait depiction, enhancing the refined management of existing members, pushing products and services to them regularly, and building stable relationships with members. The mathematical model was established to calculate the activation rate of non-active members in the life cycle of members, that is, the possibility of transforming from inactive members to active members. Based on the actual sales data, the relationship model between the activation rate and shopping mall promotion was determined. Generally speaking, the higher the commodity price is, the higher the profit will be. Joint consumption is the core of shopping center operation, if the business will plan a promotion, how to plan the promotion according to the preferences of members and the joint rate of goods. The Symbols Shows as Table 1.

1.2 Model Notations

Through the three fields of document number, cash register number and consumption time, we can uniquely identify an order (receipt), which may contain several different products of different brands.

In other words, the model assumes that there are no two customers settling accounts at the same register at the same time, so there are no identical bill numbers in the system. Suppose there are only two forms of promotional activities in shopping malls. One is direct price reduction or discount, which is reflected in the difference between the amount paid by customers and the total amount of goods; the other is store points, which is reflected in the increase of membership points.

Table 1. Symbols

Symbol	Explanation of the symbols
i	Member i
t	Time t
$P_{i,t}$	Member i purchasing power at the moment t
$M_{i,t,\Delta t}$	Member spending amount at $t - \Delta t$ to t time
$Q_{i,t,\Delta t}$	Number of items purchased by members at t − Δt to t time
$C_{i,t,\Delta t}$	Number of billing receipts member at t − Δt to t time
$S_{i,t}$	The status of member i at t times

(continued)

Table 1. (*continued*)

Symbol	Explanation of the symbols
$P_{0,2}$	Activation rate of failed members
$P_{1,2}$	Activation rate of inactive members
B_l	Brand l
$X_{l,t,total}$	Total merchandise sold by brand l to members of the store in month t
$x_{l,t,discount}$	The total number of discounted items sold by brand l to members of the store in month t
$l_a, l_b, \ldots$	Indicators for evaluating promotional activities

2   Problem Analysis

2.1   Problems to be Solved

Firstly, we determine the indicators for evaluating the promotional activities of shopping malls. According to the assumptions of the model, we will establish the evaluation indicators from the aspects of discount and points. Discount indicators provide a comprehensive measure of discount strength, such as monthly discount rates, total discounts, total number of discounted products purchased by members, percentage of discounted products out of total products sold, and average discount range for each brand. In terms of points, the total number of points issued in a month and the ratio of points to the total amount (i.e., the ratio of points) can measure the generosity of points issued by the shopping mall, which is also a method to motivate members. Then we research the correlation between the activation rate and the above indicators to determine whether promotional activities have an incentive effect on the activation rate. At the same time, we study the overall impact of indicators on the activation rate through the regression model due to the large number of indicators. In this process, considering the possible strong correlation between indicators, we screen variables through Lasso regression. In order to study the associated consumption of commodities, we can analyze the association rules by integrating the commodity records of each purchase, and find out the commodity combination that customers often buy at the same time to understand the associated consumption. Finally, we give marketing recommendations based on joint consumer preferences.

2.2   The Activation Rate of Non-active Members

The following indicators can be used to evaluate the strength of promotional activities, and the activation rate of inactive members and invalid members may be related to these indicators.

Discount rate: total sales for the current month/original selling price and price of all goods sold for the current month;

Discount number: the number of items purchased by members of the store in the current month for less than the original price;

Number of discounted items: number of discounted items/total number of items purchased by members in the current month;

Discount rate of discount brands: the collection of all discount brands purchased by members in the current month calculate the number of items sold to the store members and discount items for each brand participating in the discount, the average of the discount product ratio of each store is the discount variety ratio of the discount brand in the current month. This measure measures the degree to which stores of various brands participate in discount.

Discount merchant ratio: the ratio of the number of brands that sold discounted goods in the month to the total number of brands.

Total points distributed this month: the sum of points distributed to members of the store this month.

Ratio of bonus points issued this month: total bonus points issued this month/total sales amount of members of this month. The correlation is not significant, $P_{0,2}(5)$ Is negative correlation.

Discount Variables are shown in Table 2.

Table 2. Discount variables

The discount	The quantity of discount	Percentage of discounted packages	Discount brand discount variety ratio	Percentage of discount merchants	Bonus points issued this month	The bonus points ratio will be issued this month
I_a	I_b	I_c	I_d	I_e	I_f	I_g

The analysis shows that there is a negative correlation between the rate of bonus point payment and other discount indicators. In other words, the discount is relatively low when the rate of bonus point payment is high. The incentive of points to the lost customers and inactive customers is far less than that of discount, so the high rate of point payment does not contribute to the improvement of the activation rate. On the contrary, in the months with high rate of point payment, the activation rate is low due to the low discount, which resulting in a negative correlation between the rate of point payment and the discount rate. At the same time, the positive correlation between I_f and activation rate is probably due to the higher discount rate at that time, which leads to higher sales volume and thus increases the total number of points issued, resulting in the above positive correlation. No matter which index is evaluated, the increase of discount will increase the activation rate. Among them, the discount rate is the most correlated with the activation rate of inactive members, while the number of discount pieces is the most correlated with the activation rate of invalid members, that is, the scale of discount products. Relevance matrix are shown in Table 3. We use Lasso regression (alpha $= 0.1$) to screen variables because of the strong correlation between variables. $P_{0,2}(5)$ Loasso Model Parameters are shown in Table 4. $P_{1,2}$ Loasso Model Parameters are shown in Table 5.

Relevance matrix:

$P_{0,2}(5)$ Loasso Model Parameters:
$P_{1,2}$ Loasso Model Parameters:

2.3 Conclusions

The total quantity of discount goods and the quantity of points released are significant variables selected by Lasso regression. In general, increased discount rates, increased brand coverage and size of discount events increase activation rates for inactive and inactive members. The increase in the rate of points may have a stronger incentive effect on active members, but as the rate of points increases, the discount intensity in the

Table 3. Relevance matrix of activation rate and discount rate

	$P_{1,2}(3,5)$	$P_{0,2}(5)$	The discount	The quantity of discount	Percentage of discounted packages	Discount rate at discount stores	Percentage of discount merchants	Bonus points issued this month	The bonus points ratio will be issued this month	Total sales for this month
$P_{1,2}(3,5)$	1.0000									
$P_{0,2}(5)$	-0.0639	1.0000								
The discount	0.4557	0.2826	1.0000							
The quantity of discount	0.3946	0.4234	0.9228	1.0000						
Percentage of discounted packages	0.3864	0.3561	0.9438	0.9806	1.0000					
Discount rate at discount stores	0.3874	0.3469	0.9581	0.9405	0.9556	1.0000				
Percentage of discount merchants	0.1339	0.3292	0.7117	0.6502	0.6943	0.7856	1.0000			
Bonus points issued this month	0.2845	0.2447	0.5304	0.4599	0.4166	0.5711	0.4094	1.0000		
The bonus points ratio will be issued this month	0.0354	-0.4071	-0.2749	-0.4656	-0.4678	-0.2724	-0.1859	0.5045	1.0000	
Total sales for this month	0.2562	0.6295	0.8397	0.9051	0.8510	0.8661	0.6327	0.6268	-0.3383	1.0000

Table4. Lasso model parameter table of churn customer activation rate

The discount	The quantity of discount	Percentage of discounted packages	Discount brand discount variety ratio	Percentage of discount merchants	Bonus points issued this month	The bonus points ratio will be issued this month	Intercept	R^2
X_a	X_b	X_c	X_d	X_e	X_f	X_g	b	
0	$4.14 * 10^{-07}$	0	0	0	$9.00 * 10^{-11}$	0	0.0533	0.1825

Table 5. Lasso model parameter table for inactive customer activation rate

The discount	The quantity of discount	Percentage of discounted packages	Discount brand discount variety ratio	Percentage of discount merchants	Bonus points issued this month	The bonus points ratio will be issued this month	Intercept	R^2
X_a	X_b	X_c	X_d	X_e	X_f	X_g	b	
0	$5.94 * 10^{-06}$	0	0	0	$3.11 * 10^{-10}$	0	0.1064	0.2643

shopping mall is generally weak, so the rate of points has no incentive effect on inactive members and invalid members.

2.4 The Associated Consumption of Commodities

The associated consumption of commodities is an important phenomenon in the process of business operation. Paying attention to customers' preference in the process of consumption is beneficial to the planning of promotional activities. Establishment of association rule model: In order to analyze the associated consumption of commodities, the following definitions are given: Commodity purchase data set $T = \{T_1, T_2, ..., T_i, T_n\}$, A transaction that represents the purchase of an item by a customer, $T_i = \{I_1, I_2, ..., I_i, I_n\}$, represents an item in the T_i consumption transaction. Commodity group: let I be the set of all items in the commodity purchase data set T, and any subset of I is called the commodity group in T.

Support count: the support count of item group X is the number of times item group X appears in item purchase data set T.

Support degree: the support degree of commodity group X is the percentage of commodity group X in the commodity purchase data set T, which describes the probability of a commodity combination appearing in all commodity consumption records. The support degree of commodity group X is expressed as

$$support(X) = |\{occurency(X)|X \subseteq T\}|/occurency(T)$$

Frequent commodity group: the commodity group whose support degree is not less than the given minimum support degree is regarded as frequent commodity group. Confidence is the percentage of the goods purchase data set T that contains both goods group X and goods group Y. Write rules of $X \rightarrow Y$ confidence for the $\text{conf}(x \Rightarrow y)$

$$\text{conf}(x \Rightarrow y) = (support(X \cup Y))/(support(X))$$

Confidence means that for the association rule X Y, the higher the confidence is, the greater the probability that both X and Y of commodity group appear in the consumption. In order to mine related commodity groups that meet the minimum support degree and the minimum confidence degree, it can be divided into the following two steps:

Step 1: find out the frequent commodity group set that meets all conditions in all data of commodity purchase data set T.
Step 2: generate association rules with frequent commodity groups, that is, find the rules satisfying the minimum confidence degree from frequent commodity groups obtained in the previous step. No less than the minimum confidence, the association rules.

3 Examples and Illustration

The tables and data above shows that solution of association rules: the consumption records of members are processed, the records belonging to the same consumption are identified through the document number and time, and the commodities purchased at the same time in each consumption process are summarized and recorded in the form of code, forming the data set of commodity purchase. We found that these sets of all goods are cosmetics by observing the commodity group, which represent this category is more suitable for joint consumption. At the same time, the cosmetics is also the main item sold by the store, because the volume and sales of cosmetics are more than the volume and sales of any other category. Thus we speculate that cosmetics sales should be the main business of the store. Secondly, we found that all associated commodity combinations belong to the same brand, and customers tend to purchase multiple commodity combinations of the same brand at the same time when purchasing commodities. A common pattern is to buy sets of skincare products (for examples, a day cream with a night cream, a moisturiser with a cream, a softener with a lotion) or sets of bottom makeup products at the same time.

When only the minimum confidence in the model is changed, the generated frequent commodity portfolio and its support degree will not change. Association rules and confidence are not changed, but quantitative filtering is performed. When only the minimum support degree in the model is changed, the generated frequent commodities and their support degree will not change, but will be screened quantitatively. Association rules and confidence will change greatly. Therefore, it can be explained that the model has good robustness, and the changing of minimum support and minimum confidence will lead to frequent commodity combination and quantitative screening of association rules, while the change of the content of relatively important association rules is less. The programs explanations are shown as following:

process.py
The following packages are used:
  Pandas
  Numpy
  Matplotlib
Problem solved:
  draw histograms and data
calculate the monthly average purchasing power index and calculate the deciles of the monthly purchasing power index
calculate the optimal length of the inactive period and the active period
calculate the evaluation index of promotion
regression.py
The following packages are used:
Sklearn –Machine learning toolkit, which USES the Lasso regression and ridge regression in sklear.linear_mode
Problems solved:
ridge regression and Lasso regression, plotting, calculating statistical indicators
getRules.py
The following packages are used:
Pandas
Problems solved:
identify the code of products purchased at the same time in the same consumption behavior from the purchase record table and form the product purchase data set. Then, the commodity combination and support degree of frequent purchases in the commodity purchase data set are calculated, and then the commodity rules and confidence degree of joint purchases are calculated with frequent commodity combination.

4   Conclusions

Conclusions for shopping mall promotions: Based on the above conclusions, we give Suggestions for shopping mall promotional activities.

Conclusion 1: the main target of promotional activities should be cosmetics, and it is better to launch promotional activities for products of the same brand.
Conclusion 2: with reference to the groups of commodity combinations with associated consumption relationships, preferential package can be launched to stimulate purchase, or the sales volume of associated consumption commodities can be increased by offering discounts to actively purchased commodities.

Acknowledgements. This work was supported by Horizontal project (Grant NO. HG19012) and Humanities and Social Sciences planning fund project of the Ministry of Education (Grant NO. 20YJAZH085).

References

1. Alan, K.: The triangular purchasing power parity hypothesis: a comment. World Econ. **44**(3), 837–848 (2020)
2. Yoon, J.C., Min, D.H., Jei, S.Y.: Purchasing power parity vs. uncovered interest rate parity for NAFTA countries: the value of incorporating time-varying parameter model. Econ. Model. **90**, 494–500 (2020)
3. Jacobo, A.D., Sosvilla-Rivero, S.: An empirical examination of purchasing power parity: Argentina 1810–2016. Int. J. Finan. Econ. **26**(2), 2064–2073 (2020)
4. Kirchberger, M., Wouters, M., Anderson, J.C.: How technology-based startups can use customer value propositions to gain pilot customers. J. Bus. Bus. Mark. **27**(4), 353–374 (2020)

Analysis and FP-Growth Algorithm Design on Discount Data of Department Store Members

Jianghong Xu[1], Cherry Jiang[2], Yezun Qu[3], Wenting Zhong[1(✉)], and Zhiwang Gan[4]

[1] Beijing Polytechnic, Beijing, People's Republic of China
306191781@qq.com
[2] College of Business, Stony Brook University SUNY, Stony Brook, NY, U.S.A.
[3] Central University of Finance and Economics, Beijing, People's Republic of China
[4] Beijing Langxin Investment Consulting Co. Ltd., Beijing, People's Republic of China

Abstract. This paper mainly studies the discount data of department store members. The researsh shows that the total supply of discounted goods and the number of reward points issued have the most significant relationship with customer activation rate, the increase of discount rate and coverage scale would increase the activation rate of inactive members and invalid members. The increase of the score rate may have a stronger incentive effect on active members, but it has no obvious incentive effect on inactive and ineffective members. In addition, by integrating the commodity records of each purchase, and analyzing association rules, commodity combinations with associated consumption relationships are obtained, and the analysis model of commodity portfolio association rules is established. This paper is mainly based on the data of the member information, the sale water meter, the member consumption detailed list, the merchandise information table, through the data processing and analysis, rejects the abnormal data, prepares for the following processing. By analyzing the characteristics of member consumption and the difference between member and non-member consumption, we can provide marketing suggestions for the store manager FP-growth Algorithm is designed to evaluate the purchasing power of members based on their gender, length of membership, age and consumption frequency, and each parameter of the model is explained, so as to improve the management level of the shopping mall. On this basis, Suggestions for promotional activities in shopping malls are given.

Keywords: The score rate · FP-growth algorithm · The data dictionary · RMF model · The changing trend

1 Instructions

1.1 Question Background

The retail industry will adopt various ways to attract more consumers to become members, and try to improve the loyalty of members. At present, the development of e-commerce leads to the continuous loss of shopping mall members, which brings great losses to retail operators. At this time, operators need to implement targeted marketing

Z. Qian et al. (Eds.): WCNA 2021, LNEE 942, pp. 797–804, 2022.
https://doi.org/10.1007/978-981-19-2456-9_80

strategies to strengthen good relations with members. For example, businesses take a series of sales promotion for their members to maintain their loyalty.

Some people think that the cost of maintaining old members is too high. In fact, the investment of developing new members is much higher than taking certain measures to maintain existing members. Improve members' image, strengthen the detailed management of existing members, regularly push products and services to them, and establish a stable relationship with members is an effective way for the better development of the real retail industry.

1.2 Question Related Information

We obtain the data of the member related information from a large department store: the member information data, the sales flow table in recent years, the member consumption detailed list, the commodity information table and the data dictionary. Generally speaking, the higher the commodity price, the higher the profit. We will focus on analysing the consumption characteristics of the members of the shopping mall, compare the differences between members and non-members, and explain the value that members bring to the shopping mall. Establish a mathematical model to describe each member's purchasing power according to their consumption situation, so as to identify the value of each member. As an important resource in the retail industry, members have a life cycle. During the process from joining members to quitting, members' status, such as active or inactive will change constantly.

Therefore, it's necessary to try to establish a mathematical model of member life cycle and state division in a certain time window, so that the store managers can manage the members more effectively.

2 Model Hypothesis and Symbolic Description

2.1 Model Hypothesis

Through the data, cash register number and transaction time, an order ticket can be determined only, the small ticket may contain several different commodities of different brands. In other words, it is assumed that there are no two customers who settle accounts at the same time or at the same cash register and record the same document number in the system. It is assumed that there are only two forms of sales promotion in the market, one is direct price reduction or discount, which represents the difference between the amount paid by customers and the original price of goods, and the other is market reward points, which represents the increase of member points.

2.2 Problem Analysis

The first step, we compare the differences between members and non-members in terms of purchases quantity, purchase amount, return quantity and return amount. For some of the members from other branches, we also analyzed the differences between our members and the members from other branches in terms of purchase and return behavior.

At the same time, we analyze the different groups' consumption habits distribution according to consumption data, which can more intuitively see the differences between member groups and other customer groups in customer consumption habits and their customer value. Based on the quarterly consumption amount of members, we will establish a mathematical model to reflect how consumption amount and time affect members' purchasing power. According to purchasing power and RMF model, we can observe the change of customer value.

For the purchasing situation of members and non-members, we choose the average unit price, the total number of purchases and the total amount of purchases as three indicators. We note that the dataset provides the return records of members. We believe that returns will have a extremely important impact on the sales and personnel scheduling of the mall. The customers of the group with less quantity and amount of returns are relatively mature, resulting in relatively small profits loss and personnel loss to the shopping mall. For the returns of members and non-members, we choose the average unit price of returned goods, the total number of returned goods and the total amount of returned goods as three indicators. Most of the members are members of our store and some are members of other branches. The members from other branches also enjoy the rights of ordinary members, such as members' discounts and credits, but they are not the object of membership management in our store. Therefore, we conducted the same analysis on the purchasing and returning situation of our members and other branch members.

2.3 The Construction Model of Purchasing Power

According to the members' consumption of the characterization of every member of the purchasing power, to recognize the value of membership. According to the theory of RMF model, the RMF measure of customer value, that is, R, represents retention rate, M represents the amount of consumption, and F represents consumption times. We believe that the consumption amount of M in the RMF model, indicating the purchasing power of members. The more the amount of consumption, the higher the purchasing power. Furthermore, the shorter the last consumption time and the current time interval, the higher the value of customers. In RMF model, M represents the sum of customer's historical consumption amount, which increases over time. We believe that members' purchasing power will change over time. Considering the recent consumption amount and historical consumption amount of members, the changing trend of purchasing power can be explained.

We set the purchasing power of Member i at t Quarter as $P_{i,t}$:

$$P_{i,t} = M_{i,t} \times \frac{2}{5} + P_{i,t-1} \times \frac{3}{5}, t = 1, 2, 3, \ldots$$

$M_{i,t}$ is the Consumption at Current Quarter, $P_{i,t-1}$ is the purchasing power of the previous quarter, so $P_{i,0} = 0$.

In summary, the criteria given by the model for judging membership status are as follows:

Members are considered active members, Members have consumption records within three months, there is no consumption record in three months, but there is consumption record in five months, that is to say, it is considered to be an inactive member. Members who have no consumption records in five months are invalid members.

2.4 Trend Analysis of Purchasing Power

From the analyze, it can be seen that the purchasing power of the top 10% customers with the highest purchasing power index has been rising in the nearly 2 years, and the gap with the purchasing power of the other 90% customers has also been widening. The purchasing power of the remaining 90% of the customer base has been declining over the past two years. From this, we can see that the shopping mall's customer group presents a long tail phenomenon, 90% of the customers' consumption capacity is constantly declining, purchase intention and gradually declining. The 10% customer group with the strongest purchasing power has a more and more significant share in the development and profit of the shopping mall, and their purchasing power and willingness to buy are also increasing. This part of the customers have higher customer value.

2.5 Division of Membership Status

Members' life cycle can be defined as: membership (development) - > active period - > inactive period - > invalidation (withdrawal) period. In our opinion, how to judge that members enter the inactive period after they do not buy commodities for a period of time. And how to determine whether a member does not buy goods for a longer period of time, that is to enter the expiration period, which is very critical.

Set the status of Member i at t time as $S_{i,t}$

Let $S_{i,t}$ be the state of member i at t time.

The state $S_{i,t} = -1$ means that customer i is invalid at time t.

The state $S_{i,t} = 1$ means that customer i is inactive at t time.

The state $S_{i,t} = 2$ means that customer i is active at time t.

Let M be the symbol of the amount, Q the symbol of the quantity, and C the symbol of the number of purchases to the shopping mall. For the development state, we think that generally speaking, it can be classified as inactive state, that is, the activity of new members is not enough to enter active state. Generally speaking, we can assume that in the recent Δt_1 period, member i went to the mall more than c_1 *times*; A total payment exceeding m_1 or a purchase exceeding q_1 is considered to be active.

However, in the recent Δt_2 period, membership i goes to the mall more than c_2 times, or pays more than m_2 yuan altogether, or purchases more than q_2 goods, which is considered inactive; in other cases, membership is invalid and withdraws.

So as:

$$S_{i,t} = \begin{cases} 2, & M_{i,t,\Delta t_1} \geq m_1 \vee Q_{i,t,\Delta t_1} \geq q_1 \vee C_{i,t,\Delta t_1} \geq c_1 \\ 1, & (M_{i,t,\Delta t_1} < m_1 \wedge Q_{i,t,\Delta t_1} < q_1 \wedge \quad (M_{i,t,\Delta t_2} \geq m_2 \vee Q_{i,t,\Delta t_2} \geq q_2 \\ & \quad \wedge C_{i,t,\Delta t_1} < c_1) \quad \vee C_{i,t,\Delta t_2} \geq c_2) \\ 0, & other \end{cases}$$

Currently, members' consumption data totals three years, of which the first year is incomplete. For members' life cycle, the time of data is not long enough to support the simultaneous calculation of so many thresholds, so we simplify the model appropriately. We believe that in the recent Δt_1 period, Members i purchased at least one commodity, which is considered active; If the member has not purchased goods in the latest Δt_1 period, but has purchased at least one item in the latest Δt_2 period, the member is considered inactive. In the recent Δt_2 period, members have not purchased goods, they think that the membership has lost.

So the simplified model is

$$S_{i,t} = \begin{cases} 2, & C_{i,t,\Delta t_1} \geq 1 \\ 1, & C_{i,t,\Delta t_1} = 0 \wedge C_{i,t,\Delta t_2} \geq 1 \\ 0, & other \end{cases}$$

So now we have to determine the size of Δt_1 and Δt_2. The activation rate $P_{0,2}(t, \Delta t_2, i)$ of inactive members is defined as: at time t, member i has not purchased any products during the Δt_2 period before time t. But in the time from t to t + 1, the probability of purchasing at least one product.

The activation rate $P_{1,2}(t, \Delta t_1, \Delta t_2, i)$ of inactive members is defined as: at time t, member i has not purchased any products during the Δt_1 period before time t. At least one product has been purchased from Δt_2 to Δt_1, but the probability of purchasing at least one product from time t to time t + 1.

We assume that $P_{0,2}$ and $P_{1,2}$ are independent with the members and the current time, that is, $P_{0,2}(t, \Delta t_2, i) = P_{0,2}(\Delta t_2)$, $P_{1,2}(t, \Delta t_1, \Delta t_2, i) = P_{1,2}(\Delta t_1, \Delta t_2)$. And the probability is expressed by statistical frequency, so the following conclusions are drawn:

Conclusion 1: When the activation rate $P_{0,2}(\Delta t_2^*)$ is the minimum of $P_{0,2}(\Delta t_2)$, the Δt_2^* is the inactive period of members. That is, the longest time for members to remain inactive;

The reason is that after the Δt_2^* period, if the member does not buy, the possibility of the member resuming shopping is the lowest in next month, that is to say, the member most likely to become an invalid member. Therefore, any member who has not purchased goods in the recent Δt_2^* period is considered to be transformed from inactive state to invalid state.

Conclusion 2: When the activation rate $P_{1,2}(\Delta t_1^*, \Delta t_2^*)$ is the minimum of $P_{1,2}(\Delta t_1, \Delta t_2^*)$, the Δt_1^* is the active period of members. Similar to conclusion 1, in such a long period of time as Δt_1^* members did not shop (even if they did during the period from Δt_2^* to Δt_1^*), they were least likely to resume shopping and most likely to shift from active to inactive. First, we calculate Δt_2^*. For $\Delta t_2 = j, j$ in 2, 3, 4, ..., 11, 12 For any number in 11,12, for a month in the sample a. Calculate the number of members x_1 who did not buy in the first j months of this month, then calculate the number of customers x_2 in the next month of x_1, and record the activation rate of invalid members in the month a under the condition $\Delta t_2 = j$ that $P_{0,2}(j, a) = \frac{x_2}{x_1}$.

2.6 Sensitivity Analysis

For active period Δt_1^* and inactive period Δt_2^*, we choose 18 consecutive months as test samples to calculate Δt_1^* and Δt_2^*, in the 24-month sample length from these nearly

2 years. To evaluate the robustness of active and inactive periods. From the 24-month sample period, seven 18-month test samples can be generated. To evaluate the robustness of active and inactive periods. From the 24-month sample period, seven 18-month test samples can be generated.

3   Customer Life Cycle Model

In fact, customer activity is not constant. According to the activation rate of customers, we can get the model of customer's transition between inactive, active and loss states, that is, customer life cycle model. For each user, the probability of losing, inactive and active users in the t month is $P_{t,0}$, $P_{t,1}$, $P_{t,2}$, and $P_{t,0} + P_{t,1} + P_{t,2} = 1$. For new users, $P_{t,0} = 0, P_{t,1} = 0, P_{t,2} = 1$.

In t + 1 month, the probability that the user belongs to three types of users is respectively.

$$P_{t+1,0} = k_{0,0} P_{t,0} + k_{1,0} P_{t,1}$$

$$P_{t+1,1} = k_{1,1} P_{t,1} + k_{2,1} P_{t,2}$$

$$P_{t+1,2} = k_{0,2} P_{t,0} + k_{1,2} P_{t,1} + k_{2,2} P_{t,2},$$

$$k_{0,0} = 0.0401, k_{1,0} = 0.2807,$$

$$k_{1,1} = 0.6346, k_{2,1} = 0.2279,$$

$$k_{0,2} = 0.0509, k_{1,2} = 0.0847,$$

$$k_{2,2} = 0.7721$$

Based on the conclusion of RMF model, we find that the purchasing power of the first 10% of customers increases gradually, and their purchasing willingness becomes stronger and stronger. We believe that this part of customers have the highest customer value, so establish membership status partition model and membership life cycle model. Based on the purchasing situation of members, members can be divided into active members, inactive members and lost members. Members can switch between these three states, and the probability of conversion is activation rate.

By calculating the activation rate under different states, we find that the boundaries between the three states are that the members with consumption are active members in three months, those without consumption in three months but with consumption in five months are inactive members, and those without consumption records in five months are invalid members. Finally, we calculate the probability of transition among the three states based on historical data.

4   Model Evaluation, Improvement and Extension

Combines with the descriptive statistics of consumption habits distribution, we can roughly estimate the consumption habits of the overall customers. By establishing a purchasing power model and combining with the RMF model, the changes in customer value can be observed. RMF model can make up for the deficiency of single purchasing power index and reflect customer value more comprehensively. The relationship between member activation rate and marketing activities is studied by establishing membership status partition model and membership life cycle model. Based on the data analysis of membership status, the differences of purchase time among active members, inactive members and lost members were clarfied.

The method of determining this boundary is proved by mathematical method, which is justified by mathematical method besides traditional marketing theory. An analytical model for association rules of commodity portfolio is establish., which not only reveals the relationship between commodities, but also shows the strength of the relationship between commodities through the confidence index, which has a strong explanability. At the same time, automatic mining is more efficient and more applicable than manual mining. FP-growth algorithm is used to analyze the association rules of the problem. Compared with the traditional Apriori algorithm for computing Association rules, FP-growth algorithm has obvious advantages in the efficiency and accuracy of large-scale data processing. However, it is worth noting that FP-growth algorithm can only be used to calculate historical data, but can not operate on incremental data alone. Therefore, in the actual application process, the specific needs of market analysis may not be met, and the storage space occupied is also very large.

The purchasing power and RMF model can be further deepened, and the purchasing power can be internalized as an index in the RMF model. Clustering according to the members' retention rate and consumption frequency, dividing different customer groups. and comparing the customer value of each group, we can get more detailed customer division and clearer customer value. By using member life cycle models of the problem, we can not only monitor the member's activity, but also promote it further. Predicting the state transition of members' activity is great reference value to enterprise customer management and marketing decision-making. we assume that there are only two ways of discount: price reduction and membership points. At the same time, we are not clear about the use of membership points. If there is more detailed discount information, we can refine the relationship between the activation rate and discount activities. then the specific discount strategy will also have a clearer direction.

This paper is mainly based on the data of the member information, the sale water meter, the member consumption detailed list, the merchandise information table, through the data processing and analysis, rejects the abnormal data, prepares for the following processing. By analyzing the characteristics of member consumption and the difference between member and non-member consumption, we can provide marketing suggestions for the store manager FP-growth Algorithm is designed to evaluate the purchasing power of members based on their gender, length of membership, age and consumption frequency, and each parameter of the model is explained, so as to improve the management level of the shopping mall.

Acknowledgements. This work was supported by Beijing Municipal Commission of Education Foundation (Grant NO. AAEA2020005) and Beijing Polytechnic project (Grant NO. YZK2016012).

References

1. Junke, Z., et al.: Sphalerite as a record of metallogenic information using multivariate statistical analysis: constraints from trace element geochemistry. J. Geochem. Explor. **23**, 106883 (2021)
2. Kong, X., et al.: Patterns of near-crash events in a naturalistic driving dataset: applying rules mining. Accident Anal. Prevent. **161**, 106346 (2021)
3. Zhongfei, Z., et al.: Study on the scheme-design framework and service-business case of product service system oriented by customer value[J]. IET Collaborat. Intell. Manuf. **2**(3), 132–141 (2020)
4. Yoon, J.C., Min, D.H., Jei, S.Y.: Purchasing power parity vs. uncovered interest rate parity for NAFTA countries: The value of incorporating time-varying parameter model. Econ. Model. **90**, 494–500 (2020)
5. Zhou, J.: Customer segmentation by web content mining. J. Retail. Consum. Serv. **61**, 102588 (2021)
6. Yi, Z., Hao, X.: Customer stratification theory and value evaluation—analysis based on improved RFM model. J. Intell. Fuzzy Syst. **40**(3), 4155–4167 (2021)
7. Wu J., et al.: An Empirical study on customer segmentation by purchase behaviors using a RFM model and -means algorithm. Mathematical Problems in Engineering (2020)
8. Wei, J.T., et al.: Using a combination of RFM model and cluster analysis to analyze customers' values of a veterinary hospital. IAENG Int. J. Comput. Sci. **47**(3), 1–7 (2020)

Analysis of Subway Braking Performance Based on Fuzzy Comprehensive Evaluation Method

Hua Peng[1(✉)] and Yixin He[2]

[1] School of Mechanical and Automotive Engineering, Qingdao University of Technology, Qingdao, China
qdypenghua01@sina.com
[2] Institute of Rail Transit, Tongji University, Shanghai, China

Abstract. In the process of subway operation, the braking system is a complex system, and its state detection is for high data accuracy and state positioning accuracy According to the structure of the braking system and the principle of the braking method, the basic braking performance parameters of the system are analyzed, combined with the abnormal state of the subway brake cylinder pressure data, the braking process is divided into two stages: brake cylinder pressure establishment and peak stability. And define the six characteristic parameters of 90% brake cylinder pressure establishment time, special slope period time, stable pressure value, stable pressure standard deviation, maximum value and minimum value. Aiming at the braking process performance, a data mining theory is proposed, and software based on the fuzzy comprehensive evaluation method is written to analyze the deterioration of the braking performance of subway vehicles. The actual on-board data is used as an example to verify the reliability of the theory.

Keywords: Subway · Braking performance · Fuzzy comprehensive evaluation

1 Introduction

Urban rail transit has outstanding benefits such as large capacity, fast speed, punctuality, high economy, low environmental pollution, safety, and low energy consumption. Therefore, it has become an inevitable choice for large cities to deal with traffic congestion. As the urbanization of China continues to deepen, it is believed that more and more small and medium-sized cities will also start the era of urban rail transit [1]. For a long time in the future, China's urban rail transit will be in its golden period of development, so there is a huge market for research on urban rail transit train-related technologies. At present, for the entire braking system, the engineering has proposed an analysis method for the performance of the system, but the analysis method for the performance of the subway vehicle braking system is still relatively rough [2]. However, with the degradation of the system performance during the service time of the train and the occurrence of failures, the state of the brake system of the train changes with time and environmental changes, which in turn affects the execution of the brake command by the brake system [3]. These

© The Author(s) 2022
Z. Qian et al. (Eds.): WCNA 2021, LNEE 942, pp. 805–812, 2022.
https://doi.org/10.1007/978-981-19-2456-9_81

reflect the changes in the performance indicators of the braking system in the dynamic working state, as well as the description of the changes in the working state of the vehicle braking system. Since no clear and systematic analysis and evaluation methods are given, there is an urgent need to study the analysis of braking performance [4] (Fig. 1).

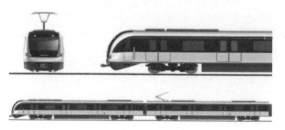

Fig. 1. Schematic diagram of subway model

2 Selection of Braking Characteristic Parameters

During the operation of subway vehicles, the brake cylinder pressure reflects the final output of the BECU and BCU, and then the brake cylinder pressure enters the basic braking device to brake the vehicle. Regarding the braking system as a black box, following the black box theory, the impact of internal changes in the braking system will affect the final output, which in turn affects the performance of the entire vehicle. Based on the data mining of the output data, the brake cylinder pressure is selected as the core observation time series data without considering the specific internal abnormality generation mechanism.

According to the simulation analysis of the abnormal characteristics of the brake cylinder pressure data in the previous section, in the actual operation of the vehicle, for example, the braking process at the initial braking speed of 80 km/h needs to cover a variety of different time series data sampling rates and large amounts of data analysis. To characterize the normal or abnormal state of the data, it is necessary to reduce the data volume of the brake cylinder pressure without losing the data characteristics. Therefore, it is necessary to extract the characteristic value of the brake cylinder pressure data to represent the complete braking process with fewer parameters.

This article divides a complete braking process into two major stages: brake cylinder pressure establishment and brake cylinder pressure stabilization. A total of six characteristic parameters are named after A, C, D, E, F, which characterize the change process of brake cylinder pressure. As shown in Table 1 below.

Table 1. Characteristic parameter table

Stage	Parameter item	Characteristic value name
Rising phase	A	90%T

(continued)

Table 1. (*continued*)

Stage	Parameter item	Characteristic value name
	B	Special slope time period
Stage two	C	Stable value
	D	Standard deviation
	E	Important maximum
	F	Important minimum

(1) A

90 % pressure build-up time of brake cylinder. The time required for the brake cylinder to start charging until the brake cylinder pressure rises to the specified pressure (90% of the target pressure) is the main indicator describing the response performance of the brake control system. The brake cylinder pressure rise time is an important performance parameter, which includes the time from when the driver's brake handle is pulled to when the air pressure of the brake cylinder rises to the start of the basic brake, which reflects idling stopping distance.

(2) B

The build-up time of brake cylinder pressure in special section. Because the subway vehicle brakes under actual working conditions, there is a small interval of braking, so the build-up of brake cylinder pressure may have been eliminated when the peak braking command is not fully reached, and the 90% brake cylinder pressure build-up time at this time is meaningless. At the same time, in the charging time of the brake cylinder pressure, the first 2 s basically belong to the action phase of the brake system. The brake cylinder pressure data at this time represents a series of actions of the brake system, and the latter part is basically the process of continuing to inflate to the target pressure. Therefore, it is set to select 50 kPa–70 kPa as the special slope section.

(3) C

Stable value. When the brake cylinder pressure is established, the brake cylinder pressure is based on the actual output pressure value of the target pressure. There is a certain difference between the actual output value of the vehicle engineering and the target set value. At this stage, due to the dynamic characteristics of the system, the actual brake cylinder pressure is real-time. Commonly used data processing methods are to take the arithmetic average of the data, geometric average, etc. During a complete braking, if the output value of the brake cylinder pressure of the vehicle is abnormally high or too low, the average value may be affected by the abnormal data. Therefore, a single value in the data segment is selected as the stable value of the brake cylinder pressure in the stable phase, that is, the most frequent data value in the stable phase is selected as the normal actual output value.

(4) D

Standard deviation. In mathematics, it can also be used as the mean square error, which is the square root of the arithmetic mean of the square of the deviation from the mean, expressed as σ. The standard deviation is the arithmetic square root of the variance. Assuming that there is a set of real number data columns: $X_1, X_2, X_3, \ldots, X_n$, the arithmetic mean value of which is μ, the standard deviation formula is as follows.

$$\sigma = \sqrt{\frac{1}{N} \sum_{i=1}^{N} (x_i - \mu)^2} \tag{1}$$

The standard deviation can reflect the discrete level of a data set. It is the most frequently used judgment that can quantify the discrete degree of a set of data, and it is also the main indicator of accuracy. Regarding the brake cylinder pressure in the stable phase, the normal state is a constant value, but the actual output results usually produce certain fluctuations. Using the standard deviation can express the degree of fluctuation of the brake cylinder pressure value, so as to monitor the stability of the system output.

(5) E

The maximum value of the stable phase. When the brake cylinder pressure is unstable and abnormal output is present, it is necessary to monitor the actual maximum output pressure. Too high brake contact surface pressure will cause the wheels to lock, which will affect the braking performance.

(6) F

The minimum value of the stable phase. When an abnormality occurs in the brake system, such as relay valve air leakage, brake cylinder air leakage, etc..Due to continuous air leakage, the brake cylinder pressure continues to drop after the brake cylinder pressure rises to the target pressure. It is necessary to pass the minimum value of the stable phase to monitor possible abnormalities.

Therefore, the feature parameter extraction table is obtained as shown in Table 2. below.

Table 2. Analysis table of six characteristic parameters

Stage	Number	Name	Meaning
Stage one	A	90%T	90% target pressure build-up time
	B	Special slope time period	Specific ascent speed
Stage two	C	Stable value	Stable stage value
	D	Standard deviation	Volatility
	E	Important maximum	Brake cylinder pressure overshoot
	F	Important minimum	Insufficient brake cylinder pressure

The graphical data of brake cylinder pressure is shown in Fig. 2 below.

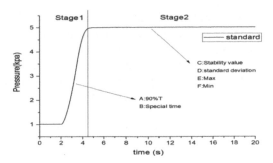

Fig. 2. The distribution of parameters on the brake cylinder pressure curve

3 Fuzzy Comprehensive Evaluation and Analysis Method Based on Characteristic Parameters

In order to analyze the train's health status from multiple angles, it is necessary to filter and analyze the indicators that characterize the train's health status. As far as the train brake system is concerned, the range of features is diverse, such as the operating time of a solenoid valve, the strategy and efficiency of the air compressor's charging and exhausting air, the degree of airtightness of the cylinder, the operating frequency of the large and small brakes, etc.. However, the first thing to consider when analyzing streaming data should be whether these variables and features exist for detection by existing sensors, and whether sensor data can be obtained through simpler streaming data acquisition channels, otherwise, just talking about multiple variables is not reasonable for realization and engineering.

The problem of state analysis is that it is difficult to establish a complete model for complex systems to analyze their failure probability, and although the operating parameters of the system and components show degradation with the increase in service time, this degradation is severely non-linear and at the same time ambiguous, without a strict boundary limit. Refined to the rail transit train braking system, due to its importance to ensure safety, there is no full life cycle database like other components. In order to realize the quantitative expression of the above-mentioned qualitative characteristics, rely on these factors to establish a stream data analysis and evaluation system, and choose the fuzzy comprehensive evaluation method.

Fuzzy comprehensive evaluation method is a comprehensive evaluation method based on fuzzy mathematics. It makes full use of the membership degree theory of fuzzy mathematics, and expresses various qualitative evaluations through quantitative evaluation, that is, uses fuzzy mathematics to make an overall evaluation of affairs or objects restricted by multiple factors. The fuzzy theory can be understood through simple examples. Water with a temperature of 0 °C can be regarded as ice water, or a mixture of ice and water, while water with a temperature of 80 °C is obviously hot water, so the properties of water at 40 °C between the two are difficult to give a clear judgment. In the process of changing properties from hot water to ice water, there is only a vague understanding of how to make accurate judgments based on temperature, and it is impossible to clearly give a reasonable judgment boundary. For example, the maximum impulse requirement for the common braking of a subway train is less than 0.75 m/s^3, so if the actual impulse is greater than this value, it is obviously a poor state, which will greatly affect the comfort of the user, and even a strong impact may cause deformation of the coupler. Then if the maximum impulse of a certain service brake is second-order derivation of the speed, the calculated value is 0.72 m/s^3. The braking performance this time is only from the perspective of impulse, which is obviously not ideal, but it does not exceed the data of 0.75 m/s^3.

The naive evaluation index is that the smaller the impulse when the train is braking, the better, and it can meet the needs within a reasonable range. When it exceeds a certain value, although it is still acceptable, it still faintly feels that there is a hidden danger, that is, the driving state of the vehicle has declined.

4 Analysis and Verification of Long-Term Vehicle Operation Status

For the braking performance degradation accompanying the long-term operation of the vehicle, the theoretical method is to conduct periodic consistency tests on the vehicle to observe the state change of the braking performance. In this article, based on the above-mentioned fuzzy comprehensive evaluation and analysis method theory, a set of software that can realize data visualization and data in-depth analysis is developed, and the braking state of the vehicle is analyzed based on the actual on-board data of many months.

4.1 Data Analysis Software Development

The development of data analysis software is based on the database as the carrier and is developed based on the Labview language, which realizes the storage and deletion of on-board data, and at the same time realizes the multi-function view of the data, and can analyze the braking state of the whole vehicle based on multi-day data. The overall structure of the software is shown in Fig. 3.

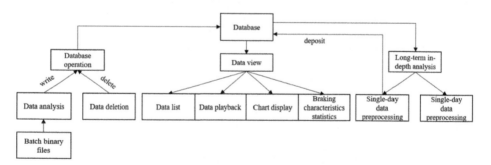

Fig. 3. Data analysis software architecture

The overall layout of the data analysis software is divided into a functional area and a working area. The functional area has database operations, data viewing, and data in-depth analysis. The working area is to implement specific operations on each functional area module. The software interface is shown in Fig. 4.

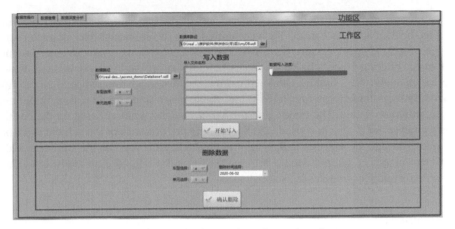

Fig. 4. Introduction to the software interface

4.2 Result Analysis

Use the software to analyze the data to get the scoring of the braking state of the vehicle. The result is shown in Fig. 5. According to the analysis method in this article, when the score is lower, it proves that the consistency of the vehicle is worse, which means that the braking performance of the vehicle has decreased.

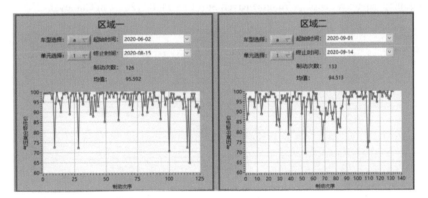

Fig. 5. Analysis of braking performance

As shown in Fig. 5, a total of 126 braking occurred from June to August, and the average braking state score was 95.592. In September, a total of 133 braking occurred, and the average value of the braking state score was 94.513. It can be seen that the braking state of the vehicle has declined over time.

5 Conclusion

First of all, this article introduces the subway brake system, which is the object of subway braking performance, analyzes its braking method and working principle, combines the output parameters of the brake system, and establishes the brake cylinder pressure data as the data analysis carrier. Then combined with the abnormal data characteristics of the brake cylinder pressure data, two major stages and six characteristic parameters of the braking process based on the brake cylinder pressure are established, which are respectively: 90% brake cylinder pressure establishment time, special slope time period, brake cylinder pressure stable value, stable phase standard deviation, maximum value, minimum value. The braking performance analysis method based on the consistency analysis method is proposed, and the braking performance of the vehicle is deeply studied. Through the analysis of data mining methods, and based on the similarity measurement model of sample data, a braking performance degradation analysis method suitable for subway vehicles is established.

References

1. Senhua, L., Chunjun, C., Lei, Y., et al.: Comprehensive comfort evaluation system for metro train passengers based on analytic hierarchy process. Sci., Technol. Eng. **019**(036), 296–301 (2019). (in Chinese)
2. Wang D.: Research on Spatial Coupling Vibration of Low and Medium Speed Maglev Train and Low Structure. Southwest Jiaotong University, (2015, in Chinese)
3. Chen C.: Measurement and Research of Urban Rail Transit Operation Comfort based on uic513 Standard. Southwest Jiaotong University, (2016, in Chinese)
4. Huang, Y., CAIDE Institute, Li M., et al.: The study on the influence of dam discharge on fish migration capacity. People's Yangtze River, **50**(008), 74–80 (2019, in Chinese)

A Method for Obtaining Highly Robust Memristor Based Binarized Convolutional Neural Network

Lixing Huang[1], Jietao Diao[1], Shuhua Teng[2], Zhiwei Li[1], Wei Wang[1], Sen Liu[1], Minghou Li[3], and Haijun Liu[1(✉)]

[1] College of Electronic Science and Technology, National University of Defence Technology, Changsha 410073, Hunan, China
liuhaijun@nudt.edu.cn
[2] Hunan Communications Research Institute Co., Ltd., Changsha 410015, Hunan, China
[3] Qingdao Geo-Engineering Sur4vering Institute, Qingdao 266100, Shandong, China

Abstract. Recently, memristor based binarized convolutional neural network has been widely investigated owing to its strong processing capability, low power consumption and high computing efficiency.However, it has not been widely applied in the field of embedded neuromorphic computing for manufacturing technology of the memristor being not mature. With respect to this, we propose a method for obtaining highly robust memristor based binarized convolutional neural network. To demonstrate the performance of the method, a convolutional neural network architecture with two layers is used for simulation, and the simulation results show that binarized convolutional neural network can still achieve more than 96.75% recognition rate on MNIST dataset under the condition of 80% yield of the memristor array, and the recognition rate is 94.53% when the variation of memristance is 26%, and it is 94.66% when the variation of the neuron output is 0.8.

Keywords: Memristor · Binarized convolutional neural network · Variation

1 Introduction

Binarized convolutional neural network [1, 2] has obtained much attention owing to its excellent computing efficiency [3] and fewer storage consumption [4]. However, when faced with complex tasks [5], the depth of the neural network will become deeper and deeper [6], increasing the demands on the communication bandwidth. And constrained by the problem of memory wall [7] in von Neumann architecture, it is difficult to realize further improvement in computing speed and energy efficiency.

Fortunately, the emerging of memristor [8] based computing system provides a novel processing architecture, viz., processing-in-memory (PIM) architecture [9], solving the memory wall problem existed in von Neumann architecture. Because the core computing component in PIM architecture, memristor array, is not only used to store weights of neural network but also to execute matrix-vector multiplier, data transferring between memory and computing units is avoided, thus decreasing the requirements of communication bandwidth and improving computing speed and energy efficiency.

© The Author(s) 2022
Z. Qian et al. (Eds.): WCNA 2021, LNEE 942, pp. 813–822, 2022.
https://doi.org/10.1007/978-981-19-2456-9_82

Nevertheless, the manufacturing technology of the memristor is still not mature, the manufactured devices existing many non-ideal characteristics [10, 11], such as yield rate of memristor array and memristance variation, which degrades the performance of application program running on the memristor based computing system. In response to this, we propose a method to keep the performance of the binarized neural network running on memristor based computing system.

2   Binarized Convolutional Neural Network and Proposed Method

2.1   Binarized Convolutional Neural Network

The architecture of the binarized convolutional neural network used for simulation only two layers, which is proposed in our previous work [12]. And the detail information of the binarized convolutional neural network is shown in Fig. 1.

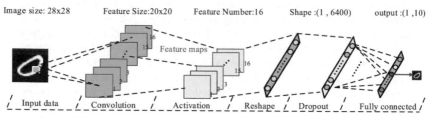

Fig. 1. Detail information of the binarized neural network

For the binarized convolutional neural network shown in Fig. 1, the input images of the network are first processed into binary, viz., the pixel value of them is processed to be 0 or 1. And the processing function is shown as follows:

$$f(x) = \begin{cases} 0 & x \leq 0.5 \\ 1 & x > 0.5 \end{cases} \tag{1}$$

The output type of the activation is the same as the input, viz., 0 or 1, and the express of the binarized function is shown as follows:

$$f(x) = \begin{cases} 0 & x \leq 0 \\ 1 & x > 0 \end{cases} \tag{2}$$

The binary form of the weight parameters in the binarized neural network is $+1$ or -1, and the processing function is shown as follows:

$$f(x) = \begin{cases} -1 & x \leq 0 \\ +1 & x > 0 \end{cases} \tag{3}$$

2.2 Proposed Method

The principle of the proposed method is to inject Gaussian noise into the binary weights and binary function of activation during the forward propagation of the training process. The purpose of injecting Gaussian noise into the weights is to improve robustness of the binarized neural network to device defects, while the counterpart of that is to improve the robustness of the network to neuron output variation.

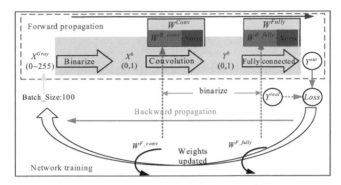

Fig. 2. Training process of binarized neural network

With Gaussian noise injected into the weights, the detail training process of the binarized convolutional neural network can be seen in Fig. 2. And it can be seen from Fig. 2 that the 'Noise' represents the random value sampled from Gaussian noise which follows normal distribution, and it is added to the binary weights, namely W^{B_conv} and W^{B_fully}, to get the weights W^{Conv} and W^{Fully}. Then, the weights W^{Conv} and W^{Fully} are used to perform convolution and vector-matrix multiply operation with inputs. In the weights updated phase of backward propagation, the weights W^{F_conv} and W^{F_fully} being float-point are updated according to the algorithm of gradient descent [13]. What should be noticed is that,since the gradient of the binary activation function at the non-zero point is 0, and the gradient at the zero point is infinite, we use the gradient of the tanh function to approximate the gradient of the binary activation function.

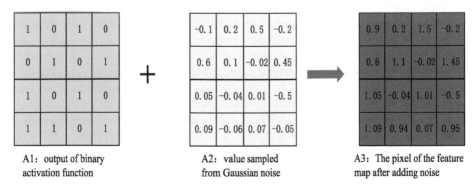

A1: output of binary
activation function

A2: value sampled
from Gaussian noise

A3: The pixel of the feature
map after adding noise

Fig. 3. Example of injecting noise into binary activation function

The implementation scheme of injecting the Gaussian noise into binary activation function can be seen in Fig. 3.

As can be seen in Fig. 3, the original outputs of the binary activation function only have two types of values, that is 0 and 1. And the value sampled from the Gaussian noise is float-point type. Therefore, the final type of the pixel value in the feature map is float-point.

3   Experiments

3.1   Simulation Settings

All the experiments in this study are conducted using a computer with 24 GB DDR4, Intel Core i7-8750H CPU (2.2 GHz), and a Nvidia GTX 1050 graphics card, and the Tensor flow [14] open-source library is used to train the binarized neural network. The simulation results are obtained using Monte-Carlo simulation method in Python. Another simulation settings are shown as following.

(1) Parameters of memristor model
    During the simulation process, two Pt/HfO2:Cu/Cu memristors [15] are used for representing one weights in the binarized convolutional neural network. And the average resistance value of the memristors with high resistance state (HRS) or low resistance state (LRS) is 1 MΩ and 1 KΩ, respectively.
(2) Parameters for training binarized convolutional neural network
    The MNIST dataset is divided into three subsets, viz., training set including 55,000 images, validation set containing 5000 images, and testing set composed of 10,000 images. The number of epoch for training network is 100, and the value of the batch size for gradient descent optimization algorithm is also 100. In addition to that, exponentially decaying learning rate is applied, and the initial learning rate is 0.01.
(3) Model of non-ideal characteristics
    The defects considered in our experiments include three types, namely yield rate of the memristor array, resistance variation of the memristor and neuron output variation.

For the problem of the yield in memristor array, meaning that there are some damaged devices in the array and each damaged device either sticks at G_{HRS} (conductance value corresponding to memristor in the state HRS) or G_{LRS} (conductance value corresponding to memristor in the state of LRS), the resistance in memristor array is randomly changed to be G_{HRS} or G_{LRS} for emulating the yield rate problem. And an assumption has been made that there is 50% possibility for each damaged device being stuck at G_{HRS} or G_{LRS}.

As for the problem of the resistance variation of the memristor, the resistance of the memristors in the state of HRS (LRS) in array is not exactly 1MΩ (1KΩ), but fluctuates around 1MΩ (1KΩ). Therefore, during the simulation process, the model of the resistance variation is depicted as Eq. (4):

$$R N(\mu, \sigma_v^2) \tag{4}$$

In Eq. (4): the parameter μ represents the average value of the memristance in HRS or LRS, viz., $1M\Omega$ in the state of HRS and $1K\Omega$ in the state of LRS. The parameter σ_v should satisfy the relation described in Eq. (5):

$$\sigma_v = \mu \times r_v \tag{5}$$

In Eq. (5): the parameter r_v denotes to the scale of the resistance variation.

With respect to the problem of neuron output variation, meaning the logical value output by the binary activation function is not corresponded to the actual voltage value output by neuron circuit, the logical value +1 (0) is not exactly mapped to the output of the neuron, viz., + VCC (0V), but fluctuates around + VCC (0V). During the simulation process, the model of the neuron output variation is depicted as Eq. (6):

$$V N(\mu_1, \sigma^2) \tag{6}$$

In Eq. (6): the parameter μ_1 represents the expected voltage value of the neuron output, viz., + VCC and 0V, and the parameter σ is the standard deviation of the normal distribution reflecting the range of the neuron output variation.

3.2 Simulatino Results

At first, the performance of the method with Gaussian noise injected into binary weights is first demonstrated. The robustness of the binarized convolutional neural network trained through the method with Gaussian noise injected into binary weights is analyzed based on the model of non-ideal characteristics.

Table 1 gives the information about the recognition rate of the network trained through method with noise injected into binary weights on MNSIT. What should be noticed is that, the parameter (σ_1) of the noise injected into binary weights is closely related to the parameter (σ_v) of resistance variation model for the reason that two memristors forming a differential pair are used to represent one weight.

Table 1. The performance of the binarized convolutional neural network trained through method with noise injected into weights.

Gaussian noise injected into binary weights (σ_1)	0.1	0.2	0.3	0.4	0.5	0.6	0.7	0.8
Accuracy (%)	98.15	98.1	98.06	98	97.87	97.57	97.18	96.83

Figure 4 shows the analysis results of network's tolerance for yieldrate of the memristor array and resistance variation of memristor when the network is trained through or not through ($\sigma_1 = 0.0$) method of injecting noised into weights. What should be noticed is that, the noise parameter $\sigma_1 = 0.0$ means that the method of injecting noise into weight is not adopted.

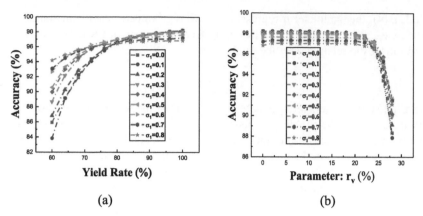

Fig. 4. Analysis results of the tolerance of binarized convolutional neural network for yield rate of the memristor array (a) and resistance variation of memristor (b).

As can be seen in Fig. 4, with the value of noise parameter σ_1 increasing, the network's robustness to yield rate of memristor array and resistance variation of memristor is improved, however, the performance of the network under ideal condition shows a gradual decline. Therefore, a reasonable noise parameter value should be given to balance the network performance and robustness. It can be noticed from table 1 that the recognition rate of the network achieves more than 97.5% when noise parameter varies from 0.1 to 0.6. And it can be seen from Fig. 4 (a) and (b), when the noise parameter is 0.6, the network not only has a good tolerance to the resistance variation of the memristor, but also has a good tolerance to the yield of the array. Therefore, the parameter value of the noise injected into weights is 0.6 in this paper. Figure 5 gives the analysis

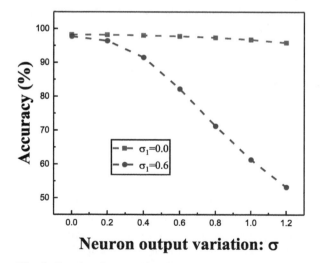

Fig. 5. Results of network's robustness to neuron variation.

results of the network's tolerance for neuron output variation when the noise parameter is 0.0 and 0.6, respectively.

What can be seen from Fig. 5 is that the network's tolerance to neuron output variation is degenerated. To improve the network's tolerance for neuron output variation, the method of injecting noise to binary activation function is also adopted during the training procedure of the network. Table 2 gives the information about the performance of the network trained with method of injecting noise into binary weights ($\sigma_1 = 0.6$) and binary activation function(σ_2).

Table 2. The performance of the network trained through method with Gaussian noise injected into weights ($\sigma_1 = 0.6$) and activation.

Gaussian noise injected into binary activation (σ_2)	0.2	0.4	0.6	0.8	1.0	1.2
Accuracy under ideal condition ($\sigma = 0.0$)	97.33%	97.13%	96.66%	96.55%	96.03%	95.66%
Accuracy when the parameter of neuron output variation ($\sigma = 1.2$)	67.99%	88.28%	91.44%	93.33%	93.62%	93.67%

What can be seen from Table 2 is that, as the noise parameter σ_1 is 0.6 and noise parameter σ_2 increase, the performance of the network under ideal condition declines continuously, but the tolerance of the network to neuron output variation increase gradually. Therefore, to keep the performance of the network excellent under ideal condition and improve the tolerance of the network to neuron output variation, we select a rough value for the noise parameter σ_2, that is 0.5. Similarity, the parameter (σ_2) is related to the parameter (σ) of the neuron output variation model for the reason that the neuron output variation follows normal distribution. Figure 6 shows the robustness of network trained through method with noise injected into weights ($\sigma_1 = 0.6$) and binary activation ($\sigma_2 = 0.5$) to non-ideal characteristics.

As can be seen in Fig. 6 (a) and (c), the robustness of the network trained through method with noise injected into binary weights ($\sigma_1 = 0.6$) and binary activation ($\sigma_2 = 0.5$) to yield of array and neuron output variation is improved. It also can be noticed from Fig. 6 (b) that the performance of the network under ideal condition declines marginally, which can be ignored.

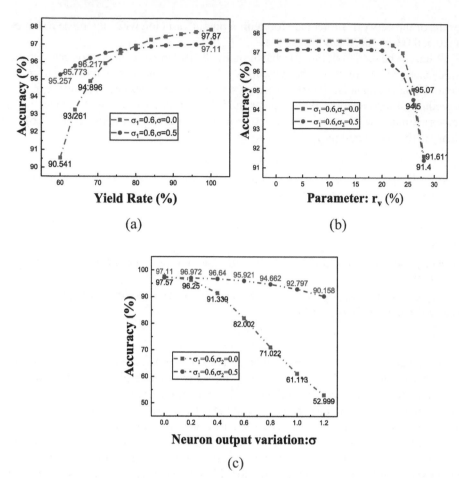

Fig. 6. The robustness of the binarized convolutional neural network trained trough method with noised injected into weights ($\sigma_1 = 0.6$) and binary activation ($\sigma_2 = 0.5$) to yield of array (a) and resistance variation of memristor (b) and neuron output variation (c).

4   Conclusion

In this paper, we propose a method for obtaining highly robust memristor based binarized convolutional neural network. By injecting Gaussian noise into binary weights and binary activation function during the training procedure, the reasonable noise parameter is selected for keeping the performance of the network and the network's tolerance to non-ideal characteristics. A binarized convolutional neural network is mapped into memristor array for simulation, and the results show that when the yield of the memristor array is 80%, the recognition rate of the memristor based binarized convolutional neural network is about 96.75%, and when the resistance variation of the memristor is 26%, it is around 94.53%, and when the neuron output variation is 0.8, it is about 94.66%.

Acknowledgements. This work was supported by the National Natural Science Foundation of China (Grant Nos. 61974164, 62074166, 61804181, 62004219, and 62004220).

References

1. Courbariaux, M., Hubara, I., Soudry, D., El-Yaniv, R., Bengio, Y.: Binarized Neural Networks: Training Deep Neural Networks with Weights and Activations Constrained to +1 or -1 (2016)
2. Courbariaux, M., Bengio, Y., David, J.-P.: BinaryConnect: training deep neural networks with binary weights during propagations. Adv. Neural Inf. Process. Syst. **28** (2015)
3. Rastegari, M., Ordonez, V., Redmon, J., Farhadi, A.: XNOR-Net: imagenet classification using binary convolutional neural networks. Computer Vision - Eccv 2016, Pt Iv (2016)
4. Qiao, G.C., Hu, S.G., Chen, T.P., et al.: STBNN: Hardware-friendly spatio-temporal binary neural network with high pattern recognition accuracy. Neurocomputing **409**, 351–360 (2020)
5. Krizhevsky, A., Sutskever, I., Hinton, G.E.: ImageNet classification with deep convolutional neural networks. Commun. ACM **60**(6), 84–90 (2017)
6. Simonyan, K., Zisserman, A.: Very deep convolutional networks for large-scale image recognition. Computer Science (2014)
7. Wulf, W.A., McKee, S.A.: Hitting the Memory Wall: Implications of the Obvious **23**(1), 20–24 (1995)
8. Strukov, D.B., Snider, G.S., Stewart, D.R., Williams, R.S.: The missing memristor found. Nature **453**(7191), 80–83 (2008)
9. Ielmini, D., Wong, H.: In-memory computing with resistive switching devices. Nature Electronics **1**(6), 333 (2018)
10. Kim, S., Kim, H.D., Choi, S.J.: Impact of synaptic device variations on classification accuracy in a binarized neural network. Sci. Rep. **9**(1), 15237 (2019)
11. Liu, B.Y., Li, H., Chen, Y.R., et al.: Vortex: variation-aware training for memristor x-bar. In: 2015 52nd Acm/Edac/Ieee Design Automation Conference; Los Alamitos (2015)
12. Huang, L., Diao, J., Nie, H., et al.: Memristor based binary convolutional neural network architecture with configurable neurons. Frontiers Neurosci. **15**, 328 (2021)
13. Lecun, Y., Bottou, L.: Gradient-Based Learning Applied to Document Recognition. 86(11), 2278-2324 (1998)
14. Abadi, M., Barham, P., Chen, J.M., et al.: TensorFlow: A system for large-scale machine learning. In: 12th USENIX Symposium on Operating Systems Design and Implementation (OSDI); Nov 02–04, Savannah, GA (2016)
15. Liu, S., Wang, W., Li, Q., et al.: Highly improved resistive switching performances of the self-doped Pt/HfO2:Cu/Cu devices by atomic layer deposition. Science China-Physics Mechanics & Astronomy. 59(12) (2016)

Real-Time Estimation of GPS Satellite Clock Errors and Its Precise Point Positioning Performance

Junping Zou[✉] and Jiexian Wang

College of Surveying and Geo-Informatics, Tongji University,
Shanghai 200092, People's Republic of China
1410902@tongji.edu.cn

Abstract. The current stochastic model in GNSS processing is constructed based on the prior experience, for example the ratio of the weight of the pseudorange and phase observations is generally determined as 1:10000. These methods ignore the precision differences of the different GNSS receivers and observation space. In this paper, the standard deviation of differenced ionosphere-free pseudorange and phase observations is computed with dual-frequency observations and then the weight ratio of the pseudorange and phase observations is obtained using the computed standard deviation. This method is introduced in satellite clock estimating and the data is processed. The results show that the presented method is feasible, with which the accuracy of the estimated satellite clock results is improved. The estimated satellite clock results are further adopted in PPP and the positioning results of the 10 users validate that the estimated satellite clock, which uses the presented method, can accelerate the convergence of PPP compared with the traditional method.

Keywords: Global Navigation Satellite System · Precise point positioning · Standard deviation · Pseudorange · Phase observations ratio of the weight

1 Introduction

The positioning accuracy of the Global Navigation Satellite System (GNSS) is affected by different kinds of error sources, and the satellite clock error is one of the most influential factors. To meet the high-precision positioning requirement of precise point positioning (PPP) users, the estimation and service for the precise satellite clock becomes an essential routine of the International GNSS Service (IGS) [1, 2]. The ionosphere-free phase and pseudorange observations (L1/L2 and P1/P2) collected from the global observation stations are used in GNSS satellite clock error estimation [3-7]. Initially, final precise satellite clock products are provided and delayed by 15 days. Considering the influence of the phase ambiguity on the estimating efficiency, some computationally efficient approaches are presented for real-time application [3, 4, 6-8]. In these computationally efficient approaches, the time-varying satellite clock correction is computed according to the epoch-differenced algorithm and phase observation, while the satellite

© The Author(s) 2022
Z. Qian et al. (Eds.): WCNA 2021, LNEE 942, pp. 823–830, 2022.
https://doi.org/10.1007/978-981-19-2456-9_83

clock error of the reference epoch is estimated with code observation. Since then, in order that high-precision satellite clock and orbit products are provided for users, with the help of multiple agencies and centers, real-time service (RTS) of IGS is published as Ratio Technical Commission for Maritime Services (RTCM) and state-space representation (SSR) streams are broadcasted on the Internet [18]. Based on the analysis for triple-frequency observations, the GPS inter-frequency clock bias is noticed [9-11] and the estimation for the triple-frequency satellite clock is developed [12, 13]. The satellite clock services for the single, dual and triple-frequency users are realized based on the IGS clock products, which is estimated with ionosphere-free phase and pseudorange combinations, and the biases between the different observations.

The reasonable stochastic model is very important for processing the GNSS data to obtain the optimal solution [14-16]. The stochastic model is generally constructed with the elevation-dependent function and standard deviations of observations. In satellite clock estimating, the code and phase observations and their corresponding stochastic models are adopted. The different weights should be applied for observations of different stations considering their precision differences [17]. The 1:10000 of the weight ratio is generally adopted for pseudorange and phase observations in satellite clock error estimating [5, 6]. It is obvious that using the same weight ratio for all observation stations ignores the precision differences of observations for different GNSS receivers. It is well known that the performance of the GNSS receiver and their observations are improved with the continuous progress of the GNSS hardware technology. Thus, the satellite clock estimation is discussed and the construction strategy of the stochastic model is presented. GPS data from 56 IGS stations on DOY 100, 2021 are processed for analyzing the quality of the estimated satellite clock errors and data from 10 user stations are used for evaluating the performance of PPP.

2   Method

Generally, undifferenced ionosphere-free carrier-phase and pseudorange observations are adopted in satellite clock estimation. During the estimation process, the biases from the satellite and the receiver are included in the estimated satellite clock and receiver clock respectively. The contribution of the biases from the pseudorange and phase observations to the clock estimations determines the used weights of observations. Thus, the strategy of the satellite clock error resolution is discussed and then the establishment of the stochastic model is presented.

2.1   The Satellite Clock Estimation

The ionosphere-free carrier-phase and pseudorange observations can be described as:

$$
\begin{aligned}
IF(L1, L2) &= \rho + \delta^r - \delta^s + \left(\frac{f_1^2}{f_1^2 - f_2^2} N_1 \lambda_1 - \frac{f_2^2}{f_1^2 - f_2^2} N_2 \lambda_2 \right) - \left(\frac{f_1^2}{f_1^2 - f_2^2} FCB_1^r - \frac{f_2^2}{f_1^2 - f_2^2} FCB_2^r \right) \\
&\quad + \left(\frac{f_1^2}{f_1^2 - f_2^2} FCB_1^s - \frac{f_2^2}{f_1^2 - f_2^2} FCB_2^s \right) - T^{r,s} + \varepsilon_{1,2} \\
IF(P1, P2) &= \rho + \delta^r - \delta^s - \left(\frac{f_1^2}{f_1^2 - f_2^2} b_1^r - \frac{f_2^2}{f_1^2 - f_2^2} b_2^r \right) + \left(\frac{f_1^2}{f_1^2 - f_2^2} b_1^s - \frac{f_2^2}{f_1^2 - f_2^2} b_2^s \right) - T^{r,s} + \omega_{1,2}
\end{aligned}
\tag{1}
$$

where ρ is the geometric distance from a satellite to a receiver, δ^r is the receiver clock error (unit: m), δ^s is the satellite clock error (unit: m), $f_i(i = 1,2)$ are carrier frequencies of signals, FCB_i^s ($i = 1,2$) are satellite FCBs of phase observations, which contain constant and time-varying parts, $FCB_i^r(i = 1,2)$ are receiver FCBs, b_i^s ($i = 1,2$) are satellite hardware delays of pseudorange observations, which also contain constant and time-varying parts, $b_i^r(i = 1,2)$ are receiver HDBs, T is tropospheric delay, $\varepsilon_{1,2}$ and $\omega_{1,2}$ are noises. During resolving, the estimated, reparameterized satellite clock error will absorb satellite-dependent biases and is written as:

$$\overline{\delta^s} = \delta^s + \left[P_P \cdot \left(\frac{f_1^2}{f_1^2 - f_2^2} FCB_1^s - \frac{f_2^2}{f_1^2 - f_2^2} FCB_2^s \right) + P_c \cdot \left(\frac{f_1^2}{f_1^2 - f_2^2} b_1^s - \frac{f_2^2}{f_1^2 - f_2^2} b_2^s \right) \right] / (P_P + P_c) \tag{2}$$

where P_p and P_c are the used weights of phase and pseudorange observations in satellite clock error estimating. Equation (2) indicates that the set weights are mainly determined by biases of pseudorange and phase observations when reparameterizing satellite clock error. Combined with the elevation-dependent weighting, the final weight function can be written as:

$$w(\theta_k) = \begin{cases} 1/\sigma^2 & 30° \leq \theta_k \leq 90° \\ 2\sin(\theta_k)/\sigma^2 & 7° \leq \theta_k < 30° \end{cases} \tag{3}$$

where θ is the elevation of the satellite; σ is the standard deviations of phase and pseudorange observations. Generally, the 1:100 standard deviations for phase and pseudorange observations are adopted and estimated, reparameterized satellite clock error contains almost all parts of FCB, since the weight of the phase observation is far greater than that of pseudorange observation. It is obvious that these weights do not consider the precision differences of the different GNSS receivers and is not beneficial for improving the estimated satellite clock results. The settings for satellite clock estimation are listed as follows. For measurements, the observation interval is 30s and the elevation cut-off angle is set as 70. In parameter correction, the least square filter is adopted and weighting is according to the presented method. Station coordinates are fixed values from IGS SINEX files and satellite orbits are based on precise ephemeris products released by IGS. Satellite and receiver clock errors are both solved as white noises at each epoch and ambiguity float solution is adopted. The troposphere delay is corrected by Saastamoinen model and residuals are estimated via piece-wise pattern. The phase center variation (PCV) is based on Absolute IGS 08 correction model. DCB(C1-P1) correction adopts monthly products released by CODE. In addition, phase windup, relativistic effects, solid tide and ocean tide corrections are also implemented.

The implementation of PPP requires dual-frequency observations and corresponding satellite clock and orbit products. Meanwhile, corrections are needed for ocean tide, solid tide, Earth rotation, phase center variation, relativistic effects and Differential Code Bias (DCB). The estimated parameters are receiver position and clock error, residual of troposphere delay and phase ambiguity. In PPP processing, the estimator of Least square filter and the corresponding stochastic model are used.

2.2 Stochastic Model

Li et al. [17] show that the reasonable stochastic model should be constructed for obtaining the optimal results. The stochastic model is established by means of the evaluation of ionosphere-free pseudorange observations. As for the presented method, the standard deviation of differenced ionosphere-free phase and pseudorange observations is described as:

$$\sigma_{dif} = \sqrt{\sigma_{ifp}^2 + \sigma_{ifc}^2} \tag{4}$$

where σ_{ifp}^2 and σ_{ifc}^2 are variances of ionosphere-free phase and pseudorange observations, respectively. In Li et al. [17], the weight ratio of the adopted phase and pseudorange observations can be obtained, once the standard deviation as Eq. 4 shown is calculated and standard deviation of ionosphere-free phase observations is determined according to the priori information.

3 Data Processing and Analysis

To validate the performance of modified satellite clock errors with presented method, GPS data of 56 IGS stations collected on DOY 100, 2021 is processed with different weights. The distribution of used stations can be seen from Fig. 1.

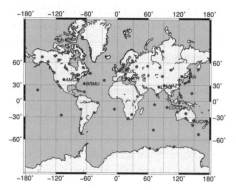

Fig. 1. Distribution of the 56 IGS stations (Red dots) and the 10 user stations (blue dots).

Based on the strategy presented in Li et al. [17], pseudorange observations are evaluated and related results can be seen from Fig. 2. It is shown that standard deviations of differenced ionosphere-free carrier-phase and pseudorange observations of different stations are different. These validate that the accuracy of different GNSS receivers is different and the different weight should be set in GNSS processing, when the different GNSS receiver are used.

To validate the presented approach, data is processed according to the settings. A simulated real-time experiment is implemented, in which streamed data from daily files are analyzed in epoch-wise pattern. The estimated satellite clock results are in comparison with that of IGS. The computed convergence time is considered as the elapsed time

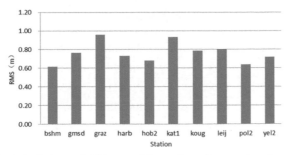

Fig. 2. The standard deviations of differenced ionosphere-free carrier-phase and pseudorange observations of different stations

when the difference between estimated clock results and corresponding IGS products is less than 0.1 ns. In difference computing, the reference satellite of PRN02 is selected:

$$RMS_{if} = \sqrt{\frac{\sum_{i=1}^{n} (\sigma_{IGS} - \sigma_E)_i}{n}} \tag{5}$$

where σ_E refers to the estimated satellite clock error and σ_{IGS} refers to the satellite clock error released by IGS. In data processing, the different strategies of #1 and #2 are used. In the #1, the traditional weights ratio of the phase and pseudorange observations is used, while the presented stochastic model is applied in #2. In the strategies of #1 and #2, the standard deviation of L1/L2 of 0.001 m is used. The convergence time can be seen from table 1. It is observed that the convergence time of #1 is longer than that of #2. The shortened time results indicate that the presented method is beneficial to build the reasonable stochastic model. This new built stochastic model considers the difference of GNSS receiver and observing environment so that the good results are obtained. Meanwhile, modified satellite clock errors are adopted in PPP positioning. In processing, convergence time of positioning refers to the elapsed time when the errors between estimated coordinates and that of IGS are all smaller than 10 cm in all directions of north, east and up. The convergence time of positioning for 10 users are shown in Table 2. The results indicate that the convergence time is shortened, when the method of #2 is used to process the GNSS data. This can be interpreted as the estimated satellite clock error values of #2 are better than of #1. When the better satellite clock error values are serviced for PPP users, high-precision PPP positioning can be obtained.

Table 1. Convergence time of satellite clock errors estimated by different strategies

Satellite	#1 (min)	#2 (min)	Satellite	#1 (min)	#2 (min)
PRN01	30	28	PRN18	40	38
PRN02	30	29	PRN19	32	32

(continued)

Table 1. (*continued*)

Satellite	#1 (min)	#2 (min)	Satellite	#1 (min)	#2 (min)
PRN03	25	23	PRN20	33	31
PRN05	30	28	PRN21	26	25
PRN06	26	24	PRN22	33	32
PRN07	23	23	PRN23	34	33
PRN08	25	24	PRN24	35	34
PRN09	26	25	PRN25	29	28
PRN10	28	27	PRN26	27	26
PRN11	33	32	PRN27	26	25
PRN12	36	35	PRN28	28	27
PRN13	32	31	PRN29	26	25
PRN14	31	30	PRN30	27	26
PRN15	26	25	PRN31	26	25
PRN16	27	26	PRN32	26	24
PRN17	22	21			

Table 2. The convergence time results of positioning for 10 users for the different strategies

Station	#1 (min)	#2 (min)	Improvement (min)
amc2	25	25	0
auck	104	102	2
brmu	63	46	17
brux	23	18	5
chan	120	118	2
coco	26	25	1
lck3	60	55	5
lck4	46	38	8
lhaz	35	32	3
mat1	67	66	1

4 Conclusion

The reasonable stochastic model is very important for obtaining the optimal estimated results. In the GNSS data processing, the weight for the GNSS observations of the phase and pseudorange is generally determined by the prior experience, for example the ratio of 1:10000. It is obvious that this neglects the precision differences of different GNSS receivers and observation space. In Li et al. [17], the standard deviation of differenced ionosphere-free carrier-phase and pseudorange observations is computed and then the weight ratio of pseudorange and phase observations are obtained using the computed standard deviation. This presented method is introduced in satellite clock estimating and the data is processed. In this processing, the results show the proposed strategy for establishing the stochastic model is feasible and it is beneficial to improve the accuracy of estimated satellite clock results. Further, improved satellite clock results are used in PPP and positioning results of 10 users demonstrate that the convergence time is shortened when satellite clock errors are estimated with the presented method.

References

1. Dow, J., Neilan, R., Rizos, C.: The International GNSS service in a changing landscape of global navigation satellite systems. J. Geod. **83**(3–4), 191–198 (2009)
2. Zumberge, J.F., Heflin, M.B., Jefferson, D.C., Watkins, M.M., Webb, F.H.: Precise point positioning for the efficient and robust analysis of GPS data from large networks. J. Geophys. Res. **102**(B3), 5005–5017 (1997)
3. Han, S., Kwon, J., Jekeli, C.: Accurate absolute GPS positioning through satellite clock error estimation. J. Geod. **75**(1), 33–43 (2001)
4. Bock, H., Dach, R., Jäggi, A., Beutler, G.: High-rate GPS clock corrections from CODE: Support of 1 Hz applications. J. Geod. **83**(11), 1083–1094 (2009)
5. Hauschild, A., Montenbruck, O.: Kalman-filter-based GPS clock estimation for near real-time positioning. GPS Solution **13**(3), 173–182 (2009)
6. Zhang, X., Li, X., Guo, F.: Satellite clock estimation at 1 Hz for realtime kinematic PPP applications. GPS Solution **15**(4), 315–324 (2010)
7. Ge, M., Chen, J., Dousa, J., Gendt, G., Wickert, J.: A computationally efficient approach for estimating high-rate satellite clock corrections in realtime. GPS Solution **16**(1), 9–17 (2012)
8. Li, H., Chen, J., Wang, J., Hu, C., Liu, Z.: Network based real-time precise point positioning. Adv. Space Res. **46**(9), 1218–1224 (2010)
9. Li, H., Li, B., Xiao, G., Wang, J., Xu, T.: Improved method for estimating the inter-frequency satellite clock bias of triple-frequency GPS. GPS Solution **20**(4), 751–760 (2015). https://doi.org/10.1007/s10291-015-0486-9
10. Montenbruck, O., Hugentobler, U., Dach, R., Steigenberger, P., Hauschild, A.: Apparent clock variations of the Block IIF-1 (SVN62) GPS satellite. GPS Solution **16**, 303–313 (2012)
11. Pan, L., Zhang, X., Li, X., Liu, J., Li, X.: Characteristics of inter-frequency clock bias for Block IIF satellites and its effect on triple-frequency GPS precise point positioning. GPS Solut **21**(2), 811–822 (2016). https://doi.org/10.1007/s10291-016-0571-8
12. Li, H., Li, B., Lou, L., Yang, L., Wang, J.: Impact of GPS differential code bias in dual- and triple-frequency positioning and satellite clock estimation. GPS Solution **21**(3), 897–903 (2016). https://doi.org/10.1007/s10291-016-0578-1
13. Li, H., Zhu, W., Zhao, R., Wang, J.: Service and evaluation of the GPS triple-frequency satellite clock offset. J. Navig. **71**(5), 1263–1273 (2018)
14. Li, B., Shen, Y., Xu, P.: Assessment of stochastic models for GPS measurements with different types of receivers. Chin. Sci. Bull. **53**(20), 3219–3225 (2008)
15. Li, B., Zhang, L., Verhagen, S.: Impacts of BeiDou stochastic model on reliability: overall test, w-test and minimal detectable bias. GPS Solution **21**, 1095–1112 (2017)
16. Wang, J., Stewart, M., Tsakiri, M.: Stochastic modeling for static GPS baseline data processing. J. Surv. Eng. **124**(4), 171–181 (1998)

17. Li, H., Xiao, J., Li, B.: Evaluation and application of the GPS code observable in precise point positioning. J. Navig. **72**(6), 1633–1648 (2019)
18. Weber, G., Dettmering, D., Gebhard, H.: Networked transport of RTCM via internet protocol (NTRIP). In: Sansò, F. (eds.) Proceedings of A Window on the Future of Geodesy. International Association of Geodesy Symposia, vol. 128. Springer, Berlin (2005)

Segmenting of the Sonar Image from an Undersea Goal Using Two Dimensional THC Entropy

Yu Liu[1], Ruiyi Wang[1], and Haitao Guo[2(✉)]

[1] College of Electronic Information Engineering, Inner Mongolia University,
Hohhot 010021, China
[2] College of Marine Science and Technology, Hainan Tropical Ocean University,
Sanya 572022, China
ghtpaper@126.com

Abstract. Sonar image segmentation is the basis of undersea goal detection and recognition. The THC (Tsallis-Havrda-Charvát) entropy can describe the statistical properties of the non-extensive systems and has a wider range of applications. There are multiple choices for the two features of the two-dimensional histograms, such as the gray value, the average gray value within a neighborhood, the median gray value within a neighborhood, the mode of gray values within a neighborhood, and so on. This paper investigates the segmentation results of a sonar image from an undersea goal using THC entropies of a variety of two-dimensional histograms, and gives the evaluation indexes for the segmentation results.

Keywords: Sonar image · Image segmentation · Entropy · Two-dimensional histogram

1 Introduction

A sonar is common equipment for undersea measurement, and in some cases it is irreplaceable. Undersea goal finding, undersea rescue, undersea manufacture, undersea robot movement, seabed treasure finding, ocean exploitation, and sea warfares often contain the goal recognition with the help of the sonar images of undersea goals [1]. In order to recognize the sonar images, segmenting the image first is usually essential. The sonar image of an undersea goal usually contains three areas: the goal light area, the goal dark one, and the seabed reverberation one. The sonar image segmentation is to obtain the above goal light area and goal dark one.

The thresholding is a well-known image segmentation method, and widely used because of its simple and fast calculation [2, 3]. There are some thresholding methods based on entropies [4]. There is a the entropy based method which uses the THC (Tsallis-Havrda-Charvát) entropy [3, 5]. The THC entropy can describe the statistical properties of the non-extensive systems and has a wider range of applications [3, 6]. The THC entropy based method in the reference [3] uses the gray value and the average gray value

© The Author(s) 2022
Z. Qian et al. (Eds.): WCNA 2021, LNEE 942, pp. 831–836, 2022.
https://doi.org/10.1007/978-981-19-2456-9_84

within a neighborhood as the features of the pixel to form the two-dimensional histogram for image segmentation. Not only the gray value and the average gray value within a neighborhood are used to form the two-dimensional histogram in this paper, but other features are also used. That is, this paper investigates the segmentation results using a variety of two-dimensional histograms.

2 Dual-Threshold Method Using the Two-Dimensional THC Entropy

There are multiple choices for the two features of the two-dimensional histograms, such as gray value, the average gray value within a neighborhood, the median gray value within a neighborhood, the mode of gray values within a neighborhood, and so on. Let $f_1(m, n)$ and $f_2(m, n)$ represent the two features of the pixel in an image, and the two-dimensional histogram is prescribed as

$$p(i,j) = \frac{n(i,j)}{M \times N} \tag{1}$$

where $n(i, j)$ denotes the pixel number when $f_1(m, n) = i$ and $f_2(m, n) = j$, $M \times N$ represents the size of the sonar image. Suppose $i = 0, 1, \cdots, i_{max}$ where i_{max} is the maximum value of $f_1(m, n)$ while (m, n) traveling across the whole image and $j = 0, 1, \cdots, j_{max}$ where j_{max} is the minimum value of $f_2(m, n)$ while (m, n) traveling across the whole image. Suppose that a sonar image of an undersea goal is divided into three areas using (t_1, s_1) and (t_2, s_2): the goal light area, the goal dark one, and the seabed reverberation one. Here t_1 and t_2 denote the thresholds of the feature $f_1(m, n)$ in the image, and s_1 and s_2 are the thresholds of the feature $f_2(m, n)$ in the image.

THC entropy with the order α related to the goal dark area is prescribed by

$$H_d^\alpha(t_1, s_1) = \frac{1}{\alpha - 1}[1 - \sum_{i=0}^{t_1}\sum_{j=0}^{s_1}(\frac{p(i,j)}{P_d(t_1, s_1)})^\alpha] \tag{2}$$

where

$$P_d(t_1, s_1) = \sum_{i=0}^{t_1}\sum_{j=0}^{s_1}p(i,j) \tag{3}$$

THC entropy with the order α related to the goal light area is prescribed by

$$H_l^\alpha(t_2, s_2) = \frac{1}{\alpha - 1}[1 - \sum_{i=t_2+1}^{i_{max}}\sum_{j=s_2+1}^{j_{max}}(\frac{p(i,j)}{P_l(t_2, s_2)})^\alpha] \tag{4}$$

where

$$P_l(t_2, s_2) = \sum_{i=t_2+1}^{i_{max}}\sum_{j=s_2+1}^{j_{max}} p(i,j) \tag{5}$$

THC entropy with the order α related to the seabed reverberation area is prescribed by

$$H_r^\alpha(t_1, s_1, t_2, s_2) = \frac{1}{\alpha - 1}[1 - \sum_{i=t_1+1}^{t_2} \sum_{j=s_1+1}^{s_2} (\frac{p(i, j)}{p_r(t_1, s_1, t_2, s_2)})^\alpha] \qquad (6)$$

where

$$P_r(t_1, s_1, t_2, s_2) = 1 - P_d(t_1, s_1) - P_l(t_2, s_2) \qquad (7)$$

The total THC entropy is given by

$$H^\alpha(t_1, s_1, t_2, s_2) = H_d^\alpha(t_1, s_1) + H_r^\alpha(t_1, s_1, t_2, s_2) + H_l^\alpha(t_2, s_2) \qquad (8)$$

Receive the value $(t_1^*, s_1^*, t_2^*, s_2^*)$ corresponding to the maximum value of the total THC entropy by means of maximizing the total THC entropy, that is

$$(t_1^*, s_1^*, t_2^*, s_2^*) = Arg \max_{t_1, s_1, t_2, s_2} [H^\alpha(t_1, s_1, t_2, s_2)] \qquad (9)$$

Here (t_1^*, s_1^*) and (t_2^*, s_2^*) are two thresholds which are used for the thresholding (segmentation) of an image.

3 Segmentation Results for the Sonar Image from an Undersea Goal

3.1 Introduction to the Sonar Image from an Undersea Goal

Figure 1(a) is a sonar image from an undersea man-made goal. The lighter part is the goal light area and the darker part is the goal dark area in the image. The goal dark area is on the goal light area and close to the goal light area. The reverberation area is around the goal light area and the goal dark one.

3.2 Segmentation Procedures and Results for the Sonar Image from an Undersea Goal

The segmentation procedures are as follows.

(1) Input the sonar image from an undersea goal.
(2) Filter the image using Wiener filter with a window size of 5×5.
(3) Calculate the two-dimensional histogram.
(4) Let $\alpha = 0.8$ [3], and calculate $H_d^\alpha(t_1, s_1)$, $H_l^\alpha(t_2, s_2)$ and $H_r^\alpha(t_1, s_1, t_2, s_2)$ using the formulas (2), (4) and (6).
(5) Calculate $(t_1^*, s_1^*, t_2^*, s_2^*)$ using the formula (9).
(6) Receive two pair of thresholds (t_1^*, s_1^*) and (t_2^*, s_2^*).
(7) Receive the thresholded image containing three gray values with the help of the thresholds (t_1^*, s_1^*) and (t_2^*, s_2^*).

In Fig. 1, Fig. 1(b) is the images after Wiener filtering, Fig. 1(c) is the image after manual segmentation which is regarded as the best segmentation result. Figure 1(d)-1(i) are the segmented images by means of the two-dimensional THC entropy. Figure 1(d)-1(i) are the segmented images corresponding to the feature combinations 1 (the gray value and the average gray value within a neighborhood), 2 (the gray value and the median gray value within a neighborhood), 3 (the gray value and the mode of gray values within a neighborhood), 4 (the average gray value within a neighborhood and the mode of gray values within a neighborhood), 5 (the average gray value within a neighborhood and the median gray value within a neighborhood), and 6(the median gray value within a neighborhood and the mode of gray values within a neighborhood). The thresholds

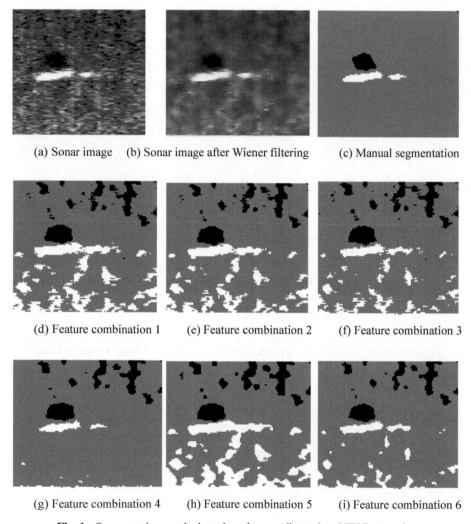

(a) Sonar image   (b) Sonar image after Wiener filtering   (c) Manual segmentation

(d) Feature combination 1   (e) Feature combination 2   (f) Feature combination 3

(g) Feature combination 4   (h) Feature combination 5   (i) Feature combination 6

Fig. 1. Segmentation results based on the two-dimensional THC entropies.

for image segmentation corresponding to Fig. 1(d)-1(i) are (38,43), (94,86); (38,42), (96,91); (39,43), (101,81); (38,41), (133,117); (40,42), (91,82); (38,40), (105,83).

It can be found out from Fig. 1 that the sonar image of an undersea man-made goal is roughly divided into a goal dark area, a reverberation one and a goal light one. However, the parts of the reverberation area are wrongly divided into the goal light area or the goal dark one. The reason for this phenomenon is that the values of the two features of each feature combination in the parts of the reverberation area are actually equal to the values of the two features of each feature combination in the goal light area or the goal dark one. Visually, although there are errors in segmentation, in comparison, Fig. 1(g), namely feature combination 4, has the best segmentation effect.

This paper attempts to give the evaluation indexes IOU (intersection over union) and FPR (false positive rate) for the above segmentation results. Table 1 gives the evaluation indexes IOU and FPR of the goal light area. Table 2 gives the evaluation indexes IOU and FPR of the goal dark area [6]. In terms of the evaluation indexes IOU and FPR, for the segmentation of the goal light area, Fig. 1(g), namely feature combination 4, has the best segmentation effect; and for the segmentation of the goal dark area, Fig. 1(i), namely feature combination 6, has the best segmentation effect. In general, the evaluation using indexes IOU and FPR is roughly the same as the visual effect.

Table 1. Evaluation indexes for the segmentation of the goal light area.

Feature combinations	1	2	3	4	5	6
IOU	0.1321	0.1520	0.1966	0.5078	0.1093	0.2552
FPR	0.8667	0.8466	0.7953	0.3973	0.8902	0.7343

Table 2. Evaluation indexes for the segmentation of the goal dark area.

Feature combinations	1	2	3	4	5	6
IOU	0.2456	0.2479	0.2293	0.2572	0.2132	0.2636
FPR	0.7520	0.7497	0.7695	0.7403	0.7847	0.7339

4 Conclusion

This paper investigates the application of the THC entropies of 6 kinds two-dimensional histograms to the sonar image segmentation. The segmentation results with different two-dimensional histograms are different. In practical applications, we can determine which two-dimensional histogram is more appropriate based on experiments. But we should also know that for a sonar image from an undersea goal, there may be mis-segmentation with any two-dimensional histogram given in the paper. That is because, for a sonar image from an undersea goal, any two-dimensional histogram given in the paper is not an ideal shape of the three peaks and two valleys.

This work is supported by Hainan Provincial Natural Science Foundation of China (No. 420CXTD439) and the National Science Foundation of China (No. 61661038).

References

1. Guo, H., Liu, L., Zhao, Y., Xu, F.: Chinese J. Sci. Instr. **34**, 2322–2327 (2013)
2. Han, S., Wang, L.: Syst. Eng. Electron. **24**, 91–94 (2002). (in Chinese)
3. Sahoo, P.K., Arora, G.: Pattern Recogn. Lett. **27**, 520–528 (2006)
4. Cao, J.: Pattern Recogn. Artif. Intel. **25**, 958–971 (2012). (in Chinese)
5. Wu, Y., Pan, Z., Wu, W.: Opto-Electron. Eng. **35**, 53–58 (2008). (in Chinese)
6. Wang, R.: Master Thesis. Inner Mongolia University, Hohhot (2018)

Optimal Decision Threshold-Moving Strategy for Skewed Gaussian Naive Bayes Classifier

Qinyuan He[1](✉) and Hualong Yu[2]

[1] Marine Design and Research Institute of China, Shanghai 200011, China
qinyuan_he@yeah.net
[2] School of Computer, Jiangsu University of Science and Technology, Zhenjiang 212100, China

Abstract. Gaussian Naive Bayes (GNB) is a popular supervised learning algorithm to address various classification issues. GNB has strong theoretical basis, however, its performance tends to be hurt by skewed data distribution. In this study, we present an optimal decision threshold-moving strategy for helping GNB to adapt imbalanced classification data. Specifically, a PSO-based optimal procedure is conducted to tune the posterior probabilities produced by GNB, further repairing the bias on classification boundary. The proposed GNB-ODTM algorithm presents excellent adaptation to skewed data distribution. Experimental results on eight class imbalance data sets also indicate the effectiveness and superiority of the proposed algorithm.

Keywords: Gaussian Naive Bayes · Class imbalance learning · Decision threshold moving · Particle swarm optimization

1 Introduction

In recent years, class imbalance learning (CIL) has become one of hot topics in the field of machine learning [1]. Also, the CIL has been widely applied in various real-world applications, including disease classification [2], software defect detection [3], biology data analysis [4], bankrupt prediction [5], etc. So-called class imbalance problem means in training data, the instances belong to one class is much more than that in other classes. The problem tends to highlight the performance of majority class, but to ignore the minority class.

There exist three major techniques to implement CIL: 1) data-level approach, 2) algorithmic-level method and 3) ensemble learning strategy. Data-level, which is called resampling, addresses CIL problem by re-balancing data distribution [6–7]. It contains oversampling that generates lots of new minority instances, and undersampling which removes a lot of majority instances. Algorithmic-level adapts class imbalance by modifying the original supervised learning algorithms. It mainly contains: cost-sensitive learning [8], and decision threshold-moving strategy [9–10]. Cost-sensitive learning designates different training costs for the instances belonging to different classes to highlight the minority class, while decision threshold-moving tune the biased decision boundary from the minority class region to the majority class region. As for ensemble

© The Author(s) 2022
Z. Qian et al. (Eds.): WCNA 2021, LNEE 942, pp. 837–843, 2022.
https://doi.org/10.1007/978-981-19-2456-9_85

learning, it integrates either a data-level algorithm or an algorithmic-level method into a popular ensemble learning paradigm to promote the quality of CIL [11–12]. Among these CIL techniques, the decision threshold-moving is relatively flexible and effective, however, it also faces a challenge, i.e., it is difficult to select an appropriate threshold.

In this study, we focus on a popular supervised learning algorithm named Gaussian Naive Bayes (GNB) [13] which also tends to be hurt by skewed data distribution. First, we analyze why the GNB tends to be hurt by imbalanced data distribution in theory. Then, we explain why adopting several popular CIL techniques could repair this bias. Finally, based on the idea, PSO optimization algorithm, we propose an optimal decision threshold-moving algorithm for GNB named GNB-ODTM. Experimental results on eight class imbalance data sets indicate the effectiveness and superiority of the proposed algorithm.

2 Methods

2.1 Gaussian Naive Bayes Classifier

GNB is a variant of Naive Bayes (NB) [14], which is used only to deal with data in continuous space. Like NB, GNB has a strong theoretical basis. GNB assumes in each class, all instances satisfy a multivariate Gaussian distribution, i.e., for an instance xi, we have:

$$P(x_i|y) = \frac{1}{\sqrt{2\pi\sigma_y^2}} e^{-\frac{(x_i-\mu_y)^2}{2\sigma_y^2}} \tag{1}$$

where μy and σy denote the mean and variance of all instances belonging to class y, respectively. P(xi| y) represents in class y, xi's conditional probability. As the prior probability P(y) is known, hence the posterior probability P(y| xi) and P(~y| xi) can be calculated as,

$$P(y|x_i) = \frac{P(x_i|y)P(y)}{P(x_i|y)P(y) + P(x_i|\sim y)P(\sim y)} \tag{2}$$

$$P(\sim y|x_i) = \frac{P(x_i|\sim y)P(\sim y)}{P(x_i|y)P(y) + P(x_i|\sim y)P(\sim y)} \tag{3}$$

We expect the classification boundary can correspond to P(xi| y) = P(xi| ~y). However, if the data set is imbalanced (supposing P(y) << P(~y), then to guarantee P(y| xi) = P(~y| xi), i.e., P(xi| y)P(y) = P(xi| ~y)P(~y), the real classification boundary must correspond to a condition of P(xi| y) >> P(xi| ~y). That means the classification boundary is extremely pushed towards the minority class y. That explains why skewed data distribution hurts the performance of GNB.

To repair the bias, data-level approaches change P(y) or P(~y) to make P(y) = P(~y), cost-sensitive learning designates a high cost C1 for class y and a low cost C2 for class ~ y to make P(y) C1 = P(~y) C2, while for decision threshold-moving strategy, it adds a positive value λ for compensating the posterior probability of class y.

2.2 Optimal Decision Threshold-Moving Strategy

As we know, decision threshold-moving is an effective and efficient strategy to address CIL problem. However, we also face a challenge that is how to designate an appropriate moving threshold λ. Some previous work adopt empirical value [9] or trial-and-error method [10] to designate the value for λ, but ignore the specific data distribution, causing over-moving or under-moving phenomenon.

To address the problem above, we present an adaptive strategy for searching the most appropriate moving threshold. The strategy is based on particle swarm optimization (PSO) [15], which is a population-based stochastic optimization technique, inspired by the social behavior of bird flocking. During the optimization process of PSO, each particle dynamically changes its position and velocity by recalling its historical optimal position (pbest) and observing the position of the optimal particle (gbest). On each round, the position of each particle is updated by:

$$\begin{cases} v_{id}^{k+1} = v_{id}^k + c_1 \times r_1 \times (\text{pbest} - x_{id}^k) + c_2 \times r_2 \times (\text{gbest} - x_{id}^k) \\ x_{id}^{k+1} = x_{id}^k + v_{id}^{k+1} \end{cases} \tag{4}$$

where v_{id}^k and v_{id}^{k+1} represent the velocities of the dth dimension of the ith particle in the kth round and the (k + 1)st round, while x_{id}^k and x_{id}^{k+1} denote their positions, respectively. c1 and c2 are two nonnegative constants that are called acceleration factors, while r1 and r2 are two random variables in the range of [0, 1]. In this study, the size of particle swarm and the search times are both set as 50, as well c1 and c2 are both set to 1. Meanwhile, the position x is restricted in the range of [0, 1] with considering the upper limit of a posterior probability is 1, and the velocity v is restricted between −1 and 1.

As for the fitness function, it should directly associate with the classification performance. We all know in CIL, accuracy is not an appropriate performance evaluation metric, thus we use a widely used CIL performance evaluation metric called G-mean as fitness function, which could be described as below,

$$\text{G-mean} = \sqrt{\text{TPR} \times \text{TNR}} \tag{5}$$

where TPR and TNR indicate the accuracy of the positive and negative class, respectively.

2.3 Description About GNB-ODTM Algorithm

Combining GNB and the optimization strategy presented above, we propose an optimal decision threshold-moving algorithm for GNB named GNB-ODTM. The flow path of the GNB-ODTM algorithm is simply described as follows:

Algorithm: GNB-ODTM.
Input: A skewed binary-class training set Φ, a binary-class testing set Ψ.
Output: An optimal moving threshold λ^*, the G-mean value on the testing set Ψ.
Procedure:

1) Train a GNB classifier on Φ;

2) Calculate the posterior probabilities of each instance in Φ, and hereby calculate the original G-mean value on Φ;

3) Call PSO algorithm and use the training set Φ to find the optimal moving threshold λ^*;

4) Adopt the trained GNB classifier to calculate the posterior probabilities of each instance in Ψ;

5) Tune the posterior probabilities in Ψ by the recorded λ^*;

6) Calculate the G-mean value on the testing set Ψ by using the tuned the posterior probabilities.

From the procedure described above, it is not difficult to observe that in comparison with empirical moving threshold setting, the proposed GNB-ODTM algorithm must be more time-consuming as it needs to conduct an iterative PSO optimization procedure. However, the time-complexity can be decreased by assigning small iterative times and population as soon as possible, which is also helpful for reducing the possibility of making classification model be overfitting. Moreover, we also note that the GNB-ODTM algorithm is self-adaptive, which means it is not restricted by data distribution, and meanwhile it can adapt any data distribution type without exploring it.

3   Results and Discussions

3.1   Description About the Used Data Sets

We collected 8 class imbalance data sets from UCI machine learning repository which is avaliable at: http://archive.ics.uci.edu/ml/datasets.php. The detailed information about these data sets is presented in Table 1. Specifically, these data sets have also been used in our previous work about class imbalance learning [16].

Table 1. Description about the used data sets

Data set	Number of attributes	Number of instances	Minority class	Majority class	Class imbalance ratio
abalone9	8	4177	Class 9	Remainder classes	5.06
abalone19	8	4177	Class 19	Remainder classes	129.53
pageblocks2345	10	5473	Class 2 ~ 5	Class 1	8.77

(*continued*)

Table 1. (*continued*)

Data set	Number of attributes	Number of instances	Minority class	Majority class	Class imbalance ratio
pageblocks5	10	5473	Class 5	Class 1 ~ 4	46.59
cardiotocographyC5	21	2126	Class 5	Class 1 ~ 4, class 6 ~ 10	28.53
cardiotocographyN3	21	2126	NSP3	NSP1, NSP2	11.08
Credit card clients	23	10000	Default payment next month1	Default payment next month 0	3.46
Wilt	5	4839	Class 1	Class 2	17.54

3.2 Analysis About the Results

We compared our proposed algorithm with GNB [13], GNB-SMOTE [7], CS-GNB [8], GNB-THR [9] and GNB-OTHR [10] in our experiments. All parameters in PSO have been designated in Sect. 2. In addition, to guarantee the impartiality of experimental comparison, we adopted external 10-fold cross-validation and randomly conducted it 10 times to provide the average G-mean as the final result.

Table 2 shows the comparable results of various algorithms, where on each data set, the best result has been highlighted in boldface.

From the results in Table 2, we observe:

1) In comparison with original GNB, no matter associating it with resampling, cost-sensitive learning or decision threshold-moving techniques could promote classification performance on imbalanced data sets. The results indicate the necessity of adopting CIL technique to address imbalance classification problem, again.
2) It is difficult to compare the quality of resampling and cost-sensitive learning as each of them performs better on partial data sets. GNB-SMOTE performs better on abalone9, pageblocks5, cardiotocographyC5 and cardiotocographyN3, while CS-GNB produces better result on rest data sets.
3) Although GNB-THR significantly outperforms to the original GNB model, it is obviously worse than several other algorithms. It indicates the unreliability of setting moving threshold by empirical approach.
4) We beleive the proposed GNB-ODTM algorithm is successful as it has produced the best result on nearly all data sets except pageblocks2345 and cardiotocographyN3. In addition, we observe on mst data sets, the performance promotion is remarkable by adopting the proposed algorithm. It should attribute to the consideration of distribution self-adaption. Although the proposed GNB-ODTM algorithm has a higher time-complexity than several other algorithms, it is still an excellent altinative for processing imbalance data classification problem.

Table 2. G-mean performance of various comparable algorithms on 8 data sets

Data set	GNB	GNB-SMOTE	CS-GNB	GNB-THR	GNB-OTHR	GNB-ODTM
abalone9	0.2793	0.6318	0.6279	0.5710	0.6329	**0.6651**
abalone19	0.0000	0.6175	0.6428	0.4930	0.6227	**0.7023**
pageblocks2345	0.8506	0.9298	**0.9441**	0.8751	0.9336	0.9420
pageblocks5	0.4716	0.9360	0.9229	0.9146	0.9322	**0.9460**
cardiotocographyC5	0.6799	0.8845	0.8736	0.7851	0.8564	**0.8991**
cardiotocographyN3	0.9077	**0.9491**	0.9256	0.8672	0.9333	0.9412
Credit card clients	0.5731	0.6885	0.6914	0.5984	0.6993	**0.7296**
Wilt	0.1026	0.9687	0.9711	0.7232	0.9704	**0.9799**

4   Concluding Remarks

In this study, we focus on a specific class imbalance learning technique named decision threshold-moving strategy. A common problem about this technique is indicated, i.e., it generally lacks adaption to data distribution, further causing unreliable classification results. Specifically, in the context of Gaussian Naive Bayes classification model, we presented a robust decision threshold-moving strategy and proposed a novel CIL algorithm called GNB-ODTM. The experimental results have indicated the effective and superiority of the proposed algorithm.

The contribution of this paper is two-folds which are described as follows:

1) In context of Gaussian Naive Bayes classifier, we analyze the hazard of skewed data distribution in theory, and indicate rationality of several popular CIL techniques;
2) Based on Particle Swarm Optimization technique, we propose a robust decision threshold-moving algorithm which can adapt various data distribution.

The work was supported by Natural Science Foundation of Jiangsu Province of China under grant No.BK20191457.

References

1. Branco, P., Torgo, L., Ribeiro, R.P.: ACM Comput. Surv. **9** (2016)
2. Dai, H.J., Wang, C.K.: Int. J. Med. Inform. **129** (2019)
3. Malhotra, R., Kamal, S.: Neurocomputing **343** (2019)
4. Qian, Y., Ye, S., Zhang, Y., Zhang, J.: Gene **741** (2020)
5. D. Veganzones, E. Severin: Decis. Support Syst. **112** (2018)
6. Ng, W.W.Y., Hu, J., Yeung, D.S., Yin, S., Roli, F.: IEEE Trans. Cybern. **45** (2015)
7. Chawla, N., Bowyer, K.W., Hall, L.O.: J. Artif. Intell. Res. **16** (2002)
8. Veropoulos, K., Campbell, C., Cristianini, N.: IJCAI (1999)
9. Lin, W.J., Chen, J.J.: Brief. Bioinform. **14** (2013)

10. Yu, H., Mu, C., Sun, C., Yang, W., Yang, X., Zuo, X.: Knowl.-Based Syst. **76** (2015)
11. Tang, B., He, H.: Pattern Recogn. **71** (2017)
12. Lim, P., Goh, C.K., Tan, K.C.: IEEE Trans. Cybern. **47** (2016)
13. Berrar, D.: Encyclopedia of Bioinformatics and Computational Biology: ABC of Bioinformatics. Elsevier Science Publisher, Amsterdam (2018)
14. Griffis, J.C., Allendorfer, J.B., Szaflarski, J.P.: J. Neurosci. Methods **257** (2016)
15. Shi, Y., Eberhart, R.C.: CEC (1999)
16. Yu, H., Sun, C., Yang, X., Zheng, S., Zou, H.: IEEE Trans. Fuzzy Syst. **27** (2019)

Some Problems of Complex Signal Representation

JingBo Xu[✉]

Information Engineering University, Zhengzhou, China
15937101761@139.com

Abstract. The time domain signal is based on the decomposition of the unit step signal, the complex signal is represented by the Heaviside Function, and the problem of the definition of the original jump time in the new function is proposed, based on the analysis and comparison of simple signal and complex signal in time domain and frequency domain, the problems needing attention in using $\varepsilon(t)$ to express signal are put forward. It is concluded that no definition or special definition of the "0" moment in the original unit step signal does not affect the composition of the composite function.

Keywords: Unit step signal · Compound signal · "0" Moment

1 The Introduction

Complex signals can be easily expressed by linear combination of step signals and delay signals. In addition [1, 2], the step function is used to represent the action interval of the signal, so that the piecewise defined function can be expressed into a unified form by the step function, and the function is cut or the piecewise defined function is unified into the function defined on the whole number line, which often makes the function representation simple and easy, and simplifies the operation, and reduces the error. The study of some characteristics of complex signals becomes convenient and easy. Using the characteristic of linear time-invariant system [3], the spectrum of complex signal can be studied and discussed through the spectrum of unit step signal and the characteristics of frequency domain, so as to reduce the calculation difficulty of complex signal spectrum.

2 Complex Functions Are Represented by Unit Step Functions

Generally, in the definition of the unit step function $\varepsilon(t)$ [4], the time of "0" is undefined or defined as "0.5" according to requirements, i.e. $\varepsilon(t) = \begin{cases} 1 & t > 0 \\ 0.5 & t = 0 \\ 0 & t < 0 \end{cases}$ or

© The Author(s) 2022
Z. Qian et al. (Eds.): WCNA 2021, LNEE 942, pp. 844–849, 2022.
https://doi.org/10.1007/978-981-19-2456-9_86

$$\varepsilon(t) = \begin{cases} 1 & t > 0 \\ \text{no definition} & t = 0 \\ 0 & t < 0 \end{cases}$$ [5, 6], Thus, when complex functions are represented by linear combinations of unit step functions [7], undefined points occur within the defined interval [8]. As shown in Fig. 1, 2 and 3, is $f(t)$ equal to the sum of $f_1(t)$ and $f_2(t)$? Since the unit step function is undefined at time "0", should the value at time "0" be added to the sum of $f_1(t)$ and $f_2(t)$ to equal f(t)? Can you express the Fourier transform of $f(t)$ using the Fourier transform of $f_1(t)$ and the linear properties of the Fourier transform?

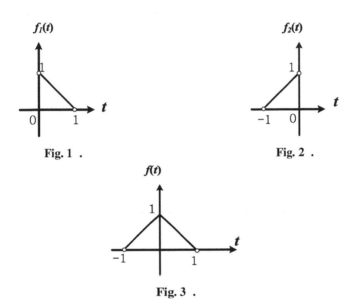

Fig. 1 .   Fig. 2 .

Fig. 3 .

2.1 $f_1(t), f_2(t)$ for $f(t)$

Using unit step signals ε(t) to describe, $f_1(t), f_2(t)$, and $f_1(t), f_2(t)$ for $f(t)$

$$f_1(t) = (-t + 1)[\varepsilon(t) - \varepsilon(t - 1)], \quad f_2(t) = (t + 1)[\varepsilon(t + 1) - \varepsilon(t)].$$

According to the definition 1 of ε(t), the functions in the above two equations are not defined at the time of "0", "1" and "-1". Is $f(t)$ properly represented by the sum of $f_1(t)$ and $f_2(t)$? The waveform shows that $f_1(t)$ and $f_2(t)$ are not defined at the "0" moment, but the value of $f(t)$ is "1". Does this mean that the value of "0" moment is missing? The following is a demonstration of the relationship between the frequency domain and the time domain.

2.2 $F_1(\omega)$ for $F(\omega)$

That's the sum of the Fourier transform of $f_1(t)$ and the Fourier transform of $f_2(t)$, compared to the Fourier transform of $F(\omega)$. Since $f_1(t) = (-t + 1)[\varepsilon(t) - \varepsilon(t - 1)]$, using the linear, time-shift, and frequency-domain differential properties of common Fourier transform, we can ge:

$$F_1(\omega) = \int_{-\infty}^{\infty} (-t+1)e^{-j\omega t}\,dt = \int_{0}^{1} (-t+1)e^{-j\omega t}\,dt$$

$$F_2(\omega) = \int_{-1}^{0} (t+1)e^{-j\omega t}\,dt$$

$$F_1(\omega) + F_2(\omega) = \int_{-1}^{0} (t+1)e^{-j\omega t}\,dt + \int_{0}^{1} (-t+1)e^{-j\omega t}\,dt$$

$$= -\int_{0}^{1} te^{-j\omega t}\,dt + \int_{0}^{1} e^{-j\omega t}\,dt + \int_{-1}^{0} te^{-j\omega t}\,dt + \int_{-1}^{0} e^{-j\omega t}\,dt$$

$$= -\int_{0}^{-1} te^{j\omega t}\,dt + \int_{-1}^{1} e^{-j\omega t}\,dt + \int_{-1}^{0} te^{-j\omega t}\,dt$$

$$= \int_{-1}^{1} e^{-j\omega t}\,dt + \int_{-1}^{0} t(e^{-j\omega t} + e^{j\omega t})\,dt$$

$$= \int_{-1}^{1} e^{-j\omega t}\,dt + 2\int_{-1}^{0} t\cos\omega t\,dt$$

Take Two integrals separately

$$\int_{-1}^{1} e^{-j\omega t}\,dt = -\frac{1}{j\omega}e^{-j\omega t}\Big|_{-1}^{1} = -\frac{1}{j\omega}(e^{-j\omega} - e^{j\omega}) = \frac{1}{j\omega}2j\sin\omega = 2\frac{\sin\omega}{\omega} \quad (2.2\text{--}1)$$

$$2\int_{-1}^{0} t\cos\omega t\,dt = \frac{2}{\omega}\int_{-1}^{0} td\sin\omega t = \frac{2}{\omega}\left(t\sin\omega t\Big|_{-1}^{0} - \int_{-1}^{0}\sin\omega t\,dt\right)$$

$$= \frac{2}{\omega}\left(-\sin\omega + \frac{1}{\omega} - \frac{1}{\omega}\cos\omega\right)$$

$$= -\frac{2}{\omega}\sin\omega + \frac{2}{\omega^2} - \frac{2}{\omega^2}\cos\omega \quad (2.2\text{--}2)$$

Add (2.2–1) and (2.2–2):

$$\frac{2}{\omega^2} - \frac{2}{\omega^2}\cos\omega = \frac{4}{\omega^2}\sin^2\frac{\omega}{2} = s_a^2\left(\frac{\omega}{2}\right) \quad (2.2\text{--}3)$$

From the Fourier transform of the commonly used signal, we can see that the Fourier transform $F(\omega) = s_a^2(\frac{\omega}{2})$ of the signal $f(t)$ in Fig. 3 is the same as formula (2.2–3), And by the one-to-one correspondence between the Fourier transform and the primitive function, we get $f(t) = f_1(t) + f_2(t)$.

2.3 The Temporal Interpretation of $f(t) = f_1(t) + f_2(t)$ Holds

From the time domain, $f(t) = f_1(t) + f_2(t)$

$$f_2(t) = (t+1)[\varepsilon(t+1) - \varepsilon(t)], f_1(t) = (-t+1)[\varepsilon(t) - \varepsilon(t-1)],$$

$$f(t) = f_1(t) + f_2(t) = (-t+1)[\varepsilon(t) - \varepsilon(t-1)] + (t+1)[\varepsilon(t+1) - \varepsilon(t)]$$

$$= (-t+1)\varepsilon(t) - (-t+1)\varepsilon(t-1) + (t+1)\varepsilon(t+1) - (t+1)\varepsilon(t)$$

$$= -2t\varepsilon(t) - (-t+1)\varepsilon(t-1) + (t+1)\varepsilon(t+1)$$

The function graph is shown below 2.3–1(a).

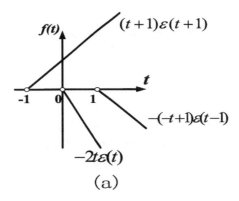

(a)

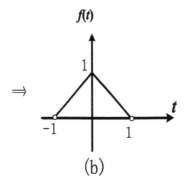

(b)

Fig. 4 .

As you can see from the figure, the value of $f(t)$ at t $= 0$ can be determined by $(t+1)\varepsilon(t+1)$, so Fig. 4(a) is the sum of three straight lines to Fig. 4(b). The fact that $f_1(t)$ and $f_2(t)$ are undefined at the "0" moment and that the value of $f(t)$ is "1" does not mean that the value of the "0" moment is missing and that it does not require $f_1(t)$

and $f_2(t)$ to add the value of "0" to get $f(t)$. When the functions defined by the step signal form a combined function, some overlapping undefined points can be naturally compensated in the process of function combination.

3 The Conclusion

Similar to the above, many functions defined by $\varepsilon(t)$ when the combination of some overlap undefined points in the process of function combination can be made up naturally, without adding. As the Common Gate Function $G_\tau(t)$ (Fig. 5).

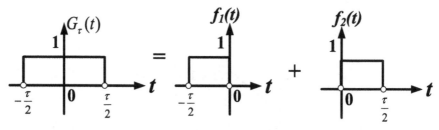

Fig. 5. Function $G_\tau(t)$

What is reasonable and right to deal with an undefined "0" moment? In this paper, two examples of Combined functions are given, and the problems needing attention in using $\varepsilon(t)$ to express signals are put forward.

References

1. Wu, D.: Signal and Linear System Analysis, 4th edn., pp. 47–49. Higher Education Press, Beijing (2008)
2. Zhang, H.: China's Science Popularization. Baidu Baike, Beijing (2018)
3. Guan, Z.: Signal and Linear System Analysis, 5th edn., pp. 26–27. Higher Education Press, Beijing (2011)
4. Junli signal system [M] Beijing: Higher Education Press, 2000
5. Oppenheim Signals and Systems. Science and Technology Press, Hangzhou (1991)
6. Tube Chih Signal and Linear System. Higher Education Press, Beijing (1992)
7. Yu, J., Shi, W.Y., Lu, C., Tang, D.Y.: Point layout optimization based on multi-signal flow graph and differential evolution algorithm. J. Instrument. **12**, 2750–2751 (2016)
8. Mao, M., Ma, Y.: Circuit gain by signal flow diagram. Ind. Instrument. Autom. **3**, 83–85 (2015)

New Principle of Fault Data Synchronization for Intelligent Protection Based on Wavelet Analysis

Zuowei Wang[1]([✉]), Hong Zhang[2], Dongchao Liu[3], Shiping E.[2], Kanjun Zhang[1], Haitao Li[3], Hengxuan Li[1], and Zhigang Chen[3]

[1] State Grid Hubei Electric Power Research Institute, Wuhan, China
48191744@qq.com
[2] State Grid Hubei Electric Power Company, Wuhan, China
[3] NR Electric Co., Ltd., Nanjing, China

Abstract. In order to eliminate the influence of the delay error of the sampled value in the data link on the longitudinal differential protection device, this paper proposes a protection data self-healing synchronization algorithm based on wavelet transform to calculate the moment of sudden change. First, calculate the mutation amount of the sampled data at each end in real time. When the mutation amount threshold is exceeded, it is determined that the multi-terminal system has a short-circuit fault. Then, according to the sudden change characteristics of the collected current waveform, the wavelet modulus maximum value is used to extract the fault sudden change time of each end data, based on the fault time at one terminal, the automatic compensation for the time differences between this terminal and others are realized, thus a new sampling sequence is formed. The resynchronized sampling sequences are used to calculate the differential current and braking current after fault to ensure the correct action of the protective device. Through theoretical analysis and simulations, the correctness and effectiveness of the proposed algorithm is verified; in addition, it is shown that this algorithm can improve the reliability of actions by the intelligent protection device, thus realizing protections such as multi-terminal differential, wide-area differential, etc.

Keywords: Mutation · Wavelet transform · Multi-terminal longitudinal differential protection · Wide-area differential protection · Synchronization algorithm · Intelligent protection

1 Introduction

With the advancement of smart grid and information technology, the research and application of the principle of multi-terminal longitudinal differential protection has received extensive attention [1–4]. The device of the multi-terminal longitudinal differential protection needs to obtain remote sampling data. These data transmission paths are different, and there will be time errors due to link blockage during transmission, so data at different collection points need to have accurate synchronization processing methods, which

© The Author(s) 2022
Z. Qian et al. (Eds.): WCNA 2021, LNEE 942, pp. 850–861, 2022.
https://doi.org/10.1007/978-981-19-2456-9_87

can ensure the synchronization of data and the accuracy of fault calculation and discrimination [5]. Data synchronization includes synchronous sampling and data window synchronization. Generally, intelligent multi-terminal longitudinal differential protection usually adopts data acquisition based on satellite and high-precision clock synchronization, and data transmission adopts high-speed optical fiber wide-area self-healing network. After adopting methods such as synchronous pulse sampling and resampling, the delay error of the data in the transformer and the sampling link can be effectively compensated, but the delay error caused by the data in the transmission link requires an effective method to realize the data window synchronization. Synchronization methods include data time-scaling method, link fixed delay compensation method, etc. [6–10]. Literature [11] proposed a fault current fundamental wave zero-crossing point identification method to solve the difficulty of protection data synchronization, and pointed out the huge cost of multi-terminal and wide-area differential protection data synchronization technology; Literature [12] analyzed the shortcomings of multiple synchronization clock methods, and proposed a network-wide time synchronization scheme based on sparse phasor measurement unit PMU; Literature [13] proposed a network sampling synchronization method based on an external reference clock source; Literature [14] proposed a data synchronization method based on clock difference to solve the problem of inconsistent data synchronization between two-way channel routing in a self-healing ring network.

According to the requirements of the specification, the relay protection device should not rely on the external time synchronization system to realize the protection function, so the data time stamping method is usually not adopted. The link fixed delay compensation method usually first measures the rated delay value of the data transmission link, and then compensates the delay error between the data according to the fixed delay value to achieve synchronization. The disadvantage is that the link delay has some uncertainties, so the delay compensation method has errors.

For the protection of the multi-terminal longitudinal differential principle, it is necessary to obtain remote sampling values to judge the fault interval. The data transmission distance is so long and the link segments are much more than we expected. During the data transmission process, link congestion and routing self-healing reconstruction may occur due to data storms. Therefore, the uncertainty of the transmission delay of the data will cause a large phase difference, calculation error and even a wrong operation of the protection [15, 16]. For the new wide-area differential protection that needs to adaptively construct the protection range according to the grid network topology, the end points and data links of the protection are not fixed, and the transmission delay of the data at each end is more uncertain. It is necessary to eliminate delay errors to ensure that the data between each end is synchronized. For UHV systems, the transmission distance is longer, the data communication volume is larger, and the data link is more complicated. The endpoints and normal communication links that constitute the multi-terminal longitudinal differential principle protection are fixed, but the end points of the wide area differential protection and the normal communication link may not be fixed. The possibility of a large delay error between the sampled data at each end is higher, and the possibility of the protection device's erroneous action is also higher.

This paper proposes a self-healing synchronization algorithm for relay protection data based on wavelet transform to calculate sudden changes. Aiming at the delay of current sampling data in the transmission process due to communication link problems in the power system, it is assumed that the sudden changes of multi-terminal faults are accurately collected. Under the premise, according to the characteristics of the waveform mutation of the sampled data during the short-circuit fault, the value of each sampled data is calculated. At the moment of the fault sudden change, the time difference is compensated by this, the sampling data synchronization is realized, and the application of the algorithm in the multi-terminal system is studied.

2 Mutation Algorithm and Data Delay Error

2.1 The Method of Calculating the Abrupt Change by Wavelet Transform

After the line fails, the waveform has abrupt and singularity. The traditional Fourier transform analysis method and the time domain analysis method will produce large errors, and the wavelet analysis has a good ability to detect the sudden change of the signal.

Let $\Psi(x)$ be the basis wavelet, $f_w(a, b)$ represents the continuous wavelet transform of the signal $f(x) \in L^2(R)$, which can be expressed as

$$f_w(a, b) = \frac{1}{\sqrt{a}} \int_{-\infty}^{+\infty} f(x) \Psi^* \left(\frac{x-b}{a} \right) = \langle f(x), \Psi_{a,b}(x) \rangle \tag{1}$$

In the formula: a is the expansion factor; b is the translation factor; $\Psi_{a,b}(x)$ is the wavelet function that selects the basis wavelet $\Psi(x)$ corresponding to a and b.

The modulus maximum point of the wavelet transform corresponds to the current fault time one-to-one. The wavelet modulus maximum point indicates that the signal has the largest rate of change at this point.

2.2 The Impact of Data Delay on the Performance of Longitudinal Differential Drotection

The multi-terminal system has m-side power supply and multi-terminal longitudinal differential protection. The differential current I_d and braking current I_r of each branch in the protection area can be expressed as

$$\begin{cases} I_d = \left| \sum_{j=1}^{m} I_j \right| \\ I_r = \sum_{j=1}^{m} \left| I_j \right| \end{cases} \tag{2}$$

In the formula, $\dot{I}_j$ is the current phasor of branch j.

In normal system operation and out-of-area faults, the differential current is 0 under ideal conditions, and the actual value is the unbalanced current caused by measurement errors and other factors, while the braking current is relatively large; when the system has an area fault, the differential current is the sum of the fault currents provided by each branch, the differential current value is larger, and the protection should satisfy the action equation for reliable action. The differential protection action equation can be expressed as

$$\begin{cases} I_d \geqslant k_r I_r \\ I_d \geqslant I_{op} \end{cases} \tag{3}$$

Where: k_r is the braking coefficient; I_{op} is the starting current.

The multi-terminal longitudinal differential protection uses the optical fiber network to transmit the sampled signal. The signal propagation speed in the optical fiber is about 2/3 of the speed of light in vacuum, the signal delay is about 5 μm/km, and the signal is converted, processed, and relayed. Additional delays are also generated in links such as relays and switches.

For multi-terminal longitudinal differential protection that needs to collect large-scale multi-point data, it is easy to sample data from each branch. But due to long data link transmission distance, channel congestion, data packet loss, route switching, etc. Loss of synchronization results in a phase difference. The relationship between the delay time difference between data Δt_{ER} and the phase difference $\Delta \varphi_{ER}$ can be expressed as

$$\Delta \varphi_{ER} = \omega_N \Delta t_{ER} \tag{4}$$

In the formula, ω_N is the power frequency angular velocity. In normal operation or an out-of-zone fault, the phase error of the two current phasors with the amplitude of I_m due to the delay error, the unbalanced differential current and the braking current are

$$\begin{cases} I_d = 2I_m sin(\Delta t_{ER}/2) \\ I_r = 2I_m \end{cases} \tag{5}$$

In the case of an out-of-zone fault, the differential protection action Eq. (3) can be expressed as

$$\frac{I_d}{I_r} = sin(\Delta t_{ER}/2) \geqslant k_r \tag{6}$$

In the case of an area fault, the delay error will also bring errors to the calculation of the differential current. The differential current and the braking current are

$$\begin{cases} I_d = 2I_m cos(\Delta t_{ER}/2) \\ I_r = 2I_m \end{cases} \tag{7}$$

In the event of a fault in the area, the differential protection action Eq. (3) can be expressed as

$$\frac{I_d}{I_r} = cos(\Delta t_{En}/2) \geqslant k_r \tag{8}$$

Table 1 shows the delay error, phase error, and the ratio of the internal and external differential current I_d to the braking current I_r of the two current phasors whose amplitudes are both Im when the fault occurs outside and inside the area.

It can be seen from Table 1 that with the increase of the delay error, the ratio shows a decreasing and increasing trend when the internal and external faults occur, and they are equal when the delay error reaches 5 ms. There is an intersection, so the delay error will bring obvious errors to the differential current calculation, and the protection device may cause the protection to malfunction or refuse to operate due to the loss of synchronization of the sampling data.

Table 1. Phase error and ratio of differential/braking current of different time delay error

Delay error /ms	Phase error /(°)	Id/Ir	
		External fault	Internal fault
0	0	0	1.000
1	18.00	0.156	0.988
2	36.00	0.309	0.951
3	54.00	0.454	0.891
4	72.00	0.588	0.809
5	90.00	0.707	0.707

For a double-ended line, the currents at each end are $\dot{I}_1$ and $\dot{I}_2$ respectively. If the differential current Id and the braking current I_r are

$$\begin{cases} I_d = \left|\dot{I}_1 + \dot{I}_2\right| \\ I_r = \left|\dot{I}_1 - \dot{I}_2\right| \end{cases} \tag{9}$$

Then the actual action equations when the fault occurs outside the zone and the zone are respectively

$$\frac{I_d}{I_r} = \tan(\Delta t_{ER}/2) \geq k_r$$

$$\frac{I_d}{I_r} = \arctan(\Delta t_{ER}/2) \geq k_r \tag{10}$$

Since the value of the tangent function is greater than the sine, it is more prone to malfunction when using this action equation in the case of an out-of-zone fault.

3   Principle of Self-healing Synchronization Algorithm for Mutation Data

In order to eliminate the influence of the delay error of the sampled value in the data link on the protection device, this paper proposes a data self-healing synchronization algorithm based on wavelet transform to calculate the moment of sudden change. The principle is that when a short-circuit fault occurs in the power system, after the protection device receives the sampling the data, first calculate the failure mutation time of each data mutation amount, and according to the data failure mutation time, compensate the transmission time error between each sampling value, realize the synchronization of the failure data sequence, and use the resynchronized sampling value to calculate the failure differential current and braking current value, realize the principle of multi-side differential and wide-area differential protection.

For m-terminal longitudinal differential protection, the received data includes m-terminal sampling data, and a fault occurs at time n, and the protection device actually receives the current data sequence at terminal j at time n as $i_j(k_j)$, j = 1, 2,..., M, as shown in Fig. 1, the data transmission delay is

$$\Delta t_j = k_j - n \tag{11}$$

In the formula: n is the time when the fault occurs; k_j is the mutation moment actually received by the protection.

By calculating the time of the sudden change of the data at each end, the time difference between the accepted current sequence $i_i(k_i)$ and $i_j(k_j)$ at the i-end can be calculated, as shown in Fig. 1, the time difference Δt_{ji} can be expressed as

$$\Delta t_{ji} = k_j - k_i \tag{12}$$

By compensating the time difference Δt_{ji} between the sequence $i_i(k_i)$ and $i_j(k_j)$, a new i-terminal current sequence $i_i(n + \Delta t_{ji})$ is obtained. Similarly, the current sequence of the other terminals after compensation is calculated, and then it is compared with the j-terminal current sequence $ij(k_j)$. Calculate the differential current, as shown in Fig. 1, the m-terminal longitudinal differential current is

$$i_d(n) = \left| \sum_{i \neq j}^{M-1} i_i\left(n + \Delta t_{ji}\right) + i_j(n) \right| \tag{13}$$

By calculating the moment of sudden change in the current sequence at each end, the time difference caused by the delay of the transmission link is compensated, the additional phase error of the current sequence at each end is eliminated, the current sequence at each end can be resynchronized, and the protection device can correctly calculate the post-fault differential current, judge the fault section, avoid the wrong operation of the protection device due to the delay error of the data transmission link.

When the system is running normally, the electrical quantity at each end does not produce a sudden change, and the sudden change method cannot be used to achieve synchronization. At this time, the phase difference of the current at each end constituting the differential protection is small, and the fixed delay compensation method and the waveform zero-crossing point detection can be used to achieve data synchronization.

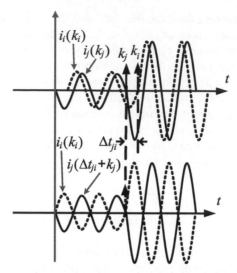

Fig. 1. Schematic of mutation data synchronization algorithm

4 Simulation Verification of Self-healing Synchronization Algorithm for Mutation Data

Use PSCAD to establish a 500 kV multi-terminal power grid system simulation model, as shown in Fig. 2, simulate the internal and external short-circuit faults under various operating conditions in the system, collect fault current signals at each end of the system, write simulation programs, and simulate sampling data is transmitted to the protection device through the optical fiber communication channel, random delay errors are generated due to factors such as channel distance, congestion, route self-healing or reconstruction, which causes the sampling data received by the protection device to lose synchronization, and this paper proposes wavelet transform to calculate the sudden change amount data self-healing synchronization algorithm resynchronizes and corrects the data to ensure that the protection device correctly judges the fault zone.

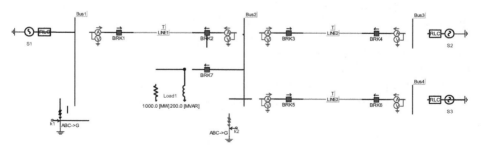

Fig. 2. PSCAD simulation principle of multi-terminal power system

4.1 External Fault Simulation Analysis

When the external fault point F1 in Fig. 2 is short-circuited, the multi-terminal longitudinal differential protection device receives the current at each end through the communication channel. For the convenience of observation, the A-phase current on each side is selected for analysis. As shown in Fig. 3(a), from the current waveform, the currents on each side that should have abrupt changes at the time of the fault are obviously out of synchronization. After eliminating the influence of the distributed capacitance of the line through current compensation and eliminating the influence of the non-periodic component in the sampled data, the phase A current on each side is shown in Fig. 3(b). It can be seen from Fig. 3 (a) and (b) that there is a significant phase difference between the short-circuit currents of phase A at each end, and a large differential current will be generated when an external fault occurs. The following calculation methods need to be used to calculate the differential current and braking current

$$\begin{cases} I_d = |\dot{I}_1 + \dot{I}_2 + \dot{I}_3| \\ I_r = |I_1| + |I_2| + |I_3| \end{cases} \tag{14}$$

Using the sudden change data self-healing synchronization algorithm proposed in this paper, the sampled data on each side can be resynchronized according to the sudden change time. After synchronization, the short-circuit current of each endpoint and the calculated differential current phase A waveform are shown in Fig. 3.

As shown in (c), it can be seen that the short-circuit current at each end after resynchronization eliminates the phase difference and only has a small differential current. Perform simulation programming on the differential current Id and braking current Ir of the multi-terminal longitudinal differential protection, calculate the effective value of the differential current Id and braking current Ir, and draw the braking curve, as shown in Fig. 3(d), including From the moment when the first fault mutation occurs on one side of the line, to the last side mutation.

Sampling data several cycles after the time, where the arrow is the direction of the order of data change over time. It can be seen from Fig. 3(d) that from the moment of the first sudden change to the sudden change on each side, the differential current action characteristic is in the action zone, indicating that in the event of an external fault, the data is out of synchronization or due to factors such as communication congestion. Part of the data is lost, guarantee.

The protective device may malfunction due to too much error in the calculated value of the differential current. The differential current action characteristic after the external fault synchronization is always in the braking zone, indicating that the out-of-synchronization data of the external fault has been resynchronized.

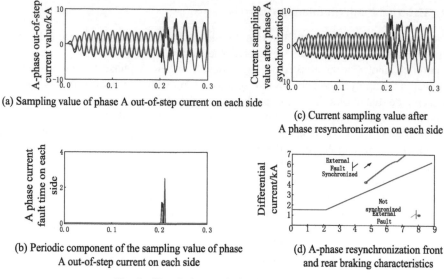

(a) Sampling value of phase A out-of-step current on each side

(c) Current sampling value after A phase resynchronization on each side

(b) Periodic component of the sampling value of phase A out-of-step current on each side

(d) A-phase resynchronization front and rear braking characteristics

Fig. 3. Simulation analysis of external faults

4.2 Ixternal Fault Simulation Analysis

When the internal fault point F2 in Fig. 2 is short-circuited, the protection device receives currents from each end point through the communication channel, and the phase A currents on each side are shown in Fig. 4(a). Obviously out of sync at the moment of sudden change in current on each side.

Phenomenon, after compensating the distributed capacitive current of the line and eliminating the influence of the non-periodic component, the phase A current on each side is shown in Fig. 4(b).

After using the sudden change data self-healing synchronization algorithm proposed in this paper to realize data resynchronization, the short-circuit current and differential current waveform diagram of each end point are shown in Fig. 4(c). It can be seen from Fig. 4(c) that the short-circuit current of each terminal after resynchronization eliminates the phase difference, and the differential current can accurately reflect the fault current.

There is an obvious phase difference in the short-circuit current of phase A at each end. The calculated differential current is greatly reduced, and the braking current is relatively large. The effective value of the current is calculated and the braking curve is drawn, as shown in Fig. 4(d). The sequence direction of the time change, and the differential current action characteristic is in the action zone, indicating that the internal fault out-of-synchronization data can ensure the correct action of the protection device after resynchronization and correction.

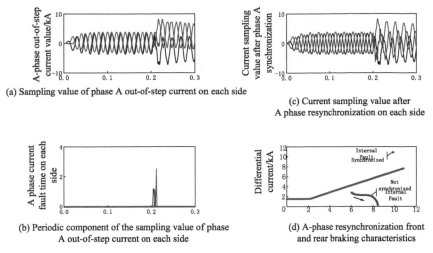

(a) Sampling value of phase A out-of-step current on each side

(c) Current sampling value after
A phase resynchronization on each side

(b) Periodic component of the sampling value of phase
A out-of-step current on each side

(d) A-phase resynchronization front
and rear braking characteristics

Fig. 4. Simulation analysis of internal faults

Figure 4(d) includes data from the moment of the first fault mutation on one side of the line to a few cycles after the last moment of mutation; the differential current action characteristic is in the braking zone, indicating that in the event of an internal fault, Using the failure sampling data that is out of synchronization or partly lost, the protection device may decrease the sensitivity of the action or even refuse to move due to the significant reduction in the calculated value of the differential current.

It can be seen from the simulation analysis that the synchronization of the sampled value is very important for the multi-terminal longitudinal differential protection. When there is a large phase error between the sampled values, it may cause errors in the calculation of the differential current and lead to protection. Misoperation or refusal of operation due to wrong judgment of the fault zone. This paper proposes a multi-terminal longitudinal differential protection mutation data synchronization algorithm based on wavelet transform that can effectively correct the transmission phase error of the sampled data, and automatically realize the multi-side sampled data. The re-synchronization ensures the accuracy of the calculation of the differential current of the multi-terminal longitudinal differential protection and the correctness of the fault interval judgment, and improves the reliability of the multi-terminal longitudinal differential protection and wide-area differential protection.

5 Conclusion

Using the sampling data of the fault current at each end, due to the complexity of the transmission link and the communication problem, and the characteristics of different sudden changes, a multi-terminal longitudinal differential protection based on wavelet transform to calculate sudden change data self-healing synchronization algorithm is proposed. Realize the resynchronization of the sampled data at each end that has lost synchronization, and ensure that the protection device correctly judges the fault interval, thereby improving the reliability of the multi-terminal longitudinal differential protection and the wide-area differential protection. The principle analysis and simulation verification prove the correctness and effectiveness of the algorithm.

This algorithm is not only suitable for multi-terminal longitudinal differential protection based on steady-state components, but also suitable for longitudinal differential protection based on transient components of sampled values. For the wide-area differential principle protection and remote backup protection center based on wide-area information, the use of mutation data self-healing synchronization algorithms or other data synchronization algorithms is even more important to ensure the reliability of protection actions.

References

1. Bao-wei, L., Chuan-kun, N., Xin-tao, D., Xu, L., Zheng, F., Ya-xin, S.: Research on the scheme of the sample synchronization scheme for optical differential protection scheme in merge unit. In: 8th Renewable Power Generation Conference (RPG 2019), pp. 1–6 (2019). https://doi.org/10.1049/cp.2019.0286
2. Wang, Q., Bo, Z., Zhao, Y., Wang, L., Ding, S., Wei, F.: Influence on the performance of multi-terminal differential protection caused by communication time desynchronizing. In: 2015 5th International Conference on Electric Utility Deregulation and Restructuring and Power Technologies (DRPT), pp. 943–946 (2015). https://doi.org/10.1109/DRPT.2015.7432364
3. Igarashi, G., Santos, J.C.: Effects of loss of time synchronization in differential protection of transformers using process bus according to IEC 61850–9–2. In: IEEE PES Innovative Smart Grid Technologies, Europe, pp. 1–6 (2014). https://doi.org/10.1109/ISGTEurope.2014.7028753
4. Ivanković, I., Brnobić, D., Rubeša, R., Rekić, M.: Line differential protection with synchrophasor data in WAMPAC system in control room. In: 2020 3rd International Colloquium on Intelligent Grid Metrology (SMAGRIMET), pp. 72–78 (2020). https://doi.org/10.23919/SMAGRIMET48809.2020.9264020
5. Aichhorn, A., Etzlinger, B., Hutterer, S., Mayrhofer, R.: Secure communication interface for line current differential protection over Ethernet-based networks. In: 2017 IEEE Manchester PowerTech, pp. 1–6 (2017). https://doi.org/10.1109/PTC.2017.7981051
6. Dahane, A.S., Dambhare, S.S.: A novel algorithm for differential protection of untransposed transmission line using synchronized measurements. In: 11th IET International Conference on Developments in Power Systems Protection (DPSP 2012), pp. 1–4 (2012). https://doi.org/10.1049/cp.2012.0025
7. Gao, H., Jiang, S., He, J.: Development of GPS synchronized digital current differential protection. In: POWERCON 1998. 1998 International Conference on Power System Technology. Proceedings (Cat. No.98EX151), vol. 2, pp. 1177–1182 (1998). https://doi.org/10.1109/ICPST.1998.729271
8. Al-Fakhri, B.: The theory and application of differential protection of multi-terminal lines without synchronization using vector difference as restraint quantity - simulation study. In: 2004 Eighth IEE International Conference on Developments in Power System Protection, vol. 2, pp. 404–409 (2004). https://doi.org/10.1049/cp:20040148
9. Villamagna, N., Crossley, P.A.: A symmetrical component-based GPS signal failure-detection algorithm for use in feeder current differential protection. IEEE Trans. Power Delivery 23(4), 1821–1828 (2008). https://doi.org/10.1109/TPWRD.2008.919035

10. Cao, T., Dai, C., Chen, J., Yu, Z.: A new method of channel monitoring for fiber optic line differential protection. In: 2008 China International Conference on Electricity Distribution, pp. 1–4 (2008). https://doi.org/10.1109/CICED.2008.5211747

11. Zhang, H., Peng, L., Xu, H., Xu, H.: Probability analysis on disoperation and misoperation of line current differential protection considering asymmetric delay. Electr. Measur. Instrument. **58**(2), 40–6 (2021)

12. Li, J., Gao, H., Zhigang, W., Bingyin, X., Wang, L., Yang, J.: Data self-synchronization method and error analysis of differential protection in active distribution network. Dianli Xitong Zidonghua/Automation Electr. Power Syst. **40**(9), 78–85 (2016)

13. Li, Z., Wan, Y., Wu, L., Cheng, Y., Weng, H.: Study on wide-area protection algorithm based on composite impedance directional principle. Int. J. Electr. Power Energy Syst. **115**(02), 119–26 (2020)

14. Huang, C., Liu, P., Jiang, Y., Leng, H., Zhu, J.: Feeder differential protection based on dynamic time warping distance in active distribution network. Diangong Jishu Xuebao/Trans. China Electrotech. Soc. **32**(6), 240–247 (2017)

15. Igarashi, G., Santos, J.C.: Effects of loss of time synchronization in differential protection of transformers using process bus according to IEC 61850–9–2. Paper presented at the 2014 IEEE PES innovative smart grid technologies conference Europe, ISGT-Europe 2014, October 12, 2014–October 15, 2014, Istanbul, Turkey (2014)

16. Yang, C.C., Song, G.B., Shen, Q.Y.: A novel principle dispensing with data synchronization for distributed bus protection. Paper presented at the 12th IET international conference on developments in power system protection, DPSP 2014, March 31, 2014–April 3, 2014, Copenhagen, Denmark (2014)

Open-Set Recognition of Shortwave Signal Based on Dual-Input Regression Neural Network

Jian Zhang$^{(\boxtimes)}$, Di Wu, Tao Hu, Shu Wang, Shiju Wang, and Tingli Li

College of Data Target Engineering, Strategic Support Force Information Engineering University, Science Avenue. 62, Zhengzhou 450001, China
gladmen@163.com

Abstract. Open-set recognition in blind shortwave signal processing is an important issue in modern communication signal processing. This paper presents a novel method for this problem. By preprocessing, the signal data matrix and vector diagram are obtained as network input. Then, the network is trained and tested with the known signal, and the upper and lower quintile algorithm is used to obtain the interval threshold for judging the known signal and the distance threshold for intercepting the length range of the unknown signal. Finally, the network is used for numerical regression in open-set range, the threshold combined with kernel density clustering algorithm is used to identify different signals. Simulation results show that the proposed method overcomes the defects of traditional algorithm, which cannot distinguish different types of unknown signals and only applicable for few signal types.

Keywords: Open-set recognition · Shortwave · Dual-input regression neural network · Data stream · Vector diagram

1 Introduction

Due to the flexibility, survivability and long-distance transmission, shortwave communication has always been a reserved and development method in the field of wireless communication. Shortwave signal automatic recognition technology [1] is an important content of signal blind processing and an important basis for subsequent signal analysis, monitoring and countermeasure. With the development of modern shortwave communication technology, shortwave communication shows a trend of diversification of types, fine differentiation of specifications and continuous emergence of new signal types. Most of the traditional signal automatic recognition technologies are concentrated in the closed-set level. When new unknown signal enter the system, the correct result cannot be obtained. Therefore, in order to meet the need of convenience, intelligence and timeliness of modern blind signal processing, it is of great value to carry out the research on efficient open-set recognition technology of shortwave signal.

At present, most traditional signal recognition algorithms as well as algorithms based on deep learning only consider the recognition of known signal types. When a new unknown signal type appears, it will be recognized as one of the known signal, resulting in

© The Author(s) 2022
Z. Qian et al. (Eds.): WCNA 2021, LNEE 942, pp. 862–873, 2022.
https://doi.org/10.1007/978-981-19-2456-9_88

discrimination error. To solve the above problem, Literature [2] proposed a support vector data description (SVDD) algorithm with density scaled classification margin (DSCM), which determines the interval between hypersphere and positive samples according to the relative density proportion of two types of positive training samples, and carries out open-set recognition in combination with support vector description, However, the algorithm can only distinguish 2 types of positive sample signals, and will classify all unknown signal types into one class. Literature [3] extends the algorithm of incremental support vector machine (ISVM) [4] combined with error correcting output codes (ECOC) [5] to multi classification for incremental learning and recognition, but this algorithm cannot solve the forgetting problem in incremental learning. Besides, designing coding matrix requires more priori information, and its multi classification ability is restricted by the coding length, as well as the model needs to be trained every time when a new signal is received, lead to its low efficiency.

The generative adversarial (GA) method is also used to solve the open-set recognition problem. Literature [6] combines the improved intra class splitting (ICS) algorithm with the genetic adversarial algorithm to obtain the boundary signal samples, then trains the boundary signal samples as unknown types of signals and realizes the open-set recognition. However, the process of constructing boundary samples is complex and the effect is unstable, and it also cannot distinguish different types of unknown signal. Literature [7] uses the generative countermeasure network theory to build a reconstruction and discrimination network (RDN) model to identify the modulation types of signals. However, the difference between the reconstructed signal data and the real unknown signal data is difficult to control, and when the known signal types is more than 2, the classification and discrimination mechanism will be very complex, which results in low operability. In addition, it is still unable to distinguish different types of unknown signals.

Some other methods, such as Literature [8] uses the extreme value-weibull distribution to fit the cut-off probability of the distance from the feature to the feature center, combines the classification cross entropy with the center loss, and modifies the output of the dual channel long-short term memory (DCLSTM) network to conduct the modulation recognition. This algorithm proposes the concepts of feature center and feature distance. In some cases, it can distinguish different unknown types of signals, but it cannot distinguish signals of different specifications with the same modulation mode.

From the above analysis, it can be concluded that the current signal open-set recognition algorithms have the following shortcomings: 1) Some algorithms are only applicable to 2 types of known signals, and no longer applicable when the number of known signal type increases; 2) The existed works focus on the signal modulation recognition, the recognition method for different specifications with the same modulation mode is hardly considered; 3) It is difficult to distinguish different types of unknown signals, unknown signals can only be distinguished into one class, called 'unknown class'.

In this paper, we propose a method to transform features of different signals into different regression values, and use these values to distinguish different signals. The contributions of proposed method are described as follow: Firstly, we design a dual-input neural network to fuse and map the feature information extracted from signal data stream and vector diagram. For better feature extraction, we design a network structure based on dense convolution theory. Secondly, different from the traditional recognition network structure, we use the hyperbolic tangent (Tanh) activation function to perform numerical regression on signal features at the end of the network, and establish a one-to-one nonlinear mapping relationship between signal feature and specific value. Thirdly, we test the network in closed-set, using the upper and lower quintile algorithm to obtain the regression discrimination threshold of each known signal and the center distance threshold for unknown signal. Finally, we perform open-set experiments to demonstrate the effectiveness of the proposed method.

2   Distinguishing Features of Shortwave Signal

2.1   Data Stream

Specific shortwave standard has unique generation algorithm and transmission specification. These rules and standards make its signal data stream presents unique information organization format. Taking MIL-STD-188-110A (110A) [9], MIL-STD-188-141B(141B) [10] and Link11 SLEW [11] as an example, the typical information transmission format is shown in Fig. 1.

4320bit Preamble Sequence	N×144bit Data Sequence	32Bit End Field	T+144Bit Flush Field

(a) 110A

768bit Protection Sequence	192bit Preamble Sequence	P×1920Bit Valid Data

(b) 141B

150bit Header Sequence	30Bit Phase Reference	60Bit Start Code	M×30Bit Tactical Data	60Bit Supervised stop code

(c)Link11 SLEW

Fig. 1. Typical transmission format for shortwave 110A, 141B and Link11 SLEW signal. The information format of 110A signal consists of preamble sequence, data sequence, end field and flush field. 141B consists of protection sequence, preamble sequence and valid data. Link11 SLEW consists of header sequence, phase reference sequence, start code, tactical data and Supervised stop code.

We can conclude that the data transmission organization structure of different signals is unique, and the bits of each sequence and field are not the same. These differences make the received 110A, 141B and Link11 data stream present the unique data characteristics of their respective signal. Based on this, if a feature extraction algorithm with high performance and strong robustness can be found for signal data, the feature extracted from signal data stream can be used as recognition criteria to distinguish the type of different shortwave signals.

2.2 Vector Diagram

Vector diagram shows the symbol track by reconstructing two channels of received signal data in time order, not only can distinguish frequency shift keying (FSK) and phase shift keying (PSK), but also can distinguish signals with different PSK modulation modes, as shown in Fig. 2. The symbols of PSK signals have a fixed phase, so the vector diagram is in the form of constellation point and symbol trajectory, while the phase of FSK signals is random during symbol conversion, so the vector diagram is in the form of circle.

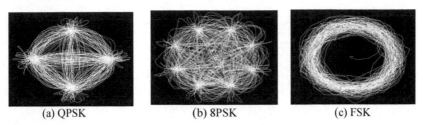

| (a) QPSK | (b) 8PSK | (c) FSK |

Fig. 2. Vector diagram of shortwave signal. It shows signal with different modulation mode has different vector diagram forms.

In this paper, the signal vector diagram is used as the supplementary feature extraction source. By powerful feature processing ability of neural network, the different feature information of signal specification represented by data flow and the modulation feature information represented by vector diagram is fused, and then learned and mapped, to further improve the performance of signal recognition.

3 Proposed Method

In this section, we first describe the dual-input neural network architecture of our method, then we present the algorithm for obtaining the discrimination threshold. Finally, we demonstrate the procedure of the proposed scheme.

3.1 Dual-Input Regression Neural Network

Regression analysis (RA) is a statistical analysis method to determine the relationship between two or more variables. We construct dual-input regression neural network to map the extracted signal feature to specific value. By using the difference of numerical regression result, we can distinguish different signals in open-set range.

The proposed dual-input regression neural network is illustrated in Fig. 3. The feature extraction is conducted by 7 feature extraction modules. The structure of feature extraction module is shown in Fig. 4. The network connects adjacent feature extraction module through the transformation module, each transformation module contains a 1 × 1 convolution and a 2 × 2 average pool. After extracting the feature via the above $(66+18) \times 2 + 5 = 173$ layers network and conduct a 7 × 7 global average pool, the acquired feature information are fused by concatenation, and then establish the nonlinear relationship between signal feature and specific value by regression processing. Except for the end of the network, the rectified linear unit (ReLu) is used in each layer. During the compilation and optimization of the network, the Adam algorithm is used to work out the optimal solution of the network structure parameters.

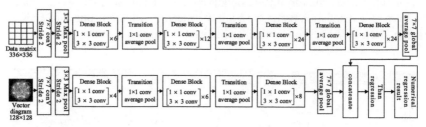

Fig. 3. Structure of dual-input regression neural network. The data matrix branch contains 4 feature extraction modules and the vector diagram branch contains 3. Each feature extraction module contains different numbers of connection nodes.

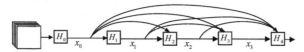

Fig. 4. Structure of the feature extraction module designed based on densely connected convolution [12], which has a better performance than residual structure [13].

At the end of the network, Tanh activation function is used for regression from signal eigenvectors to preset specific values:

$$\text{Tanh}(x) = \frac{e^x - e^{-x}}{e^x + e^{-x}}, x \in (-\infty, +\infty) \tag{1}$$

Compared with Sigmoid activation function, which is widely used in regression operation:

$$\text{Sigmoid}(x) = \frac{1}{1 + e^{-x}}, x \in (-\infty, +\infty) \tag{2}$$

The Sigmoid activation function may change the distribution of original data to some extent, as shown in Fig. 5, while Tanh does not. Moreover, Tanh has a larger gradient, so that the convergence speed is faster in regression operation, which can achieve better training effect.

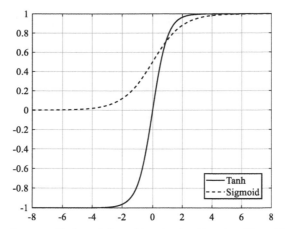

Fig. 5. Comparison between Tanh and sigmoid activation function. Sigmoid is non-zero mean, its output range is (0,1). Non-zero mean data will be mixed during output, which will change the distribution of original data to a certain extent. The Tanh activation function is zero mean and the output range is (−1,1), which solves the above problem.

3.2 Discrimination Threshold

After regression of a specific signal with several signal samples, the result values will fall into a small range. In this paper, the upper and lower quintile algorithm is used to work out the interval threshold and center distance threshold of known signal, in which the interval threshold is used as the basis to distinguish known and unknown signals, the center distance threshold is taken as the length when intercepting the numerical cluster of unknown signals. Suppose that after regression processing of a known signal S, the numerical distribution of several samples is shown in Fig. 6.

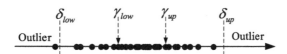

Fig. 6. Diagram of upper and lower quintile algorithm. The outliers of the numerical regression results are removed through this algorithm, and the appropriate threshold is obtained.

Define γ_{low} as the lower quintile of the data set, indicating that there is only 1/5 of all data, which value is less than γ_{low}. Similarly, define γ_{up} as the upper quintile of the data set, which means that only 1/5 of all data has a value greater than γ_{up}. According to the upper and lower quintile algorithm, the interval threshold of regression value for signal S is defined as:

$$\begin{cases} \delta_{low} = \gamma_{low} - \mu(\gamma_{up} - \gamma_{low}) \\ \delta_{up} = \gamma_{up} + \mu(\gamma_{up} - \gamma_{low}) \end{cases} \tag{3}$$

where δ_{low} is the lower bound threshold of regression value for signal S, δ_{up} is the upper bound threshold, and μ is the scale factor, which is 1.5 in this paper. In addition, $\delta_{up} - \delta_{low}$

is the upper and lower distance threshold of the regression for signal S. After regression test of known signals in the closed-set, use:

$$D = \lambda \frac{1}{2J} \sum_{j=1}^{J} (\delta_{up}^{(n)} - \delta_{low}^{(n)}) \tag{4}$$

To calculate the center distance threshold D, which is used as the length of subsequent center-distance interception of unknown signals numerical clusters. In Eq. (4), J is the number of known signal types, $\delta_{up}^{(n)}$ and $\delta_{low}^{(n)}$ represent the upper bound threshold and lower bound threshold of the j-th known signal, λ is the grace factor, the value we use is 1.38.

3.3 Algorithm Scheme

According to the above discussion, the open-set recognition process is as follows:

1) Preprocess known shortwave signals and construct training signal data sets;
2) Use the training data set to train the network, when the network's loss value falls below the preset threshold, the training is terminated and the network is saved;
3) Since the network cannot conduct zero-error regression, the trained network is used to test the known signal. With the upper and lower quintile algorithm, the interval threshold and center distance threshold of each known signal are obtained as the standard to distinguish between known and unknown signals and the subsequent interception of the unknown signal;
4) In the open-set range, use the network to recognize the preprocessed signals. For the regression value of a specific signal, if it falls within the threshold of a known signal interval in step 3), it is judged as such known signal, and if it falls outside the threshold of all known signal intervals, it is judged as unknown signal;
5) Use the kernel density clustering algorithm [14] to cluster all regression values identified as unknown signals to obtain the number of categories, regression numerical clustering clusters and corresponding density center coordinate. For each numerical clustering cluster, use the density center coordinate combined with the center distance threshold to intercept, the signal samples represented by the regression numerical points falling within the interception range are identified as such unknown signal, so as to complete the open-set recognition.

4 Experimental Results

In this section, the recognition performance of proposed method is simulated and tested. The experimental platform is configured with Intel (R) Xeon (R) e-2276m processor, NVIDIA Quadro RTX 5000 GPU and 32 GB DDR4 memory.

Signal used in the experiment includes 6 types: 110A, MIL-STD-188-110B (110B) [15], MIL-STD-188-141A(141A) [16], 141B, Link11 SLEW, PACTOR [17]. The signal setting of the experiment is shown in Table 1. During experiment, 110A, 141B, Link11 SLEW and PACTOR are used for network training as known signals, and are set to regress to the value of 0, 1, 2, and 3. 110B and 141A as unknown signals are not used for training. After obtaining the discrimination threshold according to Sect. 3.2, 110B and 141A are used as network input together with the 4 known signals in the open-set test stage.

Table 1. Attributes of experimental signal samples

Signal	Modulation	As known/unknown	Training regression value
110A	8PSK	Known	0
110B	8PSK	Unknown	–
141A	8FSK	Unknown	–
141B	8PSK	Known	1
Link11 SLEW	8PSK	Known	2
PACTOR	2FSK	Known	3

For generating vector diagram, the size is set to 128×128 to fit the structure of the network. For data stream, as the network's performance will be affected by the change of data statistical distribution, resulting in the inconsistency of calculation dimensional dynamic range and the decline of learning performance. Therefore, the normalization algorithm is adopted as:

$$\text{Norm}(data) = \frac{data - \frac{\max(data)+\min(data)}{2}}{\max(data) - \min(data)} + 0.5 \qquad (5)$$

which $data$ represents the signal data before normalization, $\text{Norm}(data)$ is the data after normalization processing. With normalization, the network can process data at the same scale, gaining better learning and regression performance. In addition, considering that the neural network can perform efficient operation on two-dimensional data structure, so the normalized data is constructed as 336×336 data matrix to obtain the high efficiency of data structure.

4.1 Recognition Performance

Table 2 shows the open-set recognition result of proposed method, The signal-to-noise ratio (SNR) of the experiment is 6dB. It is shown that after regression operation of 4 known signals 110A, 141B, Link11 SLEW and PACTOR, it does not completely regressed to the preset value, but have slight deviation. Therefore, according to the upper and lower quintile algorithm in Sect. 3.2, the upper bound and lower bound thresholds of regression for each known signals are obtained to distinguish known and unknown signal. At the same time, the center distance threshold obtained for center-distance interception of unknown signals is 0.0581. The experiment result indicates that when the SNR is 6dB, the recognition accuracy of known signals reaches more than 96%, which verifies the feasibility of the proposed method.

Table 2. Open-set recognition results of the proposed method

Signal	Lower bound of regression	Upper bound of regression	Density center	Center distance threshold	Recognition accuracy
110A	−0.1132	−0.0589	–	–	99.3%
141B	0.9226	0.9894	–	–	98.9%
Link11 SLEW	1.9132	2.1197	–	–	99.5%
PACTOR	2.9972	3.0065	–	–	96.7%
Unknow 1(110B)	–	–	−0.2923	0.0581	90.1%
Unknow 2(141A)	–	–	2.3072	0.0581	99.20%

Once regression processing is completed, use the kernel density clustering algorithm to obtain the numerical clustering clusters and density centers of unknown signal, and then intercepts them by using the center distance threshold. The proposed method can distinguish the unknown signal 1 (110B) with a recognition accuracy of 90.1%, and the unknown signal 2 (141A) with a recognition accuracy of 99.20%.

Overall, compared with the traditional open-set recognition method, which has few applicable signal types, difficult to distinguish signals of different specifications with same modulation mode and difficult to distinguish different unknown signals, the proposed method can effectively deal with the open-set signal data set, of which 4 signals are 8PSK modulation mode, and can distinguish different types of unknown signals.

4.2 Influence of Numerical Scale on Regression

This section discusses the influence of different training regression scale on network performance through comparative experiments. Table 3 shows the training regression value of 2 experiments on the known signals 110A, 141B, Link11 SLEW and PACTOR. During the training stage, 4 known signals are regressed to the value of 0, 1, 2, 3 and 0, 100, 200, 300.

Table 3. Training regression value of each experiment

Signal	Experiment 1	Experiment 2
110A	0	0
141B	1	100
Link11 SLEW	2	200
PACTOR	3	300

In order to better observe the result, signal samples are input into the network in the order of signal type during the test stage. The corresponding relationship between signal sample type and signal serial number is shown in Table 4.

The number of each signal type is 1000. The regression result of each experiment is shown in Fig. 7. It can be seen that when different scale of regression is set, the network will carry out numerical regression according to the preset scale, and the result of both experiment have good discrimination.

Table 4. Corresponding relationship between signal sample type and serial number

Sample type	Sample serial number
110A	1–1000
110B	1001–2000
141A	2001–3000
141B	3001–4000
Link11 SLEW	4001–5000
PACTOR	5001–6000

This is because, although the numerical scales are different, once the network completes the training under this scale, a nonlinear mapping relationship matching this scale is formed. In other words, the training of different scale will only lead to the difference in the numerical dimension of regression result, and will not affect the discrimination performance between signals.

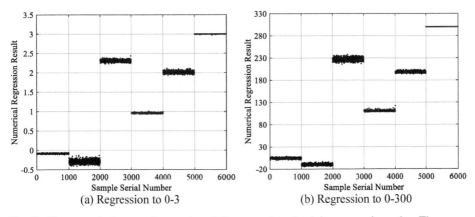

Fig. 7. The numerical regression result at different scales of training regression value. The experimental results show that different regression numerical scale will not affect the discrimination of signals.

5   Conclusions

By combining the feature information of shortwave signal data stream and vector diagram, an open-set signal recognition method is proposed. Using the good feature extraction ability of densely connected convolution and the excellent feature processing and regression performance of dual-input regression neural network, the open-set signal recognition task is well completed. Experimental results show that compared with the traditional method, the proposed method can distinguish different type of unknown signals while maintaining the open-set recognition accuracy, and can effectively distinguish signals of different specifications with same modulation mode. In addition, this paper proposes to establish the regression relationship between signal feature and specific value, and embody the feature of different signal types as different regression values. This idea of transforming feature information for processing provides a new approach for further research in this field.

References

1. Jondral, F.: Automatic classification of high frequency signals. J. Signal Process. **9**, 177–190 (1985)
2. Zhenxing, L., Shichuan, C., Xiaoniu, Y.: Two-class SVDD algorithm for open-set specific emitter identification. J. Commun. Countermeas. **36**, 1–6 (2017)
3. Ying, Y., Lidong, Z.: Method for efficiently recognize satellite interference signals via incremental support vector machine. In: 15th Annual Conference of Satellite Communications, pp.163–171. China Academic Journal Electronic Publishing House, Beijing (2019)
4. Diehl, C.P., Cauwenberghs, G.: SVM incremental learning, adaptation and optimization. In: The International Joint Conference on Neural Networks, pp. 2685–2690. IEEE Press, Piscataway(2003)
5. Escalera, S., Pujol, O., Radeva, P.: Error-correcting output codes library. J. J. Mach. Learn. Res. **11**, 661–664 (2010)
6. Yujie, X., Xiaowei, Q., Xiaodong, X., Jianqiang, C.: Open-set interference signal recognition using boundary samples: a hybrid Approach. In: 12th International Conference on Wireless Communications and Signal, pp. 269–274. IEEE Press, Piscataway (2020)
7. Yunfei, H., Zhangmeng, L., Fucheng, G., Ming, Z.: Open-set recognition of signal modulation based on generative adversarial networks. J. Syst. Eng. Electron. **41**, 2619–2624 (2019)
8. Youwei, G., Hongyu, J., Jing, W.: Open set modulation recognition based on dual-channel LSTM model. J. arXiv Preprint, arXiv: 2002.12037 (2020)
9. Hector, S., Santiago, Z., Ivan, P., Ivana, R., et al.: Special issue on MC-SS validation of a HF spread spectrum multi-carrier technology through real-link measurements. J. Eur. Trans. Telecommun. **17**, 651–657 (2012)
10. Johnson, E.E.: Simulation results for third-generation HF automatic link establishment. J. Proc. IEEE Milit. Commun. Conf. **2**, 984–988 (1999)
11. Zhu, C.: Non-cooperative demodulation of LINK11_SLEW. J. Telecommun. Eng. **54**, 1378–1384 (2014)
12. Gao, H., Zhuang, L., Laurens, V., Kilian, Q.W.: Densely connected convolutional networks. In: 2017 IEEE Conference on Computer Vision and Pattern Recognition (CVPR), pp. 2261–2269. IEEE Press, Piscataway (2017)
13. Xiong, Z., Mankun, X., Hua, P., Xin, Q., Tianyun, L.: Specific protocol signal recognition based on deep residual network. J. Acta Electronica Sinica. **47**, 1532–1537 (2019)

14. Fagui, L., Yufei, C.: An Energy aware adaptive kernel density estimation approach to unequal clustering in wireless sensor networks. J. IEEE Access. **7**, 40569–40580 (2019)
15. Nieto, J.W., Furman, W.N.: Constant-amplitude waveform variations of US MIL-STD-188–110B and STANAG 4539. In: 2016 IET International Conference on Ionospheric Radio Systems and Techniques (IRST), pp. 212–216. IET Press, London (2006)
16. Baker, M., Beamish, W., Turner, M.: The use of MIL-STD-188–141A in HF data networks. In: IEEE Military Communications Conference, pp. 75–79. IEEE Press, Piscataway (2002)
17. Mohd, Y.R., Zainal, N., Abd, M.S.: Performance of 8FSK base on PACTOR I protocol over AWGN channels. In: 5th International Conference on Information Technology, Computer, and Electrical Engineering, pp. 1–5. IEEE Press, Piscataway (2018)

Deep Person Re-identification with the Combination of Physical Biometric Information and Appearance Features

Chunsheng Hua[✉], Xiaoheng Zhao, Wei Meng, and Yingjie Pan

Liaoning University, No. 66 Chongshan Middle Road, Huanggu District, Shenyang, Liaoning, China
huachunsheng@lnu.edu.cn

Abstract. In this paper, we propose a novel Person Re-identification model that combines physical biometric information and traditional appearance features. After manually obtaining a target human ROI from human detection results, the skeleton points of target person will be automatically extracted by OpenPose algorithm. Combining the skeleton points with the biometric information (height, shoulder width.) calculated by the vision-based geometric estimation, the further physical biometric information (stride length, swinging arm.) of target person could be estimated. In order to improve the person re-identification performance, an improved triplet loss function has been applied in the framework of [1] where both the human appearance feature and the calculated human biometric information are utilized by a full connection layer (FCL). Through the experiments carried out on public datasets and the real school surveillance video, the effectiveness and efficiency of proposed algorithm have been confirmed.

Keywords: Computer vision · Deep learning · Person re-identification

1 Introduction

How to identify a person through long distance, where the facial features of target will be blurred due to the low resolution of face region, has been an important task in many fields such as surveillance, security and recommendation system. Since the outbreak of COVID-19, it has drawn more and more attention from numerous researchers because the performance of conventional face recognition algorithms will degrade greatly due to the request of wearing mask, therefore, people need other methods to identify the target person regardless of their facial masks. On the other hand, close contacts are often found in busy areas (shopping streets, malls, restaurants, etc.), the appearance of people tends to change significantly. Compared with the physical biometric information, appearance is more sensitive to clothing and lighting changes. On the contrary, people's physical information is less affected by external factors.

The methods of Person Re-identification (Re-ID) can roughly be divided into part-based Re-ID, mask-based Re-ID, pose-guided Re-ID, attention-model Re-ID, GAN Re-ID and Gait Re-ID [7].

© The Author(s) 2022
Z. Qian et al. (Eds.): WCNA 2021, LNEE 942, pp. 874–887, 2022.
https://doi.org/10.1007/978-981-19-2456-9_89

Part-based Re-ID: global features and local features of target are extracted and calculated to achieve Person Re-ID. McLaughlin [2] using color and optical flow information in order to capture appearance and motion information for Person Re-ID. Under different cameras, Cheng [3] present a multi-channel parts-based convolutional neural network (CNN) model. To effectively use features from a sequence of tracked human areas, Yan [4] built a Long Short-Term Memory (LSTM) network. To establish the correspondence between images of a person taken by different cameras at different times, Chung [5] proposed a weighted two stream training objective function. Inspired by the above studies, Zheng [6] proposed AlignedReID that extracts a global feature and first surpass human-level performance. These methods are fast, but performance will be affected when facing background clutter, illumination variations or obstacle blocking.

Mask-base Re-ID: masks and semantic information is used to alleviate the problem of Part-based Re-ID. Song [8] first designed a mask-guided contrastive attention model (MGCAM) to learn features from the body and background to improve robust during background clutter. Kalayeh, M [9] proposed an adopt human semantic parsing model (SPReID) to further improve the algorithm. To reduce the impact of the appearance variations, Qi [10] added multi-layer fusion scheme and proposed a ranking loss. The accuracy of mask-base Re-ID is improved compared with part-based Re-ID, but it usually suffers from its expensive computational cost and its segmentation result lacks of more accurate information for Person Re-Id proposes.

Pose-guided Re-ID: When extracting features from person, part-based Re-ID and mask-base Re-ID usually simply divide the body into several parts. In Pose-guided Re-ID, after prediction the human pose, the same parts of the human body features are extracted for Re-ID. Su [11] proposed a Pose-driven Deep Convolutional (PDC) model to match the features from global human body and local body parts. To capture human pose variations, Liu [12] proposed a pose-transferrable person Re-ID framework. Suh [13] found human body parts are frequently misaligned between the detected human boxes and proposed a network that learns a part-aligned representation for person re-identification. Considering of people wearing black clothes or be captured by surveillance systems in low light illumination, Xu [15] proposed head-shoulder adaptive attention network (HAA) that is effective in dealing with person Re-ID in black clothing. Pose-guided Re-ID has a good balance between speed and accuracy. But the performance is influenced by skeleton points detection algorithm, especially when pedestrians are blocking each other.

Attention-model Re-ID: using attention model to determine attention by globally considering the interrelationships between features for Person Re-ID. The LSTM/RNN model with the traditional encoder-decoder structure suffers from a problem: it encodes the input into a fixed-length vector representation regardless of its length, which makes the model cannot performing well for long input sequences. Unfortunately, Person Re-ID always working in long input sequences. Many researches chosen to use attention-based model and reached the state-of-the-art. Xu [16] proposed a spatiotemporal attention model for Person Re-ID. The model is assumed the availability of well-aligned person bounding box images, W. Li [17] and S. Li [18] proposed two different spatiotemporal attention to complementary information of different levels of visual attention re-id discriminative learning constraints. In study, researchers found the methods based on a single feature vector are not sufficient enough to overcome visual ambiguity [19] and proposed Dual Attention Matching networks [20]. Compared with above methods, attention-model re-ID method has better performance in accuracy, but it is computationally intensive.

GAN Re-ID: using generative adversarial network (GAN) to generate more training data only from the training set and reduce the interference of lighting changes. A challenge of Person Re-ID is the lacking of datasets, especially in the complex scenes and view changes. To obtain more training data only from the training set and improve performance during different datasets, semi-supervised models using generative adversarial network (GAN) such as LSRO [21], PTGA [22] and DG-Net [23] was proposed. GAN Re-ID works well in different environments, but there are still some problems in stability of training.

Gait Re-ID: using skeleton points of human to extract gait features for person Re-ID. This type of method does not focus on the appearance of a person, but requires a continuous sequence of frames to identify a person by the changes in appearance caused by motion. Gait Re-ID method exploit either two-dimensional (2D) or 3D information depending on the image acquisition methods.

For 3D methods, depth-based person re-identification was proposed [24, 25], which works on Kinect or other RGBD cameras to obtain human pose information. This method is fast and show better robustness to a variety of factors such as clothing change or carrying goods. However, not many surveillances use RGBD cameras in real-life and this method can only maintain accuracy at close distance (usually less than 4 0 or 5 m).

For 2D methods, Carley, C [26] proposed an autocorrelation-based network. Rao [27] proposed a self-supervised method CAGEs to obtain better gait representations. This method provides a solution of "Appearance constancy hypothesis" in appearance-based method, but it is more computationally expensive and require higher-quality data.

In this paper, we propose a person Re-ID algorithm with the combination of physical biometric information and appearance features. To get appearance features, we modified the ResNet-50 proposed by framework of [1] and design a new triplet loss function, trained it on Market1501 and DUKEMTMC. On the other hand, Re-ID is often used in surveillance video, where the camera's view is often fixed. By calibrating the camera and measuring the camera's position information, combined with human skeletal point, we can calculate physical biometric information such as human height, shoulder width and stride length, which is useful for the person Re-ID. In the end, we calculate the Euclidean distance between target person and others, reranking the results. In order to improve the person re-identification performance, both the human appearance feature and the calculated human biometric information are utilized by a full connection layer (FCL).

Since most of the conventional Person Re-ID datasets do not contain physical biometric information and the intrinsic matrix of cameras, we built our dataset by using real surveillance video in school and evaluate our combine method.

2 Algorithm Description

2.1 System Overview

The framework of proposed algorithm is shown in Fig. 1. To reduce calculation errors, the camera needs to be calibrated and positioned before running the person Re-ID. Our algorithm works on video streams, we need to mark the target ROI manually for query sets. The algorithm will use object detection algorithm to predict human ROI for gallery set. The method consists of two parts named global appearance features part and physical biometric information part. The first part extracts the global feature from the person image and distance from each target. The other part is designed to predict physical biometric information by using human skeleton points and calculate triple loss. The losses of these two parts are sent to a fully connected layer classified and re-ranking to match the target person. More details of this work will be described in the following sections.

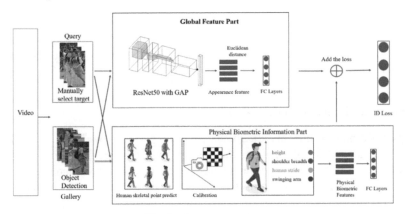

Fig. 1. System overview

2.2 Query Sets and Gallery Sets Data Collection

To simplify the operation, we need to select the target ROI manually. In this part, we mark the target in multiple video frames. Marking multiple angles of the same target can improve the accuracy of the subsequent algorithm.

To collect the gallery data, we use object detection to predict human ROI for next part. In comparison experiments, we found out that using the larger model can hardly improve the accuracy of prediction, but would greatly increase the computational cost. Consider the balance between speed and accuracy, we choose YOLOV5S, the smallest and fastest model of YOLOV5, as our detector.

After collecting Data of Query Sets and Gallery Sets, these images will be sent into the Global Appearance Part and Physical Biometric Information Part to extract features for person re-identification.

2.3   Global Appearance Part

In this research we using a modified model of framework in [1] to extract global appearance. The backbone of this model is ResNet-50 with the span of the last spatial downsampling set to 2. After extracting features by the backbone, the model uses a GAP layer to obtain the global feature. During prediction, the model will calculate the Euclidean distance of global feature between Gallery sets and Query sets. During training, the framework will calculate triplet loss based on the distance between positive pair and negative pair of global features. To improve the performance of the model, we use RKM (reliability-based k-means clustering algorithm) [33] modified the loss function. After applied the new triplet loss function (1) in the framework, we retraining and evaluated our model on Market1501 [34] and DukeMTMC [35]. The experimental results will be described in the EXPERIMENT section.

Our triplet loss (F_t) is computed as:

$$F_t = R^*[dp - dn + \alpha] \tag{1}$$

where d_p and d_n are feature distances of positive pair and negative pair. α is the margin of triplet loss. In this paper we set α as 0.2. R represents the reliability to classify a gallery sample into the query or other clusters. Detailed information about how to compute R could be found in [33].

2.4   Physical Biometric Information Part

The physical biometric information calculated by this part is shown in Fig. 1. To calculate the physical biometric information, the position information, intrinsic matrix of the camera and the skeleton points of target are needed. For getting human skeleton points we using OpenPose [29], a bottom-up algorithm, which first detect 25 human skeleton points of the human body in the whole image and then correspond these points to different individual people. The human skeleton points predicted by OpenPose are shown in the Fig. 2 By using human ROI that we get by object detection, the computation required by OpenPose decreases significantly.

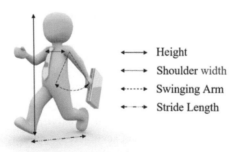

Fig. 2. Physical biometric information

In this paper, every result predicted by OpenPose will be stored in an array of 25 lengths skeleton points. The human physical biometric information is calculated by dividing ROI on human body pictures. When the whole human body is in the camera, we use the y-point coordinate at the top of the target detection frame as the top coordinate y1. The lowest point coordinates of the target ankle, max (skeleton points [24] [1], skeleton points [21] [1]), is used as the bottom coordinate y2. In order to calculate shoulder breadth, we use the skeleton points of human shoulder x coordinates, skeleton points [2] [0] and skeleton points [5] [0], as X-axis coordinates x1 and x2. By using x1, y1, x2, y2 into the Formula (2), the distance between human head, heel, left shoulder and right shoulder in the realistic reference system can be calculated, as further calculate the information of human height and shoulder width.

When the camera is on the side of the person, we can calculate the stride length and arm swing length of the person from the skeleton points of the arms and toes in consecutive video frames.In this part, we still use the (y1, y2) coordinates calculated in the height. The difference is that we take the maximum and minimum values of the left toes and right feet toes in a sequence as the x-coordinate (x3, x4). By substituting (y1, y2, x3, x4) into the Formula (2), we can obtain the stride length. Similarly, using the coordinates of the target's left elbow and right elbow we can calculate the swing arm. We use 0 fill the physical information when we can't calculate physical information because of the orientation of person or the obstruction.

This part is based on single-view metrology algorithm by obtaining the distance of object between two parallel planes. With distortion compensation processing, the images can be used to measure human physical biometric information. We use the traditional pinhole model to transform camera reference frame to world reference frame, and this model is defined as (2):

$$\begin{bmatrix} x_b - C_x \\ y_b - C_y \\ -f_k \end{bmatrix} = \begin{bmatrix} R_{11} & R_{12} & R_{13} \\ R_{21} & R_{22} & R_{23} \\ R_{31} & R_{32} & R_{33} \end{bmatrix} \cdot \begin{bmatrix} X_w - X_0 \\ Y_w - Y_0 \\ Z_w - Z_0 \end{bmatrix} \tag{2}$$

where x_b, y_b is a point on the image, C_x, C_y is the centric point of the image plane coordinates, f_k is the distance from the center of projection to the image plane, R is the extrinsic matrix of the camera, X_W, Y_W, Z_W is a point in the world reference frame, and X_0, Y_0, Z_0 is the centric point in the world reference frame.

In the experimental, we found that when human body moved, the posture changes would lead to data fluctuation, which affected the stability of calculation body height. Therefore, we used a simplified Kalman filter to solve the problem. The simplified Kalman filter formula is given by the following:

$$P_t = P_{t-1} + Q \tag{3}$$

$$K_t = \frac{P_{t-1}}{(P_{t-1} + R)} \tag{4}$$

$$X_t = X_{t-1} + K_t(HZ_tH^T - Hx_{t-1}) \tag{5}$$

$$P_t = (E - K_t)P_{t-1} \tag{6}$$

where P is the predicted matrix, X is the estimate matrix, K is the Kalman gain matrix, P is covariance matrix, Z is measurement result, Q is process noise matrix, R is measurement error covariance matrix, t, $t-1$ is current time and previous time. E is identity matrix. H is measurement matrix.

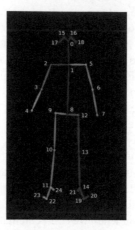

Fig. 3. OpenPose predict result

Human height and shoulder width are numerically independent, we simplify the control matrix and use the $\begin{bmatrix} h & 0 \\ 0 & w \end{bmatrix}$ as the input of Kalman filter, where the h is height of target and w is the shoulder width of target. Kalman filter takes a weighted average (5) of the predicted result of the current state (t) and the previous state (t−1) with the measurement result. The weighted mean named Kalman gain is defined by the covariance matrix of the previous state, the measurement noise covariance and the system process covariance (4). In this work (Q, R) are hyperparameters, which Q was set to 0.0001 and R was set to 1. The covariance matrix is determined by the previous moment's covariance, the process noise matrix Q and Kalman gain (3) (6). The effect of Kalman filtering for height measurement will be shown in Fig. 3. Kalman Filter Comparison Chart, where the 'truth' line refers the real height of the person, the 'original' line refers each predicted result, the 'filtered' line refers the result after Kalman filtering. As shown in the Fig. 3, after Kalman filtering, the max error predicted by our method is reduced from ± 10 cm to ± 4 cm.

In the end, the features calculated by global appearance part and physical biometric information part will be sent into a network to classification and re-ranking to find the target person.

2.5   Classification and Re-ranking

In this part, we designed a network (Fig. 4) to utilize physical biometric information and human appearance features for person re-identification. In order to ensure the independent robustness, we first use relatively independent networks and loss functions to

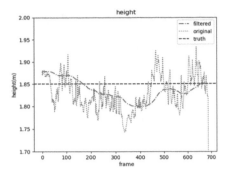

Fig. 4. Kalman Filter comparison chart

process the two features separately and score the results obtained from each in a consistent manner. We use a fully connected network with two hidden layers to jointly compute the triplet loss and SoftMax loss, at the same time, optimizing the ratio of both. We introduce Dropout into the fully connected layer to prevent overfitting. After processing the physical biometric feature information by the fully connection network, the output dimension will be consistent with the appearance features. Finally, we add two feature losses to calculate the ID loss. For comprehensive consideration, we introduce a sigmoid function and trainable parameters λ to give appropriate activation intensity, to control the weight of the two kinds of features.

During prediction, we obtain the feature vectors of query sets and gallery sets respectively to calculate the Euclidean distance between them, re-ranking the data of gallery through the distance difference, and select the top five IDs as the final result.

3 Experiment

3.1 Evaluation of Human Height Prediction

In this part, we requested 3 persons with different heights to walk in same trajectory for evaluating the accuracy of our human height prediction method. Each person was requested to walk in the circle, where the range between the camera and human varied between 5 to 10 m. Before the experiment, the true heights of each target person were manually recorded. Then we recorded a ten-minute video of each person. Table 1 shows the accuracy and max error of our prediction algorithm, where 'Truth' refers to the truth height of the person, 'Average' refers the average height of predicted person, 'Max Error' refers to the maximum error between 'truth' and predicted human height.

Table 1. Evaluation results of human height prediction

Person	Truth	Max error	Average
Person 1	183 cm	3.79 cm	183.67 cm
Person 2	178 cm	2.296 cm	178.24 cm
Person 3	180 cm	3.12 cm	179.22 cm

3.2 Evaluation on Public Dataset

In this section, we trained our modified models on Market1501 [34] and DukeMTMC [35] datasets. Market1501 [34] dataset collected 32,668 images of 1,501 identities using 6 video cameras at different perspectives distances. Due to the openness of the environment, images of each identity were captured by at least of two cameras. In this dataset, 751 of these individuals were classified as the training set, which contains 12,936 images. The remaining 750 individuals were classified as the test set, which contains 19,732 images. DukeMTMC [35] dataset is recorded by 8 calibrated and synchronized static outdoor cameras, it has over 2700 identities, with 1404 individuals appearing on more than two cameras and 408 individuals appearing on one camera. This dataset randomly sampled 702 individuals containing 17,661 images as the training set and 702 individuals containing 17,661 images as the test set.

Since most of the Person Re-ID datasets do not contain human physical information or camera location information, we evaluated our global appearance part on public dataset. The results of the evaluation are shown in Table 2. The Rank1 accuracy and mean Average Precision (mAP) are reported as evaluation metrics.

Table 2. Comparison of other methods

Type	Method	Market1501		DukeMTMC	
		Rank1	mAP	Rank1	mAP
Mask-guided	MGCAM [8]	83.79	74.33	–	–
	SPReID [9]	94.63	90.96	88.96	84.99
	MaskReID [10]	92.46	88.13	84.07	79.73
Pose-guided	PDC [11]	84.14	63.41	–	–
	PT [12]	79.75	57.98	68.64	48.06
	PABR [13]	95.4	93.1	88.3	83.9
	HAA [14]	95.8	89.5	89.0	80.4

(*continued*)

Table 2. (*continued*)

Type	Method	Market1501		DukeMTMC	
		Rank1	mAP	Rank1	mAP
Attention-based	HA-CNN [18]	91.2	75.7	80.5	63.8
	DuATM [19]	91.42	76.62	81.82	64.58
	EXAM [20]	95.1	85.9	87.4	76.0
Gan-ReID	LSRO-GAN [21]	83.97	66.07	–	–
	DG-Net [23]	94.8	86.0	86.6	74.8
Part-based	AlignedReID [6]	94.4	90.7	–	–
	IDE [30]	79.5	59.9	–	–
	TriNet [31]	84.9	69.1	–	–
	AWTL [32]	89.5	79.7	79.8	63.4
	Strong Baseline [1]	95.4	94.2	90.3	89.1
	Ours	**96.1**	**94.2**	**90.9**	**89.1**

3.3 Evaluation on Surveillance Dataset

To train and evaluate our method, we build our dataset by using real surveillance video in school. We took several videos of 30 people walking at different angles by using 3 calibrated cameras. Before recording, we calibrated and measured position of the camera. The cameras were placed horizontally and measured by a laser rangefinder to get the height and pitch angle for composition extrinsic matrix. Then, we use a checkerboard calibration plate to calibrate the camera and get intrinsic matrix. The information will be used to calculate human physiological information and reduce calculation errors.

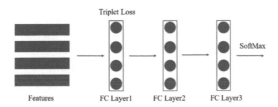

Fig. 5. Fully connection network

We randomly intercept 100 consecutive frames for each video and use object detection algorithm to obtain the bounding box of each person, then manually label each target as our dataset. For getting human physical biometric information, we measured each person's height, shoulder width, stride length and swing arm manually. Totally, we labeled 9000 images and the dataset was divided equally according to identity as our test and training set. Because of privacy problem, we put part of people images in Fig. 5 and blur them (Fig. 6).

Fig. 6. Some examples of surveillance dataset

In this part we conducted comparative experiments on the surveillance dataset, the appearance feature method (without Physical Biometric part) and the combining method (with Physical Biometric part) respectively, bold number denote the better performance (Table 3).

Table 3. Comparison on our dataset

Method	Ranking1	mAP
Ours (Only Appearance Features)	97.62%	94.93%
Ours (Appearance Features + Physical Biometric Information)	**98.68%**	**95.46%**

4   Conclusion

In this research, we propose a person re-identification algorithm that combines physical biometric information and human appearance features. We calculate human physiological parameters by human skeletal point prediction algorithm combined with camera single-view metrology algorithm. The human appearance features are extracted by a modified ResNet50. To combine appearance features and physiological biometric information, we introduce a feature-weighted fusion model to learn both feature information. By evaluating on a public dataset, we demonstrate the effectiveness of the new loss function. Since it is not feasible to conduct comparative experiments of combining methods on public datasets, we produced our own dataset to train and evaluate our improved global appearance method and combination method, confirmed the effectiveness of the combining method.

5   Future Work

In our experiments, we found that when using the object detector to predict the human body, the ROI changes also lead to incorrect prediction of human height. This situation seriously reduces the accuracy of the algorithm. We will try to use mask-based methods to predict persons and calculate biometric information in future work. On the other hand, our physical biometric part relies heavily on camera position information, which makes our method not so compatible, we will try to solve these problems in our future work.

References

1. Hao, L., et al.: Bag of tricks and a strong baseline for deep person re-identification. In: Proceedings of the IEEE/CVF Conference on Computer Vision and Pattern Recognition Workshops (2019)
2. McLaughlin, N., Del Rincon, J.M., Miller, P.: Recurrent convolutional network for video-based person re-identification. In: Proceedings of the IEEE Conference on Computer Vision and Pattern Recognition (2016)
3. Cheng, D., Gong, Y., Zhou, S., Wang, J., Zheng, N.: Person re-identification by multi-channel parts-based cnn with improved triplet loss function. In Proceedings of the IEEE Conference on Computer Vision and Pattern Recognition, pp. 1335–1344 (2016)
4. Yan, Y., et al.: Person re-identification via recurrent feature aggregation. In: European Conference on Computer Vision. Springer, Cham (2016). https://doi.org/10.1007/978-3-319-46466-4_42
5. Chung, D., Tahboub, K., Delp, E.J.: A two stream Siamese convolutional neural network for person re-identification. In: Proceedings of the IEEE International Conference on Computer Vision (2017)
6. Zheng, L., Zhang, H., Sun, S., Chandraker, M., Tian, Q.: Person re-identification in the wild. In: IEEE Conference on Computer Vision and Pattern Recognition (CVPR), pp. 1367–1376 (2017)
7. Ye, M., et al.: Deep learning for person re-identification: a survey and outlook. IEEE Transactions on Pattern Analysis and Machine Intelligence (2021)
8. Song, C., Huang, Y., Ouyang, W., Wang, L.: Mask-guided contrastive attention model for person re-identification. In: Proceedings of the IEEE Conference on Computer Vision and Pattern Recognition, pp. 1179–1188 (2018)
9. Kalayeh, M.M., Basaran, E., Gökmen, M., Kamasak, M.E., Shah, M.: Human semantic parsing for person re-identification. In: Proceedings of the IEEE Conference on Computer Vision and Pattern Recognition, pp. 1062–1071 (2018)
10. Qi, L., Huo, J., Wang, L., Shi, Y., Gao, Y.: Maskreid: a mask based deep ranking neural network for person re-identification. arXiv preprint arXiv:1804.03864 (2018)
11. Su, C., Li, J., Zhang, S., Xing, J., Gao, W., Tian, Q.: Pose-driven deep convolutional model for person re-identification. In: Proceedings of the IEEE International Conference on Computer Vision, pp. 3960–3969 (2017)
12. Liu, J., Ni, B., Yan, Y., Zhou, P., Cheng, S., Hu, J.: Pose transferrable person re-identification. In: Proceedings of the IEEE Conference on Computer Vision and Pattern Recognition, pp. 4099–4108 (2018)
13. Suh, Y., Wang, J., Tang, S., Mei, T., Lee, K.M.: Part-aligned bilinear representations for person re-identification. In: Proceedings of the European Conference on Computer Vision (ECCV), pp. 402–419 (2018)
14. Qian, X., et al.: Pose-normalized image generation for person re-identification. In: Ferrari, V., Hebert, M., Sminchisescu, C., Weiss, Y. (eds.) ECCV 2018. LNCS, vol. 11213, pp. 661–678. Springer, Cham (2018). https://doi.org/10.1007/978-3-030-01240-3_40
15. Xu, B., et al.: Black re-id: a head-shoulder descriptor for the challenging problem of person re-identification. In: Proceedings of the 28th ACM International Conference on Multimedia (2020)
16. Xu, S., Cheng, Y., Gu, K., Yang, Y., Chang, S., Zhou, P.: Jointly attentive spatial-temporal pooling networks for video-based person re-identification. In: Proceedings of the IEEE International Conference on Computer Vision, pp. 4733–4742 (2017)
17. Li, S., et al.: Diversity regularized spatiotemporal attention for video-based person re-identification. In: Proceedings of the IEEE Conference on Computer Vision and Pattern Recognition (2018)

18. Li, W., Zhu, X., Gong, S.: Harmonious attention network for person re-identification. In: Proceedings of the IEEE Conference on Computer Vision and Pattern Recognition, pp. 2285–2294 (2018)

19. Si, J., et al.: Dual attention matching network for context-aware feature sequence-based person re-identification. In: Proceedings of the IEEE Conference on Computer Vision and Pattern Recognition, pp. 5363–5372 (2018)

20. Qi, G., et al.: EXAM: a framework of learning extreme and moderate embeddings for person re-ID. J. Imaging **7**(1), 6 (2021)

21. Zheng, Z., Zheng, L., Yang, Y.: Unlabeled samples generated by gan improve the person re-identification baseline in vitro. In: Proceedings of the IEEE International Conference on Computer Vision, pp. 3754–3762 (2017)

22. Wei, L., Zhang, S., Gao, W., Tian, Q.: Person transfer gan to bridge domain gap for person re-identification. In: Proceedings of the IEEE Conference on Computer Vision and Pattern Recognition, pp. 79–88 (2018)

23. Zheng, Z., Yang, X., Yu, Z., Zheng, L., Yang, Y., Kautz, J.: Joint discriminative and generative learning for person re-identification. In: Proceedings of the IEEE/CVF Conference on Computer Vision and Pattern Recognition, pp. 2138–2147 (2019)

24. Karianakis, N., Liu, Z., Chen, Y., Soatto, S.: Reinforced temporal attention and split-rate transfer for depth-based person re-identification. In: Ferrari, V., Hebert, M., Sminchisescu, C., Weiss, Y. (eds.) ECCV 2018. LNCS, vol. 11209, pp. 737–756. Springer, Cham (2018). https://doi.org/10.1007/978-3-030-01228-1_44

25. Nambiar, A.M., Bernardino, A., Nascimento, J.C., Fred, A.L.: Towards view-point invariant person re-identification via fusion of anthropometric and gait features from kinect measurements. In: VISIGRAPP (5: VISAPP), pp. 108–119, February 2017

26. Carley, C., Ristani, E., Tomasi, C.: Person re-identification from gait using an autocorrelation network. In: Proceedings of the IEEE/CVF Conference on Computer Vision and Pattern Recognition Workshops (p. 0) (2019)

27. Rao, H., et al.: A self-supervised gait encoding approach with locality-awareness for 3D skeleton-based person re-identification. IEEE Trans. Patt. Anal. Mach. Intell. (2021)

28. Jocher, G.: ultralytics. "YOLOV5". https://github.com/ultralytics/yolov5

29. Cao, Z., et al.: "OpenPose: realtime multi-person 2D pose estimation using Part Affinity Fields. IEEE Trans. Patt. Anal. Mach. Intell. **43**(1), 172–186 (2019)

30. Zheng, Z., Zheng, L., Yang, Y.: A discriminatively learned cnn embedding for person reidentification. ACM Trans. Multim. Comput. Commun. Appl. (TOMM) **14**(1), 1–20 (2017)

31. Hermans, A., Beyer, L., Leibe, B.: In Defense of the Triplet Loss for Person Re-Identification (2017)

32. Ristani, E., Tomasi, C.: Features for multi-target multi-camera tracking and re-identification. In: Proceedings of the IEEE Conference on Computer Vision and Pattern Recognition, pp. 6036–6046 (2018)

33. Hua, C., Chen, Q., Wu, H., Wada, T.: RK-means clustering: K-means with reliability. IEICE Trans. Inf. Syst. **91**(1), 96–104 (2008)

34. Zheng, L., Shen, L., Tian, L., Wang, S., Wang, J., Tian, Q.: Scalable person re-identification: a benchmark. In: Proceedings of the IEEE International Conference on Computer Vision, pp. 1116–1124 (2015)

35. Ristani, E., Solera, F., Zou, R., Cucchiara, R., Tomasi, C.: Performance measures and a data set for multi-target, multi-camera tracking. In: European Conference on Computer Vision, pp. 17–35. Springer, Cham, October 2016. https://doi.org/10.1007/978-3-319-48881-3_2

36. Lin, Y., et al.: Improving person re-identification by attribute and identity learning. Patt. Recogn. **95**, 151–161 (2019)

Automatic Modulation Classification Based on One-Dimensional Convolution Feature Fusion Network

Ruipeng Ma[1,2]([✉]), Di Wu[2], Tao Hu[2], Dong Yi[2], Yuqiao Zhang[1,2], and Jianxia Chen[2]

[1] School of Cyber Science and Engineering, Zhengzhou University, Wenhua Road 97, Zhengzhou 450002, China
13164351610@163.com

[2] College of Data Target Engineering, Strategic Support Force Information Engineering University, Science Avenue 62, Zhengzhou 450001, China

Abstract. Deep learning method has been gradually applied to Automatic Modulation Classification (AMC) because of its excellent performance. In this paper, a lightweight one-dimensional convolutional neural network module (Onedim-CNN) is proposed. We explore the recognition effects of this module and other different neural networks on IQ features and AP features. We conclude that the two features are complementary under high and low SNR. Therefore, we use this module and probabilistic principal component analysis (PPCA) to fuse the two features, and propose a one-dimensional convolution feature fusion network (FF-Onedimcnn). Simulation results show that the overall recognition rate of this model is improved by about 10%, and compared with other automatic modulation classification (AMC) network models, our model has the lowest complexity and the highest accuracy.

Keywords: Automatic modulation classification · Feature fusion · FF-onedimcnn · Deep learning · Low-complexity · Lightweight

1 Introduction

Automatic modulation classification has broad application value in both commercial and military applications. On the business side, The number of connected devices has been growing exponentially over the past decade. Cisco [1] predicts that machine-to-machine (M2M) connections will account for half of the connected devices in the world by 2023, and the massive number of devices will put great pressure on the spectrum resources, signaling overhead and energy consumption of base stations [2, 3]. To address these challenges, software defined radio (SDR), cognitive radio (CR) and adaptive regulation systems have been extensively studied. In the military aspect, especially in the process of unmanned aerial vehicle system signal reconnaissance, how to accurately and quickly judge the modulation type of the received signal under the condition of non-cooperative communication is very important for the real-time processing of the subsequent signal.

© The Author(s) 2022
Z. Qian et al. (Eds.): WCNA 2021, LNEE 942, pp. 888–899, 2022.
https://doi.org/10.1007/978-981-19-2456-9_90

Deep learning (DL) can automatically learn advanced features. It has received much attention for its excellent performance in complex and deep architecture identification tasks. O'Shea [4] first proposed the use of CNNs to classify the modulation of raw signal samples generated using GNU radio, and their later publication [5] introduced a richer radio (OTA) data set that included a wider range of modulation types in real-world environments. To cope with a more complex realistic environment and reduce the influence of channels on transmitted signals, an improved CNN method is proposed in [6] to correct signal distortion that may occur in wireless channels. In [7], a channel estimator based on neural network is designed to find the inverse channel response and improve the accuracy of the network by reducing the influence of channel fading [8]. Based on the theoretical knowledge of signal parameter estimation, a parameter estimator is introduced to extract information related to phase offset and transform phase parameters. In terms of lightweight network design, [9] proposed a lightweight end-to-end AMC model lightweight deep neural network (LDNN) through a new group-level sparsity induced norm. [10] proposed convolutional neural network (CNN) and convolutional Long and short Term Deep neural Network (CLDNN),Reduce the parameters in the network while maintaining reasonable accuracy. One-dimensional convolutional neural network is utilized in [11], and one-dimensional convolutional neural network achieves good performance only through original I/Q samples.In terms of feature fusion, [12] proposed two ideas of feature fusion. Firstly, the received radar signal is fused with the image fusion algorithm of non-multi-scale decomposition, The image of a single signal is combined with different time-frequency (T-F) methods. Using the convolutional neural network (CNN) based on transfer learning and stacked autoencoder (SAE) based on self-training, the sufficient information of fusion image is extracted [13]. Combining the advantages of convolutional neural network (CNN) and long and short term memory (LSTM), features are extracted from the I/Q stream and A/P stream to improve performance.

The contributions of this paper are summarized as follows:

- A lightweight one-dimensional convolutional neural network module is proposed. The one-dimensional convolutional neural network can better extract the features of data flow. Experiments show that this single module can achieve recognition accuracy comparable to other network models, but with the most minor parameters.
- The performance of different neural network models on I/Q time series and A/P time series is explored. Two conclusions can be drawn from the experimental results. First, it verifies that the proposed network module performs best in two input features. Second, the input features of the I/Q time series and the A/P time series can complement each other at low SNR and high SNR.
- According to the proposed one-dimensional convolutional neural network module and the method of probabilistic principal component analysis (PPCA) to fuse the two features, we designed a one-dimensional convolutional feature fusion network model (FF-OnedimCNN). Experimental results show that this model has more advantages in both accuracy and complexity.

2 Signal Model and Preprocessing

2.1 Signal Model

After the signal passes through the channel and is sampled discretely, the equivalent baseband signal can be expressed as follows:

$$r(n) = e^{j2\pi f_0 Tn + j\theta_n} \sum_{u}^{L-1} s(u)h(nT - uT - \varepsilon T) + g(n) \tag{1}$$

where $s(u)$ is the transmitting symbol sequence, $h(nT)$ is the channel response function, T is the symbol interval, ε represents synchronization error, f_0 represents frequency offset, θ_n represents phase jitter, $g(n)$ represents noise, and $\sum_{u}^{L-1} s(u)h(nT - uT)$ represents symbol interference.

2.2 Signal Preprocessing

In this paper, the I/Q format of the original complex sample is mainly converted to A/P format; in other words, the original sample is converted from I/Q coordinates to polar coordinates [7]. In literature [15], the author directly mapped the received complex symbols to the constellation map on the complex plane as features and achieved good performance. Although this method is practical and straightforward, learning features from images on the I-Q plane loses the domain knowledge and available features of the communication system. Obviously, the constellation of QPSK can be regarded as a subgraph of 8PSK, as shown in Fig. 1(a) and (b), which will lead to their wrong classification. Therefore, preprocessing the original sample can improve the recognition accuracy. We define r as a signal segment, and the receiving and sampling period T is described in the previous section. The I/Q symbol sequence can be regarded as a sampling sequence with time step, $n = 1, ..., N$, which can be expressed as:

$$r(nT) = r[n] = r_I[n] + jr_Q[n], \quad n = 1, ..., N . \tag{2}$$

The instantaneous amplitude of the signal is defined as:

$$A[n] = \sqrt{r_I^2[n] + r_Q^2[n]} . \tag{3}$$

The instantaneous phase of the signal is defined as:

$$P[n] = \arctan(\frac{r_I[n]}{r_Q[n]}) \tag{4}$$

Although the I/Q components have been normalized, it is still necessary to normalize them after the amplitude and phase data are obtained from the I/Q components through the standard formula; otherwise the model will perform poorly. The I/Q component of the original sample is transformed into an A/P component, as shown in Fig. 1(c) and (d).

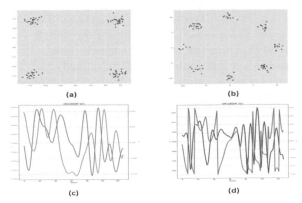

Fig. 1. (a) QPSK constellation diagram; (b) 8PSK constellation diagram; (c) 16QAM I/Q sequence waveform at SNR 12; (d) 16QAM A/P sequence waveform at SNR 12

3   The Proposed Modulation Classification Method

3.1   The Proposed One-Dimensional Convolutional Neural Network Module

The one-dimensional convolutional neural network module proposed is shown in Fig. 2. We train eight kinds of modulation signals on the RadioML data set. After the original data is preprocessed, we get two characteristic data sets, namely the I/Q sequence data set and the A/P sequence data set. For the I/Q sequence dataset, each data sample is an I/Q sampling sequence with 128-time steps, represented by a 2×128 matrix. The specific process of each layer is as follows:

- **Input layer:** The input layer of the network needs to transform the original 2×128 matrix into a 128×2 matrix, so as to input it into the one-dimensional convolution layer.
- **The first 1D CNN layer:** the first layer defines a filter (also called feature detector) of height 4 (also called convolution kernel size). We defined eight filters. So we have eight different features trained in the first layer of the network. The output of the first neural network layer is a 128×8 matrix. Each column of the output matrix contains the weight of a filter. When defining the kernel size and considering the length of the input matrix, each filter will contain 72 weight values.
- **Second 1D CNN layer:** The output of the first CNN will be input into the second CNN layer. We will define 16 different filters again on this network layer for training. Following the same logic as the first layer, the output matrix is 128×16 in size. Each filter will contain 528 weight values.
- **Third and fourth 1D CNN layers:** To learn higher-level features, two additional 1D CNN layers are used here. The output matrix after these two layers is a 128×64 matrix.
- **Global average pooling layer:** After passing four 1D CNN layers, we add a GlobalAveragePooling1D layer to prevent over-fitting. The difference between GlobalAveragePooling and our average pooling is thatGlobalAveragePooling averages each feature map internally.

- **Dropout layer:** The Dropout layer randomly assigns zero weights to neurons in the network. If we choose a ratio of 0.5, 50% of the neurons will be weighted to zero. By doing this, the network is sensitive to small changes in data.
- **The full connection layer is activated by Softmax:** finally, after two full connection layers, the number of filters is 256 and 8, respectively. After the global average pooling layer, a vector with a length of 64 is obtained. After the full connection layer, the probability of occurrence of each type in 8 modulation types is obtained.

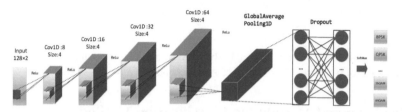

Fig. 2. Module structure of one-dimensional convolutional neural network

3.2 Datesets and Implementation Process

The three RF radioML datasets are available here: https://www.deepsig.ai/datasets. 2016.04C and 2016.10A data sets contain 11 types of modulation schemes with SNR ranging from −20 dB to 18 dB. Each data sample is an I/Q time series with 128-time steps, and the modulation signal is stored as a 2 × 128 I/Q vector. Data sets are simulated in real channel defects (generated by GNU radio), and the detailed process of data set generation can be found in O'Shea et al.'s paper [16]. There are eight digital modulation classes (BPSK, QPSK, 8PSK, PAM4, QAM16, QAM64, GFSK, CPFSK) and three analog modulation classes (WBFM, AM-DSB, AM-SSB).After a detailed exploration of three data sets. We found defects in 2016.04C and 2018.01A data sets. 2016.04C data sets were not normalized correctly. QAM16 and QAM64 occupied A larger range in value than other modulation types, while 2016.10A data were within ±0.02 on both axes.2018.01A contains 24 modulation types, but some of them are incorrectly marked. In addition, the analog modulation of the three data sets is almost impossible to distinguish between the analog modulation because the voice recording is paused. Therefore, digital modulation in 2016.10A dataset was selected for training and testing.

We divide the digital modulation data set in 2016.10A data set into training set (67%), verification set (13%) and test set (20%). Due to the limitation of memory, the batch size of time series data input is 512 and the training period is 200. In this paper, Adam optimizer is used to optimize the network, and the initial learning rate is set to 0.001. GPU environment of all programs is NVIDIA Quadro P4000.Other deep learning models include CNN [4], Resnet [5] and CLDNN [17]. Table 1 compares the performance and complexity of several indicators, including the number of parameters, training time, overall classification accuracy and classification accuracy under different signal-to-noise ratios.

Table 1. Performance comparison under different models with different features

Model	Feature	Parameters	Training time	Classification Accuracy SNR (−20 db,18 db)	Classification Accuracy SNR (−10 db, 5 db)	Classification Accuracy SNR (6 db,18 db)
CNN	IQ	2,665,816	115	55.76%	58.85%	82.69%
	AP	2,665,816	114	55.39%	54.23%	87.20%
ResNet	IQ	141,632	82	55.81%	59.26%	82.23%
	AP	141,632	81	53.91%	50.90%	86.95%
CLDNN	IQ	163,462	210	54.82%	58.46%	80.85%
	AP	163,462	210	57.93%	56.94%	92.94%
OnedimCNN (Ours)	IQ	29,632	29	58.39%	60.98%	87.59%
	AP	29,632	28	60.46%	59.61%	95.49%

As shown in Table 1, compared with other benchmark models, the one-dimensional convolutional network module is superior to other network models in all aspects of indicators. Among the benchmark models, CLDNN performs best, with a classification accuracy of 93% at a high SNR. Compared with CLDNN, the proposed one-dimensional convolutional network module has more obvious advantages, The classification accuracy of the model is slightly 3% higher than that of CLDNN, but the parameters of the model are only 1/5 of that of CLDNN. The classification accuracy within the whole SNR range is shown in Fig. 3. As can be seen from Fig. 3, among all the models, the classification accuracy of the A/P feature at high SNR is about 7.25% higher than that of the I/Q feature, and I/Q data is more resistant than the A/P feature at low SNR. As seen from the confusion matrix of OnedimCNN-IQ and OnedimCNN-AP, as shown in Fig. 4, the

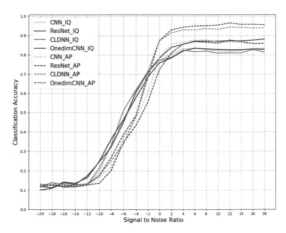

Fig. 3. Classification accuracy of the time series model within the overall SNR range

A/P feature is better than the I/Q feature to help the model distinguish between QAMs and PSKs. OnedimCNN-AP can completely distinguish 8PSK from QPSK, while QAM still confuses. This shows that amplitude-phase time series are more prominent features of modulation classification, but they are more susceptible to noise conditions.

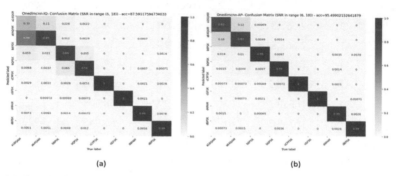

(a)   (b)

Fig. 4. (a) OnedimCNN-IQ and (b) OnedimCNN-AP confusion matrices in the SNR range of 6db to 18db

3.3 Feature Fusion

Feature fusion is divided into two steps. Firstly, we apply probabilistic principal component analysis to reduce the dimension of high-dimensional features extracted by one-dimensional convolution module. Then, we use the method of sequence fusion for feature fusion. Our one-dimensional convolution feature fusion network model (FF-Onedimcnn) is shown in Fig. 5 features are extracted from the two feature fusion networks through two convolution modules. The input size of both components is 128×2, and ReLu is selected as the activation function. In the one-dimensional convolutional feature fusion network structure, after the features are extracted through Block1, the main parts of the two segments are screened by combining the method of probabilistic principal component analysis. Then the features are fused by sequence splicing. In addition, A/P data after normalization increases the risk of overlap with the I/Q data. In order to prevent network model fitting, we have two kinds of feature extraction of regularization is introduced after the operation, L2 regularization to make the network more tend to use all the input characteristics, rather than rely heavily on the input features in some small part. L2 penalizes smaller, more diffuse weight vectors, which encourages classifiers to eventually use features from all dimensions, rather than relying heavily on a few of them. We introduce L2 regularization in the fully connected layer to improve the generalization ability of the model and reduce the risk of overfitting.

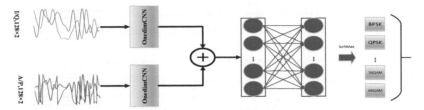

Fig. 5. Model structure of one-dimensional convolution feature fusion network

4 Experimental Results and Discussion

This section illustrates the effectiveness of the one-dimensional convolution feature fusion network model through some comparative experiments. We still conduct training and testing on the previous datasets, and first verify whether the feature fusion method can inherit the advantages of the two features. Secondly, we compared this model with the latest automatic modulation classification algorithms based on deep learning, including CNN-1 [4], CNN-2 [10], CLDNN-1 [10], CLDNN-2 [13], MCLDNN [18], and PET-CGDNN [8]. The evaluation is also carried out from four aspects: the number of parameters, training time, overall classification accuracy and classification accuracy under different SNR, as shown in Table 2. Among the classification models mentioned above, CLDNN-2 and MCLDNN both involve the idea of feature fusion. The model proposed by us is comparable to the two models in accuracy, but our model is superior in model complexity.

4.1 Classification Accuracy

As can be seen from Fig. 6, after the fusion of the two features, the classification accuracy of the one-dimensional convolution module proposed by us is consistent with that of A single module on I/Q features at low SNR, and roughly the same as that of A single module on A/P features at high SNR. We verify that the advantages of both can be inherited by the method of feature fusion. In addition, the overall recognition rate is 10% better than Resnet's A/P feature recognition rate, 5% better than the I/Q feature recognition rate of individual modules, and 3% better than the A/P feature recognition rate. Meanwhile, as shown in Fig. 6, the recognition rate of the FF-OnedimCNN model proposed is significantly higher than that of other network models starting from -4dB. When the SNR reaches 2 dB, the recognition rate of the model tends to be stable. The average recognition accuracy from 6 dB to 18 dB reaches 94.95%, which is almost equal to the recognition rate of the A/P feature of A single module. It can also be seen from Table 2 that, compared with other network models, the FF-OnedimCNN model proposed by us has the highest classification accuracy in terms of both overall classification accuracy and high SNR classification accuracy. Figure 7 shows the confusion matrices of the FF-OnedimCNN model under different SNR. For the confusion matrices, each row represents the real modulation type, and each column represents the predicted modulation type. From the confusion matrices from −20 dB to 18 dB, the confusion mainly focuses on the classification of 8PSK and QPSK, 16QAM and 64QAM. From the second section, we know that there are two reasons for the significant classification error. The first one is influenced by the channel. To simulate the real scene, the channel has interfered with frequency offset, center frequency offset, selective fading and Gaussian white noise. Second, they have overlapping constellation points, which leads to the decline of recognition rate. However, according to the confusion matrix from 6 dB to 18 dB, the FF-OnedimCNN model proposed can completely distinguish 8PSK from QPSK, and 16QAM and 64QAM are also greatly improved.

4.2 Computational Complexity

In order to better deploy the model to edge devices, we should consider not only the accuracy of the model, but also the complexity of the model. The most intuitive evaluation criteria for model complexity are the training parameters and training time of the model, as shown in Table 2. The training parameters of CNN-2 and PET-CGDNN are similar to those of the FF-OnedimCNN model, among which PET-CGDNN has the least training parameters. However, from the perspective of training time, The training time of FF-OnedimCNN model was only 1/3 of that of PET-CGDNN. The sum of model parameters of CNN-2 is almost equal to that of the FF-OnedimCNN model, from the perspective of accuracy, the FF-OnedimCNN model proposed by us has a higher classification accuracy. In addition, both CLDNN-2 and MCLDNN adopt the idea of feature fusion. Both combine two network models of convolutional neural network (CNN) and long and short-term memory (LSTM) for classification. In terms of classification accuracy, the two models both reach more than 92% at high SNR, indicating that multi-feature fusion is better than single-feature fusion. However, from the perspective of training parameters, the training parameters of the two models increased more than seven times than that of the FF-OnedimCNN model. At the same time, we also found that the LSTM network would increase the training time of the network. In summary, we can conclude that the

Table 2. Performance comparison under different models

Model	Parameters	Training time	Classification Accuracy SNR (−20 db, 18 db)	Classification Accuracy SNR (−10 db, 5 db)	Classification Accuracy SNR (6 db, 18 db)
CNN-1	2,665,816	115	55.76%	58.85%	82.69%
CNN-2	73,588	40	57.89%	60.22%	86.38%
CLDNN-1	97,864	368	58.77%	62.20%	86.57%
CLDNN-2	557,212	668	62.38%	65.64%	93.37%
MCLDNN	405,812	523	58.44%	56.93%	92.96%
PET-CGDNN	71,484	210	56.66%	60.60%	83.13%
FF-OnedimCNN (Ours)	73,176	71	63.40%	67.10%	94.95%

FF-OnedimCNN model proposed has more significant advantages in both accuracy and complexity, and has more potential in future model deployment.

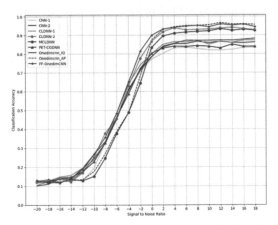

Fig. 6. Comparison between the proposed method and deep learning based method under different SNR

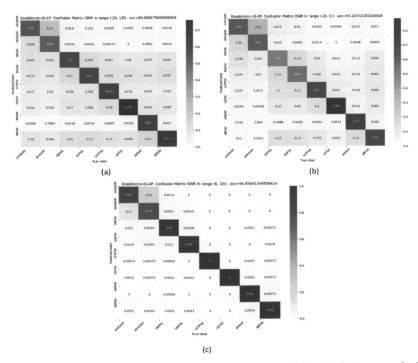

Fig. 7. Confusion matrices for the proposed method at different SNRs. (a) SNR range of −20 db to 18 db; (b) SNR range of −10 db to 5 db; (c) SNR range of 6 db to 18 db

5 Conclusions

In this article, we first proposed a lightweight one-dimensional convolutional neural network module. We compared the modules and other network model in the I/Q performance features and A/P, we found that the A/P characteristics under high signal to noise ratio of classification accuracy are about 7.25% higher than the I/Q characteristics, I/Q data under low SNR more resistance than A/P characteristics, We conclude that the I/Q feature and A/P feature can complement each other at high and low SNR. Therefore, a one-dimensional convolution feature fusion network structure (FF-OnedimCNN) is proposed by using one-dimensional convolution neural module combined with probabilistic principal component analysis (PPCA) to fuse the two features. We discuss the validity of the proposed model from two aspects of classification accuracy and complexity. Experimental results show that compared with the newly proposed network model for automatic modulation classification, our model has obvious advantages in both classification accuracy and complexity.

References

1. Cisco Annual Internet Report (2018–2023) White Paper. https://www.cisco.com/c/en/us/sol utions/collateral/executiveperspectives/annual-internet-report/white-paper-c11-741490.html
2. Qualcomm Inc.: mMTC KPI evaluation assumptions .3GPP R1-162195, April 2016
3. Evans, D.: The Internet of Things: How the next evolution of the internet is changing everything. In: Proceedings of Cisco White Pap, pp. 1–11 (2011)
4. O'Shea, T.J., Corgan, J., Clancy, T.C.: Convolutional radio modulation recognition networks. In: International Conference on Engineering Applications of Neural Networks, pp. 213–226. Springer, Heidelberg (2016)
5. O'Shea, T.J., Roy, T., Clancy, T.C.: Over-the-air deep learning based radio signal classifification. IEEE J. Sel. Top. Signal Process. **12**, 168–179 (2018)
6. Yashashwi, K., Sethi, A., Chaporkar, P.: A learnable distortion correction module for modulation recognition. J. IEEE Wirel. Commun. Lett. **8**, 77–80 (2018)
7. Teng, C.F., Chou, C.Y., Chen, C.H., et al.: Accumulated polar feature-based deep learning for efficient and lightweight automatic modulation classification with channel compensation mechanism. J. IEEE Trans. Vehicular Technol. **69**, 15472–15485 (2020)
8. Zhang, F., Luo, C., Xu, J., Luo, Y.: An efficient deep learning model for automatic modulation recognition based on parameter estimation and transformation. J. IEEE Commun. Lett. **25**, 3287–3290 (2021)
9. Liu, X., Wang, Q., Wang, H.: A two-fold group lasso based lightweight deep neural network for automatic modulation classification. In: 2020 IEEE International Conference on Communications Workshops (ICC Workshops), pp. 1–6. ICCE, Ireland (2020)
10. Pijackova, K., Gotthans, T.: Radio modulation classification using deep learning architectures. In: 2021 31st International Conference Radioelektronika (RADIOELEKTRONIKA), pp. 1–5 (2021)
11. Wang, Y., Liu, M., Yang, J., Gui, G.: Data-driven deep learning for automatic modulation recognition in cognitive radios. J. IEEE Trans. Veh. Technol. **68**, 4074–4077 (2019)
12. Gao, L., Zhang, X., Gao, J., You, S.: Fusion image based radar signal feature extraction and modulation recognition. J. IEEE Access. **7**, 13135–13148 (2019)
13. Zhang, Z., Luo, H., Wang, C., Gan, C., Xiang, Y.: Automatic modulation classification using CNN-LSTM based dual-stream structure. J. IEEE Trans. Veh. Technol. **69**, 13521–13531 (2020)

14. Meng, F., Chen, P., Wu, L., et al.: Automatic modulation classification: a deep learning enabled approach. IEEE Trans. Veh. Technol. **67**(11), 10760–10772 (2018)
15. Peng, S., Jiang,H. ,Wang, H., Alwageed, H., Yao,Y.D.: Modulation classifification using convolutional neural network based deep learning model. In: 26th Wireless Optical Communication Conference, Newark, NJ, USA, pp. 1–5 (2017)
16. O'Shea,T., West, N.: Radio machine learning dataset generation with gnu radio. In: Proceedings of the GNU Radio Conference 1 (2016)
17. West, N.E., O'Shea, T.J.: Deep architectures for modulation recognition. CoRR, vol.abs/1703.09197 (2017)
18. Xu, J., Luo, C., Parr.G., Luo,Y.: A spatiotemporal multi-channel learning framework for automatic modulation recognition. IEEE Wirel. Commun. Lett. **9**, 1629–1632 (2020)

Design and Finite Element Analysis
of Magnetorheological Damper

Yunyun Song and Xiaolong Yang[✉]

School of Mechanical and Automotive Engineering, Guangxi University of Science and
Technology, Liuzhou 545006, China
yangxiaolong@gxust.edu.cn

Abstract. In order to solve the problem that the output damping force of magnetorheological Damper is not large enough and the adjustable range is small, a
bypass magnetorheological Damper is designed in this paper. The Valve is connected in a hydraulic cylinder with pipes to form a controllable magnetorheological
Damper device. Two structures are designed by adding non-magnetic materials to
the structure so that the magnetic field lines pass vertically through the damping
gap as much as possible. One is to use two coils and add a non-magnetic material above the coil, and the other is to use only one coil and add a non-magnetic
material above the coil. The finite element method is used to simulate and analyze
the parameters of two structures which affect the damping performance, and the
results are discussed. The results show that more magnetic force lines can pass
through the damping channel vertically by adding non-magnetic material to the
structure, which can increase the damping force and adjustable coefficient.

Keywords: Magnetorheological damper · Single coil · Double coil · The finite
element

1 Introduction

Magnetorheological fluid (MRF) is a new kind of intelligent material, which is generally
composed of magnetizable particles at micron or nanometer scale, carrier fluid and
additives. When there is no external magnetic field, the Magnetorheological Fluid is a
fluid with good fluidity. When magnetic field is applied, the Magnetorheological Fluid
can be converted into a viscoelastic solid in millisecond level, and the yield shear stress
increases with the increase of magnetic field intensity until saturation. Moreover, the
transformation process of Magnetorheological Fluid and solid is controllable, rapid and
reversible. Magnetorheological damper has excellent dynamic characteristics of fast
response and low power consumption, and has outstanding functions in semi-active
vibration control [1].

Mazlan improved its performance by designing its structure and extending the path
length of the magnetorheological damper [2]. Hu and Liu studied the dual-coil magnetorheological damper, built a model to study its performance by studying different

© The Author(s) 2022
Z. Qian et al. (Eds.): WCNA 2021, LNEE 942, pp. 900–911, 2022.
https://doi.org/10.1007/978-981-19-2456-9_91

piston configurations, and optimized it by using Ansys parameter language to obtain the best damping performance [3]. Kim and Park proposed a new type of adjustable damper and analyzed its damping force characteristics by studying four cylinders with different shapes [4]. Nie and Xin analyzed the performance of the Magnetorheological Damper with different piston configurations, and optimized its structural parameters by combining particle swarm optimization and finite element method [5]. The magnetorheological damper designed by Wang and Chen can improve its performance under a certain volume [6]. Choi et al. [7] designed a new magnetorheological damper and installed the serpentine valve on the bypass channel of the damper, but in order to reduce the volume, the installation position was consistent with the cylinder shaft. Liu and Gao [8] verified the advantages of multi-slot dampers through experiments, which have large damping force and adjustable range, and can further improve their performance by increasing the number of multi-slots.

2 Theoretical Formula and Structural Design

2.1 Theoretical Formula

According to Bingham model, when the magnetorheological fluid flows through the damping gap with volume Q, the pressure difference at both ends is:

$$\Delta P = \frac{12\eta LQ}{bh^3} + \frac{CLT_y}{h} \tag{1}$$

$$Q = A_P V \tag{2}$$

The damping force in flow mode is:

$$F = \Delta P A_P = \frac{12\eta LA_P^2 V}{bh^3} + \frac{CLT_y}{h}A_P \tag{3}$$

Adjustable coefficient:

$$\beta = \frac{bh^2 T_y}{4\eta A_p V} \tag{4}$$

A_P Is the effective area of the piston, Q is the flow rate, V is the movement speed of the piston, η is the viscosity, L is the length, h is the radial height of the damping hole, D is the inner diameter of the cylinder, C is 2–3. $b = \pi D$.

2.2 Structural Design

To a preliminary magnetorheological damping hydraulic design, first of all, based on maximum damping force to calculate the diameter of the piston rod, according to the inner diameter and the relationship between the piston rod diameter, estimate the cylinder inner diameter and thickness of the cylinder block, cylinder diameter, according to the material maximum pressure and the allowable stress at work, calculate the thickness at the bottom of the end cover of magnetorheological damper. The smaller the damping clearance, the greater the damping force, but too small damping clearance may lead to plugging phenomenon, temporarily set the damping clearance as 1 mm, the piston rod line is temporarily set as plus or minus 50, hydraulic cylinder specific parameters are shown in the following Table 1.

Table 1. Parameter table

Piston rod diameter (mm)	16
Cylinder inner diameter (mm)	40
Outer diameter of cylinder (mm)	60
Cylinder thickness (mm)	10
End cap thickness (mm)	10
Damping gap (mm)	1
Stroke (mm)	±50

By adding non-conductive materials, more magnetic field lines can pass vertically through the damping channel. The structure and parameters of the two structures are shown in the figure below (Figs. 1 and 2).

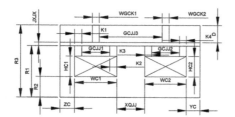 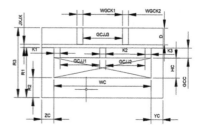

Fig. 1. The structure of the double coil is shown in this figure   **Fig. 2.** The structure of the single coil is shown in this figure

It can be deduced from the previous formula:
There are eight hysteresis drops in the first configuration (Table 2)

$$P1 = \frac{2 * T1 * YC}{JXJX} \tag{5}$$

$$P2 = \frac{2 * T2 * (GCJJ2 - WCGK2) * 0.5}{JXJX} \tag{6}$$

$$P3 = \frac{2 * T3 * (GCJJ2 - WCGK2) * 0.5}{JXJX} \tag{7}$$

$$P4 = \frac{2 * T4 * (0.5 * XQJJ)}{JXJX} \tag{8}$$

Table 2. Size parameters

Double coil	The Size (mm)	Single coil	Size(mm)
WC1	12	WC	32
HC1	6	HC	6
WC2	12	R2	6
HC2	6	ZC	4
R2	6	YC	4
ZC	4	GCC	3
YC	4	K1	2
XQJJ	8	K2	2
GCC	3	K3	2
K1	2	WGCK1	2
K2	2	WGCK2	2
K3	2	R1	15
K4	2	JXJX	1
WGCK1	2	D	5
WGCK2	2	R3	21
R1	15	GCJJ1	13
JXJX	1	GCJJ2	13
D	5	GCJJ3	13
R3	21		
GCJJ1	8		
GCJJ2	8		
GCJJ3	18		

$$P5 = \frac{2 * T5 * (0.5 * XQJJ)}{JXJX} \tag{9}$$

$$P6 = \frac{2 * T6 * (GCJJ1 - WGCK1) * 0.5}{JXJX} \tag{10}$$

$$P7 = \frac{2 * T7 * (GCJJ1 - WGCK1) * 0.5}{JXJX} \tag{11}$$

$$P8 = \frac{2 * T8 * YC}{JXJX} \tag{12}$$

Viscous pressure drop

$$P0 = \frac{12Q\eta * (ZC + WC1 + XQJJ + WC2 + YC)}{\pi * JXJX * JXJX * JXJX * 2 * R1} \tag{13}$$

The total pressure drop of the first structure is.

$$P = P0 + P1 + P2 + P3 + P4 + P5 + P6 + P7 + P8 \tag{14}$$

There are six hysteresis drops in the first configuration

$$P1 = \frac{2 * T1 * YC}{JXJX} \tag{15}$$

$$P2 = \frac{2 * T2 * (GCJJ2 - WGCK2) * 0.5}{JXJX} \tag{16}$$

$$P3 = \frac{2 * T3 * (GCJJ2 - WGCK2) * 0.5}{JXJX} \tag{17}$$

$$P4 = \frac{2 * T4 * (GCJJ1 - WGCK1) * 0.5}{JXJX} \tag{18}$$

$$P5 = \frac{2 * T5 * (GCJJ1 - WGCK1) * 0.5}{JXJX} \tag{19}$$

$$P6 = \frac{2 * T6 * ZC}{JXJX} \tag{20}$$

Viscous pressure drop

$$P0 = \frac{12Q\eta * (ZC + WC + YC)}{\pi * JXJX * JXJX * JXJX * 2 * R1} \tag{21}$$

The pressure drop of the second structure is.

$$P = P0 + P1 + P2 + P3 + P4 + P5 + P6 \tag{22}$$

3 Finite Element Analysis

3.1 Model Diagram and Magnetic Field Line Distribution Diagram of the Two Structures

Two kinds of structure modeling in ANSYS, give material properties respectively and then the simulation, observe two lines of magnetic force distribution of the structure, it can be seen that due to the structure by adding non-magnetic materials, and then make more lines of magnetic force can be vertically through the damping clearance, two-dimensional model diagram and the lines of magnetic force distribution as shown in the figure below (Figs. 3, 4, 5 and 6).

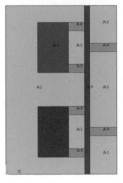

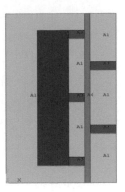

Fig. 3. Double coil as shown in this figure   **Fig. 4.** Single coil as shown in this figure

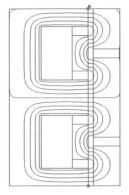

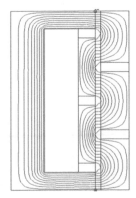

Fig. 5. Double coil as shown in this figure   **Fig. 6.** Single coil as shown in this figure

3.2 Influence of Each Parameter on Magnetic Flux Density

Influence of Radial Damping Clearance. As can be seen from the figure, the magnetic induction intensity increases with the decrease of the damping gap. As the gap becomes smaller, the magnetic resistance becomes smaller. As the total magnetic flux remains unchanged, the magnetic induction intensity increases. The output damping force is the sum of viscous damping force and hysteresis damping force, and the viscous damping force is inversely proportional to the third power of the gap. The controllable damping force is also inversely proportional to the size of the gap, so it decreases with the increase of the gap. When the clearance increases, the decrease of controllable damping force is much smaller than that of viscous damping force, resulting in a rapid increase of adjustable ratio (Figs. 7 and 8).

Influence of Current Size. By the figure can be seen when the current increases, the magnetic induction intensity is increasing, this is because the increase in the current process, other relevant size remains the same, lead to reluctance has not changed, this is increase current, equivalent to increase magnetic flux, magnetic induction intensity increasing, further influence the shear stress, leading to large damping force. The increase

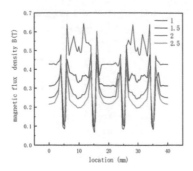

Fig. 7. Double coil as shown in this figure

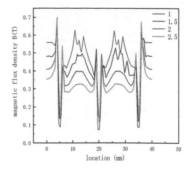

Fig. 8. Single coil as shown in this figure

of hysteresis drop indirectly leads to the increase of adjustable coefficient (Figs. 9 and 10).

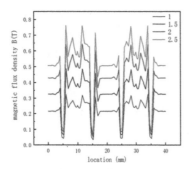

Fig. 9. Double coil as shown in this figure

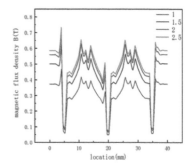

Fig. 10. Single coil as shown in this figure

Influence of Coil Turns. By the figure can be seen when the coil number of turns increases, the magnetic induction intensity is increasing, it is because the increase in the number of turns in the process, other relevant size remains the same, lead to reluctance has not changed, then increase the coil number of turns, equivalent to increase magnetic flux, magnetic induction intensity increasing, further influence the shear stress, leading to large damping force. The increase of hysteresis drop indirectly leads to the increase of adjustable coefficient (Figs. 11 and 12).

The Influence of the Width of the Magnetic Isolation Ring above the Coil. It can be seen from the figure that when the width of the magnetic isolation ring on the coil increases, the magnetic flux density decreases. This is because the increase of the width indirectly leads to the shortening of the vertical passage length of the magnetic field line, and ultimately reduces the hysteresis drop. When the hysteresis drop becomes smaller, the output damping force becomes smaller and the adjustable coefficient decreases (Figs. 13 and 14).

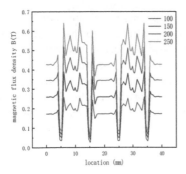

Fig. 11. Double coil as shown in this figure

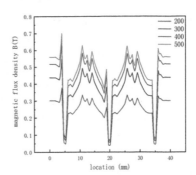

Fig. 12. Single coil as shown in this figure

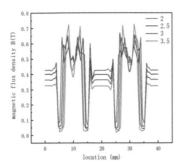

Fig. 13. Double coil as shown in this figure

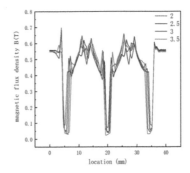

Fig. 14. Single coil as shown in this figure

3.3 Influence of Each Parameter on Damping Performance

Influence of Radial Damping Clearance. As the radial clearance increases from 1 mm to 2.5 mm, the effect on the output damping force and adjustable coefficient is shown below (Figs. 15 and 16).

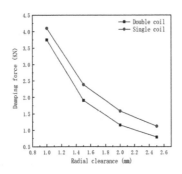

Fig. 15. Double coil as shown in this figure

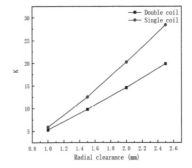

Fig. 16. Single coil as shown in this figure

When the radial clearance increases, the damping force decreases, because when the radial clearance increases, the pressure drop decreases, and then the damping force decreases. The adjustable coefficient increases with the increase of the radial clearance. This is because the increase of the radial clearance will lead to the decrease of the viscous pressure drop, and the hysteresis pressure drop also decreases with the increase of the clearance, but the decrease speed is smaller than the viscous pressure drop, so the adjustable coefficient becomes larger. It can also be seen from the figure that the damping force and adjustable coefficient of a single coil are larger than those of a double coil, possibly because the magnetic flux density along the path of a single coil is more evenly distributed than that of a double coil, and part of the two coils in a double coil will cancel out.

Influence of Current Size. As the current increases from 1A to 2.5a, the effect on the output damping force and adjustable coefficient is shown in the figure below (Figs. 17 and 18).

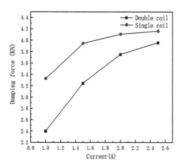
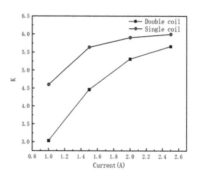

Fig. 17. Double coil as shown in this figure   **Fig. 18.** Single coil as shown in this figure

It can be seen from the figure that when the current increases, the output damping force and the adjustable coefficient are increasing. This is because when the current increases, the magnetic resistance does not change, which causes the magnetic flux to increase, the magnetic induction intensity increases, and the magnetic The stagnant pressure drop increases, and the viscous pressure drop is constant at this time, so the overall pressure drop increases, the damping force becomes larger, and the adjustable coefficient becomes larger. It can also be seen from the figure that the damping force and adjustable coefficient of the single coil are larger than that of the double coil, which may be due to the offset between the two coils in the double coil.

Influence of Coil Turns. When the number of turns of the coil increases from 200N to 500N, the influence on the output damping force and adjustable coefficient is shown in the figure below (Figs. 19 and 20).

As can be seen from the figure, when the number of turns of the coil increases, both the output damping force and the adjustable coefficient increase. This is because when the number of turns of the coil increases, the hysteresis pressure drop also increases indirectly. At this time, the viscous pressure drop is constant, so the adjustable coefficient increases. It can also be seen from the figure that the damping force and adjustable coefficient of a single coil are larger than that of a double coil, possibly because the two coils in a double coil will cancel out part of the middle.

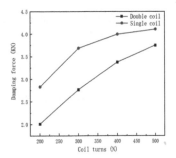

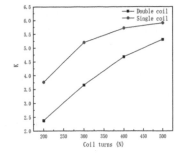

Fig. 19. Double coil as shown in this figure   **Fig. 20.** Single coil as shown in this figure

Influence of the Width of the Magnetic Isolation Ring above the Coil.

When the width of the width of the magnetic isolation ring above the coil increases from 2 mm to 3.5 mm, the effect on the output damping force and adjustable coefficient is shown in the figure below (Figs. 21 and 22).

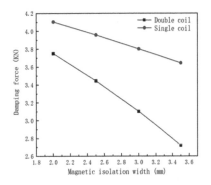

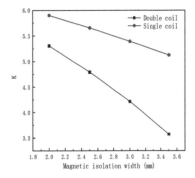

Fig. 21. Double coil as shown in this figure   **Fig. 22.** Single coil as shown in this figure

It can be seen from the figure that when the width of the magnetic isolation ring increases, the output damping force and the adjustable coefficient are both reduced. This is because when the width of the magnetic isolation ring increases, the length of the magnetic field lines passing through vertically decreases indirectly. This causes the hysteresis pressure drop to decrease. At this time, the viscous pressure drop is certain, so the adjustable coefficient becomes smaller. Due to the decrease of the hysteresis voltage drop, the output damping force is indirectly reduced. It can also be seen from the figure that the damping force and adjustable coefficient of the single coil are larger than that of the double coil, which may be due to the fact that the magnetic flux density on the path of the single coil is more uniform than that of the double coil, and the double coil Part of the two coils in the middle will cancel out.

4   Conclusion

Based on hydraulic design, two kinds of structures of magnetorheological damper are designed, and the finite element analysis is carried out on the relevant parameters that affect the damping performance. Through the analysis, the influence of each parameter on the damping performance is studied. The following conclusions are drawn:

1) As the gap increases, the damping force of the magnetorheological damper decreases and the adjustable coefficient increases.
2) When increasing the number of coil turns and current, both the damping force and the adjustable coefficient of the magnetorheological damper increase.
3) When the width of the magnetic isolation ring increases, the damping force and the adjustable coefficient of the magnetorheological damper decrease.
4) In the same volume, the damping performance of the single coil is better than that of the double coil when keeping the current and the number of turns of the coil the same.

Acknowledgements. The authors gratefully acknowledge the support of the National Nature Science Foundation of China (Grant No. 51905114), the support of the Science and Technology Project of Guangxi Province (Grant No. 2020GXNSFAA159042), and the support of the Science and Technology Project of Liuzhou (Grant No. 2017BC20204), and the support of Innovation Project of Guangxi University of Science and Technology Graduate Education (Grant No. GKYC202111).

References

1. Zhu, X., Jing, X., Cheng, L.: Magnetorheological fluid dampers: a review on structure design and analysis. J. Intell. Mater. Syst. Struct. **23**(8), 839–873 (2012)
2. Imaduddin, F., et al.: Design and performance analysis of a compact magnetorheological valve with multiple annular and radial gaps. J. Intell. Mater. Syst. Struct. **26**(9), 1038–1049 (2015)
3. Hu, G., et al.: Design, Analysis, and Experimental Evaluation of a Double Coil Magnetorheological Fluid Damper. Shock and Vibration (2016)
4. Kim, W.H., et al.: A novel type of tunable magnetorheological dampers operated by permanent magnets. Sensors Actuators a-Physical **255**, 104–117 (2017)
5. Nie, S.-L., et al.: Optimization and performance analysis of magnetorheological fluid damper considering different piston configurations. J. Intell. Mater. Syst. Struct. **30**(5), 764–777 (2019)
6. Wang, M., Chen, Z., Wereley, N.M.: Magnetorheological damper design to improve vibration mitigation under a volume constraint. Smart Mater. Struct. **28**(11) (2019)
7. Idris, M.H., et al.: A Concentric design of a bypass magnetorheological fluid damper with a serpentine flux valve. Actuators **9**(1) (2020)
8. Liu, G., Gao, F., Liao, W.-H.: Magnetorheological damper with multi-grooves on piston for damping force enhancement. Smart Materials Struct. **30**(2) (2021)

A High-Efficiency Knowledge Distillation Image Caption Technology

Mingxiao Li[✉]

International School, Beijing University of Posts and Telecommunications, Beijing, China
lmx_bupt2018@163.com

Abstract. Image caption is wildly considered in the application of machine learning. Its purpose is describing one given picture into text accurately. Currently, it uses the Encoder-Decoder architecture from deep learning. To further increase the semantic transmitted after distillation by feature representation, this paper proposes a knowledge distillation framework to increase the results of the teacher section, extracting features by different semantic levels from different fields of view, and the loss function adopts the method of label normalization. Handle unmatched image-sentence pairs. In order to achieve the purpose of a more efficient process. Experimental results prove that this knowledge distillation architecture can strengthen the semantic information transmitted after distillation in the feature representation, achieve a more efficient training model on less data, and obtain a higher accuracy rate.

Keywords: Image captioning · Knowledge distillation · Encoder-decoder · CNN-LSTM

1 Introduction

Image Captioning is very useful in the field of big data and a great advance for computers to quickly extract information from images. In addition, Image captioning actually generates a comprehensive and smooth descriptive sentence automatically by the computer based on the content of the Image. For example, the user searches for the desired items through a paragraph, or find a paper or article source through a picture, multi-object recognition in images or videos, automatic semantic annotation of medical images, object recognition in automatic driving and so on.

The original image captioning technology is mainly derived from machine learning algorithms. For example, after extracting image operators and using classifiers to obtain targets, the target and attributes are used to generate captions. In recent years, it has many kinds of methods in the model [1]: one of them is statistical method to have features with NN model based on encode decode. HAF model is the baseline based on RL [2]. In a generating caption, REN for CIDEr by assigning different weights to each of importance and its weight is word-level. It is proposed to use the language model as a large label space to complete image caption [3], and it also includes using the Attention area to generate words. But there is a problem of attention drift.

© The Author(s) 2022
Z. Qian et al. (Eds.): WCNA 2021, LNEE 942, pp. 912–917, 2022.
https://doi.org/10.1007/978-981-19-2456-9_92

This paper proposes a knowledge distillation architecture to increase our performance of an autoregressive teacher model with good generalization performance. The purpose is to provide more data for training as a reference, and introduce more unlabeled data to achieve soft target and true value as much as possible correspond. Comparing this method with two Encoder-Decoder architectures, the results implied that the model has certain improvements in calculation accuracy.

The rest of this paper includes: The second part is an overview of Image Caption; the third part is an introduction to the Encoder-Decoder architecture; the fourth part is the proposed knowledge distillation structure; the fifth part is the experiments and results, and finally is the summary.

2   Overview of Image Caption

Image caption is the automatic generation of image descriptions by human's language, which has attracted more and more attention in the AI industry. Image captioning can be said to be a huge challenge for the core problem of CV, because image understanding is much more difficult than image classification. It requires not only CV technology, but also natural language processing technology to generate meaningful language for images [4].

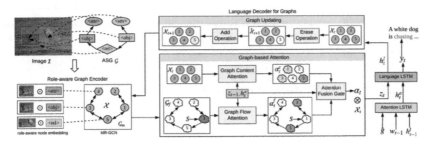

Fig. 1. ASG2Caption model [7].

A novel ASG2Caption model [5] was proposed and shown in Fig. 1, which is able to recognize the graph structure. They let encoder to encode basic information with embedding and then propose a role-aware graph encoder, which contains a role-aware node embedding to distinguish node intentions by MR-GCN. The attention model with CNN over images and LSTM sentences was proposed with three stimulus-driven: Color/Dimension/Location. The CNN-LSTM model combining with the attention principle was considered in paper [6]. The image caption generation with an LSTM was proposed by Verma [7]. The paper [8] propose a lightweight Bifurcate-CNN.

3   Encoder-Decoder Architecture

According to the output and input sequence, in order to serve different application fields, different numbers of RNNs are designed into a variety of different structures. Encoder-Decoder is one of the most important structures in the current AI industry. Since the input and output of the sequence conversion model are variable in length, in order to deal with this type of input and output, the researcher designed a structure consisting of two main parts: the first is the encoder, which is the other to the content. A representation, which is used to output a feature vector network, using a variable-length sequence as input and converting it into a coded state with a fixed shape. The second is a decoder with the same network structure as the encoder but in the opposite direction, which maps a fixed-shape encoding state to a variable-length sequence. An encoder-decoder architecture was employed for captions generation [9]. Seq2Seq can overcome the shortcomings of RNN. For example, applications such as machine translation and chatbot need to achieve direct conversion from one sequence to another. The problem with RNN is that the size of the input and output is mandatory, and the Seq2Seq model does not need to have these restrictions, so the length of the input and output is variable for any occasions.

The encoder-decoder based on fusion methods can be adopted to finish subtitle text task [10]. In the post extraction part, use the VGG16 + Faster R-CNN framework and use the fusion method to train BLSTM. Gated Recurrent Unit is used for effective sentence generation [11]. When the time interval is too large or too small, the gradient of the RNN is more likely to decay or explode. Although deleting gradients can cope with gradient explosions, it cannot solve the difficulty of gradient attenuation. The root cause of RNN's difficulty in practical applications is that RNNs always have gradient attenuation for problems with large processing time distances. LSTM allows RNN to selectively forget some past information through gating, with the purpose of establishing a more global model for long-term conditions and relationships, and retaining useful past memories. GRU believes that it is necessary to further reduce the disappearance of gradients while retaining the advantages of long-term sequences.

4   Knowledge Distillation Structure

Conceptual Captions is a data set proposed in the paper [12]. Compared with the classic COCO data set, Conceptual Captions contains more images, image styles and image annotation content. The method of obtaining Conceptual Captions is to extract and filter the target information content on the internet web page, such as image data, images Image captions and other related information are used as search and filtering tools.

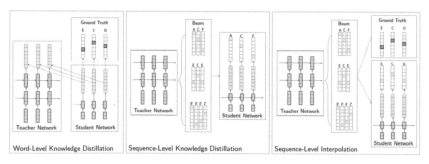

Fig. 2. The different knowledge distillation [13].

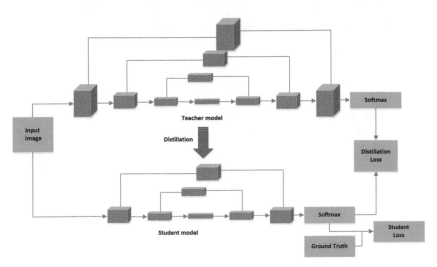

Fig. 3. The proposed knowledge distillation structure.

Like all other artificial intelligence methods, image caption mainly relies on multiple layers of deep neural networks, which introduces high computational costs. How to reduce this high computational cost, consider migrating the large-scale model used to describe large-scale knowledge to the small-scale model. The former is regarded as a teacher and the latter as a student [13, 14], as shown in Fig. 2. The problem that needs to be solved is to determine to integrate certain knowledge into the teacher model and transfer it, and also to solve the problem of the transfer process. This method is called knowledge distillation. The main principle is to map the core knowledge as the learning goal. What needs to be retained in the latter small-size model is the output layer of the previous larger-size model.

For further improving the semantic transmitted after distillation in the feature representation, this paper proposed one knowledge distillation architecture to increase the results of the autoregressive teacher model with good generalization performance, as shown in Fig. 3. The teacher model and student model use a network structure similar to U-net, which is conducive to training the model with higher efficiency on less data, and can achieve features to obtain higher results. Meanwhile, in the loss section, the label normalization method is used to deal with the unmatched image-sentence pairs. To achieve the purpose of more efficient distillation process. In addition, you can also provide more data for training as a reference, and introduce more unlabeled data to achieve the soft target and the ground truth as much as possible.

5   Experimental Results

To analysis and compare some results of the structure proposed in our paper, we selected a part of the data based on Microsoft COCO Caption and Flickr8K. Each image includes five corresponding sentence. All Backbone and Detector adopt VGG16. The multiple descriptions of the image are independent of each other and use different grammars.

These descriptions describe different aspects of the same image, or simply use different grammars [15].

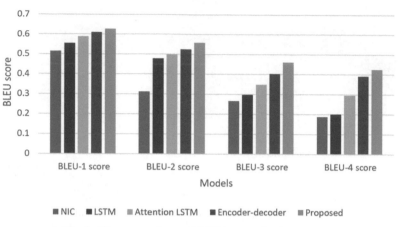

Fig. 4. The comparison of BLEU scores for five models.

In our evaluation process, BLEU is generally used as the evaluation index. BLEU calculates the number of matches between each candidate and ground truth by comparing the n-gram matches between the two. The more matches, the better the candidate gets. Figure 4 shows the test results of BLEU1–BLEU4 with five different ways. These results show that the feature extraction of different semantic levels of images has a good impact on increasing the results of image subtitles. When information is entered into the frame in different ways, such as LSTM and Attention LSTM, it may also affect the results. Comparing the method proposed in this paper with the Attention LSTM and Encoder-Decoder algorithms, the experimental results show that this knowledge distillation architecture can strengthen the semantic information transmitted after distillation in the feature representation, and achieve higher efficiency training models on less data to obtain a higher accuracy rate.

6   Conclusions

Image captioning technology is the comprehensive technology of image generation and description in real life. The recent image captioning is primary belong to the DNN Encode Decode architecture. The teacher-student knowledge distillation framework proposed in this paper can train the model with higher efficiency on less data, and can achieve features of different levels in different fields to increase the indicator of a teacher model with good generalization performance. The next step will be to study how to improve the mapping capabilities of multimodal spaces.

References

1. Smith, T.F., Waterman, M.S.: Identification of common molecular subsequences. J. Mol. Biol. **147**, 195–197 (1981)

2. Wang, H., Zhang, Y., Yu, X.: An overview of image caption generation methods. Computational intelligence and neuroscience (2020)
3. Wu, C., Yuan, S., Cao, H., et al.: Hierarchical attention-based fusion for image caption with multi-grained rewards. IEEE Access **8**, 57943–57951 (2020)
4. Xu, K., Ba, J., Kiros, R., et al.: Show, attend and tell: neural image caption generation with visual attention. In: International Conference on Machine Learning. PMLR, pp. 2048–2057 (2015)
5. You, Q., Jin, H., Wang, Z., et al.: Image captioning with semantic attention. In: Proceedings of the IEEE Conference on Computer Vision and Pattern Recognition, p. 4651–4659 (2016)
6. Chen, S., Jin, Q., Wang, P., et al.: Say as you wish: fine-grained control of image caption generation with abstract scene graphs. In: Proceedings of the IEEE/CVF Conference on Computer Vision and Pattern Recognition, pp. 9962–9971 (2020)
7. Ding, S., Qu, S., Xi, Y., et al.: Stimulus-driven and concept-driven analysis for image caption generation. Neurocomputing **398**, 520–530 (2020)
8. Verma, A., Saxena, H., Jaiswal, M., et al.: Intelligence Embedded Image Caption Generator using LSTM based RNN Model. In: 2021 6th International Conference on Communication and Electronics Systems (ICCES), pp. 963–967. IEEE (2021)
9. Zhao, D., Yang, R., Guo, S.: A lightweight convolutional neural network for large-scale Chinese image caption. Optoelectron. Lett. **17**(6), 361–366 (2021)
10. Singh, A., Singh, T.D., Bandyopadhyay, S.: An encoder-decoder based framework for hindi image caption generation. Multimedia Tools and Applications, 1–20 (2021)
11. Duan, M., Liu, J., Lv, S.: Encoder-decoder based multi-feature fusion model for image caption generation. J. Big Data **3**(2), 77 (2021)
12. Parikh, H., Sawant, H., Parmar, B., et al.: Encoder-decoder architecture for image caption generation. In: 2020 3rd International Conference on Communication System, Computing and IT Applications (CSCITA), pp. 174–179. IEEE (2020)
13. Sharma, P., Ding, N., Goodman, S., et al.: Conceptual captions: a cleaned, hypernymed, image alt-text dataset for automatic image captioning. In: Proceedings of the 56th Annual Meeting of the Association for Computational Linguistics (Volume 1: Long Papers), pp. 2556–2565 (2018)
14. Kim, Y., Rush, A.M.: Sequence-level knowledge distillation. arXiv preprint arXiv:1606.07947 (2016)
15. Park, W., Kim, D., Lu, Y., et al.: Relational knowledge distillation. In: Proceedings of the IEEE/CVF Conference on Computer Vision and Pattern Recognition, pp. 3967–3976 (2019)

Machine Learning in Medical Image Processing

Ahmed Elmahalawy[1,2(✉)] and Ghada Abdel-Aziz[3]

[1] Computer Science and Engineering Department, Faculty of Electronic Engineering, Menofia University, Al Minufiyah, Egypt
ahmed.elmahalawy@el-eng.monufia.edu.eg
[2] Higher Institute of Engineering, El Shorouk Academy, Cairo, Egypt
[3] Department of Communications and Computer Engineering, Electrical Engineering Department is Part of the Faculty of Engineering, Benha University, Banha, Egypt
Ghada.abdelaziz@bhit.bu.edu.eg

Abstract. Medical images provide information that can be used to detect and diagnose a variety of diseases and abnormalities. Because cardiovascular disorders are the primary cause of death and cancer is the second, good early identification can aid in the reduction of cancer mortality rates. There are different medical imaging modalities that the radiologists use in order to study the organ or tissue structure. The significance of each imaging modality is changing depending on the medical field. The goal of this research is to give a review that shows new machine learning applications for medical image processing and gives a review of the field's progress. The classification of medical photographs of various sections of the human body is the focus of this review. Additional information on methodology developed using various machine learning algorithms to aid in the classification of tumors, non-tumors, and other dense masses is available. It begins with an introduction of several medical imaging modalities, followed by a discussion of various machine learning algorithms to segmentation and feature extraction.

Keywords: Machine learning · Feature extraction · Segmentation · Cancer classification · Image processing · Histopathological images · HI · Magnetic resonance imaging (MRI) · Mammogram images · Supervised ML · Unsupervised ML

1 Introduction

Medical imaging makes use of emerging technology to improve people's health and quality of life. Computer-assisted diagnostic (CAD) systems in medicine are a good example. Scientists are increasingly using X-rays, magnetic resonance imaging (MRI), cardiac magnetic resonance imaging (CMRI), computed tomography (CT), Mammography, and histopathology images (HIs).

Despite major breakthroughs in diagnosis and medical treatment, cardiovascular diseases (CVDs) remain the leading cause of death worldwide. According to a World Health Organization report, there were 17.9 million deaths attributed to CVDs in 2016. Cancer is another disease with a high mortality rate, with 9 million deaths. Both developed and developing countries are affected by cancer. Because of the increase in risk

Z. Qian et al. (Eds.): WCNA 2021, LNEE 942, pp. 918–927, 2022.
https://doi.org/10.1007/978-981-19-2456-9_93

factors and late detection of diseases, death rates in low and middle-income nations are high. The early and precise detection of tumors and CVDs is the key point of treatment and diagnostic decision making [1, 2].

Prior diagnostic data should be reviewed then valuable information from previous data is obtained. Artificial intelligence (AI) applications in medical imaging have advanced exponentially in recent years as a result of technological advancements and increased computer capacity. In the image-based diagnosis procedure, machine learning (ML) is applied. It depends on previous clinical models through explicit programming identification of complex imaging data patterns. As ML technique ingest training data, it is then possible to produce more precise models depending on those training patterns. Existing review declares the incremental value of image-based diagnosis using ML methods [3, 4].

1.1 Medical Imaging

Rapid tumor detection and diagnosis using image processing and machine learning techniques can now be an important tool in increasing cancer diagnostic accuracy. Medical imaging is used for clinical diagnosis, therapy, and identifying problems in various body parts.

The goal of a medical imaging the purpose of this research is to establish the location and scale of the project, and features of the tissue or organ in question. This classification is thought to be a good technique to get useful information out of a vast volume of data. As a result, some scientists have focused their efforts in creating and interpreting medical images in order to diagnose the vast majority of diseases. As a result, medical images aid in illness identification, the detection of pathogenic abnormalities and the treatment of patients in a clinical setting.

The techniques and methods used to acquire images of various parts of the human body for diagnostic purposes are referred to as medical imaging. Different radiological imaging techniques are included in medical imaging such as:

X-Ray. The brighter areas on the X-ray are solid tissues, while the darker areas include air or normal tissues. On an X-ray film of the chest, for example, Many organs that separate the chest cavity from the abdominal cavity, such as the heart, ribs, thoracic spine, and diaphragm, are readily visible. This can be used in lung infection detection [5].

CT/CMRI. Significant aspects of the bodily organ, such as shape and size, must be understood in order to categorize the various disorders. Image processing tools such as CT or CMRI are used to **develop** the diagnosis of cardiac disease. This can be used in CVDs diagnosis [6–8].

Mammography. Mammography is regarded as the simplest approach for early breast cancer diagnosis, using only a small amount of radiation. It aids radio-graphic Breast cancer examination to detect any growth or lump in the early stages, even before it becomes obvious to the doctor or the woman herself, and that these rays are not dangerous if used at yearly intervals, as recommended by the National Guidelines for early breast cancer diagnosis. The only method that has been proved to be effective in reducing breast cancer mortality by detecting the disease early on is mammography. Mammography is the most successful approach for early detection of breast cancer, despite the fact that it cannot prevent cancer [2, 9].

Histopathological Images (HI). Despite fast developments in medical field research, the gold standard for tumor identification remains histology. HI is a type of medical imaging in which tissues from microscopy biopsies are shown. The pathologists can use these images to study tissues characteristics in a cell basis. Because HIs contain complicated geometric shapes and textures, they can be utilized to identify, monitor, and treat cancer in various organs such as the breast, lung, liver, lymph nodes, and so on... [10, 11].

1.2 Motivation

The purpose of this study is to show radiologists how to use machine learning techniques to enhance the rate of rapid and accurate cancer detection and CVD diagnosis and categorization. This research seeks to provide a review of novel applications of machine learning for the analysis of medical pictures, as well as an overview of progress in this field. This paper focuses on segmentation and feature extraction in multi-modal medical images of various areas of the human body that have lately been employed.

1.3 Paper Structure

The following is a breakdown of the paper's structure. Section 2 presents a taxonomy for categorizing medical image analysis machine learning algorithms. Section 3 displays several supervised segmentation methodologies as well as supervised ML that was used for the segmentation methods. Section 4 introduces unsupervised machine learning (ML), which is used for segmentation, and then displays various unsupervised segmentation algorithms that aim to find essential structures in medical images, which may aid diagnosis. The feature extraction methods used to describe HIs for further categorization using ML are presented in Sect. 5. Finally, in Sect. 6, the conclusions are stated.

2 Machine Learning

Machine learning (ML) is a type of data analysis that automates the generation of analytical system models. It's a subset of AI that governs how a machine learns from data, recognizes patterns, and makes decisions with little or no human assistance. ML is used to provide a pathological diagnosis of malignancy in a variety of tissues and organs (breast, prostate, skin, brain, bones, liver, and others). Machine learning methods have been widely used in segmentation, feature extraction, and classification [12].

Unsupervised and supervised machine learning methods are the two types of machine learning methods. Unsupervised learning organizes and interprets data based solely on input data, whereas supervised learning (classification and regression) creates prediction models based on both input and output data (clustering).

ML Medical analysis methods can be classified as illustrated in Fig. 1. Typically, pathology specialists are interested in tissue regions that are related to the condition being identified. The goal of medical segmentation is to Label pixels with the structure that they could represent.

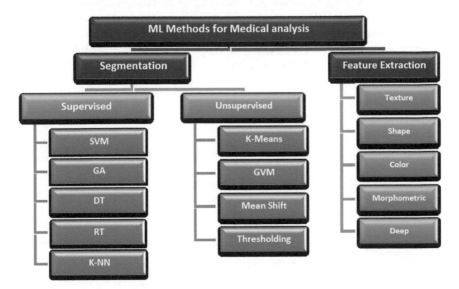

Fig. 1. ML for medical images analysis

Nucleus structure identification, for example, can be used to extract morphological information like the number of nuclei per region, their size, and format, which can be particularly useful in evaluating a tumor's diagnosis. Several segmentation methods are based on supervised or unsupervised machine learning techniques.

3 Supervised ML for Segmentation

Support vector machine (SVM), genetic algorithm (GA), decision trees (DT), regression trees (RT), and k-nearest neighbors algorithm are some of the supervised machine learning algorithms used for segmentation (k-NN).

SVMs are a type of learning machine that can recognize patterns and predict time series, among other things. The support vectors of selected samples map samples into feature space, and the greatest margin hyperplane separates feature vectors [12, 13].

GA is a search-based optimization technique depends on the idea of genetics and natural selection. Optimal or near-optimal solutions to complicated problems are found while the typical solution will consume a very long time to find out. GA searches a space of potential solutions to find one which solves the problem [14].

One of the predictive modelling methods is DT. DT moves from observations, which are represented the tree's branches lead to inferences about the target value, which the tree's leaves reflect. Classification trees are DT when the target variable is a discrete set of values. Class labels are the leaves in these tree structures, and feature combinations that lead to those class labels are the branches. When the target variable is a set of continuous values, such as real numbers, RT is DT [15].

A non-parametric classification method is the K-NN classification method. Data categorization and regression are both done with it. The k closest training samples from the data set are used as the input in all cases. Depending on the mode (classification or regression), the output changes. The outcome of k-NN classification is a class membership. The most common class of its neighbours is used to name an object based on a majority vote of its neighbours. The algorithm determines the value of a property for an object in k-NN regression. This property's value is the average of its k closest neighbours' values [12, 16].

The supervised segmentation algorithms are shown in Table 1 along with the ML methods they employ.

Table 1. Supervised segmentation approaches using different machine learning methods

Segmentation Approach	Tissue/ Organ	ML Method	Year	Paper
Supervised Prostates Segmentation Approach	Prostate	k-NN	2014	[16]
Supervised Breast Segmentation Approach1	Breast	SVM	2015	[17]
Supervised Colon Segmentation Approach	Colon	QDA	2015	[18]
Supervised Breast Segmentation Approach2	Breast	SVM	2015	[19]
Supervised Epithelium Segmentation Approach	Epithelium	SVM	2015	[20]
Supervised Breast Segmentation Approach3	Breast	GA + SVM	2016	[21]
Supervised Breast segmentation Approach4	Breast	SVM	2016	[22]
Supervised General Segmentation Approach	General	RT	2017	[24]
Supervised Medical Segmentation Approach	Breast, prostate, kidney, liver, stomach, bladder	DT	2019	[25]

4　Unsupervised Ml Segmentation

Unsupervised machine learning (ML) segmentation should discover patterns from untagged data and can be divided into several types, such as k-means, general vector machine (GVM), mean shift, and thresholding. The k-means technique is an unsupervised machine learning clustering approach that has been used to segment pixel regions. The K-means technique, which is an unsupervised clustering method, is used to separate the item from the background. It divides the input data into K-clusters, or groupings, based on the K-centroids. When unlabeled data, i.e. data with no established categories or groupings, the method is employed. The purpose is to locate specific groups based on some form of data similarity, with K being the number of groups [12].

The GVM is used to replace the SVM, which are support vectors of selected samples separated by the greatest margin hyper-plane. The support vectors are substituted by general project vectors chosen from the normal vector space, and the general vectors are found using the Monte Carlo (MC) process. GVM improves the capacity to extract features [26].

When a set of data points is given, the mean shift approach labels each data point towards the nearest cluster centroid iteratively, with the direction to the closest cluster centroid defined by where the majority of the neighbor points are. Each iteration brings each data point gets closer to the cluster centre, which contains the most data points. Each point is assigned to a cluster when the algorithm finishes [27, 28].

Decision scores, which are the output of the decision function that is used to produce the prediction, are employed in the thresholding approach. The best score from the output of the decision function can be chosen as the value of the decision threshold All decision score values less than this decision threshold value are considered negative, and all decision score values more than this decision threshold value are considered positive [29].

Table 2 depicts the unsupervised segmentation methodologies as well as the machine learning methods employed.

Table 2. Unsupervised segmentation approaches using different machine learning methods

Segmentation Approach	Tissue/ Organ	ML Method	Year	Paper
Unsupervised Prostate Segmentation Approach	Prostate	Mean shift, Similarity	2014	[27]
Unsupervised Breast Segmentation Approach	Breast	Dictionary, Thresholding	2015	[29]
Unsupervised Cardiac Segmentation Approach	Cardiac	k-means	2016	[30]
Unsupervised Lymph nodes Segmentation Approach	Lymph nodes	k-means	2016	[32]
Unsupervised Lung Segmentation Approach	Lung	k-means	2016	[33]
Unsupervised Liver Segmentation Approach	Liver	k-means	2017	[34]

5 Feature Extraction ML

Before doing classification, some methods rely on feature extraction from raw data. Feature extraction methods aim to reduce the granularity of the input and highlight relevant information related to the problem, such as the presence or absence of a specific element, the amount of that element, texture, shape, histogram, and so on, while providing a form that is unaffected by changes like translation, scaling, and rotation.

Prior to categorization, these issues necessitate the translation of picture pixels into meaningful features. Feature extraction methods take photographs and extract a reasonable number of characteristics from them that summarize the information they contain.

Several different types of characteristics, such as shape, size, texture, fractal, and even a combination of these, have been used.

1- Feature Extraction Approaches
2- Deep Learning Feature Extraction.

In conclusion, due to the nature of medical images, particularly HIs, which contain complex geometric structures and textures, multiple types of characteristics need be merged in many cases for further description. As shown in Table 3, different approaches extract several types of characteristics metamorphic characteristics are useful for identifying geometric structures, but they are more difficult to obtain due to the extensive pre-processing required. Texture, on the other hand, is one of the most significant features for identifying items or regions of interest in a photograph.

Table 3. Feature extraction approaches including deep learning approaches applied on histological images of different parts in human body to extract different features

Approach	Feature	Year	Paper
Colorectal Approach	Local object pattern	2014	[35]
Esophagus Approach	Morphometric, LBP, SIFT, color histograms	2014	[36]
Prostate cancer classification Approach	LBP	2015	[37]
Liver Approach	Morphometric, GLCM, LBP, fractal dimension, graph-based	2015	[38, 39]
Skin Approach	Z-transform coefficients	2016	[40]
Breast Cancer Approach1	Fractal dimension	2016	[41]
Breast Cancer Approach2	Deep	2017	[42]
Breast Cancer Approach3	Deep	2019	[43]

Finally, the most recent techniques rely on deep feature extraction. They're similar to a set of filters that extract geometric and textural features. As a result, deep features and deep approaches for medical image analysis appear to be quite promising.

6 Conclusion

There are different imaging modalities that the radiologists use in order to study the organ or tissue structure. The significance of each imaging modality is changing depending on the medical field. This review provides a brief description of the medical images significance using multi-modalities of different parts in human body; X-ray, CT, MRI, CMRI, Mammography and HI.

This review divides ML applications into supervised segmentation, unsupervised segmentation and feature extraction approaches and describes the various methods in ML was used to offer a summary of development in this area.

There are several supervised ML methods is used for segmentation such as SVM, GA, DT, RT and k-NN. On the other hand, unsupervised ML segmentation methods can be divided into methods such as K-means, GVM, mean shift and thresholding.

Textural characteristics, on the other hand, are crucial in segmentation and are more difficult to collect due to the extensive pre-processing required. Morphometric characteristics are crucial for identifying geometric structures, but they are more difficult to collect due to the need for extensive pre-processing. Finally, the most current feature extraction techniques use deep features to describe organ or tissue details. They're like a series of filters for detecting geometric structures and textures. This research also demonstrates that some deep feature extraction algorithms for medical picture analysis appear to be extremely promising.

References

1. World Health Statistics 2020: monitoring health for the SDGs, sustainable development goals. World Health Organization, Geneva (2020)
2. American Cancer Society. Breast Cancer Facts & Figures 2019–2020. American Cancer Society, Inc., Atlanta (2019)
3. Saxena, S., Gyanchandani, M.: Machine learning methods for computer-aided breast cancer diagnosis using histopathology: a narrative review. J. Med. Imaging Radiat. Sci. **51**, 182–193 (2020)
4. Yassin, N.I., Omran, S., El Houby, E.M., Allam, H.: Machine learning techniques for breast cancer computer aided diagnosis using different image modalities: a systematic review. Comput. Methods Programs Biomed. **156**, 25–45 (2018)
5. X-rays: Radiography. U.S. National Library of Medicine (12 April 2021). https://medlineplus.gov/xrays.html. Accessed May 2021
6. Judice, A., Geetha, K.: A novel assessment of various bio-imaging methods for lung tumor detection and treatment by using 4-D and 2-D CT images. Int. J. Biomed. Sci. (IJBS) **9**(2), 54–60 (2013)
7. Pennell, D.S.U., et al.: Clinical indications for cardiovascular magnetic resonance (CMR): consensus panel report. Eur. Heart J. **25**(21), 1940–1965 (2004)
8. Patient safety: Magnetic resonance imaging (MRI): American College of Radiology, Radiological Society of North America (June 2013). http://www.radiologyinfo.org/en/safety/index.cfm?pg=sfty_m. Accessed May 2021
9. Dheeba, J., Singh, N.A., Selvi, S.T.: Computer-aided detection of breast cancer on mammograms: a swarm intelligence optimized wavelet neural network approach. J. Biomed. Inform. **49**, 45–52 (2014)
10. Sudharshan, P., Petitjean, C., Spanhol, F.: Multiple instance learning for histopathological breast cancer image classification. Expert Syst. Appl. **117**, 103–111 (2019)
11. Aresta, G., Araújo, T., Kwok, S., Chennamsetty, S., et al.: Grand challenge on breast cancer histology images. Med. Image Anal. **56**, 122–139 (2019)
12. Mahesh, B.: Machine learning algorithms: a review. Int. J. Sci. Res. (IJSR) **9**(1), 381–386 (2020)
13. Kotsiantis, S.B.: Supervised machine learning: a review of classification techniques. Informatica **31**, 249–268 (2007)

14. Shapiro, J.: Genetic algorithms in machine learning. In: Paliouras, G., Karkaletsis, V., Spyropoulos, C.D. (eds.) Machine Learning and Its Applications. ACAI 1999. Lecture Notes in Computer Science(), vol. 2049. Springer, Heidelberg (2001). https://doi.org/10.1007/3-540-44673-7_7

15. Amin, R.K., Sibaroni, Y.: Implementation of decision tree using C4.5 algorithm in decision making of loan application by debtor (Case study: Bank pasar of Yogyakarta Special Region). In: 2015 3rd International Conference on Information and Communication Technology (ICoICT), Yogyakarta (2015)

16. Salman, S., et al.: A machine learning approach to identify prostate cancer areas in complex histological images. Adv. Intell. Syst. Comput. **283**, 295–306 (2014)

17. Chen, J., et al.: New breast cancer prognostic factors identified by computer-aided image analysis of HE stained histopathology images. Scientific Reports 2015 (2015)

18. Geessink, O., Baidoshvili, A., Freling, G., Klaase, J., Slump, C., Van Der Heijden, F.: Toward automatic segmentation and quantification of tumor and stroma in whole-slide images of H&E stained rectal carcinomas. In: Progress in Biomedical Optics and Imaging - Proceedings of SPIE (2015)

19. Zarella, M., Breen, D., Reza, M., Milutinovic, A., Garcia, F.: Lymph node metastasis status in breast carcinoma can be predicted via image analysis of tumor histology. Anal. Quant. Cytopathol. Histopathol. **37**, 273–285 (2015)

20. Santamaria-Pang, A., Rittscher, J., Gerdes, M., Padfield, D.: Cell segmentation and classification by hierarchical supervised shape ranking. In: IEEE 12th International Symposium on Biomedical Imaging, pp. 1296–1299 (2015)

21. Wang, P., Hu, X., Li, Y., Liu, Q., Zhu, X.: Automatic cell nuclei segmentation and classification of breast cancer histopathology images. Signal Process. **122**, 1–13 (2016)

22. Arteta, C., Lempitsky, V., Noble, J., Zisserman, A.: Detecting overlapping instances in microscopy images using extremal region trees. Med. Image Anal. **27**, 3–16 (2016)

23. Arteta, C., Lempitsky, V., Noble, J., Zisserman, A.: Learning to detect cells using non-overlapping extremal regions. Med. Image Comput. Comput.-Assist. Interv. **7510**, 348–356 (2012)

24. Brieu, N., Schmidt, G.: Learning size adaptive local maxima selection for robust nuclei detection in histopathology images. In: IEEE 14th International Symposium on Biomedical Imaging (2017)

25. Song, J., Xiao, L., Molaei, M., Lian, Z.: Multi-layer boosting sparse convolutional model for generalized nuclear segmentation from histopathology images. Knowl.-Based Syst. **176**, 40–53 (2019)

26. Zhao, H.: General vector machine. arXiv preprint. arXiv:1602.03950 (2016)

27. Yang, L., Qi, X., Xing, F., Kurc, T., Saltz, J., Foran, D.: Parallel content-based sub-image retrieval using hierarchical searching. Bioinformatics **30**(7), 996–1002 (2014)

28. Demirovic, D.: An implementation of the mean shift algorithm. Image Process. On Line **9**, 251–268 (2019)

29. Sirinukunwattana, K., Khan, A., Rajpoot, N.: Cell words: modelling the visual appearance of cells in histopathology images. Comput. Med. Imaging Graph. **42**, 16–24 (2015)

30. Mazo, C., Trujillo, M., Alegre, E., Salazar, L.: Automatic recognition of fundamental tissues on histology images of the human cardiovascular system. Micron **89**, 1–8 (2016)

31. Mazo, C., Alegre, E., Trujillo, M.: Classification of cardiovascular tissues using LBP based descriptors and a cascade SVM. Comput. Methods Programs Biomed. **147**, 1–10 (2017)

32. Shi, P., Zhong, J., Huang, R., Lin, J.: Automated quantitative image analysis of hematoxylin-eosin staining slides in lymphoma based on hierarchical k-means clustering. In: 8th International Conference on Information Technology in Medicine and Education (2016)

33. Brieu, N., Pauly, O., Zimmermann, J., Binnig, G., Schmidt, G.: Slide-specific models for segmentation of differently stained digital histopathology whole slide images. In: Medical Imaging 2016: Image Processing, Proceedings of SPIE (2016)
34. Shi, P., Chen, J., Lin, J., Zhang, L.: High-throughput fat quantifications of hematoxylin-eosin stained liver histopathological images based on pixel-wise clustering. Sci. China Inf. Sci. **60**, 1–12 (2017)
35. Olgun, G., Sokmensuer, C., Gunduz-Demir, C.: Local object patterns for the representation and classification of colon tissue images. IEEE J. Biomed. Health Inform. **18**, 1390–1396 (2014)
36. Kandemir, M., Feuchtinger, A., Walch, A., Hamprecht, F.: Digital pathology: multiple instance learning can detect Barrett's cancer. In: IEEE 11th International Symposium on Biomedical Imaging, pp. 1348–1351 (2014)
37. Gertych, A., et al.: Machine learning approaches to analyze histological images of tissues from radical prostatectomies. Comput. Med. Imaging Graph. **46**(2), 197–208 (2015)
38. Coatelen, J., et al.: A feature selection based framework for histology image classification using global and local heterogeneity quantification. In: 36th Annual International Conference of the IEEE Engineering in Medicine and Biology Society (2014)
39. Coatelen, J., et al.: A subset-search and ranking based feature-selection for histology image classification using global and local quantification. In: International Conference on Image Processing Theory, Tools and Applications (IPTA) (2015)
40. Noroozi, N., Zakerolhosseini, A.: Computer assisted diagnosis of basal cell carcinoma using Z-transform features. J. Vis. Commun. Image Represent. **40**, 128–148 (2016)
41. Chan, A., Tuszynski, J.: Automatic prediction of tumour malignancy in breast cancer with fractal dimension. R. Soc. Open Sci. **3**, 160558 (2016)
42. Spanhol, F., Oliveira, L., Cavalin, P., Petitjean, C., Heutte, L.: Deep features for breast cancer histopathological image classification. In: IEEE International Conference on Systems, Man, and Cybernetics (2017)
43. Vo, D., Nguyen, N., Lee, S.: Classification of breast cancer histology images using incremental boosting convolution networks. Inf. Sci. **482**, 123–138 (2019)
44. Akkus, Z., Galimzianova, A., Hoogi, A., Rubin, D.L., Erickson, B.J.: Deep learning for brain MRI segmentation: state of the art and future directions. J. Digit. Imaging **30**, 449–459 (2017)
45. Maier, A., Syben, C., Lasser, T., Riess, C.: A gentle introduction to deep learning in medical image processing. Z. Med. Phys. **29**(2), 86–101 (2019)

A New Prefetching Unit for Digital Signal Processor

Rongju Ji and Haoqi Ren[✉]

College of Electronics and Information Engineering, Tongji University, Shanghai 201804, China
renhaoqi@tongji.edu.cn

Abstract. In this paper, a new structure of instruction prefetching unit is proposed. The prefetching is achieved by building the relationship between the branch source and its branch target and the relationship between the branch target and the first branch in its following instruction sequence. With the help of the proposed structure, it is easy to know whether the instruction block of branch target blocks exist in the instruction cache based on the recorded branch information. The two-level depth target prefetching can be performed to eliminate or reduce the instruction cache miss penalty. Experimental results demonstrate that the proposed instruction prefetching scheme can achieve lower cache miss rate and miss penalty than the traditional next-line prefetching technique.

Keywords: Cache prefetching · Digital signal processor · Branch predictor

1 Introduction

Digital Signal Processors (DSPs) are widely used in communication, high performance computing, internet of things, artificial intelligence and other fields. In order to achieve extraordinary data processing ability, VLIW and SIMD are the most common techniques. The former is instruction level parallelism and the latter is data level parallelism. A VLIW instruction package contains several instructions (e.g.: 4 instructions), which will be issued in the same clock cycle [1]. On the one hand, in order to utilize the locality of executed instruction, the size of cache block should be at least 4 times the size of the instruction package [2]. On the other hand, the application program running on the DSP usually have small code amount, so that the capacity of instruction cache is not too large.

Combined with the above two factors, the number of instruction cache blocks will be relatively small, especially when way-set associative organization is used. If the program is executed following the instruction sequence, there will be no instruction cache miss with the help of next-line prefetching scheme [3]. According to statistics, however, there is one branch instruction in every seven instructions [4]. Once a branch is taken, chances are that instruction block of the branch target is not in the cache, which causes an instruction cache miss and leads to severe miss penalty.

© The Author(s) 2022
Z. Qian et al. (Eds.): WCNA 2021, LNEE 942, pp. 928–935, 2022.
https://doi.org/10.1007/978-981-19-2456-9_94

Branch target buffer (BTB) is a structure to facilitate the performance by recording the target address [5]. With BTB, it is easy to fill the instruction block of branch target into the cache in advance. It is recommended to check whether the instruction block is already in the cache before filling, to avoid unnecessary filling. Tag matching is the simplest way to check the existence, but resulting in higher power consumption [6]. From the view of power consumption and implementation cost, using an indication bit may be a better solution to label the existence of the instruction block.

The other cache miss problem caused by branch is the beginning of branch target prefetching may not early enough [7]. For a 5-stage pipeline architecture, the branch decision is generally made in the second or third stage [8]. If the target prefetching is started in the first stage of the branch instruction, prefetching can only start one or two cycles in advance. How to prefetching the target instruction block much earlier is then a problem to affect the performance of the processor.

2   Proposed Structure of the Prefetching Unit

To realize the proposed architecture and eliminate the penalty of cache miss, there is one primary problem need to be solved: how to obtain the branch target instruction early enough in advance? The traditional branch predictor can provide the clue of branch target address, or the branch target instruction. An improved prefetching unit may fetch the branch target instruction from the external memory and store it in the cache before the processer core executes the branch instruction (hereinafter referred to as '1st level branch'), so that the cache miss penalty can be reduced if the branch is taken. However, if there is another branch instruction (hereinafter referred to as '2nd level branch') in the instruction flow of the target instructions, the prefetching unit is unable to fetch the target instruction of the 2nd branch instruction before the 1st branch instruction is executed, due to difficulty of getting the target information of 2nd branch at that stage. That is, how to perform a two-level depth target prefetching?

The proposed prefetching unit solves the problem with a novel structure which builds the relationship between the 1st branch and the 2nd branch. It mainly bases on the classic N-bit branch predictor [9], as shown in Fig. 1. In the diagram, the columns 'Source' and 'N-bit' form the N-bit branch predictor. Take the most widely used 2-bit branch predictor for example [10], the source addresses of the branch instructions are recoded in the column 'Source'. Without losing generality, we can use the lower part of the source address as the index of the rows and recode the upper part of the source address as the content of the first column, to form a direct-mapped structure. That is, each row of this column corresponds to a branch instruction. The column 'N-bit' in the same row is then used to contain the prediction value of the branch instruction. In 2-bit branch prediction scheme, '11' and '10' indicate that the branch is likely to be taken, while '01'and '00' indicate that the branch is likely to be not-taken.

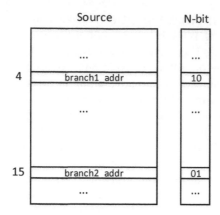

Fig. 1. Classic branch predictor

Now let's build the relationship between the 1st branch and its branch target. A simply way to achieve this is adding a new entry for each row to save the address of branch target instruction. In this case, once a branch instruction is processed by the prefetching unit, the value of branch prediction and the target address can be reached directly.by the same row indexing. Further, it is easy to adding an indication bit for each row, to distinguish whether the branch target instruction is already in the instruction cache. Thus, when processing a branch instruction, if the indication bit is '0', the prefetching unit may start to fill the instruction block addressed by the target address, to guarantee that the target instruction is in the cache or being filled into the cache when the branch is taken. The cache miss penalty is then eliminated or reduced. However, there is a problem in the scheme. Because the target address is stored as a content instead of an index, it is difficult to maintain the value of the indication bit when the target instruction block is moved in or out of the instruction cache.

To solve the above problem, the target addresses are also organized with direct-mapped structure which is similar with the structure of the source addresses, as shown in the column 'Target' in the right part of Fig. 2. A pair of pointers are adopted to connect the branch source and the corresponding branch target. On the side of branch source, 'Pointer_A' contains the row number of the branch target and two valid bits indicating whether the branch target address and the branch target instruction are valid in the prefetching unit and in the instruction cache, respectively. On the side of branch target, 'Pointer_B' contains the line number of the corresponding branch source. Therefore, it is easy to find out the target address and fill the target instruction in advance if necessary based on the branch address, while the valid bits in 'Pointer_A' can be easily modified as soon as the target address is updated or the target instruction is filled into or moved out of the instruction cache.

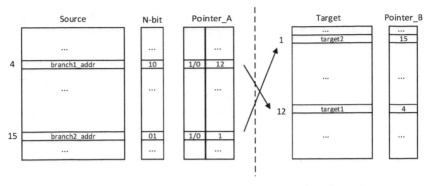

Fig. 2. Improved prefetching structure based on the branch predictor

Let's then build the relationship between the 1st branch and the 2nd branch. The 2nd branch itself is an ordinary branch source, so it is also recorded in another row of column 'source'. As shown in Fig. 3, 'Pointer_C' is used to save the row number of the 2nd branch. According to this structure, when processing a branch instruction (i.e.: the 1st branch), the target address and the source address of the 2nd branch can be obtained consequently. The corresponding instruction blocks can then be filled into the instruction cache much early in advance to eliminate or reduce the cache miss penalty.

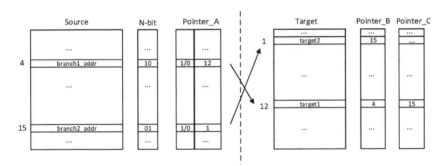

Fig. 3. Structure of the proposed prefetching unit

3 Simulation Model and Experimental Results

To estimate the performance of the proposed prefetching unit, a simulation model is built based on a 5-stage pipelined RISC processor of PISA instruction set architecture (PISA-ISA) in simplescalar simulator [11]. In the original simulator, the first stage is instruction fetching, the second stage is instruction decoding, the third stage is executing and making the branch decision, the fourth stage is memory accessing and the last stage is register writing back. Obviously, the proposed prefetching unit is placed in the first stage. Once the program counter (PC) is updated, it is used to fetch instruction from the instruction cache, as well as being sent to the prefetching unit. If the content read from the column 'source' indexing by the lower part of the PC equals to the upper part of the PC, which means the instruction according to the PC is a branch instruction recorded in the prefetching unit, the valid bits in the same row is checked to determine whether to fill its target instruction block or not. In the next clock cycle, the target address is obtained based on 'Pointer_A', which means the filling process of the 1st branch target can be started if necessary while the 1st branch instruction is in the second stage. Two more clock cycles later, the target address of the 2nd branch address can be obtained based on 'Pointer_C' and a new 'Pointer_A'. That is, the filling process of the target of the 2nd branch can be started if necessary while the 1st branch instruction is in the fourth stage and the its branch decision is just made one clock cycle before.

In PISA-ISA, there is a branch delay slot after each branch instruction, so the worst case is one branch instruction for every two instructions, and the target instruction of the 1st branch happened to the 2nd branch unfortunately. In this case, the target instruction of the 1st branch is to be filled when the 1st branch is in the second stage, and to be fetched when the 1st branch is in the third stage; the target instruction of the 2nd branch is to be filled when the 1st branch is in the fourth stage, and to be fetched when the 1st branch is in the fifth stage. The timing diagram is illustrated in Fig. 4. Besides that, the prefetching timing requirements in other cases are more relax.

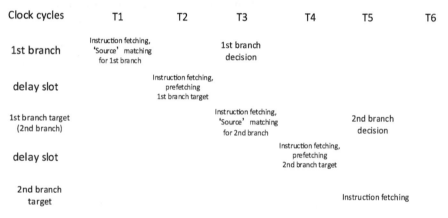

Fig. 4. The timing diagram of the worst case when using the proposed structure

With the different experimental configurations, we get several sets of simulation results. Instruction cache may have a capacity from 16k bytes to 64k bytes. The organization of instruction cache may be direct-mapped, 2-way-set associative or 4-way-set associative. Since VLIW is widely used in DSP architecture, the block size should be large enough to contain at least 4 VLIW instruction packages. Therefore, the block size is set as 64 bytes. Assume that there is a level 2 cache with a reasonable capacity, the instruction cache miss penalty may vary from 2 cycles to 6 cycles. Further, the branch prediction unit has 64 entries with direct-mapped organization, so that the prefetching unit also has 64 rows in total.

Table 1 shows the instruction cache miss rates of the traditional next-line prefetching with 2-bit branch prediction and the proposed structure. Each data row corresponds to a kind of instruction cache configuration. From the top to the bottom, the configurations are 16k bytes with direct-mapped, 16k bytes with 2-way-set-associative, 16k bytes with4-way-set-associative, 32k bytes with direct-mapped, 32k bytes with 2-way-set-associative, 32k bytes with4-way-set-associative, 64k bytes with direct-mapped, 64k bytes with 2-way-set-associative, and 64k bytes with 4-way-set-associative, respectively. In the traditional structure, miss rate has nothing to do with miss penalty, so we only need to use the leftmost data column to represent miss rate of traditional structure. The remaining columns correspond to different cases of miss penalty for the proposed structure. From the left to the right, the miss penalties are 2 cycles, 3 cycles, 4 cycles, 5 cycles, and 6 cycles, respectively.

It is obvious from the table that no matter what the miss penalty is, the proposed structure has a significant decrease in miss rate. Further, the smaller the miss penalty is, the more significant the decrease of miss rate becomes. This is because when miss penalty is small, more cache miss can be completely concealed by the two-level depth target prefetching. When the miss penalty become larger, some of the branch target is to be accessed before the prefetching is completed. This is still a cache miss, but the actual penalty of this miss can be reduced.

Table 1. Cache miss rate comparison between the next-line prefetching and the proposed one

	Next-line prefetching	Proposed, 2 cycles	Proposed, 3 cycles	Proposed, 4 cycles	Proposed, 5 cycles	Proposed, 6 cycles
16k, direct-mapped	3.45%	0.46%	0.64%	1.15%	1.84%	2.71%
16k, 2-way-set	2.20%	0.27%	0.38%	0.71%	1.15%	1.7%
16k, 4-way-set	1.84%	0.22%	0.31%	0.59%	0.95%	1.41%
32k, direct-mapped	2.25%	0.38%	0.39%	0.72%	1.17%	1.74%
32k, 2-way-set	1.48%	0.17%	0.24%	0.47%	0.76%	1.13%
32k, 4-way-set	1.33%	0.15%	0.22%	0.42%	0.68%	1.02%
64k, direct-mapped	1.40%	0.16%	0.23%	0.44%	0.72%	1.07%
64k, 2-way-set	1.34%	0.15%	0.22%	0.42%	0.69%	1.02%
64k, 4-way-set	1.33%	0.15%	0.22%	0.42%	0.68%	1.01%

Table 2 shows the instruction cache miss penalty reduction in total of the proposed structure. Comparing Table 1 and Table 2, we can see that the miss rate and the total miss penalty reduction are both relative high for the bigger penalty cases. This is because although part of the penalty is covered by the prefetching, even if there is only one penalty cycle left, it will be treated as a cache miss and increase the miss rate. In some cases, although the miss rate is still relatively high, the total miss penalty has already reduced to a very low level.

Table 2. Reduction of cache miss penalty in total

	Proposed, 2 cycles	Proposed, 3 cycles	Proposed, 4 cycles	Proposed, 5 cycles	Proposed, 6 cycles
16k, direct-mapped	86.55%	88.47%	88.85%	91.09%	94.77%
16k, 2-way-set	87.80%	89.25%	89.27%	91.30%	94.85%
16k, 4-way-set	88.16%	89.48%	89.39%	91.36%	94.88%
32k, direct-mapped	87.75%	89.22%	89.25%	91.29%	94.85%
32k, 2-way-set	88.52%	89.70%	89.51%	91.42%	94.90%
32k, 4-way-set	88.67%	89.79%	89.56%	91.45%	94.91%
64k, direct-mapped	88.60%	89.75%	89.53%	91.43%	94.91%
64k, 2-way-set	88.66%	89.79%	89.55%	91.44%	94.91%
64k, 4-way-set	88.67%	89.80%	89.56%	91.45%	94.91%

4   Conclusion

In this paper, we have described a new structure of DSP prefetching unit based on the two-level depth target prefetching scheme. The instruction block of the branch source is already in the instruction cache. The instruction blocks of the 1st branch target and the 2nd branch target are filled into the instruction cache before the branch decision is made, so that the possible instructions following the branch source are also in the cache, or being filled into the cache, which eliminate or reduce the total cache miss penalty. The performance of the proposed structure has been demonstrated by experimental results.

Acknowledgement. This work was supported by the Key-Area Research and Development Program of Guangdong Province Projects under Grant 2018B010115002, National Natural Science Foundation of China under (NSFC) Grant 61831018 and 61631017.

References

1. Ren, H., Zhang, Z., Wu, J.: SWIFT: a computationally-intensive DSP architecture for communication applications. Mob. Netw. Appl. **21**, 974–982 (2016)

2. Hennessy, J., Patterson, D.: Computer Architecture: A Quantitative Approach, 6th edn. Morgan Kaufmann Publishers, Burlington (2017)
3. Xu, J.: Research on prefetch technique of cache. Popular Science & Technology (2011)
4. Xiong, Z., Lin, Z., Ren, H.: BI-LRU: a replacement strategy based on prediction information for Branch Target Buffer. J. Comput. Inf. Syst. 11(20), 7587–7594 (2015)
5. Monchiero, M.: Low-power branch prediction techniques for VLIW architectures: a compiler-hints based approach integration. VLSI J. 38, 515–524 (2005)
6. Sun, Y., Yuan, Y., Li, W., et al.: An aggressive implementation method of branch instruction prefetch. J. Phys. Conf. Ser. 1769(1), 012062 (7pp) (2021)
7. Rostami-Sani, S., Valinataj, M., Chamazcoti, S.: Parloom: a new low-power set-associative instruction cache architecture utilizing enhanced counting bloom filter and partial tags. J. Circuits Syst. Comput. 28(6) (2018)
8. Patterson, D., Hennessy, J.: PattersnComputer Organization and Design: The Hardware Software Interface RISC-V Edition, Morgan Kaufmann Publishers (2018)
9. Lee, J., Smith, A.: Branch prediction strategies and branch target buffer design. IEEE Comput. Mag. 17, 6–22 (1984)
10. Zhang, L., Tao, F., Xiang, J.: Researches on design and implementations of two 2-bit predictors. Adv. Eng. Forum 1, 241–246 (2011)
11. Burger, D., Austin, T.: The SimpleScalar Tool Set Version 2.0, Department Technical Report, University of Wiscon-Madison (1997)

Optimizing Performance of Image Processing Algorithms on GPUs

Honghui Zhou[1], Ruyi Qin[1], Zihan Liu[2], Ying Qian[2], and Xiaoming Ju[2(✉)]

[1] Ningbo Power Supply Company of State Grid, Zhejiang Electric Power Co., Ltd., Ningbo, China

[2] Zhejiang Jierui Electric Power Technology Co., Ltd., Ningbo, China

10152130243@stu.ecnu.edu.cn, xmju@sei.ecnu.edu.cn

Abstract. The application of machine learning algorithms in the field of power grid improves the service level of power enterprises and promotes the development of power grid. NVIDIA Volta and Turing GPUs powered by Tensor Cores can accelerate training and learning performance for these algorithms. With Tensor Cores enabled, FP32 and FP16 mixed precision matrix multiplication dramatically accelerates the throughput and reduces AI training times. In order to explore the cause of this phenomenon, we choose a convolutional neural network (CNN), which is widely used in computer vision, as an example and show the performance characteristics with tensor core on general matrix multiplications and convolution calculations as benchmark. Building a CNN based on cuDNN and TensorFlow, we analyze the performance of CNN from various aspects and optimize performance of it by changing the shape of convolution kernel and using texture memory, etc. The experimental results prove the effectiveness of our methods.

Keywords: Machine learning algorithms · Convolution neural network · Computer vision · Convolution kernel · Texture memory

1 Introduction

Electricity has become an indispensable part of people's life. The application of Artificial Intelligence technology in the field of power grid improves the service level of power enterprises and promotes the development of power grid. With the in-depth application of intelligent technology in power grid, a large number of image data are produced. At this time, with the help of big data image processing technology, enterprises can solve the problem of processing and saving massive data. It can reduce the workload of the enterprise, improve the efficiency and accuracy of the staff, promote the development of the enterprise and enhance the core competitiveness of the enterprise. Among the Artificial Intelligence technologies, machine learning is a research hot spot in many research organizations. Machine learning techniques, especially deep learning such as recurrent neural networks and convolutional neural networks have been applied to fields including computer vision 1, speech recognition 2, natural language processing 3 and drug discovery 4. Deep Learning requires substantial computing power. Graphics Processing Unit (GPU) can accelerated computing.

© The Author(s) 2022

Z. Qian et al. (Eds.): WCNA 2021, LNEE 942, pp. 936–943, 2022.

https://doi.org/10.1007/978-981-19-2456-9_95

Recently, NVIDIA published Turing architecture 5 as the successor to the Volta architecture 6 with tensor cores 7 which can accelerate general matrix multiplication (GEMM). GEMM is at the heart of deep learning. Here's a diagram from 8, where the time's going for a typical deep convolutional neural network doing image recognition using Alex Krizhevsky's Imagenet architecture 1. All of the layers that start with fc (for fully-connected) or conv (for convolution) are implemented using GEMM, and almost all the time (95% of the GPU version, and 89% on CPU) is spent on those layers.

In order to construct the machine learning models conveniently, various high-performance open-source deep learning frameworks emerge these years such as tensorflow 9 and caffe 10. These frameworks support running computations on a variety of types of devices, including CPU and GPU (Fig. 1).

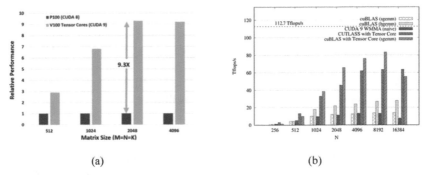

Fig. 1. Performance improvement in GEMM given by the official white paper and practical application

In some tasks of image processing, CNN can be applied to image recognition, classification and enhancement, etc. CNN used a special structure for image recognition and can be trained quickly. In order to explore the reasons for such huge difference, we will implement a typical CNN named LeNet-5 23, which is commonly used in deep learning.

2 Related Work

AI computing has become the driving force of the NVIDA GPU, as a computing accelerator, it integrates built-in hardware and software for machine learning. Some studies have investigated the tensor core by programing111213. Sorna et al. proposed a method that can use computational capability of tensor core without degrading the precision of the Fourier Transform result 14. Carrasco et al. applied a reduction strategy based on matrix multiply-accumulate with tensor core. Their found showed that tensor core can promote the arithmetic reductions15. Markidis et al. evaluated performance of NVIDIA Tensor core with Tesla V100 using GEMM operating 16. They tested the capability with tensor Core using naive implementation with CUDA 9 WMMA, CUTLASS and cuBLAS. Martineau et al. analyzed and evaluated the tensor core through optimization a GEMM benchmark 11, finding similar conclusion of V100 GPU presented by14. Different from previous studies, we will make use of neural network parallel library to further evaluate the performance of GPU on the basis of benchmark.

In deep learning, CNN is a class of artificial neural network structure gradually emerging in recent years. A representative CNN involves convolutional layer, pooling layer and full-connected layer. The convolutional layer extracts feature by convolving input with a group of kernel filters, which contains plenty of matrix operations. The pooling layer contains average, max and stochastic pooling, which contributes to invariance to data variation and perturbation. The fully connected layer in a CNN combines the results of convolutions. It performs the weights which represent the relationship between the input and output and the input multiplication and generates the output.

3   Experiment

The following experiment environment is: AMD Ryzen CPU, NVIDIA Geforce RTX 2080TI (Turing) GPU, Microsoft Windows 10 64-bit, CUDA SDK 10.0, CUTLASS 1.3. Nvprof is selected to evaluate from instruction running time to number of calls. The performance of experiment uses TFlops/s to statistics with operand divided by operation time.

General Matrix Multiplication (GEMM) defined in BLAS 18 and cuBLAS 19 is a matrix multiplication and accumulation routine as fllows:

$$C \leftarrow \alpha A \times B + \beta C$$

where $A \in R^{M \times K}$, $B \in R^{K \times N}$, and $C \in R^{M \times N}$ are matrices, and α and β are scalars. GEMM is the heart of deep learning and is mainly used in neural networks of specific structures such as CNN/RNN. The main purpose of the Tensor core in the Volta architecture and Turing architecture is to accelerate the calculation of GEMM. Many optimization efforts have also been incorporated to the widely used GEMM libraries: MAGMA 20, CUTLASS 21 and cuBLAS.

3.1   Performance of GEMM with Matrix Dimension

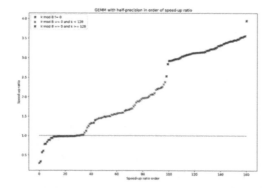

Fig. 2. Performance of GEMM at half-precision with k.

When calculate GEMM, the dimensions of matrix are m, n and k respectively in (1). Each cell is multiplied by a $1 \times K$ matrix and a $K \times 1$ matrix, this operation will be split and distributed to the tensor core for processing with tensor core on. We try to investigate the effect of m, n, k dimension on the speed-up ratio and the shared size K has a greater impact on performance. In order to find the optimal size k, the GEMM is performed with half-precision in Fig. 3. It can be seen that the speed-up ratio of the test sample that cannot be divisible by 8 is relatively low, close to 1; Most of samples which can be divisible by 8 can be effectively accelerated by the tensor core; and as the k value increases, the speed-up ratio also shows an upward trend, indicating that the tensor core is more sensitive to the value of k (Fig. 2).

3.2 Performance Analysis of GEMM with Tensor Core on and off

A series of self-written cases supplemented by the deep learning test suite DeepBench 22 are tested the performance with the tensor core on or off in the new architecture. Table 1 shows the results of running GEMM using Nvprof with the tensor core turned on and off, including the number of calls and running time of each API.

With the tensor core on, since the matrix multiplication operation that originally required multiple dot product instructions is replaced by only one wmma instruction, the calculation is more dense and the time of device synchronization become less, the performance is improved significantly.

Table 1. Performance analysis of GEMM with Tensor Core on and off (API Calls)

API	Tensor Core on			Tensor Core off		
	RT (ms)	CN	ART (ms)	RT (ms)	CN	ART (ms)
cudaDeviceSynchronize	**186156**	322	**578.12**	543509	322	168792
cudaFree	45565.9	811	56.185	47447.6	811	58.185
cudaMalloc	1835.06	805	2.2796	1838.33	805	2.2796
cudaLaunchKernel	557.250	64961	0.0086	693.12	64961	0.0086
cudaMemsetAsync	130.150	**27268**	0.0082	230.83	40501	0.0082

[*]RT-running time, CN- the number of calls, ART-average running times.

3.3 Convolution Calculation

In the CNN model, the fully connected layer is often served as the last layer, and the body of the network is composed of convolutional layers. Therefore, it is critical to speed up the calculation of convolution for the performance of the entire network.

There are several methods developed to efficiently implement the convolution operation besides directly computing the convolution named direct convolution. One is based on Fast Fourier Transform (FFT) named FFT convolution to reduce computational complexity, computing the convolution in the frequency domain 错误! 未找到引用源。.

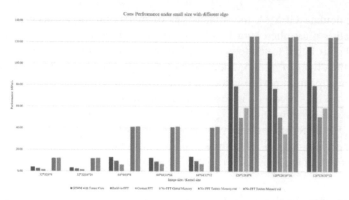

Fig. 3. Performance of convolution based on different algorithm (Small images).

Another is based on matrix multiplication (e.g., GEMM) which is one of the most widely used algorithms for convolution. Figure 4 shows the performance of each method when the image size is less than 128 * 128. When the input image size become smaller, the performance of the two methods mentioned above drops sharply, while the direct method calculates the convolution performance is stable. For the direct method using texture memory, the row and column convolutions are not much different.

3.4   Convolutional Neural Networks (CNN) Based on cuDNN

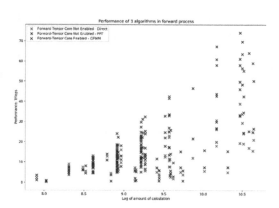

Fig. 4. Performance of DIRECT, FFT, GEMM algorithm in cuDNN.

The construction of CNN refers to LeNet-5 23, and the pooling layer is omitted for the reason that GEMM concentrated on the full-connected layer and convolutional layers, leaving only the input/output layer, convolution layer, and fully connected layer.

The results are shown in Fig. 5. In the forward process of the convolutional neural network, in addition to the convolution calculation, the forward propagation according to the weight is also the main calculation. The performance advantage with tensor core on is still obvious, except in the case of the image size is very small (such as 10^1), which also corresponds to the phenomenon in convolution calculation.

3.5 Convolutional Neural Networks (CNN) Based on Tensor Flow

Table 2. Optimization result in CNN based on TensorFlow.

Convolution Kernel	Convolution	Time(s)
5 * 5	GEMM	54.016
5 * 5	**Texture**	**50.775**
5 * 5	FFT	59.957
8 * 8	GEMM	52.395

We use TensorFlow framework to build CNN based on the LeNet-5 with cifar-10 as the dataset, which contains 50,000 images with 32×32 pixel and can be divided into ten different categories. The latest ·version of TensorFlow is enabled by default with tensor core on. We change the size of the convolution kernel and the convolution calculation method and in TensorFlow. The result is shown in Table 2. When the size of the convolution kernel was changed to 8×8, the performance improved significantly, proving the conclusions that the tensor core is more sensitive to the value of K in the GEMM experiment.

4 Conclusion

We make a series of experiments based on GEMM, convolution calculations and CNN and analyze the improvement of performance on tensor core. Based on the analysis of the above experimental results, it can be concluded that the new architecture can indeed bring significant performance improvements to a large number of GEMM in machine learning under certain circumstances and improving the performance of overall machine learning applications. However, in some cases the improvement of performance is limited for the shape of matrix and other operation except GEMM and traditional calculation methods still have higher performance.

Acknowledgement. The work is supported by State Grid Zhejiang Electric Power Co., Ltd., science and technology project (5211nb200139), the key technology and terminal development of lightweight image elastic sensing and recognition based on AI chip.

References

1. Krizhevsky, A., Sutskever, I., Hinton, G.E.: Imagenet classification with deep convolutional neural networks. In: Advances in Neural Information Processing Systems, pp. 1097–1105 (2012)
2. Abdel-Hamid, O., Mohamed, A., Jiang, H., et al.: Convolutional neural networks for speech recognition. IEEE/ACM Trans. Audio Speech Lang. Process. **22**(10), 1533–1545 (2014)
3. Conneau, A., Schwenk, H., Barrault, L., et al.: Very deep convolutional networks for natural language processing. arXiv preprint arXiv:1606.01781, February 2016
4. Segler, M.H.S., Kogej, T., Tyrchan, C., et al.: Generating focused molecule libraries for drug discovery with recurrent neural networks. ACS Cent. Sci. **4**(1), 120–131 (2017)
5. NVIDIA: Nvidia turing architecture whitepaper. Technical report, NVIDIA Corp., August 2018. https://www.nvidia.com/content/dam/en-zz/Solutions/design-visualization/technolog ies/turing-architecture/NVIDIA-Turing-Architecture-Whitepaper.pdf
6. NVIDIA Volta GPU Architecture (2017). https://images.nvidia.com/content/volta-architect ure/pdf/volta-architecture-whitepaper.pdf
7. NVIDIA: NVIDIA TENSOR CORES, The Next Generation of Deep Learning (2019)
8. Jia, Y.: Learning semantic image representations at a large scale. UC Berkeley (2014)
9. Abadi, M., Barham, P., Chen, J., et al.: Tensorflow: a system for large-scale machine learning. In: 12th {USENIX} Symposium on Operating Systems Design and Implementation ({OSDI} 2016), pp. 265–283 (2016)
10. Jia, Y., et al.: Caffe: convolutional architecture for fast feature embedding. In: Proceedings of the 22nd ACM International Conference on Multimedia, pp. 675–678. ACM (2014)
11. Martineau, M., Atkinson, P., McIntosh-Smith, S.: Benchmarking the NVIDIA V100 GPU and tensor cores. In: Mencagli, G., et al. (eds.) Euro-Par 2018. LNCS, vol. 11339, pp. 444–455. Springer, Cham (2019). https://doi.org/10.1007/978-3-030-10549-5_35
12. Jia, Z., Maggioni, M., Staiger, B., Scarpazza, D.P.: Dissecting the NVIDIA volta GPU architecture via microbenchmarking. arXiv preprint arXiv:1804.06826
13. Jia, Z., Maggioni, M., Staiger, B., Scarpazza, D.P.: Dissecting the NVidia Turing T4 GPU via microbenchmarking. arXiv preprint arXiv:1903.07486 (2019)
14. Sorna, A., Cheng, X., D'Azevedo, E., Won, K., Tomov, S.: Optimizing the fast Fourier transform using mixed precision on tensor core hardware. In: 2018 IEEE 25th International Conference on High Performance Computing Workshops (HiPCW), Bengaluru, India, pp. 3–7 (2018)
15. Carrasco, R., Vega, R., Navarro, C.A.: Analyzing GPU tensor core potential for fast reductions. In: 2018 37th International Conference of the Chilean Computer Science Society (SCCC), Santiago, Chile, pp. 1–6 (2018)
16. Markidis, S., Chien, S.W.D., Laure, E., Peng, I.B., Vetter, J.S.: NVIDIA tensor core programmability, performance & precision. In: 2018 IEEE International Parallel and Distributed Processing Symposium Workshops (IPDPSW), Vancouver, BC, pp. 522–531 (2018)
17. Chetlur, S., Woolley, C., Vandermersch, P., et al.: cuDNN: efficient primitives for deep learning. arXiv preprint arXiv:1410.0759 (2014)
18. Lawson, C.L., Hanson, R.J., Kincaid, D.R., et al.: Basic linear algebra subprograms for Fortran usage (1977)
19. Nvidia, C.: Cublas library. NVIDIA Corporation, Santa Clara, California, 15(27): 31 (2008)
20. Nath, R., Tomov, S., Dongarra, J.: An improved MAGMA GEMM for fermi graphics processing units. Int. J. High Perform. Comput. Appl. **24**(4), 511–515 (2010)
21. NVIDIA. CUTLASS: Fast Linear Algebra in CUDA C++ (2018). https://devblogs.nvidia. com/cutlass-linear-algebra-cuda/

22. Narang, S., Diamos, G.: Baidu DeepBench (2017)
23. LeCun, Y.: LeNet-5, convolutional neural networks (2015). http://yann.lecun.com/exdb/lenet. 20: 5

Nakagami Parametric Imaging Based on the Multi-pyramid Coarse-to-Fine Bowman Iteration (MCB) Method

Sinan Li, Zhuhuang Zhou, and Shuicai Wu$^{(\boxtimes)}$

Department of Biomedical Engineering, Faculty of Environment and Life,
Beijing University of Technology, Beijing, China
wushuicai@bjut.edu.cn

Abstract. Nakagami-m parametric imaging has been used for imaging and detection of coagulation zone in microwave ablation. In order to improve the image smoothness and accuracy of coagulation zone detection, the multi-pyramid coarse-to-fine bowman iteration (MCB) method was proposed and compared with traditional moment-based estimator (MBE) method. Phantom simulations showed that the MCB method could obtain better image smoothness and higher accuracy in lateral target size detection than the MBE method. Experimental results of porcine liver *ex vivo* ($n = 18$) indicated that the m parameter obtained by the MCB method was more accurate than that obtained by the MBE method in detecting the coagulation zone. Nakagami-m parametric imaging based on MCB method can be used as a potential tool for microwave ablation monitoring.

Keywords: Multi-pyramid · Nakagami imaging · Moment-based estimator · Microwave ablation

1 Introduction

Microwave ablation is one of the important methods for clinical treatment of hepatic tumors. As a way of image guidance, ultrasound can make full use of its advantages of real-time, non-radiation and cheapness. However, traditional B-mode ultrasound image cannot accurately display the boundary of the coagulation zone after tumor ablation. Parametric imaging methods based on statistical distribution models of ultrasonic backscattered signals were proposed to improve imaging and tissue characterization.

Wagner et al. [1] first applied Rayleigh distribution to B-mode imaging to show that the set of scatterers was full of high density of random scatterers. The Rice model proposed by Wang et al. [2] can represent not only random scatterers in the set of scatterers, but also periodic scatterers. The K-distribution corresponded to a variable density of random scatterers with no coherent signal component and was introduced in ultrasound imaging by Shankar et al. [3]. The homodyned K (HK) distribution corresponded to the case of random scatterers with or without coherent signal component [4]. The Nakagami distribution [5] was an approximation of HK.

© The Author(s) 2022
Z. Qian et al. (Eds.): WCNA 2021, LNEE 942, pp. 944–954, 2022.
https://doi.org/10.1007/978-981-19-2456-9_96

Some of these models and improved methods have been applied to ultrasound parametric imaging. Tsui et al. [6] applied Nakagami distribution to thermal lesions monitoring of radiofrequency ablation. Rangraz et al. [7] used HIFU-intensity focused ultrasound Nakagami imaging for thermal lesions monitoring. Tsui et al. [8] proposed the window-modulated compounding (WMC) Nakagami imaging for ultrasound tissue characterization, which improved the image smoothness. The coarse-to-fine Bowman iteration method (CTF-BOW) was used by Han et al. [9] for plaque characterization, which provided better accuracy of parameter estimation and image smoothness compared with traditional method [10].

In this paper, we proposed a Nakagami-m parametric imaging method based on multi-pyramid compound, then applied it to the coagulation zone imaging and detection. We performed phantom simulations on this new method and compared the smoothness and resolution of images obtained by the new method with those obtained by the traditional moment-based estimator (MBE) [11] method. Microwave ablation experiments were carried out on porcine liver *ex vivo* ($n = 18$), and the receiver operating characteristic (ROC) curve was drawn to assess the accuracy of the proposed method for coagulation zone detection.

2   Theoretical Algorithms

The Nakagami statistical model was proposed to express the statistics of the envelope of ultrasonic backscattered signals. The probability density function (PDF) of the envelope, $f(r)$, is given by [5]

$$f(r) = \frac{2m^m r^{2m-1}}{\Gamma(m)\Omega^m} \exp(-\frac{m}{\Omega}r^2)U(r). \tag{1}$$

where $m > 0$ is the shape parameter and $\Omega > 0$ is the scaling parameter. Values of m parameter can be calculated by the MBE method, which is expressed as [11]

$$m_{MBE} = \frac{[E(R^2)]^2}{E[R^2 - E(R^2)]^2}. \tag{2}$$

$$\Omega = E(R^2). \tag{3}$$

where E $(\cdot)$ stands for the expectation, and R is a sequence of envelope data.

Figure 1(a) illustrates the Nakagami-m parametric imaging. Firstly, the raw ultrasonic backscattered signals were acquired from the tissue. Secondly, a Hilbert transform was performed to obtain the envelope data. Lastly, the MBE method and the MCB method were used to construct Nakagami-m parametric images, respectively. The latter is detailed as below.

The basis of this method is the CTF-BOW method [9], which is shown in Fig. 1(b). Envelope data were divided into 3 layers to build a pyramid model. Original envelope matrix was the zeroth layer, which was given a Gaussian blur operation and down-sampling to get the first layer data matrix. Both rows and columns were reduced by half. Repeated the above to get the second layer data matrix. The maximum likelihood

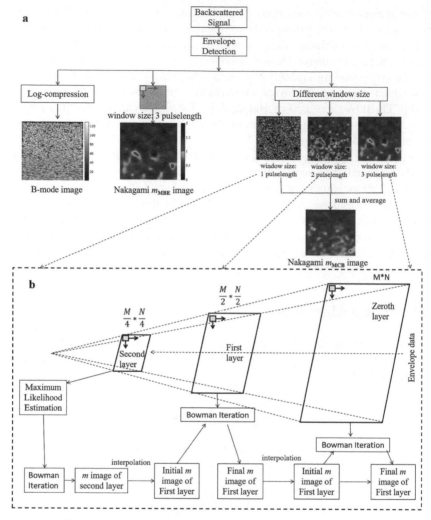

Fig. 1. Flow chart for the algorithm of MCB Nakagami imaging.

estimation was performed on the second layer data matrix to obtain the initial values for Bowman iteration [12].

$$m_{MLE} = \frac{6 + \sqrt{36 + 48\Delta}}{24\Delta}. \tag{4}$$

$$\Delta = \ln[\frac{1}{N}\sum_{i=1}^{N} r_i^2] - \frac{1}{N}\sum_{i=1}^{N} \ln r_i^2 \tag{5}$$

The Bowman estimator is defined by

$$m_j = \frac{m_{j-1}\{\ln(m_{j-1}) - \psi(m_{j-1})\}}{\Delta}. \tag{6}$$

where $\psi(x) \equiv \frac{d\ln\Gamma(x)}{dx}$ is digamma function. Through first Bowman iteration, the Nakagami-m parametric image corresponding to the second layer was obtained. It was interpolated to get the same size of the first layer envelope data and used as the initial value for another Bowman iteration. The second Bowman iteration was used to obtain the corresponding m parametric image from the first layer envelope data. Performing the above process again, and the zeroth layer m parametric image was the final image.

In the CTF-BOW method, each layer uses a sliding window of the same size for iterative calculation of m parameters. However, the window sizes should be different when the detection targets are different. In order to improve the universality of the method, three Nakagami parametric images obtained by using CTF-BOW method with different window sizes are summed and averaged in this study, which constituted the MCB method.

3 Materials and Methods

3.1 Phantom Simulations

In order to evaluate the performance of the MCB method, we used the Field II Toolbox [13, 14] to simulate the ultrasonic backscattered signals. We used it to simulate a 5-MHz Gaussian pulse (pulse length = 0.924 mm) as the incident wave, with the sampling rate of 40 MHz and sound speed of 1540 m/s. Two types of phantoms were generated: homogeneous phantom and heterogeneous phantom. 10 phantoms were produced in each kind of densities.

The volume of homogeneous phantoms was $30 \times 30 \times 1$ mm^3, and the concentrations were 2, 4, 8 and 16 scatterers/mm^3, respectively. The MCB method and MBE method were used to build the Nakagami-m parametric images. For the MBE method, a window size of 3 pulse lengths was adopted, which corresponded to the conclusion of Tsui et al. [10]. For the MCB method, the sliding windows of the three pyramid models were 2 times, 3 times and 4 times the pulse length, respectively. We used the full width at half-maximum (FWHM) to evaluate the smoothness of Nakagami parametric image. A smaller FWHM value indicated that the image smoothness was improved. The autocorrelation function (ACF) was also calculated to compare the resolution effect of the images. The parametric images were adjusted to 256×256 image data to calculate the ACFs. The smaller the widths of the ACF along the X and Y axes, the smaller the resolution of the image.

The volume of the heterogeneous phantom was also $30 \times 30 \times 1$ mm^3, with a circular target zone in the middle. The scatterer densities in the inclusion and surrounding tissues were 40/mm^3 and 4/mm^3, respectively. In order to test the ability of the MCB method to recognize the target boundary of different sizes, we used two kinds of dense circles with diameters of 10 mm and 6 mm, respectively. The diameters of the circle region in the Nakagami parametric images obtained by the MCB and MBE methods were measured and compared along the axial and lateral directions.

3.2 Porcine Liver *ex vivo* Experiment

The experimental platform for microwave ablation consists of a portable ultrasound scanner (Terason t3000); a 128 linear-array transducer (Terason 12L5A); a water-cooled ablation needle (KY-2450B) and a microwave ablation device (KY-2000). Fresh porcine livers *ex vivo* were purchased form the market. Before the experiments, the liver was placed into a $6 \times 6 \times 6\ cm^3$ acrylic box with appropriate size and was inserted horizontally through a circular hole of the acrylic box with an ablation needle. The backscattered signals of porcine liver tissues ($n = 18$) during microwave ablation were collected by this platform. For each ablation experiment, the power was set at 80 W and the ablation duration was 60 s. The backscattered signals were recorded into .bin files with 2 frames/s for the following Nakagami imaging on MATLAB. After each collection, the tissue was cut along the scanning plane of the ultrasound transducer, and the gross pathology image was taken as the reference standard of the coagulation zone.

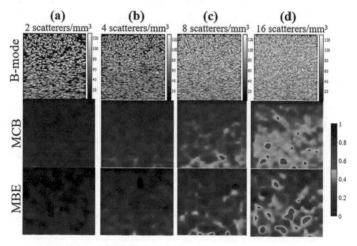

Fig. 2. B-mode image, Nakagami-*m* parametric images obtained by the MCB and MBE methods for homogeneous phantom with different densities: (a) 2 scatterers/mm^3; (b) 4 scatterers/mm^3; (c) 8 scatterers/mm^3; (d) 16 scatterers/mm^3.

4   Results

4.1 Phantom Simulations

Figure 2 shows the B-mode images, Nakagami-*m* parametric images using the MCB and MBE methods for different scatterer concentrations. With the increase of scatterer concentration, the Nakagami parametric images obtained by two methods became brighter, which corresponded to the larger values of *m* parameters.

Figure 3 illustrates the FWHMs of m-parameter distributions obtained by the MCB and MBE methods. At each scatterer concentration, the FWHM obtained by the MCB method was smaller than that of the MBE method, which indicated the MCB method could improve the image smoothness.

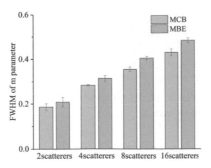

Fig. 3. Comparison of full width at half-maximum (FWHM) of the m parameters distribution between Nakagami images based on MCB and MBE methods.

The autocorrelation functions (ACFs) of Nakagami-m images were obtained by the MCB method and MBE methods for various scatterer densities. The X-axis and Y-axis widths corresponding to 10% of the peak height of ACF surfaces were taken as indicators to measure image resolution. It could be seen from Fig. 4 that the width calculated by the MCB method was larger than that of the MBE method at low concentration (≤ 4 scatterers/mm^3). When scatterer concentration was high (≥ 8 scatterers/mm^3), there was no significant difference between the width obtained by the MCB and MBE methods ($p > 0.05$).

Figure 5 shows B-ultrasound images and Nakagami parametric images corresponding to the heterogeneous phantom. In order to compare the accuracies of two methods in detecting the target boundary, the white dotted lines in the images were taken as the reference, and the length of the bright strong reflecting region was measured along axial and lateral directions, respectively. Figure 6 shows the measured results. No matter how large the diameter of the central strong reflecting region was, the axial width estimated by the MCB method was smaller than that of the MBE method, while the lateral width was larger than that of the MBE method. This indicated that the MCB method was inferior to the MBE method in axial detection capability, but superior to the MBE method in lateral detection capability.

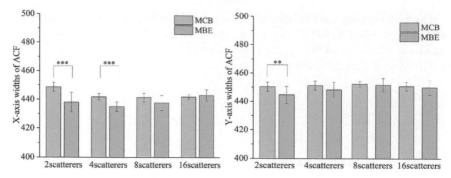

Fig. 4. Comparisons of the *X*-axis and *Y*-axis widths of autocorrelation function (ACF) among the MCB and MBE Nakagami images.

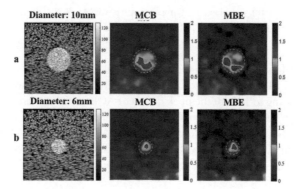

Fig. 5. B-mode image, Nakagami parametric images obtained by MCB and MBE methods for heterogeneous phantom with a circular target of (a) 10 mm diameter; (b) 6 mm diameter.

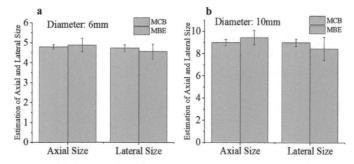

Fig. 6. Axial and lateral size estimations of strong scattering region (a) 6 diameter and (b) 10 diameter using MCB and MBE methods.

4.2 Porcine Liver *ex vivo* Experiments

Figure 7(a) shows the gross pathology image of a porcine liver after ablation; Fig. 7(b–d) are the corresponding B-mode image, Nakagami m_{MCB} image and Nakagami m_{MBE} image, respectively. The coagulation zone in the middle of the parametric image obtained by the MCB method was brighter and the contour was more obvious. In order to further quantitatively evaluate the accuracy of coagulation zone identification, the squares enclosed by red dotted lines were used to select regions of interest (ROIs) with a size of 30×30 mm^2 from all the images. Figure 8 shows the ROC curves of coagulation zone detected using m_{MCB} and m_{MBE} parametric images, which corresponds to the case shown in Fig. 7. The AUCs of the m_{MCB}, and m_{MBE} parametric images were 0.8696 and 0.8655, respectively. Table 1 shows the average AUCs for detecting coagulation zone of porcine liver *ex vivo* ($n = 18$) by using m_{MCB} and m_{MBE} parametric imaging. The performance of the MCB method in the detection of coagulation zone is slightly higher than that of the MBE method.

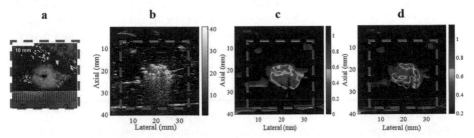

Fig. 7. Region of interest (ROI) of the gross pathology image (a), B-mode image (b), Nakagami m_{MCB} parametric image (c), Nakagami m_{MBE} parametric image (d). The squares enclosed by red dotted lines in (a)–(d) are ROI$_{GP}$, ROIm_{MCB}, ROIm_{MBE}, respectively.

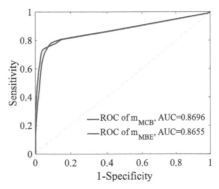

Fig. 8. Receiver operating characteristic (ROC) curves of Nakagami m_{MCB} parametric imaging (blue) and Nakagami m_{MBE} parametric imaging (red) to detect the coagulation zone of microwave ablation.

Table 1. The area under the receiver operating characteristic curve (AUC) for Nakagami m_{MCB} parametric imaging and Nakagami m_{MBE} parametric imaging to detect coagulation zones of porcine liver *ex vivo* ($n = 18$) with the binarized gross pathology image as the reference.

	m_{MCB} prarmetric imaging	m_{MBE} parametric imaging
AUC (mean ± SD)	0.8464 ± 0.07	0.8353 ± 0.07

5 Discussion

According to the results, the MCB method can improve the image smoothness of Nakagami parametric imaging and accuracy of coagulation zone detection compared with the MBE method. The Nakagami-m_{MCB} image was obtained from summing and averaging three parametric images with sliding window sizes of 2 times, 3 times and 4 times the pulse length, respectively. When performing Nakagami images compounding, the small window contains more local information to maintain the image resolution, while the large window contains more global information improve the smoothness of the image.

The results also showed that the MCB method lose more axial resolution than the MBE method in the case of low number densities of scatterers. This is due to the use of Gaussian pyramid decomposition. Half of the envelope data is lost in each decomposition layer, resulting in the loss of some local information. Although the image becomes smoother, the axial resolution is lower. It can be seen that the image smoothness and the image resolution are two variables that restrict each other.

Han et al. [9] proposed Nakagami imaging based on single Gaussian pyramid decomposition, and verified that it was better than the MBE method in m parameter estimation. However, they used a fixed sliding window, and the window size needs to be adjusted to the size of the detection target. Tsui et al. [10] also used a fixed 3 pulse lengths window for thermal lesions Nakagami imaging based on the MBE method. In our work, we used three pyramid decomposition models with different window sizes to sum and average. Heterogeneous phantom simulations have proved that the MCB method could obtain more accurate lateral size estimation in the target contour detection of different sizes compared with the MBE method. The results of microwave ablation experiment showed that the smoothness improved by MCB method made red shadings in Fig. 7(c) increase obviously. Because the 4 pulse lengths window used in the MCB method is larger than the 3 pulse lengths window used in the MBE method, which brings better smoothness. Meanwhile, the 2 pulse lengths window used in the MCB method reduce the loss of image resolution as much as possible.

6 Conclusions

In our work, we proposed the MCB method for ultrasound Nakagami imaging. Phantom simulations showed the MCB method could not only improve image smoothness, but also improve the detection ability of lateral target contour. However, the axial resolution of images obtained by MCB method at low scatterer concentration was weaker than that of MBE method, and there was no significant difference at high concentration. The result of microwave ablation of porcine liver *ex vivo* ($n = 18$) showed that the average AUC of coagulation zone detection based on the MCB method was 0.8464 ± 0.07, which was higher than that of the MBE method.

Acknowledgments. This work was supported by the National Natural Science Foundation of China (Grant Nos. 61871005, 11804013, 61801312, and 71661167001), Natural Science Foundation of Hebei Province (Grant No. F2020101001), the Beijing Natural Science Foundation (Grant No. 4184081).

References

1. Wagner, R.F., Smith, S.W., Sandrick, J.M.: Statistics of speckle in ultrasound B-scans. IEEE Trans. Son. Ultrason. **30**, 156–163 (1983)
2. Insana, M.F., Wagner, R.F., Garra, B.S.: Analysis of ultrasound image texture via generalized Rician statistics. Opt. Eng. **25**, 743–748 (1986)
3. Shankar, P.M., Reid, J.M., Ortega, H.: Use of non-Rayleigh statistics for the identification of tumors in ultrasonic B-scans of the breast. IEEE Trans. Med. Image **12**, 687–692 (1993)
4. Dutt, V., Greenleaf, J.F.: Ultrasound echo envelope analysis using a homodyned K distribution signal model. Ultrason. Imaging **16**, 265–287 (1994)
5. Shankar, P.M.: A general statistical model for ultrasonic backscattering from tissues. IEEE Trans. Ultrason. Ferroelectr. Freq. Control **47**(3), 727–736 (2000)
6. Wang, C.Y., Geng, X., Yeh, T.S.: Monitoring radiofrequency ablation with ultrasound Nakagami imaging. Med. Phys. **40**(7), 072901 (2013)
7. Rangraz, P., Behnam, H., Tavakkoli, J.: Nakagami imaging for detecting thermal lesions induced by high-intensity focused ultrasound in tissue. J. Eng. Med. **228**(1), 19–26 (2004)
8. Tsui, P.H., Ma, H.Y., Zhou, Z.: Window-modulated compounding Nakagami imaging for ultrasound tissue characterization. Ultrasonics **54**(6), 1448–1459 (2014)
9. Han, M., Wan, J.J., Zhao, Y.: Nakagami-m prarametric imaging for atheroscerotic plaque characterization using the coarse-to fine method. Ultrasound Med. Biol. **43**(6), 1275–1289 (2017)
10. Tsui, P.H., Chang, C.C.: Imaging local scatter concentrations by the Nakagami statistical model. Ultrasound Med. Biol. **33**(4), 608–619 (2007)
11. Lin, J.J., Cheng, J.Y., Huang, L.F.: Detecting changes in ultrasound backscattered statistics by using Nakagami parameters: comparisons of moment-based and maximum likelihood estimators. Ultrasonics **77**, 133–143 (2017)
12. Bowman, K.O., Shenton, L.R.: Maximum likelihood estimators for the gamma distribution revisited. Commun. Stat. Simul. Comput. **12**(6), 697–710 (1983)
13. Jensen, J.A.: FIELD: a program for simulating ultrasound systems. Med. Biol. Eng. Comput. **34**(1), 351–352 (1996)
14. Jensen, J.A., Svendsen, N.B.: Calculation of pressure fields from arbitrarily shaped, apodized, and excited ultrasound transducers. IEEE Trans. Ultrason. Ferroelectr. Freq. Control **39**, 262–267 (1992)

Building Machine Learning Models for Classification of Text and Non-text Elements in Natural Scene Images

Rituraj Soni[1([⊠])] and Deepak Sharma[2]

[1] Department of CSE, Engineering College Bikaner, Bikaner, Rajasthan, India
`rituraj.soni@gmail.com`
[2] Department of Computer Science Engineering, School of Engineering and Applied Sciences,
Bennett University, Greater Noida, Uttar Pradesh, India

Abstract. Computer vision aims to build autonomous systems that can perform some of the human visual system's tasks (and even surpass it in many cases)among the several applications of Computer Vision, extracting the information from the natural scene images is famous and influential. The information gained from an image can vary from identification, space measurements for navigation, or augmented reality applications. These scene images contain relevant text elements as well as many non-text elements. Prior to extracting meaningful information from the text, the foremost task is to classify the text & non-text elements correctly in the given images. The present paper aims to build machine learning models for accurately organizing the text and non-text elements in the benchmark dataset ICDAR 2013. The result is obtained in terms of the confusion matrix to determine the overall accuracy of the different machine learning models.

Keywords: Natural scene images · Machine learning models · Text and non-text components · Classifiers

1 Introduction

Computer Vision, often abbreviated as CV, can be formally defined as a field of study that seeks to develop techniques to help computers visualize and understand the content of digital images such as photographs and videos. It aims to develop some computational models for the human visual system concerning the biological view. Whereas, if the Engineering view is considered, it seeks to establish an autonomous system that will perform similarly to a human. Thus, Computer Vision (CV) has numerous applications in various domains of Engineering and medical sciences [1]. It finds application in the automotive, manufacturing, retail industry like Walmart and Amazon Go, financial services, health care, agriculture industry, surveillance, navigation by robots, automatic car driving sign translation, etc. Researchers are also developing an autonomous system to automatically extract the information from the old documents and help form digitized versions of such records. One of the most important uses of computer vision is to extract

© The Author(s) 2022
Z. Qian et al. (Eds.): WCNA 2021, LNEE 942, pp. 955–968, 2022.
https://doi.org/10.1007/978-981-19-2456-9_97

the text regions [2] from the natural scene images and born digital images, which will further assist in language and sing translation and tourist navigation. Thus, with such a vast domain of applications, CV plays an essential role in improving the quality of humanity.

1.1 Natural Scene Images

Natural Scene images [3] are images captured with the help of cameras or other hand-held devices in pure natural conditions. These images may be incidental images or non-incidental images. These natural scene images contain images from Advertisement boards, billboards, notices, various boards from shops, hotels, and other public offices & buildings. Such type of images often contains non-text as well as text components within them. The text present in such images includes essential information about those images. Such data can be used for implementing different applications like tourist navigation, assistance in-car driving, etc. Figure 1 displays the samples from the many natural scene images datasets, such as ICDAR 2003 [4], ICADR 2011 [5], ICDAR 2013 [6], available for research works. The research in this domain is carried out with the help of these datasets only.

Fig. 1. Examples of natural scene images [5]

The natural scene images contain various types of text, as shown in Fig. 1. The font of the text can be fancy or regular. It may prevent fonts of different orientations, colors, and different languages. In this paper, we are focusing on ICDAR datasets, which mainly contain the English language. The significant hurdles [7] in extracting the text regions apart from the variation in the font are the other non-text elements present in the images. The images contain various further details apart from the text regions. There may be natural scenery like trees, plants, and objects like chairs, tables, fencing, etc. These non-text elements must be removed from the images to get the proper text regions for extracting information from the text. This requires classifying the text and non-text features from the scene images, which is the paper's main aim.

1.2 Classification in Machine Learning

Machine learning is a domain of Computer Vision (CV) and Artificial intelligence (AI) that uses data and algorithms to work similarly as humans learn, thus gradually improving its accuracy. Therefore, it can be stated that machine learning uses computer programs and data that can be used for its learning. The aim is to make the computer or given machine learn itself. The learning process requires observation or data that is available on various internet sources for the given problem.

The learning process requires the classification among the different types of sample spaces available for a given problem. Thus, category deals with providing labels to different objects or samples. The classification process requires training on the datasets, and those results are evaluated on the given testing sets. For this work, it is necessary to build different classification machine, learning models. The machine learning models are different supervised or unsupervised types of machine learning algorithms. These machine learning models are the pre-trained models that can be further used for testing purposes.

The present paper aims to build different machine learning models [8] to classify the text and non-text elements in the natural scene images. The machine learning models are evaluated based on the confusion matrix obtained and overall accuracy. The rest of the paper is organized as follows; Sect. 1 describes the basic introduction, Sect. 2 covers the literature review related to the problem, Sect. 3 demonstrates the proposed methodology with experiments, Sect. 4 discusses the results, Section 5 discusses the conclusion, and the future work.

2   Literature Review

The importance of the various applications like contents-based image retrieval, license plate recognition, language translation from the scene, word detection from document images encourages the researchers to work in text detection and recognition from the scene images. There are various categories [9] of the method available on which work has been carried out in the past, such as Region-based, Texture based, connected components based and Stroke based methods. Each method has one thing in common: text-specific features are required to classify the text and non-text elements present in the image. Thus, to identify the text and non-text elements correctly, one of the important tasks is the choice of the classifier, that will give maximum accuracy to the selected features.

The classification of the text & non-text elements is one of the crucial processes in text detection from scene images. Researchers have used different features and classifiers for classification purposes using machine learning algorithms. Iqbal et al. [10] propose using four classifiers, Adaboost M1, Regression, Bayesian Logistic, Naïve Bayes, & Bayes Net, to classify text & non-text components. The sample space taken consisted of only 25 images. Zhu et al. [11] use a two-stage classification process to separate the text & non- txt elements that increase time complexity. Lee et al. [12] and Chen and Yullie [13] discuss the utility of the AdaBoost classifiers, but the selection of the inappropriate features gives less efficient results. Pan et al. [14] propose implementing boosted classifier & polynomial classifier to separate the text & non-text components. MA et al. [15] insist on using a linear SVM and LBP & HOG & statistical features. Pan et al. [16] use a CRF using single perceptron & multi-layer perceptron classifier. Minori Maruyama et al. [17] propose implementing the classification work using SVM (RBF kernel) and stump classifier in the second stage. Fabrizio et al. [14] use K-NN in first stage & RBF kernel with SVM classifier in the second stage. Ansari et al. [18]

insists a method for classifying components with the assistance of T-HOG & LBP (SVM) classifier. The drawback is the high computation cost.

There is no method mentioned for selecting the classifiers in the previous work done by the researchers. Most of the work is carried out using SVM classifiers and Adaboost Classifiers. There is no such method discussed in earlier work in this domain for selecting any classifier. They are chosen arbitrarily. Some of the methods have used two-stage classification that has increased the computation cost. The method in [19] uses SVM classifiers and thus takes a long time due to detailed segmentation. In some of the previous works [20], the inclusion of the deep learning architecture for classification purposes increases the computation time to a great extent.

Moreover, it requires a significant amount of time to train and give accurate results. The choice of the suitable classifier is one of the critical tasks in classification using machine learning algorithms. It will increase the accuracy of the results & reduce the time taken to give results. Therefore, choosing a classifier that will give high accuracy for classification of text & non-text elements in natural scene images is required.

3 Proposed Methodology

This section introduces the proposed methodology for building the machine learning models used in the paper to classify the text and non-text elements. The benchmark dataset ICDAR 2013 is used for the same. The images from the ICDAR dataset undergoes the modified WMF-MSER method to remove the connected characters and text present in the images. Further, then the classification is performed using the ground truth available for the images. The flowchart for the proposed method is shown in Fig. 2.

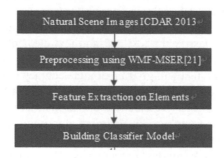

Fig. 2. Flowchart for the proposed methodology

3.1 Introduction to MSER & WMF-MSER

The domain of Computer Vision involves one of the majorly used techniques for blob detection termed Maximal Stable Extremal Regions (MSERs). It was developed by Matas et al. [22], and therefore used extensively in the domain of the text region detection. The main principle of the method is to detect the similarity between the same images when viewed from two different angles. The MSERs remain stable throughout thresholds, which may be darker or brighter than their close areas. The pixels present in those extremal regions have either higher or lower intensity corresponding to those present on the boundary regions. Therefore, it helps identify the areas with a considerable variation of the intensity in the given images. The text present in the natural scene images has different intensity (higher or lower) compared to the background, and thus it helps in resembling the text with human eyes. Since the MSER works on the principle of the variation of the intensity, it motivates us to use the MSER method in our method for separating the interconnected text or characters.

Fig. 3. WMF-MSER [21] a) original image b) original MSER [22] c) WMF-MSER

From our previous work, stated in [21], we use the WMF-MSER algorithm for separating the interconnected characters. The results obtained by the WMF-MSER algorithm can be shown in Fig. 3. The resultant images in Fig. 3(c) have properly separated characters compared to the original images in Fig. 3(b). Thus, the main advantage of using WMF-MSER is that the features can be extracted accurately on these properly separated text elements. The features extracted will then be used for building the classification model using machine learning algorithms. In the next section, we will discuss the features used in the paper.

3.2   Extraction of the Features

The text elements present in the images have significant variations among themselves. The non-text elements are different from the text elements. The naked human eye can quickly identify this as we humans have complete information about the alphabets and text used in our native language. But certainly, machines cannot recognize such text or characters until they are trained for the same. The training process requires features to make a proper difference between two entities. In the same way, in this domain, it is inevitable to have appropriate mutually exclusive features for differentiating between the text and non-text elements.

Fig. 4.   Example of text elements [23]

Fig. 5. Examples of non-text Elements [23]

Figures 3 and 4 display a few examples of the text and the non-text elements obtained after applying WMF-MSER. The researchers, over the years, extracted many features for the above-said work. In this paper, we prefer to choose three features: Maximum Stroke Width Ratio, Color Variation, and Solidity. The text elements present in the images have different sizes, colors, shapes as well orientations. So, we have considered three mutually exclusive features to differentiate between text & non-text elements properly. The definitions of the feature are as follows:

a) **Maximum Stroke Width Ratio (MSWR):** The stroke width [24] of any text is one of its unique features. The stroke width of the text always remains uniform, and thus it is one of the prominent features to identify between the text and non-text elements. The non-text elements do not have uniform text width due to their irregular structure. So, the stroke width obtained for the non-text elements has many variations compared to the text elements. It is evident from the Figs. 4 and 5 that the text elements have uniform text width. On the other hand, non-text elements do not possess uniformity. So MSWR can be chosen as one of the features for separating the non-text & text elements.

b) **Color Variation:** Color is one of the essential traits of any element that assists in differentiating objects. The text present in the images possesses different colors as compared to the non-text elements. The background present around the text also helps in identifying the text correctly from the images. Therefore, the variation in the color is taken as one of the features for classification purposes. The color variation is calculated by the Jenson-Shanon divergence (JSD) [25]. It calculates the difference between the color using the probability distribution of the text and its background.

c) **Solidity:** The text elements in the images have a very uniform structure, and the non-text elements have a non-uniform stricture. Therefore, to differentiate the elements at the structural level, we choose solidity as the third feature in our work. It is the ratio of the area covered by total pixels in the region R to the area of the convex (smallest) hull surrounding that region.

Thus, we consider these three features mentioned above to build the classification models. These three features are mutually exclusive to each other. The mutually exclusive condition is essential as we must consider different aspects of the text for its discrimination with non-text elements. It will help us to identify the text more accurately as each feature is distinct. The MSWR is related to the uniformity present in the stroke width, and the color variation contributes to the different backgrounds of the elements (Text & non-text). In contrast, the solidity feature contributes to making a difference based on the uniformity of the area occupied by the elements. In the next section, the machine learning classifications models are built using the training dataset of ICDAR 2013.

3.3   Building Classification Models

Machine Learning includes classification, which predicts the class label for a given set of input data. The classification model provides a conclusion to assign a label to the object based on the input values given for the training and machine learning algorithm used. The classification problems are binary and multi-classification. The binary classification refers to labeling one out of two given classes, whereas it refers to one out of many classes in multi-classification. In this paper, we have a binary classification problem, in which the label is to give as text or non-text elements by the classification algorithm. The classification is performed based on the features extracted in the previous section. We have chosen four classifiers for the purpose, and experiments are performed using MATLAB [26] classifiers Learner Application. The dataset used for the training and building classification model is ICDAR 2013 dataset. It consists of 229 images from the natural scene images. These 229 images consist of 4786 text characters. We applied WMF-MSER algorithms and obtained 4549 non-text elements. After that, we calculated the three features on both texts and non-text elements, as mentioned in Sect. 3.2. The four classifiers chosen for the building classification model using the dataset and the three features are Bagged Trees [27], Fine Trees [28], K-Nearest Neighbor [29], Naïve Bayes [30]. There can be two possibilities for an element present in the images, text and non-text. The following parameters for classification are used in the paper:

a)   True Positives (TP): Text is discovered as text.
b)   True Negative (TN): Non-text is discovered as non-text.
c)   False Positive (FP): Non-text is discovered as text.
d)   False Negative (FN) Text is discovered as non-text.

Therefore, the overall accuracy (A) of the classifiers is interpreted as mentioned in the equation

$$\text{Accuracy}(A) = \frac{TP + TN}{TP + TN + FP + FN}$$

The accuracy calculated in the equation is used as the final parameter for the overall accuracy of the classifiers.

4   Experiments and Results

The experimental setup and the results obtained are discussed in the given section. The three features are calculated on both txt (4786) and non-text (4549) elements are combined to make a feature vector (FV). There will be two classes, text (1) or non-text (0), so the class or response vector (R) consists of two values, 1 and 0. Thus the feature vector and class vector is shown as

$$FV = \{SWV, CV, S\}$$

$$R = \{0, 1\}$$

For building the classification model, we prefer to use Matlab classification learner application for classification purposes. This application is a part of Matlab, which trains the model to classify the data. There are many classifiers based on the supervised machine learning algorithms available in this application. The data can be explored, trained, validated, and assessed using this application, which is very easy to use and gives accurate results. The detailed experimental set-up is displayed in Table 1.

Table 1. Experimental details for building classification models

S.n	Particulars	Value/details
1.	Classifier application	MATLAB learning application
2.	Preprocessing	WMF-MSER
3.	Cross fold	10
4.	Data set	ICDAR 2013
5.	Text elements	4786
6.	Non-text elements	4549

The 10-fold cross-validation is used in the experiments to obtain good accuracy in this paper. The feature vector is passed as an input to the four classifiers mentioned in Sect. 3.3, and the accuracy for the different classifiers is obtained.

The results obtained are displayed in Table 1. It is evident from the Table that the highest accuracy is obtained for the Bagged Tree classifier. Bagging is an entirely data-specific algorithm. The bagging technique eliminates the possibility of over-fitting. It also performs well on high-dimensional data. Moreover, the missing values in the dataset do not affect the performance of the algorithm. The bagged tree combines the performance of the many weak learners to outperform the strong learner's performance.

Therefore, the accuracy obtained from Bagged Tree is highest using the feature vector consists of three features due to the advantages mentioned above. The Confusion matrix, which consists of the TP, TN, FP, FN, is used to make the ROC for the classifiers and is shown in Figs. 6, 7, 8 and 9. The ROC curve is also an indicative measure of the best classifier based on the area occupied by the ROC curve (Table 2).

Table 2. Classification accuracy obtained for four classifiers

S.n	Classifiers	Text/non-text	Classification		A
			T	NT	
1.	Bagged Tree	T	4127	659	83%
		NT	914	3635	
2.	Fine Tree	T	4358	428	81.7%
		NT	1283	3266	
3.	KNN	T	4169	617	82%
		NT	1042	3507	
4.	Naive Bayes	T	4272	1703	76.3%
		NT	514	2846	

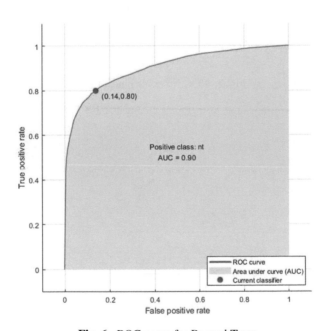

Fig. 6. ROC curve for Bagged Trees

The area under the curve in the ROC curve is shown as best in the Bagged Trees cases, indicating that the bagged trees are the best classifiers among the rest three chosen classifiers.

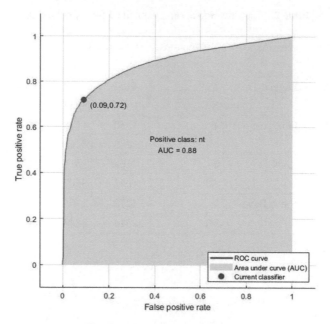

Fig. 7. ROC curve for Fine Tree

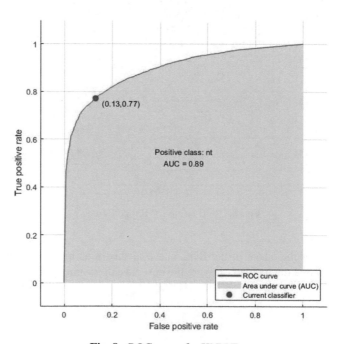

Fig. 8. ROC curve for KNN Tree

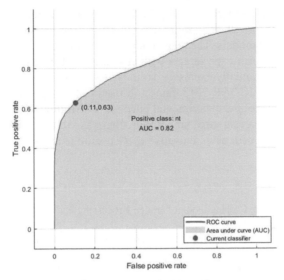

Fig. 9. ROC curve for Naïve Bayes

The choice of the classifier is necessary for the classification of the text, and non-text elements are an essential step in the classification process. It is since many classifiers exists in the domain of machine learning algorithms. The researchers had made either arbitrary choice of the classifier or focused on the traditional approach to use SVM/Adaboost classifiers. We contribute to achieving the task of selecting the classifier with the help of the Matlab Classifier Learner Application. This Matlab application is not very well explored in the classification for text & non-text elements.

In comparison with other states of the arts, Iqbal et al. [10] have considered 25 images of the ICDAR 2011 dataset for experiments, whereas we have chosen 229 images for choosing the classifier. The type of the images is very different and thus helps build a more accurate training model for handling different testing sets.

The method [31] applies CNN for classification and thus requires high computation time for evaluating the training model compared to proposed method using traditional classifiers. Mukhopadhyay et al. [32] used 100 images with one-class classifier & obtained 71% accuracy, whereas we acquired (83%) obtained in our work.

The methods using Deep learning have higher accuracy, but the issue lies in the computation cost, which is high in deep learning methods. An extensive training set [33] is required for the training process. These methods can detect the different text patterns [34, 35] in images, and the need for the GPU framework [36] increases the cost parameters. So, we choose to work on traditional machine learning classifiers and achieve results with small training sets.

5   Conclusion

The present paper demonstrates the work done to build a classifier model for the text and non-text classification present in the natural scene images. The classification of text and non-text elements is the preliminary step for detecting and extracting the text regions. The present paper explores the possibility of the existing machine learning algorithms to build the classification models. The reason behind this approach is to sue the simplicity of the model and perform experiments with less time and training data. The features used in the paper are mutually exclusive, so they will contribute to identifying the text and non-text correctly. ICDAR 2013 dataset is used in the paper as it provides proper ground truth available for the experimental purpose. The future work includes using the weka tool and other relevant edge smoothing filters as well as deep learning tool for classification purposes with new innovative text-specific features.

References

1. Distante, A., Distante, C.: Handbook of Image Processing and Computer Vision: Volume 2: From Image to Pattern (2020)
2. Rainarli, E.: A decade: review of scene text detection methods. Comput. Sci. Rev. **42**, 100434 (2021)
3. Shivakumara, P., Alaei, Pal, U.: Mining text from natural scene and video images: a survey. Wiley Interdiscipl. Rev. Data Min. Knowl. Discov. e1428 (2021)
4. Lucas, S.M., et al.: ICDAR 2003 robust reading competitions: entries, results, and future directions. Int. J. Doc. Anal. Recogn. (IJDAR) **7**(2–3), 105–122 (2005)
5. Shahab, A., Shafait, F., Dengel, A.: ICDAR 2011 robust reading competition challenge 2: reading text in scene images. In: 2011 International Conference on Document Analysis and Recognition, pp. 1491–1496. IEEE (2011)
6. Karatzas, D., et al.: ICDAR 2013 robust reading competition. In: 2013 12th International Conference on Document Analysis and Recognition, pp. 1484–1493. IEEE (2013)
7. Raisi, Z., Naiel, M.A., Fieguth, P., Wardell, S., Zelek, J.: Text detection and recognition in the wild: a review. arXiv preprint arXiv:2006.04305 (2020)
8. Sullivan, E.: Understanding from machine learning models. Br. J. Philos. Sci. (2020)
9. Shiravale, S.S., Sannakki, S.S., Rajpurohit, V.S.: Recent advancements in text detection methods from natural scene images. Int. J. Eng. Res. Technol. **13**(6), 1344–1352 (2020)
10. Iqbal, K., Yin, X.-C., Yin, X., Ali, H., Hao, H.-W.: Classifier comparison for MSER-based text classification in scene images. In: The 2013 International Joint Conference on Neural Networks (IJCNN), pp. 1–6. IEEE (2013)
11. Zhu, A., Wang, G., Dong, Y.: Detecting natural scenes text via auto image partition, two-stage grouping, and two-layer classification. Pattern Recogn. Lett. **67**, 153–162 (2015)
12. Lee, J.-J., Lee, P.-H., Lee, S.-W., Yuille, A., Koch, C.: AdaBoost for text detection in natural scene. In: 2011 International Conference on Document Analysis and Recognition, pp. 429–434. IEEE (2011)
13. Chen, X., Yuille, A.L.: Detecting and reading text in natural scenes. In: Proceedings of the 2004 IEEE Computer Society Conference on Computer Vision and Pattern Recognition, CVPR 2004, vol. 2, p. II. IEEE (2004)
14. Pan, Y.-F., Liu, C.-L., Hou, X.: Fast scene text localization by learning-based filtering and verification. In: 2010 IEEE International Conference on Image Processing, pp. 2269–2272. IEEE (2010)
15. Ma, L., Wang, C., Xiao, B.: Text detection in natural images based on multi-scale edge detection and classification. In: 2010 3rd International Congress on Image and Signal Processing, vol. 4, pp. 1961–1965. IEEE (2010)
16. Pan, Y.-F., Hou, X., Liu, C.-L.: A hybrid approach to detect and localize texts in natural scene images. IEEE Trans. Image Process. **20**(3), 800–813 (2010)

17. Maruyama, M., Yamaguchi, T.: Extraction of characters on signboards in natural scene images by stump classifiers. In: 2009 10th International Conference on Document Analysis and Recognition, pp. 1365–1369. IEEE (2009)
18. Ansari, G.J., Shah, J.H., Yasmin, M., Sharif, M., Fernandes, S.L.: A novel machine learning approach for scene text extraction. Future Gener. Comput. Syst. **87**, 328–340 (2018)
19. Wei, Y., Zhang, Z., Shen, W., Zeng, D., Fang, M., Zhou, S.: Text detection in scene images based on exhaustive segmentation. Sig. Process. Image Commun. **50**, 1–8 (2017)
20. Long, S., He, X., Yao, C.: Scene text detection and recognition: the deep learning era. Int. J. Comput. Vis. **129**(1), 161–184 (2021)
21. Soni, R., Kumar, B., Chand, S.: Extracting text regions from scene images using weighted median filter and MSER. In:2018 International Conference on Advances in Computing, Communication Control and Networking (ICACCCN), pp. 915–920. IEEE (2018)
22. Matas, J., Chum, O., Urban, M., Pajdla, T.: Robust wide-baseline stereo from maximally stable extremal regions. Image Vis. Comput. **22**(10), 761–767 (2004)
23. Soni, R., Kumar, B., Chand, S.: Optimal feature and classifier selection for text region classification in natural scene images using Weka tool. Multimedia Tools Appl. **78**(22), 31757–31791 (2019). https://doi.org/10.1007/s11042-019-07998-z
24. Epshtein, B., Ofek, E., Wexler, Y.: Detecting text in natural scenes with stroke width transform. In: 2010 IEEE Computer Society Conference on Computer Vision and Pattern Recognition, pp. 2963–2970. IEEE (2010)
25. Majtey, A.P., Lamberti, P.W., Prato, D.P.: Jensen-Shannon divergence as a measure of distinguishability between mixed quantum states. Phys. Rev. A **72**(5), 052310 (2005)
26. The Math Works, Inc.: MATLAB, Version 2020a. Natick, MA: The Math Works, Inc. (2020). https://www.mathworks.com/. Accessed 28 May 2020
27. Mousavi, R., Eftekhari, M.: A new ensemble learning methodology based on hybridization of classifier ensemble selection approaches. Appl. Soft Comput. **37**, 652–666 (2015)
28. Rokach, L., Maimon, O.Z.: Data Mining with Decision Trees: Theory and Applications, vol. 69. World Scientific (2007)
29. Fix, E., Hodges, J.L.: Discriminatory analysis. Nonparametric discrimination: consistency properties. Int. Stat. Rev./Revue Internationale de Statistique**57**(3), 238–247 (1989)
30. Zhang, Y., Jatowt, A.: Estimating a one-class naive Bayes text classifier. Intell. Data Anal. **24**(3), 567–579 (2020)
31. Wu, H., Zou, B., Zhao, Y.-Q., Guo, J.: Scene text detection using adaptive color reduction, adjacent character model and hybrid verification strategy. Vis. Comput. **33**(1), 113–126 (2015). https://doi.org/10.1007/s00371-015-1156-1
32. Mukhopadhyay, A., et al.: Multi-lingual scene text detection using one-class classifier. Int. J. Comput. Vis. Image Process. (IJCVIP) **9**(2), 48–65 (2019)
33. Jaderberg, M., Simonyan, K., Vedaldi, A., Zisserman, A.: Deep structured output learning for unconstrained text recognition. arXiv preprint arXiv:1412.5903 (2014)
34. He, T., Huang, W., Qiao, Y., Yao, J.: Text-attentional convolutional neural network for scene text detection. IEEE Trans. Image Process. **25**(6), 2529–2541 (2016)
35. Ou, W., Zhu, J., Liu, C.: Text location in natural scene. J. Chin. Inf. Process. **5**(006) (2004)
36. Busta, M., Neumann, L., Matas, J.: Deep textspotter: an end-to-end trainable scene text localization and recognition framework. In: Proceedings of the IEEE International Conference on Computer Vision, pp. 2204–2212 (2)017

Point Cloud Registration of Road Scene Based on SAC-IA and ICP Methods

Yan Liu[1], Hu Su[2(✉)], Yu Lei[3], and Fan Zou[3]

[1] School of Electrical Engineering, Southwest Jiaotong University, Room 6407, 999 Xi'an Road, Pidu District, Chengdu, Sichuan Province, China
[2] School of Electrical Engineering, Southwest Jiaotong University, Room 10922, 999 Xi'an Road, Pidu District, Chengdu, Sichuan Province, China
suhu@swjtu.edu.cn
[3] School of Electrical Engineering, Southwest Jiaotong University, 999 Xi'an Road, Pidu District, Chengdu, Sichuan Province, China

Abstract. Registration of point cloud data obtained by vehicle-mounted LiDAR is necessary process to establish high-precision road scene 3D model automatically. This paper presents a set of multi-line LiDAR point cloud registration method in road scenarios. Firstly, the obtained original point cloud data are pre-processed according to the characteristics of multi-line LiDAR point cloud. Then an initial registration algorithm (SAC-IA) with sampling consistency based on fast point feature histogram (FPFH) is used to achieve the coarse registration for two frame point clouds. Lastly, ICP algorithm optimized by KD-tree is used for precise registration and global road point cloud model can be obtained by iterative registration. In order to verify the method, actual road point cloud data are collected. The experimental results show that the method is feasible and its registration accuracy can meet the requirements of road model.

Keywords: Multi-line lidar · Multi-view scene point cloud · Point cloud registration · SAC-IA · ICP

1 Introduction

Due to the limitation of measurement conditions, it is often necessary to carry out multi-view point cloud registration [1] in order to restore complete road point data when obtaining road point cloud data by LiDAR. At present, it is considered that point cloud registration is generally divided into two stages: coarse registration and precise registration. Using only Iteration Closest Point (ICP) algorithm is easy to fall into local optimal solution [2]. Though many coarse or precise registration methods based on features [3] accelerate the speed and accuracy of point cloud registration [4, 5] to some extent, most of the researches at the present stage are in the theoretical stage. Multiperspective point cloud data collected in the actual environment are more complex, and the registration process is different from that of point cloud registration of single-object. In addition, current studies believe that the parameters setting often relies on experience and requires manual intervention [6].

© The Author(s) 2022
Z. Qian et al. (Eds.): WCNA 2021, LNEE 942, pp. 969–978, 2022.
https://doi.org/10.1007/978-981-19-2456-9_98

VLP-16 LiDAR is a kind of multi-line LiDAR, which has widely applications in unmanned driving and robot navigation and obstacle avoidance [7]. According to the characteristics of VLP-16 LiDAR, this paper designs the data pre-processing model of this kind of multi-line LiDAR point cloud. Firstly, the obtained point cloud data are simplified, and the outliers are removed according to the threshold. Then SAC-IA based on FPFH [8] is applied to coarse registration and the ICP algorithm optimized by KD-tree is used for accurate registration. Lastly, the point cloud model after point cloud registration is obtained, and the setting method of searching for domain radius in road scene is given.

2　Data Pre-processing

2.1　VLP-16 LiDAR Data Characteristics

Point cloud data obtained by VLP-16 LiDAR are different from those obtained by general point cloud acquisition devices such as stereo cameras, depth cameras and laser scanners in the term of surface distribution characteristics. Point cloud data obtained by VLP-16 LiDAR are concentrated on 16 scan lines, and each line is evenly distributed along the Z-axis direction. The point cloud data have vertical field of view from $+15°$ to $-15°$ and $360°$ horizontal scan field of view. The point cloud data are dense in horizontal direction and sparse in vertical direction because the point clouds acquired by VLP-16 LiDAR are distributed on 16 scan lines.

The characteristics of the data are shown in Fig. 1 below when the VLP-16 LiDAR is mounted on the vehicles to collect data in road scenes. The viewing angle of Fig. 1(a) is the positive Z-axis. The viewing angle of Fig. 1(b) is the positive X-axis. It can be seen from the two figures that the data are distributed discretely in form, and the positions and intervals of data points are distributed irregularly in three-dimensional space. The Fig. 1(a) shows that in the vertical direction, the data density near the ground is high. The Fig. 1(b) conveys that the farther away from the collection point, the thinner the data density is. Most of the points on either side fall on buildings and trees on both sides of the road because of the limitation of laser penetration. A laser with a negative vertical angle will scan to the ground, resulting in a ring of ground points in the collected point cloud data. These ground points in the point cloud will not only affect the extraction of the point cloud features, but also bring redundant computation, so the conditional filter is adopted to filter them.

a

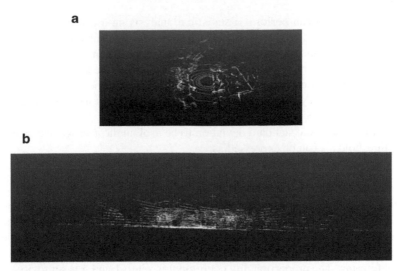

b

Fig. 1. (a) Point cloud image from the positive direction of Z-axis. (b) Point cloud image from the positive direction of X-axis.

2.2   Point Cloud Data Pre-processing Model

Due to the environment, experimental equipment, equipment accuracy and other factors, there will be noise points, outliers and holes that do not meet expectations, as well as some non-noise points that affect the experimental results when obtaining point cloud data in the field. In order to make subsequent experiments more accurate, point cloud data pre-processing should be carried out to eliminate some points that affect subsequent experimental results. Firstly, the original data are cleaned to obtain the point cloud frames which are suitable for registration. Then statistical filters based on statistical principles are used to filter outliers and noise points. Finally, conditional filters are used to filter the ground ring point clouds in road scenes, so as to improve the speed and accuracy of registration (Fig. 2).

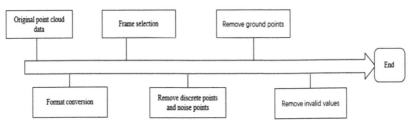

Fig. 2. Point cloud pre-processing timeline

Statistical filtering can perform a statistical analysis on a certain domain of each point and calculate the average distance from it to its adjacent points. It is assumed that the calculated results satisfy the Gaussian distribution. Then if the average distance obtained for a point is outside the standard range (defined by the global range mean and variance), such a point can be defined as an outlier or noise point removed from the original data. In this way, the influence of outliers and noise points on the registration results can be greatly reduced.

Set the mean to be and standard deviation to be σ of all the average distance and the distance threshold d_{th} can be expressed as:

$$d_{th} = l + s * \sigma \tag{1}$$

As the proportionality coefficient, the constant needs to be set according to degree of statistical filtering required. Finally, point cloud data are traversed to eliminate the points whose average distance with n neighbor points are greater than the threshold value. This paper uses the standard statistical filter of the official document of PCL to carry out statistical filtering. The proportionality coefficient is set to 1 and n is set to 50.

Conditional filters allow users to freely add and combine the range limits of XYZ axis. Compared with the simplest filter, conditional filter can be designed according to different requirements. Since point cloud data collected by vehicle-mounted LiDAR in road scenes are always in the negative direction of the Z axis, the condition for setting the Z axis of the conditional filter is: the vertical distance from the center of the LiDAR to the ground.

3   Coarse and Precise Registration of Point Cloud Data

The steps of coarse and precise registration scheme of point cloud collected by VLP-16 LiDAR in road scene are as follows: Firstly, the fast point feature histogram (FPFH) is calculated according to the point normal vector and Euclidean distance. Then, the initial registration algorithm (SAC-IA) with sampling consistency based on the fast point feature histogram (FPFH) is used for coarse registration. Finally, the precise registration of the road field is completed by using ICP algorithm with KD-tree acceleration.

3.1   Extraction of FPFH Feature Descriptor

As one of the most basic feature descriptors, FPFH is a feature descriptor of traditional Point Feature Histogram (PFH) to reduce the computational complexity and improve the computational efficiency. It captures the geometric information around a point by analysing the difference of the normal direction near each point. The result of normal estimation is important for the quality of FPFH calculation. The extraction steps of feature points are as follows:

- Set the search radius of each point as r_1, and estimate the normal vector of each point.
- Calculate the three characteristic element values between the query point and each other point within its search radius, namely the $\alpha, \varnothing, \theta$ values in PFH. Then these values are calculated into a simplified point feature histogram(SPFH).

- Determine the domain of each point in the domain of the search radius r_2 and form SPFH according to the second step.
- The SPFH of each point in the domain of the query point is weighted count. The ω_k represents the distance between the query point p and p_K. The formula is as follows:

$$FPFH(p) = SPFH(p) + \frac{1}{k} \sum_{i=1}^{k} \frac{1}{w_k}(*SPFH(p_K)) \tag{2}$$

The key to calculate the FPFH is to set the domain radius r_1 of normal estimation and the domain radius r_2 of FPFH. Search areas that are slightly too large or too small are allowed. However, if the threshold is set too small, it will lead to wrong estimation of the normal vector, resulting in the local information missing, which can not be registered. The time cost of calculating the normal vector and FPFH increases sharply when the threshold is set too much. It may occur that multiple separated objects in the scene are calculated together with the surface normal vectors, and the feature description information is inaccurate, resulting in the decline of registration quality. In previous studies, parameters setting is mostly dependent on experience. This paper presents the method of parameters setting in road scene.

In general, the point cloud data in road environment will be influenced by trees on both sides of the road and other obstacles and it is difficult to determine the normal vectors and FPFH on the surface of the trees. Therefore, buildings on both sides of the road should be regarded as key descriptors at this time. The VLP-16 LiDAR supports a vertical field angle of $\pm 15°$ and the angle between each scan line is $1.875°$ approximately. The distance between the building and the vehicle is estimated to be between 22 m and 30 m. The spacing between the two scan lines projected on buildings is calculated to be between 0.72 m and 0.98 m.

In order to satisfy the correct calculation of normal vectors on more buildings as far as possible, set r_1 to be 1 m. In the case of characteristics of point cloud data obtained by VLP-16 LiDAR in road scene, it is best that r_2 takes twice the scan line spacing, so it is set to 2 m.

3.2   ICP Precise Registration Optimized by KD-Tree

KD-tree is a data structure that divides k-dimensional data space. It is mainly applied to the search of key data in multi-dimensional space(such as range search and most recent collar search). The steps of ICP precise registration algorithm optimized by KD-tree are shown as follows (Fig. 3):

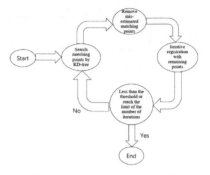

Fig. 3. Precise registration step figure

The main process of ICP algorithm is to find the best transfer matrix between source point cloud and target point cloud. For two groups of point clouds $\{X = x_1, x_2, \ldots, x_{Nx}\}$ and $\{P = p_1, p_2, \ldots, p_{Np}\}$, the rotation matrix R and the translation matrix T are solved iteratively, making the following formula minimum.

$$E(R, T) = \frac{1}{N_P} \sum_{i=1}^{N_P} \|x_i - R_{Pi} - T\|^2 \tag{3}$$

The iteration stopped when the above value is less than the set threshold, or reaches the limit of the number of iterations that is pre-set.

4   Experiment

The experimental data are collected by VLP-16 LiDAR at a location in Chengdu. The average point cloud data of each frame have about 22000 points, which are used to test the effectiveness of the model in this page. The whole algorithm model is implemented in PCL 1.8 using C++ language.

In this experiment, a section with curves, trees and general buildings on both sides of the road and good pavement condition is selected as the experimental sampling site. Placing the VLP-16 LiDAR on the roof of the car can scan the scene more stably and extensively, so a device that can fix the VLP-16 LiDAR is designed to be placed on the roof the car according to the vehicle model. After making preparations, the driver tries to keep the speed even in one direction, so as to obtain the road point cloud data of the scene.

The final experimental results are shown in the figure below. The first two images are displayed by the software--CloudCompare. The last two images are displayed using the visual portion of the PCL library. Figure 4(a) is the point cloud data of one frame with an interval of 20 frames after extraction and screening. Figure 4(b) is the result of pre-processing the point cloud data of one frame of road scenes respectively. Figure 4(c) is the point cloud image obtained after the precise and coarse registration of two frames of point clouds. The red point cloud is the point cloud data of previous frame, and the green point cloud is the point cloud data of the later frame. Figure 4(d) is the local details of the point cloud magnified in two frames after registration.

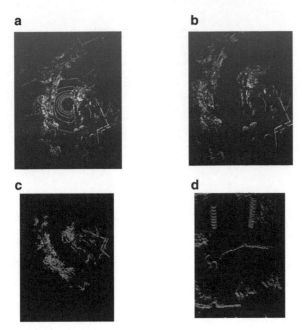

Fig. 4. (a) Original point cloud image. (b) Point cloud image after pre-processing. (c) Point cloud image after registration. (d) Local details after registration

The obtained point cloud data should be extracted at a certain interval. The interval between the two frames should not be too large, because too large will lead to too large difference between the two frames of point cloud images, and finally the two frames can not be registered. If the frame interval is too small, the workload of global point cloud registration will be increased, resulting in a great amount of redundant computation. In this experiment, sampling is conducted at an interval of 20 frames, and the point cloud coordinate system of the first frame is taken as the reference coordinate system of point cloud image sets. Using multi-thread programming, the final experimental results are obtained by registration of point cloud image sets frame by frame. The final results of global point cloud data set registration are shown in Fig. 5 below. The three pictures of Fig. 5 are the results of different perspectives. Figure 5(a) is viewed from the positive Z-axis. Figure 5(b) and (c) are viewed from the left and right sides of the driving direction. As can be seen from the picture, the trees, green belts, signs and some buildings on both sides of the road are clearly displayed. The overall results show a better picture of the road and important information on both sides. Figure 6 shows some details of trees, green belts and buildings. The desired effect has been achieved by using the experimental model.

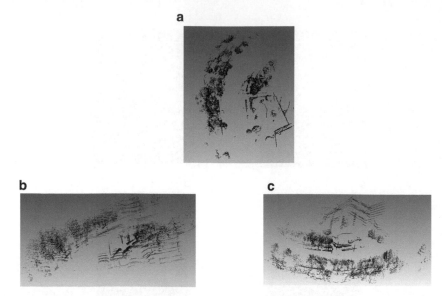

Fig. 5. (a) The final result from the upper view. (b) The final result from the right view. (c) The final result from the left view

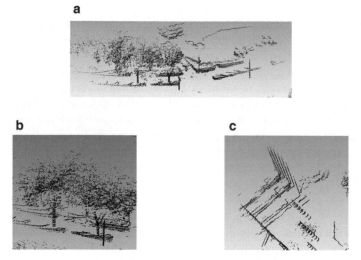

Fig. 6. (a) Details of green belts. (b) Details of trees. (c) Details of buildings

The following table can be obtained by comparing the transfer matrices calculated after coarse registration and precise registration with those measured by precise devices at the actual measuring site (Table 1).

Table 1. Registration error analysis.

	Pre-set transition matrix		Registration transition matrix		Error	
	Rotation angle	Translation angle	Rotation angle	Translation angle	Rotation angle	Translation angle
X-axis	0.00604667	0.00932314	0.00603953	0.00940045	7.13579e−06	7.731e−05
Y-axis	0.000368052	2.18737	0.000402868	2.18738	3.4816e−05	1.00136e−05
Z-axis	−0.0440608	0.0279537	−0.0440639	0.0290932	−3.09944e−06	0.0011395

It is found that the rotation angle and translation distance have not changed much except the Z-axis translation distance. The reason why the error on the Z-axis is larger than the error on the other values may be as follows: The road surface is not smooth enough and there are irresistible bumps, resulting in some errors in the translation distance of the point cloud in the vertical direction of the two frames.

5 Conclusion

In this paper, the registration algorithm of VLP-16 LiDAR point cloud is studied deeply for practical application. According to the characteristics of multi-line LiDAR point cloud in road scene, a special pre-processing model of multi-line LiDAR and a reasonable calculation method of plane normal vectors and search radius of FPFH under this scene are proposed. The SAC-IA algorithm based on FPFH is used for coarse registration, and the ICP algorithm optimized by KD-tree is used for precise registration of point clouds.

The experimental results show that this model is suitable for road scene registration, and the registration of multi-view point cloud data in road scene is completed, and the error is very small compared with the real rotation matrix. This shows that the model has applicability and effectiveness.

The deficiency of the model is that the scope of application of the parameters setting method has limitations in the calculation process of coarse registration. The results will be in a decline in registration accuracy under complex and extreme conditions. Therefore, the improvement of the adaptability of algorithm will be the goal of the next research.

References

1. Philipp, J., Ivo, K., Ralf, B., Achim, S., Floris, E.: Efficient registration of high-resolution feature enhanced point clouds. J. IEEE Trans. Pattern Anal. Mach. Intell. **41**, 1102–1115 (2019)
2. Zhang, C.J., Xu, Y.Z., Zheng, S.X., Zheng, J.G., Zhang, Y.: Application of improved weighted iterative nearest point algorithm in point cloud registration. J. Geodesy Geodyn. **39**, 417–420 (2019)
3. Wang, P., Zhu, R.Z., Sun, C.K.: Coarse registration algorithm for scene classification point cloud based on improved RANSAC. J. Laser Optoelectron. Prog. **57**, 312–320 (2020)
4. Shi, L., Yan, L.M.: Point cloud registration algorithm based on normal vector and gaussian curvature. J. Microelectron. Comput. **37**, 68–72 (2020)

5. Ma, W.: A 3D point cloud registration algorithm based on cuckoo optimization. J. Comput. Appl. Softw. **37**, 216–223 (2020)
6. Chen, Q., Yue, D.J., Chen, J.: LiDAR point cloud registration algorithm based on feature space matching. J. Geodesy Geodyn. **40**, 1303–1307 (2020)
7. Glennie, C.L., Kusari, A., Facchin, A.: Calibration and stability analysis of the VLP-16 laser scanner ISPRS J. Photogramm. Remote Sens. **XL-3/W4**, 55–60 (2016)
8. Rusu, R.B.: Semantic 3D object maps for everyday manipulation in human living environments. J. KI **24**, 345–348 (2010)

Crayfish Quality Analysis Based on SVM and Infrared Spectra

Zijian Ye and Yi Mou[✉]

School of Electrical and Electronic Engineering, Wuhan Polytechnic University, Wuhan 430023, China
mouyi@whpu.edu.cn

Abstract. Different algorithms combined with Near-infrared spectroscopy were investigated for the detection and classification of crayfish quality. In this study, the crawfish quality was predicted by partial least square-support vector machine, principal component analysis-support vector machine, BP neural network and support vector machine after pre-processing the NIR spectral data of crawfish. The result shows that the accuracy of near-infrared spectroscopy technology combined with SVM to classify crayfish quality can reach 100%, and the prediction can guide the sampling of crayfish food safety in practice, thus improving food safety and quality.

Keywords: Crayfish quality analysis · SVM · Infrared spectra

1 Introduction

The quality of crayfish is mainly determined by the three links of breeding, processing and storage, all of which are capable of significantly affecting its quality score [1]. Therefore, the use of traditional methods such as sanitary inspection, sensory evaluation, and physical and chemical analysis. They not only require professional testing, but also have the disadvantages of being too subjective and having long operation cycle [2].

The NIR spectroscopy is a green non-destructive detection with the advantages of low cost, high analytical efficiency, high speed and good reproducibility [3], and has been widely used in various fields such as food, pharmaceutical and clinical medicine [4], biochemical [5], textile [6], and environmental science. Modern NIR spectroscopy must rely on chemometric methods to complete spectral pre-processing and model building. Spectral pre-processing methods include smoothing algorithms, multivariate scattering correction, wavelet transform, etc.; Commonly used multivariate correction methods include linear correction methods such as principal component regression and partial least square, and nonlinear correction methods such as artificial neural networks and support vector machines [7].

In this study, we experimentally analyze four algorithms in crayfish quality detection and compare their prediction rates. Although PCL PLS and BP neural network have achieved better results in experiments, there is still room for improvement compared to support vector machines. Support vector machine has high generalization ability and

© The Author(s) 2022
Z. Qian et al. (Eds.): WCNA 2021, LNEE 942, pp. 979–987, 2022.
https://doi.org/10.1007/978-981-19-2456-9_99

can better handle practical problems such as small samples, nonlinearity, ambiguity, and high dimensionality [8]. The crayfish classification model with high stability and high accuracy in near-infrared spectroscopy using support vector machines, aiming to provide reference for subsequent research.

The second part of this paper gives a brief introduction of each model as well as its derivation; the third part selects the optimal model from the above machine algorithms through experimental analysis and comparison; the fourth part is an analysis of the advantages and disadvantages of the algorithms and a summary.

2   Theoretical Approach to Modeling

In this paper, four different machine classification algorithms will be used to predict crayfish quality, namely: SVM, PLS-SVM, PCA-SVM, and BP neural network. Firstly, we process the original data and divide the training set and test set according to a certain proportion. The training set is used as the input for training, and the classification model is obtained by adjusting the optimization parameters of each algorithm. Then the test set is used as the input. Finally, compare the accuracy of the four classifiers and find the appropriate optimal model.

2.1   Support Vector Machines

The basic idea of SVM is to find the support vector which constructs the optimal classification hyperplane in the training sample set which means that samples of different categories are correctly classified and the hyperplane interval is maximized. The mathematical form of the problem is:

$$y_i[(w^T x_i + b)] \geq 1, i = 1, 2, 3, ..., N \tag{1}$$

For linear indivisibility, there is a certain classification error that does not satisfy Eq. (1). Therefore, a slack variable is introduced in the optimization objective function. At this time, The problem of finding the optimal classification hyperplane will be converted into a convex optimization problem with constraints for solving:

$$\begin{cases} \min & \frac{1}{2}w^T w + C \sum_{i=1}^{N} \zeta_i \\ s.t. & y_i(w^T x_i + b) \geq 1 - \zeta_i \end{cases} \quad i = 1, 2, 3, ..., N \tag{2}$$

In the Eq. (2): C is called the penalty parameter. If the value of C is larger, the penalty for misclassification is larger. And the smaller C is, the smaller the penalty for misclassification is [9].

The classifier discriminant model function in n-dimensional space . At this time, the problem of the linear indivisible support vector machine becomes a convex quadratic programming problem. And we can use the Lagrangian function to solve it.

When the sample is non-linear, we can choose the kernel function to solve. In this paper, we mainly use RBF for SVM. The corresponding classification decision function is:

$$f(x) = \text{sgn}(\sum_{i=1}^{N} \lambda_i y_i K(x, x_i) + b) \tag{3}$$

2.2 Partial Least Square

Partial least square is a dimensionality reduction technique that maximizes the covariance between the prediction matrix composed of each element in the space and the predicted matrix [10]. It concentrates the features of principal component analysis, typical correlation analysis and linear regression analysis in the modeling process. Therefore, it can provide richer and deeper systematic information [11]. The partial least square model is developed as follows:

Pre-process the prediction matrix and the predicted matrix to make them mean and centered, and then decompose them:

$$\begin{cases} X = AP^T + B \\ Y = TQ^T + E \end{cases} \tag{4}$$

where $Y \in R^{n*m}$ and $X \in R^{n*m}$ are the predicted matrix, $A \in R^{n*a}$ and $T \in R^{n*a}$ are the score matrix, $P \in R^{m*a}$ and $Q \in R^{m*a}$ are the load matrix, $B \in R^{m*n}$ and $E \in R^{n*m}$ are the residual matrix.

The matrix product AP^T can be expressed as the sum of the products of the score vector t and the load vector P_j, then we have:

$$X = \sum_{j=1}^{a} t_j p_j^T + B \tag{5}$$

The matrix product TQ^T can also be expressed as the sum of the products of the score vector u_j and the load vector q_j, so it can be expressed as:

$$Y = \sum_{h=1}^{a} u_j q_j^T + E \tag{6}$$

Let $u_j = b_j t_j$, where b_j is the regression coefficient, then $U = AH$, $H \in R^{aa}$ is the regression matrix:

$$Y = AHQ^T + E \tag{7}$$

2.3 Principal Component Analysis Method

Principal component analysis is a mathematical transformation method in multivariate statistics that uses the idea of dimensionality reduction to transform the original multiple variables into a few integrated variables with most important information [12]. These integrated variables reduce the complexity of data processing, and reflect the maximization of the content contained in the original variable, reduce the interference of error factors, and reflect The relationship between the variables within the matter.

For the raw data, we can extract the intrinsic features among the data by some transformations, and one of the methods is to go through a linear transformation to achieve [13]. This process can be expressed as follows:

$$Y = wX \tag{8}$$

Here w is a transformation value, which can be used as a basic transformation matrix to extract the features of the original data by this transformation. Let x denote the m dimensional random vector. Assume that the mean value is zero, that is:

$$E[X] = 0 \tag{9}$$

Let w be denoted as an m dimensional unit vector x and make it project on x. This projection is defined as the inner product of the vectors x and x, it is denoted as:

$$Y = \sum_{k=1}^{n} w_k x_k = w^T x \tag{10}$$

In the above equation, the following constraints are to be satisfied:

$$\|w\| = (w^T w)^{1/2} = 1 \tag{11}$$

The principal component analysis method is to find a vector of weights $E[y^2]$, which enables the expression to take the maximum value [14].

2.4 BP Neural Network

BP neural networks simulate the human brain by simulating the structure and function of neurons. And it has the ability to solve complex problems quickly, accurately and in parallel. When the training samples are large enough, the BP neural network makes the error very small and makes the prediction result accurate enough. Compared to other neural network algorithms, BP neural networks are able to propagate the error backwards from the output to the input layer by using hidden layers. And modify the weights and threshold values during the back propagation process using the fastest descent method to make the error function converge quickly, which has fast training speed [15].

3 Experimental Results and Analysis

3.1 Support Vector Machine Classification Model

In supervised learning theory, two data sets are included. One is used to build the model, called the training sample set; the other is used to test the quality of the built model, called the test sample set. After preprocessing the data, we select half of the experimental data as the training set randomly, and use them to build the model. Finally, the remaining half of the experimental data are used as a test set and input them to the established model for classification and identification of crayfish.

LIBSVM is chosen as the training and testing tool for this model, and Gaussian kernel is chosen as the kernel function. We can search for parameters (c, g) by 10-fold cross-validation, and calculate the optimal value of 10-fold cross-validation accuracy. The set of (c, g) with the highest cross-validation accuracy is taken as the best parameter, obtaining c = 0.1, g = 4, as shown in Fig. 1.

As shown in Fig. 2, according to the comparison between the model and the actual sitution, where all samples are correctly classified with an accuracy rate of 100%, and it shows that the model has an extremely strong generalization ability and has a very high accuracy in high dimensionality.

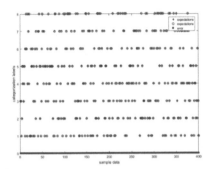
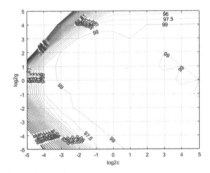

Fig. 1. Optimization parameters by grid searching technique.

Fig. 2. The Sample error in the SVM Model.

3.2 Principal Component Analysis for Clustering Crayfish

In order to remove the overlapping information in the NIR spectra and the information lacking correlation with the sample properties as much as possible, we reduced the original data matrix from 800×215 to 800×3 (3 principal components) by PCA. Since the principal component score plots of the samples can reflect the internal characteristics and clustering information of the samples, we obtained the contribution rate plots of the first three principal components as shown in Fig. 3 and the three-dimensional score distribution plots of the first three principal components as shown in Fig. 4.

Figure 6 is a plot of the scores of principal component 1, 2, 3 for 800 crayfish, where the $x\ y\ z$ axis represent the first principal component score, the second principal component score and the third principal component score respectively. From the figure, we can see that crayfish are clearly classified into 8 categories, indicating that components 1, 2, and 3 have a significant impact on crayfish with a better clustering effect. To describe the classification results quantitatively, we build a classification model for principal components using SVM.

We randomly select one-half of the standardized sample data as the training set to train the model, and the remaining one-half as the test set. The first 5 principal component score data are taken as the data features for identification. As shown in Table 1.

After that, we obtain a classification accuracy equal to 98.75% for this experiment by SVM.

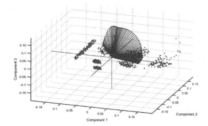

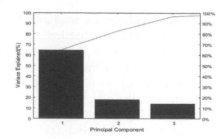

Fig. 3. Contribution of the top three principal components.

Fig. 4. 3D score distribution of the top three principal components.

Table 1. Reliabilities of principal compoents.

Principal components	Eigenvalue	Cumulative credibility
PC1	138.6437	0.985
PC2	38.4181	0.996
PC3	29.9760	0.980

3.3 Partial Least Squares Regression Analysis

It is especially important to determine the number of principal components in the PLS model. As the number of principal components increases, the degree of importance gradually decreases and represents less and less effective information. If too few principal components are selected, the characteristics of the sample are not fully reflected thus reducing the accuracy of the model prediction, this situation called under-fitting; if too many principal components are selected, some noisy information will be used as the characteristics of the sample, making the prediction ability of the model lower, this situation called overfitting [16]. Therefore, in order to reasonably determine the principal component score of the model, we derived a principal component score of 3 by taking the sum of squared prediction residuals [17] as the evaluation criterion.

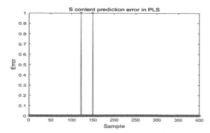

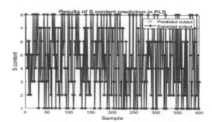

Fig. 5. Contribution of the top three Comparison of predicted values and actual values.

Fig. 6. Error analysis of S content in PLS.

The SVM model is built by the LIBSVM toolbox, and the comparison chart between predicted and reference values is shown in Fig. 5, and the error analysis is shown in Fig. 6. We came up with an accuracy rate of 99.5%.

3.4 BP Neural Network Model

The crawfish classification BP network model uses a three-layer network structure, namely input layer, implicit layer, and output layer, and the layers are interconnected. Among them, the number of neurons in the input layer is 215 features of the samples. the number of labels of the samples in the output layer is 1 layer, and the number of implicit neurons is 20 layers. The weights of the BP neural network model are set to default, the learning step is set to 0.01, the maximum number of training sessions is 1000, and the expected error is 0.001. We normalize the 800-group sample as the input term,after several training sessions, if the error meets our expectation, then the neural network model is valid and can be applied.

Figure 7 shows the performance curve of the training, indicating its variance variation. After four cycles, the network achieves convergence with a mean squared error of 0.00089, which is less than the set expectation error target of 0.001. The whole curve decreases faster, indicating the appropriate size of the learning rate.

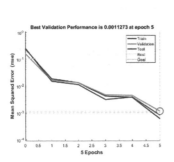

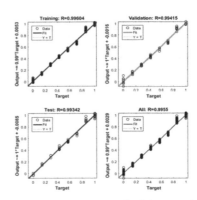

Fig. 7. Performance curve of BP neural neural network.

Fig. 8. Sample error plot in the BP network training.

Figure 8 shows the regression plot corresponding to the BP regression function, from which the fit of the training data, validation data, test data and the whole data,we know the correlation coefficient R are 0.99604, 0.99415, 0.99342 and 0.9955 respectively with high correlation, indicating the model fit. Through the above analysis, the BP neural network model has a good prediction effect with strong generalization ability in this study. Finally, we measured the accuracy rate of 97%.

4 Conclusion

Crayfish quality is affected by several factors, and it is necessary to ensure a reasonable classification of crawfish quality for objective evaluation of all aspects. This paper introduces SVM, PCA, PLS and BP neural networks in crayfish quality detection, leading to the following conclusions:

(1) To ensure the comparability of the model, 800 learning samples were selected with 215 feature vectors as input and classification label level as output. The results of the validation data show that the classification accuracy is all greater than 95%, which meets the accuracy requirement of mine environment evaluation.
(2) Compared with the BP neural network algorithm, the SVM algorithm shows more obvious superiority: the SVM model introduces the cross-validation method to program the automatic optimal selection of parameters, which overcomes the disadvantage that the neurons in the hidden layer of the BP neural network are not easily determined, and thus has a higher accuracy rate.
(3) Compared with PCA, the PLS algorithm can not only solve the problem of variable multicollinearity, but also solve the regression problem of multiple dependent variables with independent variables, reducing the influence of overlapping information.

In summary, the support vector machine model is chosen to be more suitable for the classification of crayfish quality, which has high accuracy and low time at high dimensionality and fuzziness.

Acknowledgments. This work is supported by Chutian Scholar Programm-Chutian Student of Hubei province, Hubei Provincial Department of Education (No. B2019064), the Recruitment Program of Wuhan Polytechnic University (No. 2017RZ05), Research and Innovation Initiatives of WHPU (No. 1017y27).

References

1. Jiang, H., Mengyuan, Y., Tao, H.: Food Mach. **35**, 232–236 (2019)
2. Jing, L., Xiao, G., Cuiping, Y.: Food Mach. **32**, 38–40 (2016)
3. Xu, G., Yuan, H., Lu, W.: Spectroscopy and spectral analysis, **20** (2000)
4. Watanabe, K., Mansfield, S.D., Avramidis, S.: J. Wood Sci. **57** (2011)
5. Falls, F.S.: J. Petrol. Sci. Eng. **51**, 127–137 (2006)
6. Lim, H.: Int. J. Pharm. **4**, 1–8 (2011)
7. Yang, Q.: Southwest Univ. **4**, 32 (2009)
8. Vapnik, V.N.: IEEE Trans. Netw. **10**, 988–999 (1999)
9. Li, H.: Statistical Learning Methods (Ed. by, H. Xue) , p. 125. Tsinghua University Press (2019)
10. Jiang, T., Russell, E.L., Bratz, R.D.
11. Wang, H.: Partial Least Squares Regression Methods and Their Applications, pp. 1–2. Defense Industry Press (1999)

12. Li, W.: Hu Bing and Wang Ming wei. Spectrosc. Spectral Anal. **34**, 3235–3240 (2014)
13. Chen, L.: J. Image Process. **8**, 38–41 (2007)
14. Qiang, Z.: J. Shenyang Univ. **19**, 33–35 (2007)
15. Cai, Q., Wang, J., Li, H.: J. Food Sci. Technol. **32** (2014)
16. Qiong, W., Zhonghu, Y., Xiaoning, W.: J. Shenyang Univ. **19**, 33–35 (2007)
17. Yin, C.: Revision of factor analytic theory, **6**, 27–28 (2012)

Application of Image Recognition in Precise Inoculation Control System of Pleurotus Eryngii

Xiangxiu Meng, Xuejun Zhu(✉), Yunpeng Ding, and Dengrong Qi

School of Mechanical Engineering, Ningxia University, Yinchuan 750000, China
zhuxuejunnxu@sina.com

Abstract. The traditional inoculation technology of Pleurotus eryngii is artificial inoculation, which has the disadvantages of low efficiency and high failure rate. In order to solve this problem, it is necessary to put forward the automatic control system of Pleurotus eryngii inoculation. In this paper, based on the system of high reliability, high efficiency, flexible configuration and other performance requirements, PLC is used as the core components of the control system and control the operation of the whole system. In order to improve the efficiency of the control system, the particle swarm optimization algorithm was used to optimize the interpolation time of the trajectory of the manipulator. Through simulation, it was found that the joint acceleration curve was smooth without mutation, and the running time was short. Because the position deviation of the Culture medium of Pleurotus eryngii to be inoculated will inevitably occur when it is transferred on the conveyor belt, the image recognition technology is used to accurately locate them. In order to improve the efficiency of image recognition, the genetic algorithm (GA) is used to improve Otsu to find the target region of Culture medium of Pleurotus eryngii to be inoculated, and the simulation results showed that the computational efficiency could be increased by 70%. In order to locate the center of the target region, the mean value method is used to find their centroid coordinates. At last, it is found by simulation that the centroid coordinates could be accurately calculated for a basket of 12 Pleuroides eryngii medium to be inoculated.

Keywords: Image recognition · Centroid coordinate · PLC · Robot

1 Introduction

Pleurotus eryngii is a kind of rare edible fungus, which is very popular among consumers. In the process of factory cultivation, Pleurotus eryngii should be inoculated in a sterile working environment and a highly efficient and stable inoculation process, otherwise it will lead to failure of inoculation or directly affect the quality of the mushroom [1].

At present, as shown in Fig. 1, most enterprises adopt the traditional manual inoculation method. Workers wearing protective clothing use buttons to control the operation of the conveyor belt to transfer the packed Pleurotus eryngotus medium to the appropriate location, and press the buttons to control the liquid strains to enter the syringe. However, the inoculation efficiency of Pleurotus eryngus was reduced due to the following three

© The Author(s) 2022
Z. Qian et al. (Eds.): WCNA 2021, LNEE 942, pp. 988–1005, 2022.
https://doi.org/10.1007/978-981-19-2456-9_100

reasons: after a long period of inoculation, workers could not maintain the standard inoculation operation due to physical exhaustion; Because the temperature of the inoculation room is required to be kept at 25 °C, the heat emitted by the workers themselves will cause adverse effects on the inoculation room. The optimal liquid strain content required for Pleurotus eryngii inoculation is 30 ml, so it is difficult to guarantee the precision of liquid strain injection by manual injection.

In order to solve the above problems, this paper designed a set of automatic control system for Pleurotus eryngii inoculation, which can replace manual automatic completion of Pleurotus eryngii inoculation, including PLC control, manipulator trajectory optimization and center positioning based on image recognition. By the simulation analysis, the system can not only effectively replace manual to complete the inoculation work, but also significantly improve the work efficiency.

Fig. 1. Traditional artificial Pleurotus eryngii inoculation

2   Design of Pleurotus Eryngii Inoculation Control System

The control system mainly includes four links, which are the start and stop of the conveyor belt, the opening and closing of the solenoid valve, the precise positioning of machine vision and the trajectory planning of the manipulator arm. The Culture medium of Pleurotus eryngii to be inoculated is placed in a box in groups of 12 and transported to the appropriate location by a conveyor belt. When the position sensor senses the frame, the conveyor belt stops moving. Due to the influence of external factors such as the delay of transmission signal and the skew bag of culture medium of Pleurotus eryngii, machine vision is used to collect images of culture medium of Pleurotus eryngii and find 12 central positions accurately. Next, they will be transmitted to the manipulator arm by the upper computer in turn. The function of the manipulator is to take the syringe and insert it into the culture medium of Pleurotus eryngus according to the spatial coordinates obtained from the image recognition. Finally, PLC accurately controls the injection amount of liquid strain to 30ml by controlling the start and stop time of the solenoid valve.

2.1 Design of Pleurotus Eryngii Inoculation Hardware System

It is well known that PLC has the advantages of high reliability, flexible configuration, convenient installation, fast running speed and so on [2]. Therefore, the hardware system of the automatic production line of Pleurotus eryngii inoculation designed in this paper uses PLC as the control processing unit. As is shown in Fig. 2, The hardware of the control system includes PC (personal computer), PLC, servo drives, servo motors, industrial camera, electromagnetic valve, cylinder, belt and mechanical arm device, such. According to the specified technological process, PLC controls each hardware equipment to cooperate with each other to realize the automatic production of Pleurotus eryngii inoculation.

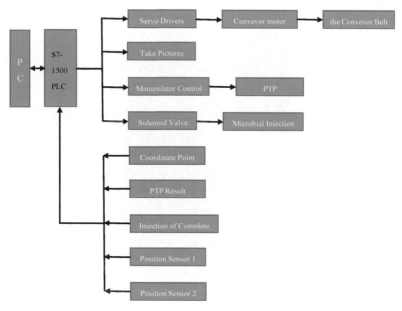

Fig. 2. Structure of Pleurotus eryngii inoculation control system

After considering the control system performance, development cost and I/O points and other factors, this paper selects Siemens S7-1500 series PLC as the controller of the equipment, and chooses the CPU as 1516. CPU 1516 has 2 PROFINET ports (X1 P1/P2 and X2 P1) and 1 PROFIBUS port. X2 P1, as the slave access port, interacts with the data of the upper computer through the industrial Ethernet bus. X1 and P1, as the access ports of the device, realize PROFINET communication with touch screen, frequency converter, distributed I/O unit, manipulator and other modules through switches.

2.2 Design of Pleurotus Eryngii Inoculation Software System

The inoculation control software of Pleurotus eratus is developed on TIA Portal platform, mainly including the design of S7-1500 PLC control program, TP1200 touch screen interface and upper computer monitoring interface. The PLC program is used for the automatic control of Pleurotus eryngii inoculation production line and the response to the monitoring request of the console, touch screen and upper computer. The program control flow is shown in Fig. 3.

The specific work steps are as follows.

- Put the baskets of Culture medium of Pleurotus eryngii to be inoculated on the running conveyor belt;
- The sensor 1 senses the basket and sends a signal to the PLC, and records the number of baskets through the PLC;
- The sensor 2 inducts the basket and sends a signal to the PLC, which stops the conveyor belt running;
- Take pictures of 12 Culture medium of Pleurotus eryngii to be inoculated inside each basket by industrial camera, and upload them to PC;
- Image processing is carried out on PC through MATLAB, all centroid coordinates are found, and coordinate values are transmitted to PLC through OPC (OLE for Process Control) protocol;
- By PROFINET protocol, PLC transmits the centroid coordinates to the manipulator successively;
- The mechanical arm accepts the centroid coordinates and drives the syringe to the centroid coordinates according to the program and sends a signal to inform the PLC;
- PLC starts the solenoid valve and records the time T. When T is equal to the preset time T, stop the solenoid valve, that is, the injection task of 30 ml liquid strain has been completed;

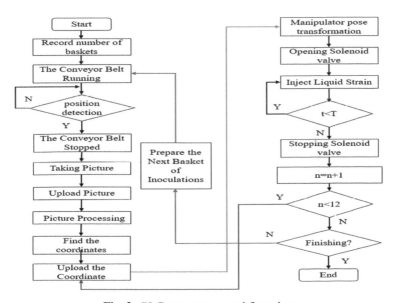

Fig. 3. PLC program control flow chart

- The mechanical arm is reset and ready to receive the next coordinate point;
- The PLC transmits the next coordinate point to the robot, and then returns to Step 6;
- When the basket's Pleurotus eryngii medium has been inoculated, the conveyor belt is started for preparing the next basket for inoculation. Then, return to Step 3;
- When the number of inoculated baskets is equal to the number of baskets sensed by sensor 1, the conveyor belt is stopped. Finally, complete the inoculation task.

2.3 Improvement of Work Efficiency

In practical application, the control system should not only meet the requirements of control performance, but also improve work efficiency as much as possible. In the whole control system, the time of trajectory planning carried out by the manipulator operated syringe occupies more than half of the whole inoculation time, so it is of great significance to select the time optimization for the trajectory of the manipulator. Because the particle swarm optimization (PSO) algorithm has the characteristics of simple structure, easy implementation, easy parameter adjustment, and can directly choose the polynomial interpolation time as the variable to optimize the PSO algorithm [3], so the PSO algorithm is selected to optimize the trajectory of the manipulator.

Analysis of Quintic Polynomial Trajectory Planning Algorithm. In order to reduce the vibration of the manipulator, the manipulator should meet the requirements of smoothness during operation. The solution of the quintic polynomial of joint Angle can satisfy the restriction of diagonally plus acceleration and avoid abrupt acceleration. Let the trajectory planning formula of joint Angle be:

$$\theta(t) = a_0 + a_1 t + a_2 t^2 + a_3 t^3 + a_4 t^4 + a_5 t^5 \tag{1}$$

In the formula (1), t represents time. $\theta(t)$ represents the Angle varying with time. $a_0, a_1, a_2, a_3, a_4, a_5$ represent the coefficients of the above formula. Set the initial time as 0, θ_0 as the initial position, t_1 as the time when the end reaches the end, and θ_1 as the end position, then the constraint conditions are as follows:

$$\begin{cases} \theta_0 = a_0 \\ \theta_0 = a_0 + a_1 t_1 + a_2 t_1^2 + a_3 t_1^3 + a_4 t_1^4 + a_5 t_1^5 \\ \dot{\theta}_0 = a_1 \\ \dot{\theta}_1 = a_0 + 2a_2 t_1 + 3a_3 t_1^2 + 4a_4 t_1^3 + 5a_5 t_1^5 \\ \ddot{\theta}_0 = a_0 \\ \ddot{\theta}_1 = 2a_2 + 6a_3 t_1 + 12a_4 t_1^2 + 20a_5 t_1^5 \end{cases} \tag{2}$$

In the formula (2), $\dot{\theta}_0$ represents velocity. $\ddot{\theta}_0$ represents initial acceleration, final velocity $\dot{\theta}_1, \ddot{\theta}_1$ represents final acceleration. According to the formula (2), there are six formulas in total, and the values of six unknowns $a_0, a_1, a_2, a_3, a_4, a_5$. The results are shown in formula (3).

$$
\begin{cases}
a_0 = \theta_0 \\
a_1 = \dot{\theta}_0 \\
a_2 = \frac{\ddot{\theta}_0}{2} \\
a_3 = \frac{20\theta_1 - 20\theta_0 - (8\dot{\theta}_1 + 12\dot{\theta}_0)t_1 - (3\ddot{\theta}_0 - 2\ddot{\theta}_1)t_1^2}{2t_1^3} \\
a_4 = \frac{30\theta_0 - 30\theta_1 - (14\dot{\theta}_1 + 16\dot{\theta}_0)t_1 - (3\ddot{\theta}_0 - 2\ddot{\theta}_1)t_1^2}{2t_1^4} \\
a_5 = \frac{12\theta_1 - 12\theta_0 - (6\dot{\theta}_1 + 6\dot{\theta}_0)t_1 - (\ddot{\theta}_0 - \ddot{\theta}_1)t_1^2}{2t_1^5}
\end{cases}
\tag{3}
$$

Trajectory Planning Simulation. Particle Swarm Optimization (PSO) trajectory planning was simulated using MATLAB software. Set the population M as 100, the range of initial position as [0.1, 4], the range of initial velocity as [−2, 2], and the number of iterations as 100. In order to reduce amount of calculation of PSO algorithm, differential time is chosen as the optimization function, its fitness function $f(t) = min \sum t$.

Shi and Eberhart studied the inertial weight W and proposed a particle swarm optimization algorithm with W decreasing linearly as the number of iterations increases. This algorithm can quickly determine the optimal target azimuth in the initial optimization process. With the increase of the number of iterations, the value of W gradually decreases and the optimization is carried out in this azimuth.

$$
w = w_{max} - (w_{max} - w_{min}) \times \frac{k}{k_{max}}
\tag{4}
$$

In the above formula, w_{max} refers to the maximum inertial weight, $w_{max} = 0.9, w_{min}$ refers to the minimum inertial weight, $w_{min} = 0.4$, and k_{max} refers to the maximum number of iterations. In order to prevent particles from running out of the solution space for optimization, a maximum value, V_{max}, is set such that $V_k \leq V_{max}$. When $V_k > V_{max}$; set $V_k = V_{max}$.

The 3-5-3 interpolation trajectory planning algorithm can not only solve the problems of polynomial interpolation, such as second-order polynomial interpolation, no convex hull and difficulty in optimization, but also reduce the computational difficulty and improve the efficiency [4]. Let the 3-5-3 polynomial be:

$$
\begin{cases}
\theta_{j1} = a_{j13}t^3 + a_{j12}t^2 + a_{j11}t + a_{j10} \\
\theta_{j2} = a_{j25}t^5 + a_{j24}t^4 + a_{j23}t^3 + a_{j22}t^2 + a_{j21}t + a_{j20} \\
\theta_{j3} = a_{j33}t^3 + a_{j32}t^2 + a_{j31}t + a_{j30}
\end{cases}
\tag{5}
$$

The angles corresponding to the initial positions, path points and end points of joints 1–3 are shown in Table 1.

Table 1. Angular interpolation points in joint space

Joint position	X_0	X_1	X_2	X_3
Joint 3	3.231	3.658	4.132	4.465

MATLAB was used to simulate joints 1, and the results were shown in Figs. 4 and 5.

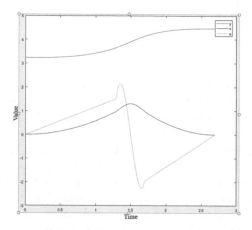

Fig. 4. Change curves of Angle, angular velocity and angular acceleration of joint 1

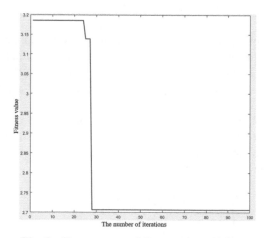

Fig. 5. Change curve of fitness value of joint 1

In Fig. 4, the change curves of Angle and angular velocity are relatively smooth, and there is no abrupt change in acceleration, indicating that the manipulator runs smoothly. In Fig. 5, the fitness value of the function decreases with the increase of the number of iterations, indicating that the total interpolation time of joint trajectory obtained at the end of iteration is the minimum.

3   Central Positioning

3.1   Image Acquisition

Figure 6 shows the image processing experimental platform built, which firstly studies a single Culture medium of Pleurotus eryngii to be processed. The industrial camera is fixed on the end of the manipulator arm through a clamp and moves along with the end of the manipulator arm, In the eye-in-hand mode. Given a camera calibration position, the manipulator moves to the calibration position before the camera takes pictures. The pictures taken by the industrial camera are uploaded to the PC, and the result after cropping is shown in Fig. 7.

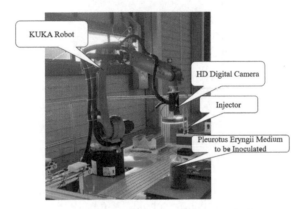

Fig. 6. Image processing experimental platform

Fig. 7. Picture of Pleurotus eryngii culture medium to be inoculated

3.2   Image Processing

Image Grayscale Processing. The grayscale processing of color images refers to the conversion of color images to grayscale images, that is, according to the color component RGB of the image into the grayscale value of the brightness range is (0, 255), so as to reflect the morphological characteristics of the image. According to the different sensitivity of human eyes to red, green and blue colors, the weighted average method is used for gray processing. According to the different sensitivity of human eyes to colors, different weights are given to RGB, and then the weighted average value of RGB brightness is taken as the gray value, as shown in formula (6).

$$gray(i, j) = 0.3R(i, j) + 0.59G(i, j) + 0.11B(i, j) \tag{6}$$

MATLAB was used for simulation, and the results were shown in Fig. 8.

Fig. 8. Grayscale processing results

Image Segmentation. Maximum inter-class variance method is a typical image segmentation method proposed by Japanese scholar Otsu in 1978 based on the principle of least square method, also known as Otsu method, abbreviated as Otsu. The measurement standard adopted in the OTSU algorithm is the maximum inter-class variance, whose principle is to obtain the inter-class variance between the target and the background through the threshold value. The larger the inter-class variance is, the greater the difference between the two parts of the image is, which means the minimum misclassification probability between the target and the background [5].

The calculation steps are as follows:

- Assume that the range of gray value I in the image is $[0, L - 1]$, the pixel of gray value i is n_i and the total number of pixels is N, then

$$N = \sum_{i=0}^{L-1} n_i \tag{7}$$

Set the probability of occurrence of grayscale value as p_i, then

$$p_i = \frac{n_i}{N} \tag{8}$$

- Assume that the range of gray value of background region and target region is $[0, T-1]$ and $[T, L-1]$ respectively, and the probability of background region and target region is P_0 and P_1 respectively, then

$$P_0 = \sum_{i=0}^{T-1} p_i \tag{9}$$

$$P_1 = \sum_{i=T}^{L-1} p_i \tag{10}$$

- Calculate the average gray scale of the background area and the target area, respectively expressed by μ_0 and μ_1, then

$$\mu_0 = \frac{1}{P_0} \sum_{i=0}^{T-1} (i \times p_i) = \frac{\mu(T)}{P_0} \tag{11}$$

$$\mu_1 = \frac{1}{P_1} \sum_{i=T}^{L-1} (i \times p_i) = \frac{\mu - \mu(T)}{1 - P_0} \tag{12}$$

- Set the average grayscale of the image to μ, then

$$\mu = \sum_{i=0}^{L-1} (i \times p_i) = \sum_{i=0}^{T-1} (i \times p_i) + \sum_{i=T}^{L-1} (i \times p_i) = P_0\mu_0 + P_1\mu_1 \tag{13}$$

- Let the total variance of the region be σ_i^2, then

$$\sigma_i^2 = P_0 \times (\mu_0 - \mu)^2 + P_1 \times (\mu_1 - \mu)^2 \tag{14}$$

MATLAB was used to simulate Fig. 8, and the results were shown in Fig. 9.

Fig. 9. Image segmentation results

Image Segmentation Optimization Based on Genetic Algorithm. Although the maximum inter-class variance method can be used to obtain an appropriate threshold for image segmentation, the need to select K value from the gray scale range $[0, L - 1]$ leads to a large amount of calculation and a long time. Genetic algorithm (GA) is used to optimize the maximal class inter-square method, which can quickly find the optimal threshold [6]. Combined with the principle of the maximum inter-class variance method in Sect. 3.1, the use of genetic algorithm is to quickly find the T value that maximizes σ_i^2.

The use of genetic algorithm is mainly divided into the following four stages:

- Population initialization
  In population initialization, n chromosomes and m genes need to be created. Each chromosome consists of m genes and represents a solution for each generation. Since the gray value range of the image is $[0, 255]$, which corresponds to 8-bit binary number, if $m = 8$, as shown in Fig. 10, the chromosome is encoded, and there are 2^8 situations on each chromosome. Let's say there are 10 solutions in each generation. Let's say $n = 10$.

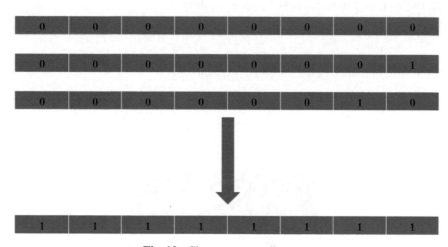

Fig. 10. Chromosome coding map

- Fitness assessment
  After population initialization, fitness function should be established to evaluate the fitness of each chromosome, that is, the performance of the solution. In this section, the maximum inter-class variance method is taken as the core, so $F_i = \sigma_i^2$ is selected as the fitness function, where $i = 1, 2, \cdots, 10$. The larger the F_i value of fitness is obtained, the more suitable the chromosome is.
- Duplication
  The process is mainly divided into three parts: selection, crossover and mutation.
  Firstly, the optimal solution from the previous generation population was copied to the next generation. According to the Roulette Wheel Selection method, the

probability of chromosome Selection was set as P_i, and the following results were obtained:

$$P_i = \frac{F_i}{\sum_i^n F_i} \qquad (15)$$

According to formula (15), a chromosome with a higher fitness F_i value has a higher probability of P_i, which means that it is more likely to be selected in the population. Finally, through 10 random screening, the next generation group was selected.

In order to speed up the solving speed of the optimal threshold, gene exchange was carried out on some chromosomes, and the selection crossover probability was 0.7. In order to avoid falling into the trap of local optimal solution, the chromosome mutation operation is selected, that is, the gene in the chromosome is changed, and the probability of selection mutation is 0.4.

- Decode

  The chromosome with the largest F_i fitness value was selected from the last generation and decoded into a decimal number, which is the optimal threshold T.

  The calculation process is shown in Fig. 11.

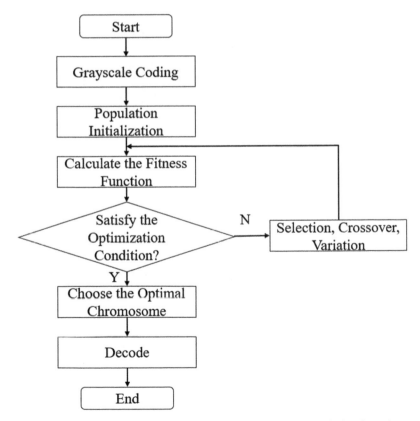

Fig. 11. Genetic algorithm to optimize the optimal threshold solution flow chart

MATLAB was used to optimize and simulate the genetic algorithm in Fig. 6, and the results were shown in Figs. 12 and 13, which were the optimal adaptive value evolution curve and the optimal threshold evolution curve respectively.

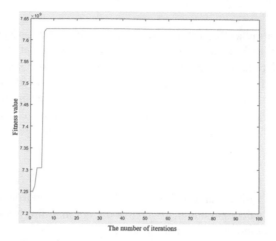

Fig. 12. Evolution curve of optimal fitness value

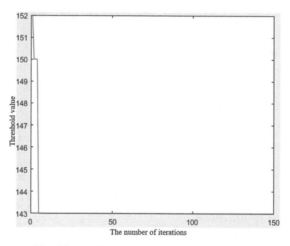

Fig. 13. Evolution curve of optimal threshold

As can be seen from Figs. 12 and 13, the fitness value of the genetic image was relatively small at the beginning of the inheritance. With continuous evolution, the unsuitable chromosomes were eliminated, and the fitness value became higher and higher, and the optimal threshold was found at the fifth generation of evolution. Through many simulations, it is found that the optimal threshold can be obtained by no more than 15 generations of evolutionary algebra. Therefore, according to the simultaneous calculation of 10 chromosomes in each generation, the optimal threshold value can be obtained within 150 threshold calculations. Compared with the traditional OTSU, which requires 256 thresholds to be calculated to compare the regional total variance, the calculation efficiency is improved by 70%.

To Solve the Center of Mass. Taking Fig. 9 as the research object, the target region we required to be solved is in the middle, but there are most interference regions outside the target region, and the centroid coordinates of the target region can be solved only if the interference region is removed.

The image connectivity domain includes four neighborhood connectivity and eight neighborhood connectivity. Since eight neighborhood connectivity is used to identify whether there are pixels (white) in eight directions of a pixel point in a binary image, eight-neighborhood connectivity is more comprehensive and has good generality [7]. In this paper, eight-neighborhood connectivity is used to remove white interference areas. The operation process is shown in Fig. 14. In the above way, imclear Border function in MATLAB was used in this paper to clear the white interference area connected with the boundary, and the result is shown in Fig. 15-a. As can be seen from Fig. 15-a, the peripheral white area of the central target area has been cleared, but many small white interference areas are still left.

Set up the image of the target area for the P, $P = \{P_1, P_2, \cdots, P_n\}$, $P_1, P_2, \cdots, P_n$ respectively represented in Fig. 15-a white area. Let the areas of $P_1, P_2, \cdots, P_n$ be $s_1, s_2, \cdots, s_n$ respectively. Through calculation, the white region with the largest area is retained. Through simulation, the results are shown in Fig. 15-b.

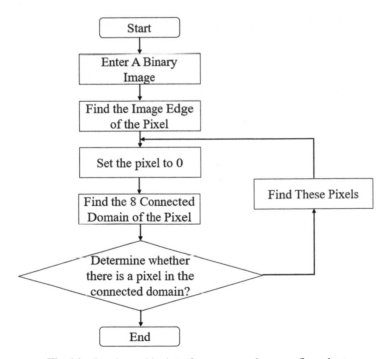

Fig. 14. Oundary white interference area clearance flow chart

As shown in Fig. 15-b, there are some small black spots inside the target region. In order to improve the accuracy of centroid solution, image expansion algorithm is used to remove the black spots.

Let A be the object to be processed and B be the structural element. The structural element B is used to scan all pixel points of image A, that is, the origin of B is used as the coordinate to scan each pixel point of A. If A pixel point in A is 1 when B covers the region of A, the corresponding pixel point of B is also 1, then the scanning point is 1;

$$A \oplus B = \{x | \hat{B}_x \cap A \neq \varnothing\} \tag{16}$$

In formula (16), $\hat{B}$ is the mapping of B to A, $\hat{B}_x$ said image B shift distance along the vector x. Through simulation, the result is as shown in Fig. 15-c, and the black spots have been removed.

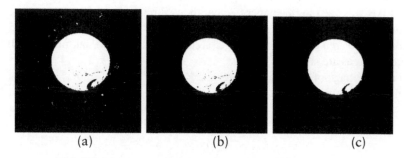

(a)   (b)   (c)

Fig. 15. Binary image interference region processing results

Calculate Coordinates by means of the Mean Value. According to Fig. 10-c, the small black points in the target area have all disappeared. Next, the coordinate of the center point of the target area is solved. Let the horizontal and vertical coordinates of the center point be X and Y respectively, and the horizontal and vertical coordinates of the target region be m and n respectively, then:

$$X = \frac{\sum_{(m,n) \in S} x_{(m,n)}}{S} \tag{17}$$

$$Y = \frac{\sum_{(m,n) \in S} y_{(m,n)}}{S} \tag{18}$$

$x_{(m,n)}$ and $y_{(m,n)}$ respectively represent the horizontal and vertical coordinates of the pixel points in the target region. S represents the area of the target region. According to Eqs. (17) and (18), the horizontal and vertical coordinates of the pixels in the target region are respectively added and then divided by the total area of the target region to obtain the horizontal X and vertical Y of the center point MATLAB is used for simulation, and the result is shown in Fig. 16. The coordinate position has been marked in the central area, and the coordinate point is (241, 191).

Fig. 16. Target area center point diagram

3.3 Multi-target Identification and Verification

In a practical production line, as shown in Fig. 17-a, in order to speed up the inoculation efficiency, 12 eryngii in a basket need to be identified simultaneously. The algorithm described above is used to simulate Fig. 17-a, and the result is shown in Fig. 17-b to obtain 12 target regions.

a   b

Fig. 17. Image processing of whole basket of Culture medium of Pleurotus eryngii

Set and the target area for $Q = \{Q_1, Q_2, \cdots, Q_{12}\}$, the area of it: $S = \{S_1, S_2, \cdots, S_{12}\}$, the center coordinates of it: $O = \{O_1, O_2, \cdots, O_{12}\}$, $O_i = (x_i, y_i)$, $i = 1, 2, \cdots, 12$. The steps for solving the central coordinates are as follows:

- Calculate the number of connected domains and mark each connected domain;
- Find the area of each connected domain S;
- Find the sum of the abscissa and ordinate of each connected domain;
- The sum of the abscissa and the sum of the ordinate of each connected domain is divided by the area to get the central coordinate of the connected domain O_i.

Through simulation, the result is shown in Fig. 18. The central coordinates of all target regions have been worked out.

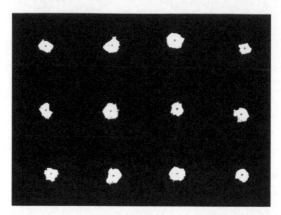

Fig. 18. Image processing results of a whole basket of Pleurotus eryngii

4 Conclusion

This paper presents a set of automatic control system for Pleurotus eryngii inoculation. The control part of the system mainly includes PLC control, mechanical arm control and visual control. Among them, PLC, as the upper computer, controls the operation of the entire Pleurotus eryngii inoculation production line to ensure that each link is completed normally according to the steps. In order to improve the running efficiency of the production line, the PSO method was used to optimize the trajectory of the manipulator. Through simulation analysis, it was found that the algorithm could reduce the running time of the manipulator. In order to improve the accuracy of inoculation by robotic arm, image recognition technology was used to accurately locate the culture medium of Pleurotus eryngii to be inoculated. Among them, the genetic algorithm is used to optimize the maximum inter-class variance method for image segmentation, and the simulation results show that the target region recognition accuracy can be reduced and the computational efficiency can be improved. Finally, the whole basket of Culture medium of Pleurotus eryngii to be processed was simulated, and 12 centroid coordinates were accurately obtained by means of the mean value method.

References

1. Suyun, Y.: Key points of industrial cultivation techniques of Pleurotus eryngii in northwest China. Northern Hortic. 7, 150 2 (2017). (in Chinese)
2. Dong, W.: Application analysis of PLC technology in electrical automatic control. J. Phys. **1533**(2), 022012 (2020)
3. Pattanayak, S., Choudhury, B.B.: An effective trajectory planning for a material handling robot using PSO algorithm. Adv. Intell. Syst. Comput. **990**, 73–81 (2020)

4. Rong, F., Hehua, J.: Time-optimal trajectory planning algorithm for manipulator based on particle swarm optimization. Inf. Control **40**, 802–808 (2011). (in Chinese)
5. Dong, Y.X.: An improved Otsu image segmentation algorithm. Adv. Mater. Res. **989–994**, 3751–3754 (2014)
6. Sun, H.: Image segmentation method based on genetic algorithm and OTSU. Boletin Tecnico/Tech. Bull. **55**, 55–61 (2017)
7. Wang, F., Zhou, G., Zhang, R., Liu, D.: A fast marking method for connected domain oriented to FPGA. Comput. Eng. Appl. **56**, 230–235 (2020). (in Chinese)

A Novel Robust Adaptive Color Image Watermarking Scheme Based on Artificial Bee Colony

Tingting Xiao and Wanshe Li[✉]

College of Mathematics and Statistics, Shaanxi Normal University, Xi'an, Shaanxi, China
liwanshe@126.com

Abstract. This paper proposes a new robust adaptive watermarking scheme based on dual-tree quaternion wavelet and artificial bee colony, wherein the host images and watermark images are both color images. Color host images and watermark images in RGB space are transformed into YCbCr space. Then, apply Arnold chaotic map on their luminance components and use the artificial bee colony optimization algorithm to generate embedding watermark strength factor. Dual-tree quaternion wavelet transform is performed on the luminance component of the scrambled host image. Apply singular value decomposition on its low-frequency amplitude sub-band to obtain the principal component (PC). Embed the watermark into the principal component. Analysis and experimental results show that the proposed scheme is better as compared to the RDWT-SVD scheme and the QWT-DCT scheme.

Keywords: Dual-tree quaternion wavelet transform (DTQWT) · Singular value decomposition · Artificial bee colony (ABC) · Color image watermarking

1 Introduction

Image watermarking is an important method to solve lots of security problems such as the authenticity of digital data, copyright protection, and legal ownership. At present, the watermarking schemes of a large number of papers take binary or grayscale images as watermarks.

In recent years, the design of watermarking schemes for embedding color watermarks into color host images has been a difficult problem. The color image watermarking scheme in the literatures [1, 2] uses grayscale images or binary images as the watermarks. Sharma et al. [3] put forward an novel color image watermarking scheme based on RDWT-SVD and ABC, in addition, the watermark images are color images. In order to improve the performance of image processing schemes, nature-inspired optimization algorithms have become an important tool. Particle swarm optimization (PSO) [4], differential evolution (DE) [5], and artificial bee colony [3] are widely used in digital image schemes. DTQWT not only provides a wealth of phase information and solves the common shortcomings of the wavelet transform, but also can consider the local characteristics of the image at different scales [6].

© The Author(s) 2022
Z. Qian et al. (Eds.): WCNA 2021, LNEE 942, pp. 1006–1017, 2022.
https://doi.org/10.1007/978-981-19-2456-9_101

This paper proposes a new color image watermarking scheme based on dual-tree quaternion wavelet transform, ABC algorithm and singular value decomposition. Apply the single level dual-tree quaternion wavelet decomposition on the host image, apply the singular value decomposition on the obtained low-frequency amplitude sub-band, and ABC algorithm is used to obtain the embedding watermark strength factor. Experimental results show that the scheme has better performance in terms of imperceptibility and robustness.

2 DTQWT and ABC

2.1 Dual-Tree Quaternion Wavelet Transform (DTQWT)

Chan et al. [6] used quaternion algebra and the two-dimensional (2D) Hilbert transform to extend the real wavelet transform and complex wavelet transform and then proposed DTQWT. In addition, the DTQWT can achieve multiresolution analyses. In digital image watermarking, the DTQWT transformation of the host image can extract the characteristics image in different frequency domains. Because the DTQWT coefficients of the host image are also quaternions, we can get the amplitude, phase, and frequency information of corresponding scales. The watermark is embedded in the stable component that has little effect on the host image, and the inverse DTQWT is applied to obtain the watermark in the host image. DTQWT not only provides rich phase information but also overcomes the common shortcomings of the wavelet transform. Taking into account the local characteristics of the image on different scales, DTQWT shows a better performance than RDWT [3], QWT [7]. We realize the DTQWT and inverse DTQWT by using the dual-tree filter bank [8] framework.

2.2 ABC Optimization

Karaboga presented an optimization algorithm about population size and called it artificial bee colony (ABC) in the year 2005 [9]. It is derived from the intelligent search for nectar source behavior of the bee colony. The ABC optimization algorithm determines the optimal value of a variable by minimizing or maximizing a given objective function in a given search space.

There are three types of bees in the ABC algorithm:employed bees, onlooker bees, and scout bees. Employed bees indicate the number of solutions. The number of initial solutions of the ABC algorithm is N, in which each solution is D-dimensional vector. An initialization solution can be expressed as $X_i = \{x_{i,1}, x_{i,2}, \cdots, x_{i,D}\}$; where $i = 1, 2, N$. The ABC algorithm optimization process includes the following steps [3]:

1) During initialization, population N is randomly selected, in which each solution $X_i = \{x_{i,1}, x_{i,2}, \cdots, x_{i,D}\}$ $(i = 1, 2, \cdots, N)$ is a D-dimensional vector. The ith food source described as in Eq. (1).

$$x_{i,j} = x_{\min,j} + rand(0, 1)(x_{\max,j} - x_{\min,j})(j = 1, 2, \cdots, D) \qquad (1)$$

2) Each employed bee useslocal information available to generate a new solution on based and then compares the fitness value of generated solution with the initial solution. Choose the better solution of the two solutions for the next iteration. Generate a new solution Y_i through Eq. (2).

$$y_{i,j} = x_{i,j} + \Phi_{i,j}(x_{i,j} - x_{k,j}) \tag{2}$$

In which $k \in \{1, 2, \cdots, N\}$ and $j \in \{1, 2, \cdots, D\}$, k is different from i. $\Phi_{i,j}$ is a random number between -1 and 1.

3) Update the fitness value. Now the onlooker bees generate a new solution by Eq. (3).

$$P_i = \frac{fitness_i}{\sum_{i=1}^{N} fitness_i} \tag{3}$$

$$fitness_i = \begin{cases} \frac{1}{F(X_i)+1}, f(x_i) > 0 \\ 1 + |F(X_i)|, otherwise \end{cases} \tag{4}$$

Where $F(X_i)$ represents the fitness value at X_i. The fitness function used in this paper is defined by Eq. (17).

4) each onlooker bee generates a random solution and the value is between zero and one; if the value of P_i is bigger than the random solution in the step 2.

5) ABC has three main control parameters: N(number of solutions), number of onlooker or employed bees, the value of limit, and the maximal iteration number. The ABC optimization algorithm circularly executes the above steps until the best solution is received.

3 Watermarking Scheme

3.1 *Watermark Embedding Process*

The watermark embedding scheme proposed in this paper is shown in Fig. 1. The specific steps are as follows:

1) Firstly convert the color host image I to a YCbCr color space, which obtains components I_Y, I_{Cb}, I_{Cr}. Apply Arnold chaotic map to I_Y to get $\tilde{I}_Y$.

2) Convert the color watermark image W to a YCbCr color space, which obtains components W_Y, W_{Cb}, W_{Cr}. Apply Arnold chaotic map to W_Y to get $\tilde{W}_Y$.

3) Perform the single level dual-tree quaternion wavelet transform on $\tilde{I}_Y$ and decompose it into sixteen sub-bands, select the low-frequency amplitude sub-band LL_Y^I as the area to embed the watermark.

4) Apply singular value decomposition on LL_Y^I to get the U_{LL}, S_{LL} and V_{LL}^T matrices.

$$LL_Y^I = U_{LL} S_{LL} V_{LL}^T \tag{5}$$

5) Use the $\tilde{I}_Y$ and $\tilde{W}_Y$ obtained in the first and second steps, and then generate an adaptive embedding watermark strength factor α according to the Sect. 3.3

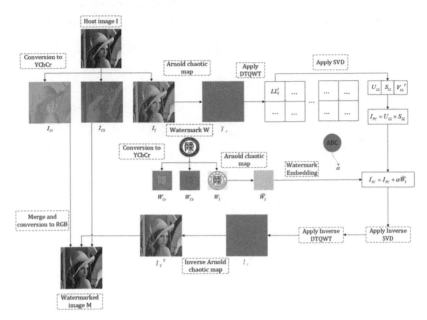

Fig. 1. The block diagram of the watermark embedding scheme

6) Using the U_{LL} and S_{LL} obtained in the fourth step, calculate the principal component I_{PC} of the host image.

$$I_{PC} = U_{LL} \times S_{LL} \tag{6}$$

7) Embed the watermark to modify the principal component.

$$I'_{PC} = I_{PC} + \alpha \tilde{W}_Y \tag{7}$$

8) Perform singular value decomposition on I'_{PC}. Save U'_{LL}, V'^{T}_{LL} matrices, for watermark extraction scheme.

$$I'_{PC} = U'_{LL} \times S'_{LL} \times V'_{LL} \tag{8}$$

9) Perform inverse SVD (ISVD) to obtain modified LL^I_W. Perform the single level inverse dual-tree quaternion wavelet transform on LL^I_W sub-band with other fifteen sub-bands to obtain I'_Y.

$$LL^I_W = U_{LL} \times S'_{LL} \times V^T_{LL} \tag{9}$$

10) Perform the inverse Arnold chaotic transform on I'_Y component to get I^N_Y.
11) Merge I^N_Y (luminance) with I_{Cb} and I_{Cr}, get the image with the watermark embedded in the YCbCr color space. Convert it to RGB color space and obtain the color watermarked image M.

3.2 Extraction Process

The watermark extracting scheme proposed in this paper is shown in Fig. 2. The specific steps are as follows:

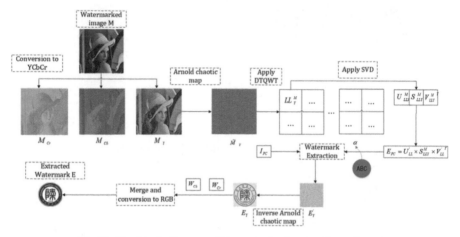

Fig. 2. The block diagram of the watermark extracting scheme.

1) Firstly convert the color watermarked image M to the YCbCr color space, which obtains components M_Y, M_{Cb}, M_{Cr}.
2) Take M_Y as the area for watermark luminance information extracting. Apply Arnold chaotic map on it and obtain $\tilde{M}_Y$.
3) Perform the single level dual-tree quaternion wavelet transform on the $\tilde{M}_Y$ and decompose it into sixteen sub-bands, select the low-frequency amplitude sub-band LL_Y^M.
4) Apply singular value decomposition on LL_Y^M.

$$LL_Y^M = U_{LLY}^M \times S_{LLY}^M \times V_{LLY}^M{}^T \tag{10}$$

5) Compute the extracted principal component E_{PC} using S_{LLY}^M generated in the forth step and $U_{LL}', V_{LL}'^T$.

$$E_{PC} = U_{LL}' \times S_{LLY}^M \times V_{LL}'^T \tag{11}$$

6) Compute the extracted watermark luminance component E_Y' using the strength factor α. Use the strength factor α to obtain the luminance component E_Y' of the extracted watermark.

$$E_Y' = (E_{PC} - I_{PC})/\alpha \tag{12}$$

7) Perform the inverse Arnold chaotic transform on E_Y' and obtain the unscrambled watermark luminance component E_Y.

8) Merge the watermarked E_Y (luminance) component with components W_{Cb} and W_{Cr}, and convert it RGB color space, finally we get the color extracted watermark image E.

3.3 Generation of Adaptive Embedding Strength Factor

It is very important to generate the watermark embedding strength factor, because it affects the imperceptibility and robustness of the watermarking scheme. The smaller the value of the embedded watermark strength factor is, the better the invisibility of the watermark scheme and the poorer robustness. On the contrary, the bigger the value of the embedding watermark strength factor, the less visibility of the watermark scheme and the better robustness. Therefore, it is necessary to find an optimal strength factor value to achieve a balance between imperceptibility and robustness. They are defined as follows [10]:

$$Imperceptiblity = correlation(H, H_W) \tag{13}$$

$$Robustness = correlation(W, W^*) \tag{14}$$

$$correlation(X, X^*) = \frac{\sum_{i=1}^{n} \sum_{j=1}^{n} \overline{X_{i,j} XOR \ X_{i,j}^*}}{n \times n} \tag{15}$$

Here H denotes the luminance component I_Y of the host image, H_W denotes the luminance component M_Y of watermarked image, W denotes the luminance component W_Y of the watermark image, W^* denotes the luminance component E_Y of the extracted watermark image, $n \times n$ is the size of the image X and XOR denotes the exclusive-OR(XOR)operation. Suppose add N type of attacks on the watermarked image M, average robustness is defined as follow:

$$Robustness_{average} = \frac{\sum_{i=1}^{N} correlation(W, W_i^*)}{N} \tag{16}$$

$$Minimizef = \frac{1}{Robustness_{average}} - Imperceptibility \tag{17}$$

The better the robustness indicates that the extracted watermark is very similar to the original watermark. In addition, the fitness function is defined as Eq. (17). Figure 3 shows the specific process of embedding strength factors optimization. Table 1 shows the control parameters optimized by ABC.

Table 1. The value of ABC optimization.

ABC optimization parameters	Values
Number of Swarms	50
Maximal iteration	30
Limit	15
Initialization range	0.001–1
Number of Employed bees	50% of the swarm
Number of Onlooker bees	50% of the swarm
Number of Scout bees	50% of the swarm
Fitness Function parameters	Noise, Filter attacks,Geometric attacks

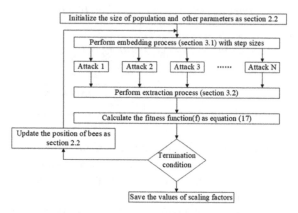

Fig. 3. Block diagram of optimization process.

4   Experimental Results and Discussion

Simulation experiments are carried out on MATLAB 2016A. To verb the performance of the proposed scheme, four RGB space color host images Lena, Plane, Pepper, and Mandrill with a size of 512×512 are selected from the database [55], as shown in Fig. 4. The color Shaanxi Normal University badge with a size of 256×256 in the RGB space is selected as the watermark image, as shown in Fig. 4. The embedding strength factor of the watermark is generated in Sect. 4. Figure 5 shows the convergence of the fitness values of different host images. The quality metrics used here include peak signal-to-noise ratio (PSNR), mean square error (MSE), normalized correlation coefficient (NCC), and structural similarity (SSIM) index.

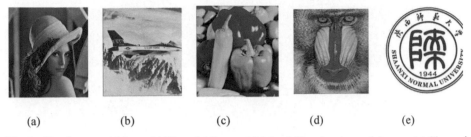

(a)   (b)   (c)   (d)   (e)

Fig. 4. Host images: (a) Lena (b) Plane (c) Pepper (d) Mandrill and watermark image: (e) Shaanxi Normal University badge.

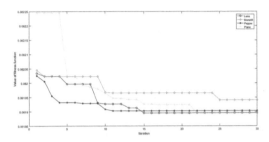

Fig. 5. Fitness value vs. iterations.

4.1 Imperceptibility Results

Figure 6 shows the watermarked and extracted watermark images applying the proposed scheme. We calculated the value of PSNR, SSIM and NCC using different host images, as shown in Table 2. The Human Visual System (HVS) shows that if the PSNR value is greater than 30 dB and the SSIM value is greater than 0.9, the imperceptibility of the watermark is better. Otherwise, the average PSNR calculated between the original color host image and the color watermarked image is 47.6349 db, which is higher than 30 db, and the average SSIM value is 0.9974, which is higher than 0.9. The high PSNR and SSIM results show that the proposed method obtains a good imperceptibility.

Table 2. Imperceptibility results without attack.

Host image	PSNR	SSIM	NCC
Lena	47.3377	0.9966	0.9974
Mandrill	47.2691	0.9987	0.9976
Pepper	47.1189	0.9973	0.9978
Plane	48.8138	0.9971	0.9979
Average	47.6349	0.9974	0.9977

Fig. 6. Obtained watermarked images: (a) Lena (b) Plane (c) Pepper (d) Mandrill and extracted watermark images: (e) (f) (g) (h)

4.2 Robustness Results

The visual result of the robustness of the image with Lena as the host, Fig. 7 shows the image obtained after adding an attack to an image embedded with a watermark. The image attacks are additive noise, filtering, rotation, cropping, blurring, and histogram equalization. Figure 7 shows the corresponding watermark images extracted from the attacked images. Table 3 shows the calculated NCC values under different image attacks. This watermarking scheme has achieved remarkable robustness results under many common signal processing attacks, especially geometric distortion.

Table 3. The NCC value of extracted watermark using Lena as the host image

Attack	Parameter	Lena	Plane	Pepper	Mandrill
Salt & pepper noise	0.05	0.9995	0.9995	0.9993	0.9994
	0.1	0.9986	0.9984	0.9986	0.9990
	0.2	0.9939	0.9925	0.9947	0.9939
Gaussian noise	0.05	0.9970	0.9975	0.9983	0.9959
	0.1	0.9919	0.9922	0.9949	0.9899
	0.2	0.9857	0.9831	0.9890	0.9818
Speckle noise	0.05	0.9992	0.9994	0.9996	0.9996
	0.1	0.9991	0.9974	0.9991	0.9985
	0.2	0.9962	0.9906	0.9959	0.9936
Gaussian fitler	[2 2]	0.9961	0.9961	0.9966	0.9917
	[3 3]	0.9954	0.9952	0.9961	0.9877
	[5 5]	0.9949	0.9943	0.9957	0.9851
Median filter	[2 2]	0.9965	0.9968	0.9969	0.9933

(continued)

Table 3. (*continued*)

Attack	Parameter	Lena	Plane	Pepper	Mandrill
	[3 3]	0.9959	0.9964	0.9968	0.9904
	[5 5]	0.9952	0.9955	0.9964	0.9851
Average filter	[2 2]	0.9961	0.9961	0.9966	0.9917
	[3 3]	0.9950	0.9947	0.9958	0.9864
	[5 5]	0.9935	0.9921	0.9945	0.9800
Histogram equalization	[3 3]	0.9500	0.9047	0.9644	0.9460
Sharpening	4	0.9965	0.9941	0.9977	0.9698
Rotation	45°	0.9879	0.9823	0.9932	0.9916
	5°	0.9977	0.9945	0.9952	0.9956
	2°	0.9974	0.9987	0.9972	0.9957
Cut	1/4	0.9923	0.9923	0.9877	0.9912
Motion blur	$\theta = 4 len = 3$	0.9924	0.9934	0.9943	0.9840
JEPG compression	$Q = 10$	0.9960	0.9968	0.9962	0.9946
	$Q = 30$	0.9962	0.9971	0.9964	0.9961
	$Q = 50$	0.9963	0.9971	0.9965	0.9968
	$Q = 80$	0.9965	0.9972	0.9967	0.9972
Brightening		0.9921	0.9930	0.9838	0.9733

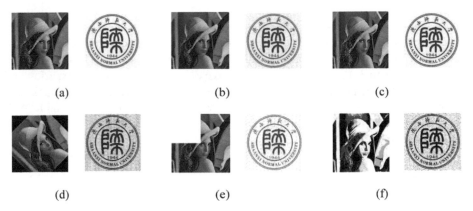

(a)   (b)   (c)

(d)   (e)   (f)

Fig. 7. The attacked watermarked image (Lena) and extracted watermark under attacks (a) Gaussian noise (b) Median filter (c) Sharpening (d) Rotation (e) Cut (f) Histogram Equalization

4.3 Comparative Analysis

Sharma et al. [3] put forward a new color image watermarking scheme based on RDWT-SVD and ABC algorithm. S. Han et al. [7] proposed a color image watermarking algorithm based on QWT-DCT, and the embedded watermark strength factor is a fixed constant. The proposed watermarking scheme is compared with the above two schemes, and the NCC values of each scheme under different image attacks are calculated. The results are shown in Table 4. Compared with the optimized and unoptimized color watermarking schemes, the robustness in this paper is significantly better.

Table 4. The comparative analysis

Attack	Parameter	Sharma et al. [3]	S. Han et al. [7]	Proposed scheme
Gaussian noise	0.001	0.9882	0.9908	0.9919
Salt&pepper noise	0.02	0.9966	0.9907	0.9986
Speckle noise	0.1	0.9813	–	0.9991
Median filter	[3 3]	0.9955	0.9859	0.9959
Average filter	[3 3]	0.9948	0.9895	0.9950
JEPG compression	50	0.9960	0.9911	0.9963
Sharpening	1	0.9931	–	0.9984
Rotation	5°	0.9914	–	0.9977
Cut	1/4	0.9648	–	0.9923

5 Conclusion

In this paper, we propose a novel color image watermarking scheme based on DTQWT-SVD and ABC optimization. The color host image is converted to YCbCr space, use the ABC optimization to generate the embedding watermark strength factor, and then modify the principal component of the host image to insert the watermark. Experimental results show that the proposed scheme has strong robustness under common attacks and geometric attacks. Compared with the adaptive watermarking scheme based on RDWT [3] and the color image watermarking scheme based on QWT [7], the scheme in this paper has better robustness.

References

1. Roy, S., Pal, A.K.: An SVD based location specific robust color image watermarking scheme using RDWT and arnold scrambling. J. Wirel. Pers. Commun. **98**, 2223–2250 (2018)
2. Kalra, G.S., Talwar, R., Sadawarti, H.: Adaptive digital image watermarking for color images in frequency domain. Multimedia Tools Appl. **74**(17), 6849–6869 (2014). https://doi.org/10.1007/s11042-014-1932-3

3. Sharma, S., Sharma, H., Sharma, J.B.: An adaptive color image watermarking using RDWT-SVD and artificial bee colony based quality metric strength factor optimization. J. Appl. Soft Comput. 2019 **84**(C), 105696

4. Thakkar, F., Srivastava, V.K.: An adaptive, secure and imperceptive image watermarking using swarm intelligence, Arnold transform, SVD and DWT. J. Multimed Tools Appl. **80**, 12275–12292 (2021)

5. Ali, M., Ahn, C.W.: An optimized watermarking technique based on self adaptive DE in DWT–SVD transform domain. J. Signal Processing **94**, 545–556 (2014)

6. Chan, W.L., Choi, H., Baraniuk, R.G.: Coherent multiscale image processing using Dual-tree quaternion wavelets. J. IEEE Trans Image Process **17**(7), 1069–1108 (2008)

7. Han, S., Yang, J., Wang, R., Jia, G.: A novel color image watermarking algorithm based on QWT and DCT. In: Yang, Jinfeng, Hu, Qinghua, Cheng, Ming-Ming., Wang, Liang, Liu, Qingshan, Bai, Xiang, Meng, Deyu (eds.) CCCV 2017. CCIS, vol. 771, pp. 428–438. Springer, Singapore (2017). https://doi.org/10.1007/978-981-10-7299-4_35

8. Selesnick, I.W., Baraniuk, R.G., Kingsbury, N.C.: The dual-tree complex wavelet transform. J. IEEE Signal Process. Mag. **22**(6), 123–151 (2005)

9. Karaboga: An idea based on honey bee swarm for numerical optimization, Tech. report TR06, Ericyes University (2005)

10. Ansari, I.A., Pant, M., Ahn, C.W.: ABC optimized secured image watermarking scheme to find out the rightful ownership. J. Optik-Int. J. Light Electron Opt. **127**(14), 5711–5721 (2016)

11. Image database. http://sipi.usc.edu/database/. Signal and Image Processing Institute, University of Southern California

Detection and Location of Myocardial Infarction from Electrocardiogram Signals Using Median Complexes and Convolutional Neural Networks

Shijie Liu, Guanghong Bin, Shuicai Wu, Zhuhuang Zhou, and Guangyu Bin[✉]

Faculty of Environmental and Life Sciences, Beijing University of Technology, Beijing, China
binguanghong@aliyun.com, {wushuicai,zhouzh}@bjut.edu.cn,
guangyubin@qq.com

Abstract. When doctors judge myocardial infarction (MI), they often introduce 12 leads as the basis for judgment. However, the repetitive labeling of nonlinear ECG signals is time-consuming and laborious. There is a need of computer-aided techniques for automatic ECG signal analysis. In this paper, we proposed a new method based on median complexes and convolutional neural networks (CNNs) for MI detection and location. Median complexes were extracted which retained the morphological features of MIs. Then, the CNN was used to determine whether each lead presented MI characteristics. Finally, the information of the 12 leads was synthesized to realize the location of MIs. Six types of MI recognition were performed, including inferior, lateral, anterolateral, anterior, and anteroseptal MIs, and non-MI. We investigated cross-database performance for MI detection and location by the proposed method, with the CNN models trained on a local database and validated by the open PTB database. Experimental results showed that the proposed method yielded F1 scores of 84.6% and 80.4% for the local and PTB test datasets, respectively. The proposed method outperformed the traditional hand-crafted method. With satisfying cross-database and generalization performance, the proposed CNN method may be used as a new method for improved MI detection and location in ECG signals.

Keywords: Electrocardiogram (ECG) · Myocardial infarction · Median complex · Convolutional neural network (CNN) · Computer-aided diagnosis (CAD)

1 Introduction

The decrease or stop of blood flow in the heart will lead to myocardial infarction (MI), resulting in myocardial damage [1]. Electrocardiogram (ECG) is often used to diagnose patients with possible or confirmed myocardial ischemia. The judgment of ECG needs the participation of professionals with certain electrophysiological knowledge, and the ECG of various patients should be considered differently. With the rapid development of ECG recording and processing equipment and analysis technology, 12-lead ECG data

© The Author(s) 2022
Z. Qian et al. (Eds.): WCNA 2021, LNEE 942, pp. 1018–1030, 2022.
https://doi.org/10.1007/978-981-19-2456-9_102

can be optimized for diagnosis of MIs and other heart diseases, especially with the use of computer-aided methods.

ECG provides information about both the presence and location of MIs. MI characteristics (MICHs) include abnormal Q wave appearance, ST-segment elevation, and T-wave inversion [2]. Abnormal Q wave on 12-lead ECG indicates previous transmural MI. The ST-segment changes related to acute ischemia or infarction on standard ECG are due to the current flowing through the boundary between ischemic and non ischemic areas, which is called injury current. Some T-wave changes are related to the stage after reperfusion.

The MI area can be located using ECG. The ECG lead of the display Mitch reflects the MI area. It should be noted that the ECG complex does not look the same in all leads of the standard 12 lead system, and the shape of the ECG component wave may vary from lead to lead. For example, the current ECG criteria for the diagnosis of acute ischemia / infarction require ST segment elevation greater than 0.2 MV in leads V1, V2 and V3 and greater than 0.1 MV in all other leads [3]. The criteria of abnormal Q waves are inconsistent in the individual leads [4].

In previous studies, linear or nonlinear ECG signal feature sets are input to a shallow classifier for MI classification. Bozzola et al. [5] extracted 96 morphologic features from 12 leads for MI classification including QRS, Q and R amplitude and duration, T amplitude and Q/R ratio. Ouyang et al. [6] measured the voltages of Q-, R-, S-, T-waveforms and ST deviation, 80 ms after point J in the I, II and V1-V6 leads of the standard 12-lead ECG, collecting 40 measurements from each case of ECG. Arif et al. [7] extracted a 36-dimensional feature vector and classified the signals with the K-nearest neighbor classifier. Kumar et al. [8] processes the segmented ECG signal and decomposes it into subband signals to extract sample entropy, which is then used as the input of different classifiers. Acharya et al. [9] extracted 47 features for MI classification and achieved an accuracy of 98.80%.

In recent years, the method based on deep learning has shown great application potential in the diagnosis of MIS and other heart diseases. Rajpurkar et al. [10] developed a-34 layer convolutional neural network (CNN), which exceeds the performance of committee certified cardiologists in detecting multiple arrhythmias through ECG recorded by single lead wearable monitor. Lodhi et al. [11] used one CNN for each lead in 12 lead ECG data, so 12 CNN constitute the voting mechanism for myocardial infarction detection. Lui and Chow [12] developed a classifier combining convolutional neural network and recursive neural network, which achieves better performance than using CNN alone. Acharya et al. [13] used CNN model and only lead II was used to automatically detect MIS, even if there was noise in ECG data. Liu et al. [14] proposed a new multi lead ECG myocardial infarction detection algorithm based on CNN. Subsequently, Liu et al. [15] proposed a multi-feature-branch CNN (MFB-CNN) to automatically detect and locate myocardial infarction using ECG. The method based on deep learning does not need early feature extraction and show many advantages.

Most of the current studies are based on the open-access PTB diagnostic ECG database [16]. The database contains 549 records from 290 subjects, among which 148 subjects are diagnosed as MIs. There are two methods for evaluating the system performance: class-based and subject-based methods [17, 18]. For the classroom based method, the data is divided into training data and test data, which is independent of patients. In the subject based method, the data from one patient is used for testing, and the other subjects are trained [18]. When using class-based approaches, the accuracy (Acc), specificity (Spe) and sensitivity (Sen) can reach more than 98.00% [7, 9, 17, 18]. However, when the subject-based method is used for evaluation, the system performance may be reduced. Sharma and Sunkaria [17] reported that the performance is Acc = 98.84%, Sen = 99.35%, and Spe = 98.29% for class-based methods, while the performance is Acc = 81.71%, Sen = 79.01%, and Spe = 79.26% for subject-based methods. Liu et al. [18] reported that the performance for class-based methods is Acc = 99.90%, Sen = 99.97%, and Spe = 99.54%, and the performance is Acc = 93.08%, Sen = 94.42%, Spe = 86.29% for subject-based methods. Note that cross-database MI detection performance have not been investigated.

We proposed a new CNN method for MI detection and location, with the CNN models trained on a local database and validated by the open PTB database. The local database was a well-labeled database of 12-lead ECG data. Doctors marked the presence of MICHs in each lead. Locations of MIs were also marked. We trained a one-dimensional (1D) CNN for each lead of ECG data, and then combined the results of each lead for discrimination and location of MIs. The proposed method showed satisfying cross-database performance in detecting and locating MIs in ECG signals.

2 Materials and Methods

2.1 ECG Dataset

Two groups of ECG datasets were used in this study: a local ECG database and the PTB diagnostic ECG database.

There are a total of 90927 records in our local database. All records were 12-lead 10 s ECG raw data collected by the GE Marquette equipment. For the sampling frequency of the ECG signals, there were 250 Hz and 500 Hz. Those signals with the sampling frequency of 500 Hz were resampled to 250 Hz. Doctors made a clinical diagnosis for all ECG records. These clinical diagnosis opinions included ECG abnormalities such as ventricular premature beats, atrial fibrillation, and MIs. The doctors also marked whether each lead presented MICHs, but they did not mark MICHs in lead aVR. We screened the clinical diagnosis opinions and selected 1146 cases of MI records. One hundred and twenty MI records and 100 non-MI records were selected from the database. These 220 records were used as a test dataset in this study. In some records, there are multiple MI locations in each single record, containing a total of 275 MIs; for these records, we asked cardiologists to review the record. The remaining records were used to train eleven 1D CNNs (MICHs vs non-MICH) for each lead, except lead aVR. Considering issues such as the balance of sample types in each lead, the number of training set, verification set and test set for each lead was finally determined, as shown in Table 1. For each lead, the ratio of the number of the training set to the number of the verification set was 3:2. The training, validation and test sets were completely independent. The validation set was used to perform hyperparameter tuning of deep neural networks. The test set was used to test the generalization performance of the CNN model.

Table 1. Number of cases in the training, validation and test sets for 11 leads.

	V1	V2	V3	V4	V5	V6	I	aVL	II	aVF	III
Training set	14600	13000	9800	6000	3800	3000	2200	2200	12200	20000	20400
Validatoin set	1825	1625	1225	750	475	375	275	275	1525	2500	2550
Test set	1460	1300	980	600	380	300	220	220	1220	2000	2040

The PTB database has been widely used for investigating MI detection. There were 148 MI patients and 52 normal subjects in the PTB database. A total of 103 cases with inferior, lateral, anterior, anterolateral, and anteroseptal MIs were included in this study, while the remaining 45 cases with infero-posterior, postero-latera, posterior, or infero-postero-latera MIs were not included. The PTB database contains 1 to 7 ECGs per patient. In this study, we only used those ECGs obtained within the first week after MI. The first 30 s ECG data were used for obtaining median complexes. Table 2 shows the statistics of the local and PTB datasets for testing MI location.

Table 2. Statistics of test sets for MI location.

MI location	Local dataset	PTB dataset
Inferior MI	40	37
Lateral MI	40	1
Anterolateral MI	15	18
Anterior MI	40	27
Anteroseptal MI	40	20
Non-MI (normal)	100	52
Total	275	155

2.2 Extraction of Median Complexes

We first extracted the median complex from the 10 s ECG. The median complex retains the characteristics of the ECG waveform morphology and can remove interference. The extraction steps of median complexes are described as follows.

QRS Detection. The Pan-Tompkins QRS detection algorithm was employed for locating QRS complexes of each lead [19]. To improve the reliability of detected QRS complexes, a method by Chen et al. [20] which combined QRS locations of 12 leads was used to determine the final QRS fiducial mark $qrs_n, n = 1, 2, \ldots, N$, where N is the number of beats.

Beats Grouping. A template matching method by Hamilton [21] was used to group beats by morphology. The segment data S_n, $S_n \in \mathbb{R}^{100 \times 12}$ around qrs_n was extracted as

$$S_n = \left[X(qrs_n - 200\,\text{ms}), \ldots, X(qrs_n + 200\,\text{ms}) \right] \tag{1}$$

where $X \in \mathbb{R}^{2500 \times 12}$ is the raw ECG data. The correlation coefficient was defined as a criterion for the similarity of two beats:

$$\rho_{n,m} = \frac{Cov(S_n, S_m)}{\sqrt{D(S_n) \times D(S_m)}} \tag{2}$$

where $Cov(.)$ is the covariance operator and $D(S_n)$ is the variance of S_n. The steps of the template matching method are shown in Algorithm 1.

Algorithm 1. Beats grouping algorithm.

1. Initialize the number of types $M = 0$.
2. Define array $[T_1, T_2, \ldots T_{Mmax}]$ to store the templates of all types.
3. **For** all segment data $S_n, n \in [1, N]$
4. Calculate the $\rho_{n,m}$ between S_n and the template of all types $T_m, m = 1, 2, \ldots M$.
5.   **If** for all $m = 1,2 \ldots M, \rho_{n,m} < thr$
6.   Add a new template, and M++
7.   The type of nth beat $G_i = M$
8.   **Endif**
9.   **If** there is only one template T_{m_0} that meets the conditions $\rho_{i,m_0} > thr$
10.   The type of nth beat $G_n = m_0$
11.   **Endif**
12.   **If** there are more than one template that meet the conditions $\rho_{n,m} > thr$, $m = m_0, m_1, \ldots$
13.   Combine the templates $m_0, m_1, \ldots$, and $G_n = m_0$
14.   **Endif**
15. **Endfor**

After steps of template matching, $G_n \in [1, 2, \ldots M]$ was obtained as the type of each beat, where M is the number of beat types in the record.

Beat Group Alignment. For each type of beats, an alignment operation was performed by

$$\max_{t_0}\{Cov(S_n(t), S_m(t - t_0))\}, \quad -50\,\text{ms} < t_0 < 50\,\text{ms} \tag{3}$$

where $S_n(t)$ and $S_m(t)$ are two beats in a same group. The time shift t_0 was found which maximized the correlation coefficient. The QRS fiducial mark was then corrected by $qrs_m = qrs_m - t_0$.

Median Complex Extraction. Firstly, we selected the primary beat group. This selection does not depend on the number of beats per beat type. More specifically, for analysis, the beat type with the largest amount of information is a popular beat type, and any beat type with three or more complexes can meet the conditions. After selecting the main beat type, each related beat is used to generate an intermediate complex for each lead. Then, a representative complex is generated using the median voltage of an aligned set of beats. In this study, -400 to 600 ms around qrs_n was extracted. The median complex was a matrix of 12×250.

Figure 1 shows the flow chart of median complex extraction, where the ECG signals came from an inferior MI record. The third beat was a premature ventricular contraction, and was grouped as type 1. The other beats were grouped as type 0. Beats of type 0 were selected as primary beats. An alignment and median operation was conducted in the primary beats to obtain the final median complex. The median complex shows abnormal Q wave appearance in leads II, III and aVF.

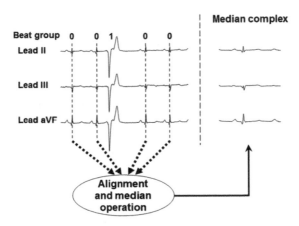

Fig. 1. Flow chart of median complex extraction.

2.3 Determination of MICH Presence

Eleven 1D CNNs were trained to determine whether each lead presented MICHs. The 11 $\times$ 1 output vector (MICH vector) was used for MI location. Lead aVR did not contribute to the MI location, so it was excluded from our analysis [4].

CNNs. Median complexes of each leads (1×250) were used as the input of CNNs. Table 3 presents the architecture of the CNN classifiers used in this study, which were constructed by four convolutional blocks and one fully connected layer. Each convolutional block contained a 1D convolutional layer, a batch normalization layer, a rectified linear unit (ReLU) layer, and a 1D max-pooling layer. The filter size of the four 1D convolutional layers were 32, 64, 128, 256, respectively; the kernel size was 11. All the max-pooling layers had a pooling size of 2. The softmax activation function was used for the output layer.

Table 3. Architecture of CNN classifiers.

Layers	Type	Output shape
0	Inputs	(1, 250)
1−4	Convolutional block	(32, 120)
5−8	Convolutional block	(64, 55)
9−12	Convolutional block	(128, 23)
13−16	Convolutional block	(256, 7)
17	Flattened	1792
18	Dropout 50%	1792
19	Full connected	2
20	Softmax output	2

Classifiers with Hand-Crafted Features. In order to compare the performance of our deep learning classifier, the traditional classifier method with manual features is also tested. Eight characteristic parameters (QRS, Q and R amplitude and duration, T amplitude and Q/R ratio) were extracted from 12 leads, and a total of 96 morphological features were obtained. Then, the Minnesota Code method was used to locate the MIs [4].

Location of MIs. The current ECG standards for diagnosing MIs require that MICHs be present in 2 or more contiguous leads. Table 4 show that the relationship between heart location and leads. The chest leads V1 through V6 are in contiguous order from right anterior (V1) to left lateral (V6); for the limb leads from left superior-basal to right inferior, the contiguous order should be aVL, I, − aVR (i.e., lead aVR with reversed polarity), II, aVF, and III. Abnormal Q waves in leads V1 and V2 are related to septal wall MIs. Those in V3 and V4 are related to anterior wall MIs. Those in V5 and V6, I, and aVL are related to lateral wall MIs. Those in II, III, and aVF are related to inferior wall MIs. Similar considerations may be applied for ECG location of ST-segment deviation. Therefore, Fig. 2 show that how the MICH vector used to locate MIs (Table 4).

Table 4. Relationship between heart location and leads.

Location	Lead
Inferior	II, aVF, III
Lateral	I, -aVL, V5, V6
Anterolateral	V3, V4, V5, V6
Anterior	V3, V4
Anteroseptal	V1, V2

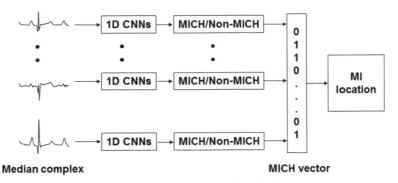

Fig. 2. The MICH vector used to locate MIs

3   Results

3.1   Classification of MICHs vs Non-MICH

Figure 3 shows the accuracy curves during the training process of 11 leads. It can be seen that the CNNs of each lead are effectively learning, and the final model is also in a good state. In this paper, the F1 score [22] was used to evaluate the performance in classification of MICHs vs non-MICH. Table 5 shows the F1 scores of each lead by the proposed CNN method, in comparison with the traditional hand-crafted method. The average F1 scores of the traditional and proposed methods were 71.32% and 94.28%, respectively. This implies that the proposed CNN method is more effective than traditional hand-crafted method in identifying the presence of MICHs.

3.2   MI Location

With the results of CNNs' discrimination of each lead and the discrimination method, we located MIs for the local and PTB datasets. The confusion matrices are shown in Fig. 4.

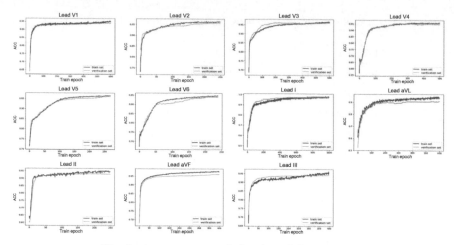

Fig. 3. Accuracy curves during the training process.

Table 5. F1 scores of the test result of the proposed CNN method and the traditional hand-crafted method.

Lead	Hand-crafted method (%)	The proposed CNN method (%)
V1	87.87	93.06
V2	83.00	95.12
V3	77.38	95.62
V4	77.33	95.19
V5	65.97	93.13
V6	58.49	93.65
I	63.35	98.17
aVL	77.16	92.24
II	53.98	91.20
aVF	60.10	95.33
III	79.89	94.38
Average	71.32	94.28

The F1 scores of MI location for the local and PTB datasets are shown in Table 6. For binary classification task (MI vs non-MI), our method achieved Sen = 94.2%, Spe = 90.0%, and Acc = 92.6% for the local dataset, and Sen = 91.2%, Spe = 90.4%, and Acc = 90.9% for the PTB dataset.

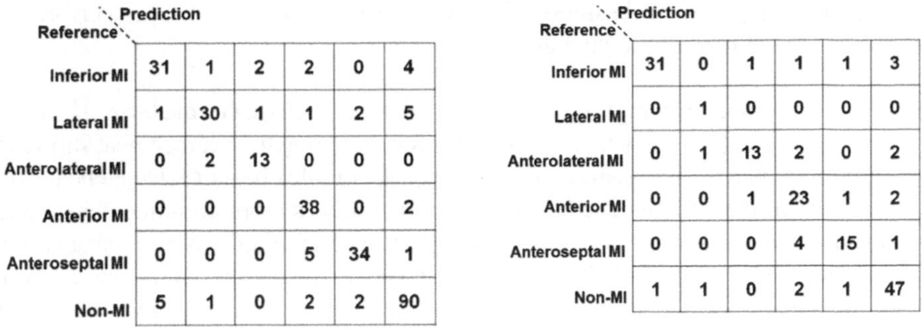

Prediction / Reference — local dataset (a):

Reference \ Prediction						
Inferior MI	31	1	2	2	0	4
Lateral MI	1	30	1	1	2	5
Anterolateral MI	0	2	13	0	0	0
Anterior MI	0	0	0	38	0	2
Anteroseptal MI	0	0	0	5	34	1
Non-MI	5	1	0	2	2	90

(a)

Prediction / Reference — PTB dataset (b):

Reference \ Prediction						
Inferior MI	31	0	1	1	1	3
Lateral MI	0	1	0	0	0	0
Anterolateral MI	0	1	13	2	0	2
Anterior MI	0	0	1	23	1	2
Anteroseptal MI	0	0	0	4	15	1
Non-MI	1	1	0	2	1	47

(b)

Fig. 4. Confusion matrices for the local (a) and PTB datasets (b).

Table 6. F1 scores of MI location for the local and PTB datasets.

MI location	Local dataset (%)	PTB dataset (%)
Inferior MI	0.805	0.899
Lateral MI	0.846	0.667
Anterolateral MI	0.813	0.813
Anterior MI	0.857	0.780
Anteroseptal MI	0.850	0.789
Non-MI	0.905	0.879
Average	0.846	0.804

4 Discussion and Conclusions

In recent years, several researchers have proposed different techniques using the PTB database to identify patients with MIs and used the subject-based method to evaluate the performance. Most of these studies implemented the binary classification task (MI vs non-MI). Keshtkar et al. [23] proposed a method based on wavelet transformed ECG signals and probabilistic neural networks to detect MIs, achieving Sen = 93%, Spe = 86%, and Acc = 89.5%. Bakul et al. [24] developed a set of features called relative frequency band coefficient to identify MIs automatically, with Sen = 85.57%, Spe = 83.97%, and Acc = 85.23%. Correa et al. [25] developed a set of features including five depolarization and four repolarization indices to detect MIs, achieving Sen = 95.8%, Spe = 94.2%, and Acc = 95.25%. Liu et al. [18] proposed a MFB-CBRNN method for MI detection, with Sen = 94.42%, Spe = 86.29% and Acc = 93.08%. However, cross-database performance of these methods have not been investigated. In this study, we trained the CNN models by using a local dataset, and tested the trained models by using the local and PTB datasets. Our method for binary classification task achieved Sen = 91.2%, Spe = 90.4% and Acc = 90.9% in the PTB database. The cross-database performance implies the robustness of the proposed CNN method. This performance may be attributed to the following aspects.

1) We used the median complex wave instead of the original ECG waveform. The results of Reddy et al. [26] show that the program for analyzing the average beat shows less variability than the program for measuring each complex beat or selected beat, while the noise of the intermediate beat is less, and produces more accurate measurement results than the analysis of the original beat. The median complex preserves the morphological characteristics of the waveform, reduces the data dimension and eliminates the noise interference. In addition, it may be helpful to automatically analyze other abnormal ECG forms, such as left bundle branch block, right bundle branch block and left ventricular hypertension.

2) Unlike other studies, we did not directly train different types of MIs, but we let the CNNs learn whether each lead presented MICHs. This discrimination method was more consistent with the doctors' clinical experience. At the same time, the CNN models of this two-category task is relatively simple, and it is not prone to problems such as over-fitting.

3) The use of 1D CNNs avoided the manual extraction of features. The extraction of hand-crafted features often brings errors, resulting in a decline in the classification performance.

4) There are some limitations in this work. Firstly, the size of test datasets is small, and the performance in more test datasets remains to be verified. Secondly, although 5 locations of MIs have been classified, there are some other locations of MIs in the clinics which have not been included in this study.

5) In conclusion, we proposed a new method based on CNNs for MI detection and location in ECG signals. Six types of MI location were accomplished by the proposed method, including inferior, lateral, anterolateral, anterior, and anteroseptal MIs, and non-MI. The CNN method achieved satisfying cross-database performance in detecting and locating MIs.

References

1. Thygesen, K., Alpert, J.S., Jaffe, A.S., et al.: Fourth universal definition of myocardial infarction (2018). J. Am. Coll. Cardiol. **72**(18), 2231–2264 (2018)
2. Das, M.K., Khan, B., Jacob, S., Kumar, A., Mahenthiran, J.: Significance of a fragmented QRS complex versus a Q wave in patients with coronary artery disease. Circulation **113**(21), 2495–2501 (2006)
3. Kligfield, P., Gettes, L.S., Bailey, J.J., et al.: Recommendations for the standardization and interpretation of the electrocardiogram: part I: the electrocardiogram and its technology a scientific statement from the American heart association electrocardiography and arrhythmias committee, council on clinical cardiology; the American college of cardiology foundation; and the heart rhythm society endorsed by the international society for computerized electrocardiology. J. Am. Coll. Cardiol. **49**(10), 1109–1127 (2007)
4. Prineas, R.J., Crow, R.S., Zhan, Z.M.: The Minnesota Code Manual of Electrocardiographic Findings. Springer, London (2010). https://doi.org/10.1007/978-1-84882-778-3

5. Bozzola, P., Bortolan, G., Combi, C., Pinciroli, F., BroHet, C.: A hybrid neuro-fuzzy system for ECG classification of myocardial infarction. In: Computers in Cardiology 1996. IEEE (1996). https://doi.org/10.1109/CIC.1996.542518

6. Ouyang, N., Ikeda, M., Yamauchi, K.: Use of an artificial neural network to analyse an ECG with QS complex in V1–2 leads. Med. Biol. Eng. Comput. 35(5), 556–560 (1997)

7. Arif, M., Malagore, I.A., Afsar, F.A.: Detection and localization of myocardial infarction using K-nearest neighbor classifier. J. Med. Syst. 36(1), 279–289 (2012)

8. Kumar, M., Pachori, R., Acharya, U.: Automated diagnosis of myocardial infarction ECG signals using sample entropy in flexible analytic wavelet transform framework. Entropy 19(9), 488 (2017)

9. Acharya, U.R., Fujita, H., Sudarshan, V.K., et al.: Automated detection and localization of myocardial infarction using electrocardiogram: a comparative study of different leads. Knowl. Based Syst. 99, 146–156 (2016)

10. Rajpurkar, P., Hannun, A.Y., Haghpanahi, M., Bourn, C., Ng, A.Y.: Cardiologist-level arrhythmia detection with convolutional neural networks. arXiv preprint arXiv:1707.01836 (2017)

11. Lodhi, A.M., Qureshi, A.N., Sharif, U., Ashiq, Z.: A novel approach using voting from ECG leads to detect myocardial infarction. In: Arai, K., Kapoor, S., Bhatia, R. (eds.) IntelliSys 2018. AISC, vol. 869, pp. 337–352. Springer, Cham (2019). https://doi.org/10.1007/978-3-030-01057-7_27

12. Lui, H.W., Chow, K.L.: Multiclass classification of myocardial infarction with convolutional and recurrent neural networks for portable ECG devices. Inform. Med. Unlocked 13, 26–33 (2018)

13. Acharya, U.R., Fujita, H., Oh, S.L., Hagiwara, Y., Tan, J.H., Adam, M.: Application of deep convolutional neural network for automated detection of myocardial infarction using ECG signals. Inf. Sci. 415, 190–198 (2017)

14. Liu, W., Zhang, M., Zhang, Y., et al.: Real-time multilead convolutional neural network for myocardial infarction detection. IEEE J. Biomed. Health Inform. 22(5), 1434–1444 (2018)

15. Liu, W., Huang, Q., Chang, S., Wang, H., He, J.: Multiple-feature-branch convolutional neural network for myocardial infarction diagnosis using electrocardiogram. Biomed. Sig. Process Control 45, 22–32 (2018)

16. Goldberger, A.L., Amaral, L.A., Glass, L., et al.: PhysioBank, PhysioToolkit, and PhysioNet: components of a new research resource for complex physiologic signals. Circulation 101(23), E215–E220 (2000)

17. Sharma, L.D., Sunkaria, R.K.: Inferior myocardial infarction detection using stationary wavelet transform and machine learning approach. SIViP 12(2), 199–206 (2017). https://doi.org/10.1007/s11760-017-1146-z

18. Liu, W., Wang, F., Huang, Q., Chang, S., Wang, H., He, J.: MFB-CBRNN: a hybrid network for MI detection using 12-lead ECGs. IEEE J. Biomed. Health Inform. 24(2), 503–514 (2020)

19. Pan, J., Tompkins, W.J.: A real-time QRS detection algorithm. IEEE Trans. Biomed. Eng. 32(3), 230–236 (1985)

20. Chen, G., Chen, M., Zhang, J., Zhang, L., Pang, C.: A crucial wave detection and delineation method for twelve-lead ECG signals. IEEE Access 8, 10707–10717 (2020)

21. Hamilton, P.: Open source ECG analysis. In: Computers in Cardiology. IEEE (2002). https://doi.org/10.1109/CIC.2002.1166717

22. Liu, F., Liu, C., Zhao, L., et al.: An open access database for evaluating the algorithms of electrocardiogram rhythm and morphology abnormality detection. J. Med. Imag. Health Inform. 8(7), 1368–1373 (2018)

23. Keshtkar, A., Seyedarabi, H., Sheikhzadeh, P., Rasta, S.H.: Discriminant analysis between myocardial infarction patients and healthy subjects using wavelet transformed signal averaged electrocardiogram and probabilistic neural network. J. Med. Sig. Sens. 3(4), 225–230 (2013)

24. Bakul, G., Tiwary, U.S.: Automated risk identification of myocardial infarction using relative frequency band coefficient (RFBC) features from ECG. Open Biomed. Eng. J. **4**, 217–222 (2010)
25. Correa, R., Arini, P.D., Correa, L.S., Valentinuzzi, M., Laciar, E.: Identification of patients with myocardial infarction. Methods Inf. Med. **55**(3), 242–249 (2016)
26. Reddy, B.R., Xue, Q., Zywietz, C.: Analysis of interval measurements on CSE multilead reference ECGs. J. Electrocardiol. **29**(Suppl), 62–66 (1996)

Sustainable Pervasive WSN Applications

Control Model Design for Monitoring the Trust of E-logistics Merchants

Susie Y. Sun[✉] [iD]

Wenhua College, Wuhan 430074, China
sunyi@whc.edu.cn

Abstract. In order to improve the subjectivity bias of the traditional autocorrelation function analysis method, this paper tries to introduce the mutual correlation criterion to establish the asymptotic control model in the e-logistics trust degree control application. By pre-constructing the resource database structure model of e-logistics, we adopt the DOI mutual correlation criterion to describe the user's trust degree evaluation of resources, and then form the user trust Simulation experiments show that the model has a good asymptotic control performance on the trust degree of e-logistics with accurate trust evaluation and high estimation accuracy. The decision making approach based on the mutual correlation criterion, in which two or more users jointly make the normalized evaluation of the mutual trust value model, can effectively improve the traditional model autocorrelation active selection bias. The new model realizes the progressive control of e-logistics trust degree based on the mutual correlation criterion, which can significantly improve the supervision of e-logistics enterprises.

Keywords: Electronic logistics · Cloud computing · Control model

1 Introduction

With the strong development of network transactions and e-commerce logistics industry, the information and data of users of network transactions of e-commerce logistics are expanding, along with the expanding information field of e-commerce logistics and the expanding information space, how to extract the information that users care from the massive information and improve the evaluation performance of merchants has become a research topic of concern [1]. The problem of accurate and effective evaluation algorithms for trust in online transactions is studied. In the open and complex network environment, factors such as randomness and ambiguity in the transaction process through the network are unpredictable, and the traditional evaluation mechanism does not make accurate judgment and quantitative assessment of them. In an open and complex network environment, buyers and sellers choose each other through a virtual network platform. For example, in Taobao, where the number of online transactions is powerful, buyers choose whether a merchant can fulfill their promises based on their needs and the reputation of similar merchants. Likewise, sellers evaluate buyers who have chosen their goods based on their trustworthiness. It is necessary to control and evaluate the trust

© The Author(s) 2022
Z. Qian et al. (Eds.): WCNA 2021, LNEE 942, pp. 1033–1040, 2022.
https://doi.org/10.1007/978-981-19-2456-9_103

degree of each other, and to improve the quantitative assessment performance of merchants by designing a logistic trust degree progressive control model for e-commerce and conducting trust degree ratings of online physical objects [2].

Traditional models have solved the trust assessment methods and evaluation degree calculation of buyers and sellers in online transactions to varying degrees under certain application conditions, but with the popularity of the Internet, the increase in online users, and the increase in user satisfaction, many shortcomings have emerged in the specific application of these models [3]. For example, in the open and complex online environment, the randomness of the communication between buyers and sellers in the process of purchasing products and the unpredictability of whether the transaction between merchants and buyers can be carried out smoothly are uncertainties that cannot be accurately predicted when using the knowledge of probability theory for estimation, and if there are malicious buyers or sellers who deliberately break the trust degree by making false evaluations, the evaluation mechanism cannot If there are malicious buyers or sellers who deliberately break the trust, the evaluation mechanism cannot make a definite judgment and eliminate them. At the same time, the above model does not give different trust degree evaluation mechanisms according to the different characteristics of entities, and lacks some flexibility [4, 5]. It can be seen that the traditional e-logistics trust degree control model adopts the model design method of autocorrelation function analysis, and the evaluation effect is not good due to the large subjectivity of autocorrelation feature analysis [6–8]. In response to the above problems, this paper proposes a progressive control model of trust degree of e-logistics based on inter-correlation criterion. Firstly, the resource database structure model of e-logistics is constructed, and based on the mutual correlation criterion, the e-logistics user recommendation model construction and network trust degree control model are carried out to realize the algorithm improvement, and the simulation experiment is carried out to demonstrate its superior performance by performance test.

2 Resource Database Structure Model and Trust Influence Parameters for E-logistics

2.1 Resource Database Structure Model for E-logistics

Design the resource database structure model of e-logistics based on cloud computing, and set the query history of e-logistics resource database users as $W = \{w_1, \ldots, w_p\}$. The query pattern $\sigma(W)$ is a two-dimensional matrix of $p \times p$. For $1 \leq i, j \leq p$, the cascade layer depth is $N_k (k = 0, 1, \ldots, L)$, denotes the number of k-layer data connections data target position location state estimation vector is

$$\alpha = (\alpha_1, \alpha_2, \ldots, \alpha_n) \neq 0 \tag{1}$$

Denote by $W_{ij}^{(k)}$, the connection weight of the jth layer of k, $x_i^{(k)}$ is the i ($i = 1, 2, \ldots, N_k$) input vector of the hidden data set in the e-logistics database. Denote the linear input and reversible invariant output of the e-logistics trustworthiness evaluation system by $s_j^{(k)}$ and $y_j^{(k)}$, Expressed as an eigenvector as:

$$x^k = \left[x_1^{(k)}, x_2^{(k)}, \ldots x_{N_{k-1}}^{(k)} \right]^T \tag{2}$$

$$s^{(k)} = \left[s_1^{(k)}, s_2^{(k)}, \ldots s_{N_k}^{(k)} \right]^T \tag{3}$$

$$y^{(k)} = \left[y_1^{(k)}, y_2^{(k)}, y_{N_k}^{(k)} \right]^T \tag{4}$$

The power spectrum optimized allocation probability density function of the resource database obtained by grid assignment (Pr):

$$\left| Pr\left(Real_{\sum,A}(k) = 1\right) - Pr\left(Simulator_{\sum,A,Sim}(k) = 1\right) \right| \leq negl(k) \tag{5}$$

In e-commerce transactions, if both parties agree, the transaction can be carried out smoothly, and the buyer evaluates the seller according to the merchant's various service attitudes, and the evaluation can be converted into the merchant's reputation to facilitate the smooth conduct of the next transaction. The above process shows that only the mutual trust mechanism between buyers and sellers can ensure the smooth transaction in the virtual network environment. The above process realizes the construction of the resource database structure model of e-logistics and lays the foundation for the progressive control of trust degree.

2.2 E-logistics Trust Influence Parameters and Cloud Preprocessing

The main parameters influencing the trust in e-logistics are: the credibility of the evaluator, the historical evaluation value accumulated by the merchant, and the price of the transacting entity. Due to the unpredictability factors such as randomness and ambiguity of the buyer and seller in conducting the transaction process, and also if there is a deliberate breaking of trust by the user, the existing evaluation mechanism does not make an exact judgment on it. Trust between subjects includes both direct trust and indirect trust. Direct trust is obtained by the subject based on his own experience, assuming the existence of n evaluations of the evaluated goods, corresponding to m characteristic attributes. If each evaluation is considered as a cloud factor, then m trust attribute clouds are obtained using the trust attribute inverse growth cloud algorithm. The data set contains n samples for n uncorrelated independent vectors, let the range e-logistics data value domain for N discrete points $A = \{a_1, \ldots, a_N\}$, and meet $a_1 < a_2 < \ldots < a_N$. The set X is divided into class c and the set of subscripts is assigned:

1) $V_1 = \{> a_1, > a_2, \ldots, > a_{N-1}\}$
2) $V_2 = \{\geq a_1, \geq a_2, \ldots, \geq a_N\}$
3) $V_3 = \{< a_1, < a_2, \ldots, < a_N\}$
4) $V_4 = \{\leq a_1, \leq a_2, \ldots, \leq a_N\}$
5) $V_5 = \{= a_1, = a_2, \ldots, = a_N\}$

Suppose U is a quantitative domain and C is a qualitative concept in U. When the quantitative value x is a random realization in the qualitative concept C and the degree of certainty $\mu(x) \in [0, 1]$ of x with respect to C is a stable random number, then the distribution of x over the quantitative domain U is called a cloud, denoted as $C(X)$. Each x is called a cloud droplet. Where the cloud droplet is described quantitatively by a standard normal function. The cloud model is described by a large number of quantitative values with certainty for qualitative quantities, and it mainly utilizes forward and inverse cloud generators for interconversion of qualitative and quantitative concepts. Suppose that U is a topological quantitative domain of trust data of God's network and C is a qualitative concept in U. When the quantitative value x is a random realization of the qualitative concept C in, The determinacy of x with respect to C, $\mu(x) \in [0, 1]$ is a stable random number. Through the above processing, the correlation analysis and cloud pre-processing of the parameters influencing the trust degree of e-logistics are realized to provide an accurate data base for conducting the trust degree of e-logistics.

3 Improvement of Trust Degree Asymptotic Control Model Based on Mutual Correlation Criterion

On the basis of the above model design, algorithm improvement is carried out, and the superior traditional e-logistics trust degree control model adopts the model design method of autocorrelation function analysis, which is more subjective in autocorrelation feature analysis and has poor evaluation effect. In this regard, this paper proposes a progressive control model of trust degree of e-logistics based on the inter-correlation criterion.

The DOI (Degree of Interest) intercorrelation criterion is used to describe the user's trust evaluation of the resource, and the posterior probability of successful negotiation between two subjects for the n + 1th time follows a Beta distribution.

$$P_{a+1} = E(Beta(P|a + 1, n - a + 1)) = \frac{a + 1}{n + 2} \tag{6}$$

Let the mutual correlation function weight function be U, where $\sum u = 1$. The trust relationship model between e-logistics user A and user B, where I_a is the resource identifier of user A. The following must be satisfied by network users for e-logistics A products:

$$v - p_1 + \rho_1 A_1 \geq 0 \tag{7}$$

$$v - p_1 + \rho_1 A_1 \geq \delta \cdot v - p_2 + \rho_2 A_2 \tag{8}$$

$$U = \begin{cases} v \geq p_1 - \rho_1 A_1 \\ v \geq \frac{p_1 - p_2 + \rho_2 A_2 - \rho_1 A_1}{1 - \delta} \end{cases} \tag{9}$$

Based on the mutuality criterion, a consumer who chooses logistics product B must satisfy:

$$\delta \cdot v - p_2 + \rho_2 A_2 \geq 0 \tag{10}$$

The above equation tabulates the rating of resource i by users A, B in the user trust network control system. The indirect trust relationship between users is obtained denoted as $A \rightarrow B, B \rightarrow C$. Launching:

$$MSD_{a \rightarrow b} = 1 - \frac{\sum\limits_{i=1}^{|I_{a,b}|} \sqrt{(d_{a,i} - \overline{d}_a)^2 + (d_{b,i} - \overline{d}_b)^2}}{|I_{a,b}| \times \sum\limits_{i=1}^{|I_{a,b}|} \left[\sqrt{(d_{a,i} - \overline{d}_a)^2} + \sqrt{(d_{b,i} - \overline{d}_b)^2} \right]} \tag{11}$$

The randomness of buyers in the process of shopping for goods and communication with sellers, merchants and buyers to conduct transactions between them meet the following constraints:

$$v - p_1 + \rho_1 A_1 < \delta \cdot v - p_2 + \rho_2 A_2 \tag{12}$$

That is:

$$U = \begin{cases} v \geq \frac{p_2 - \rho_2 A_2}{\delta} \\ v < \frac{p_1 - p_2 + \rho_2 A_2 - \rho_1 A_1}{1 - \delta} \end{cases} \tag{13}$$

Users trust the rating of resource i by users A, B in the network control system, and if there are malicious buyers or sellers who make false ratings to deliberately break the trust level, there are:

$$p_2 - \rho_2 A_2 \geq \delta \cdot (p_1 - \rho_1 A_1) \tag{14}$$

At this point, the market only has demand for product A; when the following inequality is satisfied:

$$p_2 - \rho_2 A_2 \leq p_1 - \rho_1 A_1 - Q(1 - \delta) \tag{15}$$

In the above equation, $w(k) \in R^n$ the expert rating results in an unknown perturbation in the finite energy local range. When:

$$\delta \cdot (p_1 - \rho_1 A_1) \leq p_2 - \rho_2 A_2 \leq p_1 - \rho_1 A_1 - Q(1 - \delta) \tag{16}$$

The asymptotic coefficients of user trust evaluation $\gamma > 0$, if there exist positive definite symmetric matrices Q, S, M, the asymptotic control solutions of e-logistics trust degree are:

$$\frac{p_1 - p_2 + \rho_2 A_2 - \rho_1 A_1}{1 - \delta} \leq v \leq Q \tag{17}$$

$$\frac{p_2 - \rho_2 A_2}{\delta} \leq v \leq \frac{p_1 - p_2 + \rho_2 A_2 - \rho_1 A_1}{1 - \delta} \tag{18}$$

In the above equation, $Trust_{a \rightarrow b}$ represents the trust weight value of target user A to user neighbor B. The use of using TW to increase the number of similar users in the traditional collaborative filtering recommendation method produces an uncertain time lag due to the high number of similar users in the trust network model, At this time, two users jointly make a normalized evaluation of each other's trust value model and construct a user trust assessment mechanism and network control model. This realizes the progressive control of e-logistics trust based on the mutual correlation criterion and improves the management and control benefits for e-logistics merchants.

4 Simulation Experiments and Results Analysis

In order to test the performance of the algorithm in this paper in achieving the progressive control of trust in e-logistics, simulation experiments are conducted. Experimental environment: Myeclipse 8.0 experimental simulation platform and Java platform development language and combined with swarm program package. According to the analysis, the e-logistics network trading merchants receive customer orders, through the multi-subject negotiation, the subject respectively in accordance with their role in the merchant and the sector in which they are synergistically play their role, together to serve the business objectives. The trust level of network information is modeled according to the index system described in the previous section and divided into five levels, A, B, C, D and E. The user trust perception model uses the trust level evaluation of network information on C2C websites as the index system. Suppose there are trust attribute clouds TPC_1, TPC_2 and their mathematical properties are Ex_1, En_1, He_1, Ex_2, En_2, He_2 respectively. using the algorithm of this paper, the response output of the mutual correlation function of e-logistics users is calculated as shown in Fig. 1.

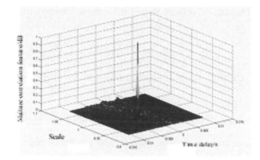

Fig. 1. e-Logistics user correlation function response output

As seen in Fig. 1, the algorithm of this paper is used for the mutual correlation function feature analysis, based on the mutual correlation criterion, the feature extraction accuracy is high, and the estimation performance of the trust degree of e-logistics is superior, for the trust attribute cloud TPC1 generates a normal random number W_1 with En_1, $He_1{}^2$ as variance, and the trust value is calculated as 6.2 by the division of the trust interval with low confidence as [3.5–6.5]. In order to compare the performance of the algorithm, the simulation experiment of the progressive control accuracy of the trust degree of e-logistics is carried out using the algorithm of this paper and the traditional algorithm, and the results are obtained as shown in Fig. 2.

In Fig. 2, assuming that the historical trust degree and the current trust degree are weighted half each, since the trust degree of the previous evaluation is 6, then the trust degree of the network transaction of this e-logistics merchant is 6 × 50% + 6.2 × 50% = 6.1. Comparing the control accuracy of this paper's algorithm and the traditional algorithm, we get that this paper's algorithm has better asymptotic control performance, accurate evaluation and higher estimation accuracy.

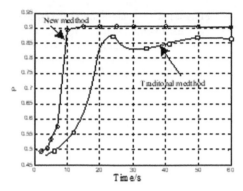

Fig. 2. Progressive control accuracy of e-logistics trust degree

5   Conclusion

The design of the logistic trust degree progressive control model for e-commerce is used to carry out the trust degree rating of network entity objects and improve the quantitative evaluation performance of merchants. The traditional e-logistics trust degree control model uses the model design method of autocorrelation function analysis, which is not effective in evaluation due to the large subjectivity of autocorrelation feature analysis. A progressive control model of trust degree of e-logistics based on inter-correlation criterion is proposed. Firstly, the resource database structure model of e-logistics is constructed, and based on the mutual correlation criterion, the e-logistics user recommendation model is constructed and the network trust degree control model is implemented to improve the algorithm, and the simulation experiments show that the algorithm in this paper has good asymptotic control performance, accurate evaluation and high estimation accuracy. The asymptotic control of e-logistics trust degree based on the mutual correlation criterion is realized to improve the management and control benefits of e-logistics merchants.

References

1. Kanagavalli, G., Azeez, R.: Logistics and e-logistics management: benefits and challenges. Int. J. Recent Technol. Eng. (IJRTE) **8**(4), 12804–12809 (2019)
2. Jaisankar, N., Priya, G.: A fuzzy based trust evaluation model for service selection in cloud environment. Int. J. Grid High Perform. Comput. **11**(4), 13–27 (2019)
3. Park, M.S., Heo, T.Y.: Seasonal spatial-temporal model for rainfall data of South Korea. J. Appl. Sci. Res. **5**(5), 565–572 (2009)
4. Liu, D., Su, Y., et al.: Customer evaluation based trust model in cloud service. Commun. Netw. **42**(9), 99–102 (2016)
5. Zhong, L., Zhang, J., Liang, J.: Multidimension integrated electronic commerce trust model based on interval-valued intuitionstic fuzzy. Comput. Eng. **45**(4), 316–320 (2019)
6. Kamarianakis, Y., Prastacos, P.: Space–time modeling of traffic flow. Comput. Geosci. **31**(2), 119–133 (2005)

7. Halim, S., Bisono, I.N., Sunyoto, D., et al.: Parameter estimation of space-time model using genetic algorithm. In: IEEE International Conference on Industrial Engineering and Engineering Management 2009. IEEM 2009, pp. 1371–1375. IEEE (2009)
8. Teacy, W.T., Luck, M., Rogers, A.: An efficient and versatile approach to trust and reputation using hierarchical Bayesian modelling. Artif. Intell. **193**(12), 149–185 (2012)

Experimental Performance Analysis of Machine Learning Algorithms

Ganesh Khekare[1]([✉]) [iD], Anil V. Turukmane[1] [iD], Chetan Dhule[2] [iD], Pooja Sharma[3] [iD], and Lokesh Kumar Bramhane[4] [iD]

[1] Parul University, Vadodara, India
khekare.123@gmail.com, {ganesh.khekare19325,
anil.turukmane21100}@paruluniversity.ac.in
[2] G H Raisoni College of Engineering, Nagpur, India
[3] Indira College of Engineering and Management, Pune, India
[4] National Institute of Technology, Goa, India
lokesh.bramhane@nitgoa.ac.in

Abstract. Machine Learning models and algorithms have become quite common these days. Deep Learning and Machine Learning algorithms are utilized in various projects, and now, it has opened the door to several opportunities in various fields of research and business. However, identifying the appropriate algorithm for a particular program has always been an enigma, and that necessitates to be solved ere the development of any machine learning system. Let's take the example of the Stock Price Prediction system, it is used to identify the future asset prediction of a industry or other financial aspects traded on a related transaction. Now, it is a daunting task to find the right algorithm or model for such a purpose that can predict accurate values. There are several other systems such as recommendation systems, sales prediction of a mega-store, or predicting what are the chances of a driver meeting an accident based on his past records and the road they've taken. These problem statements require to be built using the most suitable algorithm and identifying them is a necessary task. This is what the system does, it compares a set of machine learning algorithms while determining the appropriate algorithm for the selected predictive system using the required data sets. The objective is to develop an interface that can be used to display the result matrix of different machine learning algorithms after being exposed to different datasets with different features. Besides that, one can determine the most suitable (or optimal) models for their operations, using these fundamentals. For experimental performance analysis several technologies and tools are used including Python, Django, Jupyter Notebook, Machine Learning, Data Science methodologies, etc. The comparative performance analysis of best known five time series forecasting machine learning algorithms viz. linear regression, K – nearest neighbor, Auto ARIMA, Prophet, and Support Vector Machine is done. Stock market, earth and sales forecasting data is used for analysis.

Keywords: Best known machine learning algorithms · Survey · Experimentation · Performance analysis · Stock market prediction · Earth and sales forecasting

Z. Qian et al. (Eds.): WCNA 2021, LNEE 942, pp. 1041–1052, 2022.
https://doi.org/10.1007/978-981-19-2456-9_104

1 Introduction

The system mainly concentrates on machine learning algorithms that are used in predic-tion modeling. Machine learning algorithms are self-programming methods to deliver better results after being exposed to data. The learning portion of machine learning sig-nifies that the models which are build changes according to the data that they encounter over the time of fitting.

The idea behind the building of this system was to determine which one among the chosen time series forecasting algorithms are the most suitable for these operations. The uniqueness of this work is specified using the help of the literature review section of this study. The five algorithms that were chosen are Linear Regression, K-Nearest Neighbor, Auto ARIMA, Support Vector Machine, and Facebook's Prophet, which were never compared altogether on a common platform. Also, several datasets were extracted for building and testing these models, along with the evaluation metrics.

Since the extracted datasets are time-series forecasting types, that's why algorithms that are most suitable for these kinds of works are chosen in this system. The term time series forecasting means that the system is going to make a prediction based on time-series data. Time series data are those where records are indexed on the basis of time, that can be anything like a proper date, a timestamp, quarter, term, or year. In this type of forecasting, the date column is used as a predictor/independent variable for predicting the target value.

A machine learning algorithm builds a model with the help of a dataset by getting trained and tested. The dataset is split into two parts as train and test datasets, and generally, the record of these two do not overlap, and there are different mechanisms around machine learning for this task. After fitting/training the model on the basis of the train portion, it must be tested, and for that, the test dataset comes into play. Further, the results that are generated are matched with the desired targets with the help of evaluation metrics. The two-evaluation metrics viz., the Mean Absolute Percentage Error and the Root Mean Squared Error are considered for comparison purpose is broadly discussed in the Methodology chapter.

2 Literature Review

This section delivers the opinion and conclusion of several researchers who contributed their works to the field of machine learning algorithms. Also, this section manifests the comparative outcomes of the machine learning algorithms.

Vansh Jatana mentioned in his paper Machine Learning Algorithms [1] that Machine Learning is a branch of AI which allows System to train and learn from the past data and activities. Also, it explores a bunch of regression, classification, and clustering algorithms through several parameters including the memory size, overfitting tendency, time for learning, and time for predicting. In the comparison of Random Forest, Boosting, SVM, and Neural Networks, the time for learning is weaker in the case of Linear Regression. Also, like Logistic Regression and Naive Bayes [2], the overfitting tendency of Linear Regression is low. However, in the research Linear regression is the only pure regression model, as else are Classification as well as Clustering model too.

Ariruna Dasgupta and Asoke Nath [3] discuss the broader classification of a prominent machine learning algorithm in their journal and also, specifies the new applications of them. In supervised learning, priori is necessary and always produces the same output for specific input. Similarly, Reinforcement learning requires priori too, but the output changes if the environment doesn't remain the same for a specific result. Nevertheless, Unsupervised Learning doesn't require priori.

Talking about Auto ARIMA, Prapanna Mondal, Labani Shit, and Saptarsi Goswami [4] in their paper carried a study on 56 stocks from 07 divisions. Stocks that are registered in the National Stock Exchange (NSE) are considered. The authors have chosen 23 months of information for the observational research. They've calculated the perfection of the ARIMA model in prediction of stock costs. For all the divisions, the ARIMA model's accuracy in anticipating stock costs is higher than eighty fifths, which symbolizes that ARIMA provides sensible accuracy.

A work by Kemal Korjenić, Kerim Hodžić, and Dženana Đonk [5] evaluates its performance in very real-world use cases. The prophet model has inclinations of generating fairly conventional monthly as well as quarterly forecasts. Also, as an enormous potential for classification of the portfolio into many classes consistent with the expected level of statement authenticity: some five-hundredths of the merchandise portfolio (with large amount of dataset) will be projected with MAPE < 30% monthly, whereas around 70% can be predicted with MAPE < 30% quarterly (out of that 40% with MAPE <15%).

Sibarama Panigrahi and H.S. Behra [6] used FTSF-DBN, FTSF-LSTM, and FTSF-SVM models as comparative algorithms for their Fuzzy Time Series Forecasting (FTSF) in their journal. These Machine learning algorithms are used model FLRs (Fuzzy Logic Relationships [7]. The paper concluded that FTSF-DBN outperformed DBN (Deep Belief Network) method. But it also reported that the statistical difference between FTSF-LSTM and LSTM is insignificant.

Talking about K-Nearest Neighbour (KNN), it has been stated in a paper [8] that KNN as a data mining algorithm has a broad range of use in regression and classification scenarios. It is mostly used for data Mining or data categorization. In Agriculture, it can be applied for simulating daily precipitations and weather forecasts. KNN can be used efficiently in determining required patterns and correlations between data. Along with those other techniques such as hierarchical clustering and k-means, regression models, ARIMA [9], and decision tree analysis can also be applied over this massive field of exploration. Also, KNN [10] can be applied medical field to predict the reason for a patient's admission to the hospital.

In the end, the whole analysis of the different journals published in recent years features a broad perspective of different machine learning algorithms specifically time series and prediction algorithms, that are about to be featured in the implementation of this system. Also, from the above study, it can be concluded that each algorithm belongs to different categories and have significant applications. Further, some of the comparative studies define the best machine learning techniques based on several parameters. Nevertheless, in this whole process of encountering the brilliant works, team never came across any work where five algorithms that they've chosen being compared in on one platform with common dataset, and that's why the team saw this as an opportunity to compare these five algorithms that are different nature but also share some similarities so that they can be used for time series forecasting as well.

3 Methodology

The idea was to create an interface that could display result matrix and multiple analysis with words, numbers, statistics, and pictorial representations. The visual interface created by the team should not deviate from the topic for the audience and should only include limited and necessary items such as what algorithms are used, what dataset are used, their data analysis and respected comparative results. Anyway, the construction of the interface was the ultimate concern in the entire research and system construction campaign.

3.1 Linear Regression

Linear regression [11] is a simplistic and well-known Machine Learning algorithm. It is a mathematical procedure that is applied for the prognosticative analytical study. Simple Linear regression delivers forecasts for continuous or numeric variables like trades, wages, span, goods worth, etc.

Mathematically, it can be represented as shown in "Eq. (1)",

$$y = \theta 0 + \theta 1 \times 1 + \theta 2 \times 2 + \ldots + \theta n \times n \tag{1}$$

Here, y is the target variable and x1, x2, …, xn are predictive variables that represents every other feature in a dataset. $\theta 0$, $\theta 1$, $\theta 2$, …, θn represent the parameters that can be calculated by fitting the model.

In the case of using two variables i.e., 1 independent and 1 dependent variable, it can be represented as shown in "Eq. (2)":

$$y = \theta 0 + \theta 1 x \tag{2}$$

where parameters $\theta 0$ is said to be the intercept that forms on y-axis, and $\theta 1$ can be generated once the model is trained.

3.2 K Nearest Neighbour

K-Nearest Neighbour [12] calculates the similarity among the recent data and recorded cases and sets the new records into the section where alike data exists.

It computes the length between the input and the test data and provides the prognostication subsequently as shown in "Eq. (3)".

$$d(p,q) = d(q,d) = \sqrt{(q_1 - p_1)^2 + (q_2 - p_2)^2 + \ldots + (q_n - p_n)^2}$$
$$= \sqrt{\sum_{i=1}^{n} (q_i - p_i)^2}. \tag{3}$$

The n number of specifications are taken into consideration. The marking that is situated at the merest position from marking is in similar class. Here q and p are new and existing data-points respectively.

3.3 Auto ARIMA

ARIMA [13] is a standard word that refers to Auto-Regressive Integrated Moving Average. It is a mere and efficient ML algorithm used to perform time-series forecasting. It consists of two systems Auto Regression and Moving average.

It takes past values into account for future prediction. There are 3 essential parameters in ARIMA:

p => historical data used for predicting the upcoming data
q => historical prediction faults i.e., used for forecasting the Upcoming data
d => Sequence of variation

3.4 Prophet

The prophet [14] is an open-source library by FB company made for predicting time series data to learn and likely forecast the exchange. Seasonality variations occur over a short duration and aren't notable enough to be described as a trend. The equations related to the terms are defined as shown in "Eq. (4)",

$$fn(t) = g(t) + s(t) + h(t) + e(t) \tag{4}$$

where,

g(t) => trend
s(t) => seasonality
h(t) => forecast effected by holidays
e(t) => error term
fn(t) => the forecast

The variation of the given terms is maths dependent. And if not studied properly it might lead them to make the wrong prediction which may be very problematic to the customer or for business in practice.

3.5 Support Vector Machine

The SVM [15] is a machine learning algorithm that is employed for both regressions and classifications depending upon the enigmas. In Linear SVM, features are linearly arranged [16] that can utilize a simple straight line to implement SVM in this case. The formula for obtaining hyperplane in this case is as shown in "Eq. (5)":

$$y = mx + c \tag{5}$$

If the feature that is being used is of non-linear type, then more dimensions are needed to be added to it. And in that case, one need to use a plane. The formula for obtaining hyperplane in this case is as shown in "Eq. (6)":

$$z = x2 + y2 \tag{6}$$

In this system, to determine the accuracy, 2 evaluation metrics that are used for generating results are Mean Absolute Percentage Error and Root Mean Squared Error, and both depend on the obtained values and actual value.

The Root Mean Squared Error a.k.a. RMSE value is obtained by taking the square root of the addition of the individually calculated mean squared errors. The formula for the same is given in "Eq. (7)":

$$RMSE = \sqrt{\sum_{i=1}^{n} \frac{(\hat{y}_i - y_i)^2}{n}} \tag{7}$$

Here, $\hat{y}1$, $\hat{y}2$, $\hat{y}3$, …, $\hat{y}n$ are the actual value and $y1$, $y2$, $y3$…yn are respective obtained value and n here is the number of iterations performed.

In MAPE or Mean Absolute Percentage Error, the value is calculated by taking absolute subtraction of obtained value from actual value divided by the actual value, later the individual value to obtain the result were added as shown in "Eq. (8)"

$$M = \frac{1}{n} \sum_{t=1}^{n} \left| \frac{A_t - F_t}{A_t} \right| \tag{8}$$

Here, $A1$, $A2$, $A3$, …, An represents actual value, while $F1$, $F2$, $F3$, …, Fn represents the obtained data, and n is the number of iterations taken under consideration.

4   System Design

The design of the whole system depends on the flow of modules. The work is segregated into six modules, and the team developed the whole system going through these six modules that are discussed in this section of the study. Figure number 1 describes the modules and processes that are going to be involved in the long process of implementation of the required interface (Fig. 1).

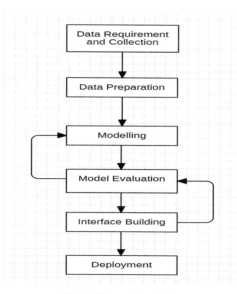

Fig. 1. Flow of modules

4.1 Data Requirements and Collection

In this phase of the whole implementation, the main objective is to understand what kind of datasets are required in the massive process. Understanding the data requirements plays a vital role in upcoming modules in this long process. Further, after understanding the data requirements, the next step is to focus on the collection of the required datasets.

4.2 Data Preparation

This phase of the implementation is the most crucial. It let the implementer determine the bruises in the collected data. To operate with the data, it needs to be developed in a way that inscribes abstaining or fallacious values and eliminates copies and ensures that it is accurately formatted for modeling.

4.3 Modelling

In this module, the implementation of algorithms is done as per the requirement in Python with the help of some Python libraries. It is the phase, that allows to decide how the information can be envisioned to find the solution that is required. All five algorithms which are either predictive or descriptive that are mentioned in the previous section were implemented here.

4.4 Model Evaluation

Model's assessment will probably assess the calculations that are actualized in the past module. It is intended to decide the right logical methodology or strategy to take care of the issue. With the help of RMSE, and MAPE, it can be determined which model is most suitable for a particular time series dataset. The closer the value of RMSE and MAPE towards zero, the better the model for that dataset.

4.5 Interface Building

In this module, the work went under the interface development of the system. Also, the team established a connection between the interface and the models that were implemented in previous phases. Also, as per the requirement, the team can also revert to the fourth phase of the implementation. Django was used as the web-framework for this phase of the implementation.

4.6 Deployment

Once the models are evaluated and the interface is developed, it is deployed and put to the ultimate test. It showed required comparative results and satisfied the objectabletive the team has taken prior to initiating the hands-on working on this system.

5   Results

As per the discussion, the results that need to generate were nothing else but the comparative results of the evaluation metrics value of the respective dataset. First in that trail was the Stock Prediction dataset, and table number 1 describes shows the comparative values for the same (Table 1).

Table 1. Results of stock prediction dataset.

Algorithms	RMSE	MAPE
Linear regression	47.51609	11.32705
K - nearest neighbor	65.11185	16.92529
Auto ARIMA	3.74366	0.72129
Prophet	53.01529	13.01318
Support vector machine	69.81082	12.44615

Table 2. Results of earthquake forecasting dataset.

Algorithms	RMSE	MAPE
Linear regression	0.43306	2.49101
K - nearest neighbor	0.46377	2.86797
Auto ARIMA	0.41603	2.58689
Prophet	0.43047	2.71666
Support vector machine	0.43734	2.78535

Table 3. Results of sales forecasting dataset.

Algorithms	RMSE	MAPE
Linear regression	2.22990	23.76444
K - nearest neighbor	2.35999	24.30888
Auto ARIMA	2.23614	23.97399
Prophet	2.24678	24.12586
Support vector machine	2.33276	22.56927

Auto ARIMA has been the best performer with the lowest value of RMSE and MAPE. However, SVM and KNN are the worst performers according to the RMSE and MAPE respectively. Similarly, Table 2 shows the output generated for the Earthquake dataset, and here reader can observe that Auto ARIMA and Linear Regression are the best performers with the lowest value of RMSE and MAPE respectively. However, KNN was the worst performer according to both RMSE and MAPE. But the numerals were so much close in this case (Table 3).

The results of the Sales forecasting dataset are described in table number 3, where it can be observed that Linear Regression and SVM turns out to be the best performer with the lowest value of RMSE and MAPE respectively. However, KNN was the worst performer according to both RMSE and MAPE (Fig. 2).

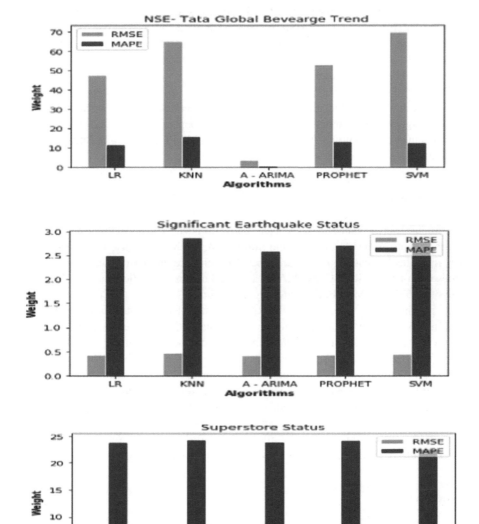

Fig. 2. Model performance comparative graph

The graphs in figure number 2 shows the comparison of the value attains by the Evaluation metrics. The Tata Global Beverage graph signifies that RMSE has higher values than MAPE; however, the other two datasets say otherwise. Ultimately, it all depends on the target variable and dataset.

The trend of First Dataset says Auto ARIMA has a significantly lower value of RMSE (3.74366) and MAPE (0.72129) than other models. However, talking about the worst performer, KNN beats other algorithms according to RMSE (16.92529), and SVM according to RMSE (69.81082).

Looking at the trend of the second dataset one can say that there is minimal difference between models according to RMSE; however, among all Auto ARIMA (0.41603) gave a bit better satisfying result. But according to the MAPE, Linear Regression (2.49101) went on top followed by Auto ARIMA (2.58689). White RMSE and MAPE both signified that KNN wouldn't be a good choice for this dataset.

The third dataset i.e., for Sales prediction had very difficult in choosing an optimal algorithm according to the graph. Nevertheless, Linear Regression became the more favorable algorithm than others according to the numbers of RMSE (2.22990). Similarly, SVM became a more optimal algorithm according to MAPE (22.56927). But again, KNN significantly became not a good choice.

6   Conclusion

Experimental performance analysis of five algorithms viz., linear regression, K – Nearest Neighbor, Auto ARIMA, Prophet, and Support Vector Machine is done. Stock market, earth and sales forecasting data is analyzed. To compare the performance and accuracy of these algorithms, RMSE and MAPE are used as the evaluation metrics. Lower the value of RMSE and MAPE, the better the algorithm.

As per the results, according to the RMSE, Auto ARIMA is the most optimal algorithm in two cases out of three. However, MAPE states that the Auto ARIMA is suitable for only one case. Taking it all in determination, it can be said that Auto ARIMA jostled all the other four algorithms, followed by Linear regression in the second place. Also, KNN is going to be the worst choice for Time-Series Forecasting. In the end, it won't be wrong to say that everything depends upon the trends and variables of the dataset, and that's why choosing an appropriate machine learning model becomes priority before going for a business idea. Here, one can observe that there is small difference between results of the evaluation metrics of earthquake and sales dataset. Yet, the numeral gaps between Auto ARIMA and other models in Stock Prediction dataset is clearly observed.

References

1. Panesar, A.: Machine learning algorithms. In: Machine Learning and AI for Healthcare, pp. 119–188. Apress, Berkeley, CA (2019). https://doi.org/10.1007/978-1-4842-3799-1_4
2. Abdualgalil, B., Abraham, S.: Applications of machine learning algorithms and performance comparison: a review. In: International Conference on Emerging Trends in Information Technology and Engineering (ic-ETITE), pp. 1–6, Vellore, India, (2020). https://doi.org/10.1109/ic-ETITE47903.2020.490

3. Dasgupta, A., Nath, A.: Classification of machine learning algorithms. Int. J. Innov. Res. Adv. Eng. (IJIRAE). **3**, 6–11 (2016). ISSN: 2349–2763, https://doi.org/10.6084/M9.FIGSHARE. 3504194.V1

4. Mondal, P., Shit, L., Goswami, S.: Study of effectiveness of time series modeling (Arima) in forecasting stock prices. Int. J. Comput. Sci. Eng. Appl. **4**, 13–29 (2014). https://doi.org/10. 5121/ijcsea.2014.4202

5. Korjenić , K., Hodžić, K., Đonk, D.: Application of Facebook's prophet algorithm for successful sales forecasting based on real-world data. Int. J. Eng. Data Technol. (IJCSIT). **12**(2), ten.5121/ijcsit.2020.12203 (2020)

6. Panigrahi, S., Behera, H.: A study on leading machine learning techniques for high order fuzzy time series forecasting. Eng. Appl. Artif. Intell. **87**, 103245 (2020)

7. Roondiwala, M., Patel, H., Varma, S.: Predicting stock prices using LSTM. Int. J. Sci. Res. **6**(4), 1754–1756 (IJSR) (2017)

8. Joosery, B., Deepa, G.: Comparative analysis of time-series forecasting algorithms for stock price prediction. 1–6 (2020)

9. Ariyo, A.A., Adewumi, A.O., Ayo, C.K.: Stock price prediction using the ARIMA model. In: UKSim-AMSS 16th International Conference on Computer Modelling and Simulation, Cambridge, pp. 106–112 (2014)

10. Khekare, G., Verma, P.: Prophetic probe of accidents in Indian smart cities using machine learning. In: Bhateja, V., Satapathy, S.C., Travieso-González, C.M., Aradhya, V.N.M. (eds.) Data Engineering and Intelligent Computing. AISC, vol. 1407, pp. 181–189. Springer, Singapore (2021). https://doi.org/10.1007/978-981-16-0171-2_18

11. Imandoust, S.B., Bolandraftar, M., Imandoust, S.B., et al.: Application of K-nearest neighbor (KNN) approach for predicting economic events theoretical background. Int. J. Eng. Res. Appl. **3**(5), 05–661 (2013)

12. Ayyub, K., Iqbal, S., Munir, E.U., Nisar, M.W., Abbasi, M.: Exploring diverse features for sentiment quantification using machine learning algorithms. IEEE Access **8**, 142819–142831 (2020)

13. Khekare, G.: Internet of everything (IoE): intelligence, cognition. Catenate. MC Eng. Themes **1**(2), 31–32 (2021)

14. Zhang, Y., Cheung, Y.L.: Learnable weighting of intra-attribute distances for categorical data clustering with nominal and ordinal attributes. In: IEEE Transactions on Pattern Analysis and Machine Intelligence (2021)

15. Kumar, N., Kumar, U. Diverse analysis of data mining and machine learning algorithms to secure computer network. Wireless Pers. Commun. 1–27 (2021). https://doi.org/10.1007/s11 277-021-09393-0

16. Pant, M., Kumar, S.: Fuzzy time series forecasting based on hesitant fuzzy sets, particle swarm optimization and support vector machine-based hybrid method. Granul. Comput. 1–19 (2021)

From Information Resources Push to Service Aggregation: The Development Trend of Mobile Government Service

Jinyu Liu[✉], Dongze Li, and Yongzhao Wu

School of Politics and Public Administration, South China Normal University, Guangzhou 510631, China
liujinyu@m.scnu.edu.cn

Abstract. Whether the mobile government service in this paper can meet the use standard of "4-b". The mobile government service can be divided into four development stages since its emergence: one is to solve the main problem of how to build the basic framework of mobile government service, which is based on the push stage of government information resources of information offline browsing system; the second is to solve the main problem of how to identify the user's identity conveniently, which is based on the user identity authentication stage of the mobile client; the third is to solve the main problem of fast interaction between server and client, which is based on the intelligent document processing stage of QR code; fourthly, it solves the main problem of fast access to services, which is based on the service aggregation stage of "App + applet". These four stages are inherited from each other, which is a process of continuous improvement. With the solution of service aggregation, the mobile government service will fully meet the "4-b" usage standard and become the mainstream form of e-government.

Keywords: Mobile government service · "4-b" use standard · Service aggregation

1 Introduction

Mobile government service is a kind of practice form of "Internet + government service", which is oriented to the public, with mobile phone, PDA, wireless network, Bluetooth, RFID and other technologies as its main application forms, mobile client terminals as its intermediary, and providing information and services based on mobile Internet as its main content [1–3]. To investigate the development trend of mobile government service, we can adopt the "4-b" standard [4], that is, whether it meets the standard that users can use conveniently in "beach, buses, bathroom and beds". Essentially, this standard is a method to test whether the existing electronic public service is convenient, comprehensive and reliable. Mobile government service has been highly concerned and widely used by governments of various countries [5]. In 2012, the U.S. government issued the "Digital Government" strategy, the primary goal of which is to ensure that

© The Author(s) 2022
Z. Qian et al. (Eds.): WCNA 2021, LNEE 942, pp. 1053–1058, 2022.
https://doi.org/10.1007/978-981-19-2456-9_105

the American citizens and the increasing number of mobile e-government, "4-b" use standard and service aggregation.

According to the "4-b" standard: "beach, buses, bathroom and beds", and two dimensions that are solving problem and instrument, we use the Document analysis method, empirical analysis and the Model method, and draw the conclusion that the mobile government service can be divided into four development stages.: The push stage of government informantion resources, the user identity authentication stage, the intelligent document processing stage, and the service aggregation stage (Fig. 1).

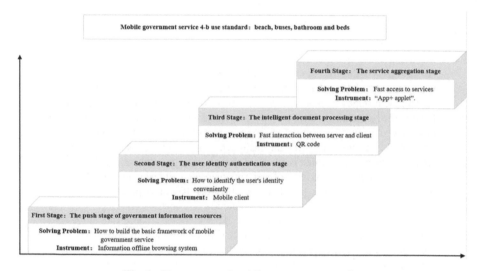

Fig. 1. Four stages of mobile government service

Users can obtain high-quality digital government information and services anytime, anywhere on any terminal [6]. Some EU countries have positioned mobile government as the main link to promote the strategy of "multi-channel delivery of public services" [7]. It can be said that the practice of mobile government affairs service in contemporary mainstream countries is developing towards meeting the "4-b" use standard.

2   Government Information Resources Push Based on Information Offline Browsing System

At this stage, we mainly solve the problem of how to build the basic framework of mobile government service. Generally speaking, mobile government service is an extension of traditional e-government [8]. However, handheld devices, such as smart phones and tablet computers, are greatly different from the platforms serving desktop computers in the aspects of information organization, including information transmission, storage, presentation and user reading habits. Therefore, we can't simply copy the traditional government service model to build the basic framework of mobile government service. Therefore, at this stage, our main task is to explore the push mode of mobile government information resources, and build the basic framework by building the mobile government information portal.

Mobile government information, as an information resource that users can browse with handheld devices, is the so-called "mobile newspaper" in its early form. Essentially, this is an information push method based on information offline browsing system. Its main feature is that the system can download relevant information to the mobile terminal when the mobile network is idle, and the offline device does not affect the storage of information, and users can browse the information at their own convenience. At this stage, the personalized information service system for users' needs also needs to be put forward, and it will dominate the future development form of mobile government service. Because the storage capacity of early mobile terminals is very small, in order to save storage space and improve user experience and satisfaction, "accurate delivery of information resources" has received extensive attention from the industry. Therefore, combined with the characteristics of government information resources, on the basis of continuously compressing information resources catalogue and simplifying traditional webpage elements, the basic framework of mobile government service has been gradually established, and it has taken a different development path from traditional e-government and mobile business services.

3 User Authentication Based on Mobile Client

At this stage, we mainly need to solve the problem of how users can get services by real-name registration system. Identity authentication is the basic function of various network applications. As long as users log in to each website, they need to provide corresponding identity and authentication information [9]. There are usually two ways to authenticate the identity of government users, one is anonymous registration system, the other is real-name registration system. With the national basic databases such as population basic information database and legal entity basic information database put into use one after another, real-name registration system registration has become the main way of identity authentication in e-government system. Technically, the mobile phone number is exclusive. Its Subscriber Identity Module (SIM) number can be in one-to-one correspondence with the user ID number. Mobile government process can interface with the real-name management system and technology of mobile communication services: when any mobile client terminal loaded with SIM card authenticates the identity of government service users, mobile government process can access the management data of mobile communication service providers through the management mechanism of SIM card, thus improving the authenticity and reliability of registered information, and further improving the security of user data by using the security mechanism provided by mobile communication equipment and service providers.

4 Smart Document Processing Based on QR Code

At this stage, we mainly solve the problem of information interaction between server and client. Because of the limited interface width of the mobile terminal, it is impossible to simply apply the spreadsheet and document technology of traditional websites to realize the interaction between the client and the server. With the emergence of smart document and QR code technology, the above problems will be solved easily. The so-called intelligent document is a problem of information processing between structured data and unstructured documents. Its main technical feature is that it embeds the logical connection function of database, such as data verification and routing instructions [10]. This will enable the electronic form to exchange data with the back-end database, and make the user's information, process and business model merge to the maximum extent, thus greatly improving the efficiency of business management.

The QR code is actually a barcode. Barcode refers to a graphic identifier that arranges a number of black bars and spaces with different widths according to certain coding rules to express a group of information. A standard article (commodity) barcode can load various types of information such as country of production, manufacturer, article name, production date, model, specification, etc. The idea of barcode technology was born in the 1940s, but it was not widely used until more than 30 years later when laser technology and computer technology matured. Different from the traditional bar code, the new type of QR code is a kind of black-and-white figure which is distributed in the plane (two-dimensional direction) according to a certain rule and records data symbol information by using a certain geometric figure. Compared with one-dimensional bar code, two-dimensional code is a barcode composed of multiple lines. The QR code itself can store a large amount of data without connecting to a database. The application of mobile phone QR code can make data exchange more convenient, which is widely recognized by users.

5 Service Aggregation Based on "App + Applet"

Service aggregation refers to integrating electronic service items scattered in different government departments and presenting them to users in an integrated image and personalized way, that is, to solve problems such as how to make users get services quickly. With the emergence of the "App + Applet" model, there is a more appropriate way to solve the above problems.

APP(Application) mainly covers software for mobile terminals, that is, mobile clients. Generalized mobile terminal software, combined with industrial cellphone device, has already been widely used in scientific research, production and transportation fields such as geological exploration, warehousing and logistics. The narrow sense of mobile terminal software mainly refers to the application programs that are emerging in recent years and can be used by handheld devices such as smart phones and tablets. Correspondingly, the government affairs APP refers to the mobile client or application program whose main content is to provide government affairs services. Common government apps can be divided into professional and general types. Professional government affairs APP refers to the mobile client application that serves specific groups and specific industries. General-purpose government APP refers to a mobile client application that provides comprehensive services for companies, social organizations and individuals based on the integrated platform of government service network. Compared with the traditional government affairs portal, the government affairs APP can integrate more functions, especially the easy integration of government affairs services. Therefore, the development speed of government affairs APP is very fast, and there is a tendency to replace the traditional government affairs portal. However, because of the high development, maintenance and upgrading costs of government APP, it will face some problems such as the limited flexibility of the mobile government service it loads [11, 12].

Fortunately, the maturity of "applet application" in recent years has greatly alleviated the above problems. The so-called "applet" refers to the development of small-scale packages. Taking the WeChat Mini Program as an example, this is an application software that can be used without downloading and installing. Wechat APPlets can be used directly in WeChat app. When users want to use small programs with specific functions, such as paying subway and bus tickets, they only need to use the corresponding programs in WeChat without downloading software packages. Since WeChat launched the small program in 2015, it has been upgraded and revised several times. Now, it has realized data sharing and process docking with many public utilities and government services. With more and more software platforms paying attention to the development and application of applets, the application mode based on "App + Applets" is covering all fields of government services at an unprecedented speed, which accelerates the integration progress of mobile government services.

6 Conclusion

From the developer's point of view, the applet architecture is simple, the development threshold is much lower than that of APP, and it can satisfy simple basic applications. From the manager's point of view, mini programs have short development cycle and low cost, which can meet the needs of low-frequency use. From the user's point of view, the applet embodies the idea of "putting it aside after use", and it has the convenience charactcristics of no installation and no desktop resources. For e-government apps, they can use the advantages of social users, business circle users and entertainment circle users of commercial apps to spread e-government services in the form of mini programs among clients of various social, business and entertainment applications. For social, business and entertainment APP software, they can also use small programs as an intermediary to attract more users and keep users sticky with the help of the resource advantages of government APP. To sum up, "App + Applet" is a mode of "integration and symbiosis" that can truly meet the requirements of "4-b" usage standard, and it represents the development trend of mobile government service.

Acknowledgments. This work was supported by the following items: National Social Science Fund Project total "community-level data based on the authorization of a major community-level public health emergencies coordinated prevention and control mechanisms of innovative research" (20BGL217).

References

1. Guozhang, F.Y.: Development and prospect of mobile government. E-Government **12**, 11–21 (2010)
2. Chanana, L.F., Agrawal, R.S., Punia, D.K.T.: Service quality parameters for mobile government services in India. Glob. Bus. Rev. **17**(1), 136–146 (2016)
3. Liu, S.F., Hua, Z.S., Yuan, Q.T.: Mobile government and urban governance in China. E-Government **6**, 2–12 (2011)
4. Giussani, B.F.: Roam: Making Sense of the Wireless Internet, 1st edn. CITIC Publishing House, Beijing (2002)
5. Song, G.F., Li, M.S.: Reinventing public management by mobile government. Off. Informatization **11**, 10–13 (2006)
6. Chen, L.F.: Are the government websites mobile? Informatization Construct. **6**, 24–26 (2013)
7. Kushchu, I.F., Kuscu, M.H.S.: From E-government to M-government: facing the Inevitable. In: 3rd European Conference on eGovernment, pp. 1–13 (2004)
8. Lin, S.F.: Mobile E-government construction based on the public requirements. Chin. Public Admin. **4**, 52–56 (2015)
9. Jian, L.F., Changxiang, S., Han, Z.T.: Survey of research on identity management. Comput. Eng. Des. **30**(6), 1365–1370+1375 (2009)
10. Zhang, C.F.: Application analysis of electronic form system. East China Sci. Technol. **9**, 60–63 (2021)
11. Wei, P.F., Su, L.S.: Research on mobile government affairs and the construction of intelligent-service-government. J. Shanxi Youth Vocat. Coll. **34**(02), 42–45 (2021)
12. Chen, Z.F.: Analysis of typical problems of China mobile government app client. E-Government **3**, 12–17 (2015)

Performance Analysis of Fault Detection Rate in SRGM

Zhichao Sun[✉], Ce Zhang, Yafei Wen, Miaomiao Fan, Kaiwei Liu, and Wenyu Li

School of Computer Science and Technology, Harbin Institute of Technology, Weihai, China
szc20160365@outlook.com

Abstract. The fault detection rate is one of the main parameters of the software reliability model. Different forms of fault detection rates have different functions. This paper focuses on the influence of fault detection rate on software reliability, proposes a single reliability model multi-failure data set multi-fault detection rate analysis plan, and analyses the impact of fault detection rate on SRGM. After experimental analysis, the performance of the software reliability model corresponding to the power function and the S-type fault detection rate is better, the performance of the software reliability model corresponding to the constant fault detection rate is acceptable, and the comprehensive performance of the software reliability model corresponding to the exponential fault detection rate is poor. The research in this paper has a certain guiding role in the selection of parameter models in software reliability modelling and the determination of the optimal release time.

Keywords: Failure detection rate · Reliability modeling · Software reliability growth model · Empirical analysis

1 Introduction

With the development of information technology and networks, the application of computers has become more and more extensive. As the main carrier and function provider for users to use computers, computer software plays an important role in production and life. In order to meet people's expectations for the improvement of software functions, the scale and complexity of software continue to increase. When the scale of software gradually increases, maintaining software quality is an important part of the software development and testing process. Software reliability is an important factor in software quality, and high-quality software must be highly reliable. The software reliability growth model SRGM is an important method of software reliability research, and it is also the current mainstream research method. In the general SRGM model, there are two types of basic parameters [1], one is the total software failure, which is the abstraction of the overall number of failures in the software system, and the other is the failure detection rate, which is a description function of the test capability in the software test environment. In the process of software testing, testers will continue to find and repair faults. In order to better grasp the reliability of the software and meet the expected (release) requirements, it is necessary to study the function of FDR in reliability research.

© The Author(s) 2022
Z. Qian et al. (Eds.): WCNA 2021, LNEE 942, pp. 1059–1066, 2022.
https://doi.org/10.1007/978-981-19-2456-9_106

The fault detection rate characterizes the comprehensive ability of the test environment, test technology, test resource consumption and tester skills [2]. Objectively, the difference in the test environment and the difference in the test strategies implemented by the testers make different system projects show different external characteristics in the test. From the perspective of establishing a mathematical model, the difference between different models is closely related to the fault detection rate FDR. In this way, FDR portrays the test effect as a whole, making it the main evaluation point that affects the performance of SRGM. It is of great significance to build models for software reliability, predict the number of software failures, determine the optimal release time, and control test costs.

This paper mainly starts from the fault detection rate, proposes a single SRGM, multiple FDS and multiple FDR schemes, the correlation between reliability model, FDS and FDR, based on the experimental results of FDR on the reliability model and FDS, combines different actual scenarios to implement, and analyses the effect of fault detection rate on the efficacy of SRGM.

2   Modelling the Influence of Failure Detection Rate on Reliability Model

First, give the hypothesis for establishing SRGM in this article:

- Software failure satisfies the NHPP process [3, 4];
- The number of faults detected within $(t + \Delta t)$ is proportional to the number of faults remaining in the current software;
- There is no new fault introduction phenomenon in the software repair process [5];

So far, hundreds of SRGMs have been proposed. Assumption (1) mentioned above are included in the assumptions of all these models and based on this, different forms of differential equations have been established. In order to facilitate the observation of $b(t)$ performance, this article gives the more a general form based on the basic establishing process of many SRGMs [3–11]:

$$\frac{dm(t)}{dt} = b(t) \cdot [a - m(t)]$$

In this formula, $b(t)$ is the fault detection function, whose value is between (0, 1); a is the total number of faults in the software system. a is set to be a constant in this article. Based on the model mentioned above, the $b(t)$ function can be set as needed to get software reliability model corresponding to different fault detection rate.

This article will proceed from the following three steps to gradually determine FDR, SRGM and FDS.

Step 1: Based on our previous research results and a large number of experiments, select the set of SRGMs with excellent performance on the scheduled FDS. These SRGM sets include the reliability model established from the FDR perspective obtained above;

Step 2: Establish the set of FDRs to be observed. Although they cannot be derived from previous experiments, they can be selected by collecting b(t) that appear frequently in the current research;

Step 3: Establish the FDR for observation and the set of observation points at which the FDR may have impact on SRGM.

Based on the determined correlation model, an empirical analysis is carried out based on the proposed scheme to explore the impact of fault detection rate on the performance of SRGM.

3   Single SRGM Multiple FDS Multiple FDR Model

Under certain SRGM (i.e. $m(t)$) conditions, you can observe the SRGM performance at this time by changing the FDR, that is, substituting multiple $b(t)$ functions into $m(t)$. This situation is called single SRGM multiple FDS multiple FDR mode.

At this time, for the selected SRGM and FDS, the former has good fitting and predictive capabilities for the latter. Therefore, in this good situation, different FDRs are brought into SRGM for experiments. The dashed line of fitting and prediction obtained by observation and decision-making can give the FDR ranking result (i.e. partial order set). Figure 1 and Fig. 2 respectively describe the basic process of this scheme and the corresponding execution algorithm EvaluateFDREffectOnSRGM—SSSFMF.

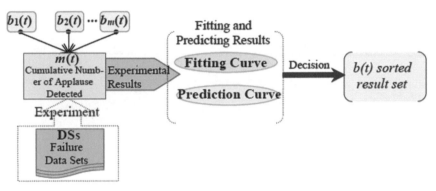

Fig. 1. $b(t)$ Evaluation and decision-making process

In this single SRGM, multiple FDS, and multiple FDR modes, since the research is based on a single SRGM sum on many failure data sets FDS, each FDS can be regarded as a specific software test environment. Therefore, based on the analysis results, it is convenient to improve the test strategy, so that the fault detection rate can be improved in the direction of meeting the test requirements.

4   Experiment and Analysis

4.1   Experiments Settings

For the single SRGM multi-FDS multi-FDR scheme, different $b(t)$ function forms are substituted based on the above formula, and the model expressions obtained are shown in Table 1.

	FDR Evaluation algorithm *EvaluateFDREffectOnSRGM—SSSFMF*
	Input: (Through a lot of experiments) select SRGM model m(t) and failure data set DSs, failure detection rate vector **FDRSet** $= [bt_1, bt_2, ..., bt_m]$
	Output: FDR partially ordered set **FDRSet**

1:	*EvaluateFDREffectOnSRGM—SSSFMF:*
	For each DS in (DSs) {
	    For each $b(t)$ **in (FDRSet) {**
	        MT[i]=*Fitting*(m(t), DS, b(t))
	        Draw the fitted curve.
	        RE[i]=*CalculateRE*(MT[i])
	        Draw the prediction curve.
	    }
	}

2:	**FDRSet**=*SortFDR*(**MT, RE**) //Obtain the partially ordered set.
3:	**Return FDRSet**

Fig. 2. Execution algorithm evaluateFDREffectOnSRGM—SSSFMF

Table 1. Formatting sections, subsections *and* subsubsections

Models	FDR type	$b(t)$ function	$m(t)$
M-1	*Constant type*	$b_1(t) = b$ [6]	$m_1(t) = a(1 - e^{-bt})$
M-2	*Power function type*	$b_2(t) = b^2 t/(1 + bt)$ [7, 8]	$m_2(t) = a \cdot (1 - (1 + bt)e^{-bt})$
M-2	*S type*	$b_3(t) = \frac{b(1+\sigma)}{1+\sigma e^{-b(1+\sigma)t}}$ [9,10]	$m_3(t) = \frac{a(1-e^{-(1+\sigma)bt})}{1+\sigma e^{-(1+\sigma)bt}}$
M-4	*Complex exponential type*	$b_4(t) = b\alpha\beta e^{-\beta t}$ [11]	$m_4(t) = a[1 - e^{-b\alpha(1-e^{-\beta t})}]$

The above is the SRGM model corresponding to different $b(t)$ functions under the perfect hypothesis, and then the fitting and prediction were carried out on several published real failure data sets to observe the influence of different FDRS on the SRGM model.

4.2 Experiment and Analysis

Fitting Performance Analysis. This section mainly analyses the fitting performance of different models under the real failure data set. Based on a series of real failure data sets, we draw the fitting curve of different models for the data sets, as shown in Fig. 3. The closer the fitting curve is to the real failure curve, the better the fitting performance of the model.

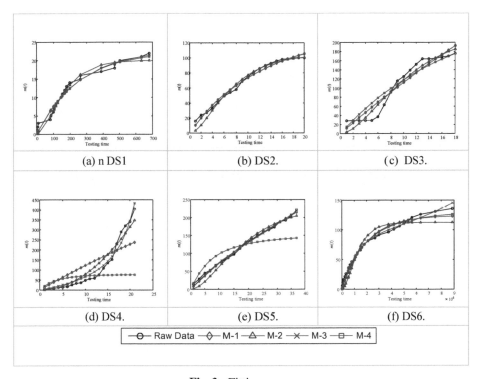

Fig. 3. Fitting curves

It can be seen in Fig. 3 in this paper, the choice of most real failure data growth curve for convex type growth form, which suggests that most of the real situation of software testing, DS3 is S type growth curve, showed more complex software systems and test environment, DS4 growth curve is concave type growth forms, corresponds to the part of the real software test scenarios. On the whole, the fitting curve of most models is consistent with the growth trend of the data set except for some models with serious deviation. For the real failure data set with convex growth, most of the models have good fitting effect. Only some models have a large deviation from the real failure data set (such as M-4 on DS5). For the concave growth data set DS4, the SRGM models (M-3 and M-2) corresponding to s-type and power function $b(t)$ function have good fitting effect, indicating that S-type and power function $b(t)$ function have better applicability. For the S type DS3 which has both concave and convex growth forms, the fitting performance of M-3 corresponding to S type $b(t)$ function is better, which further indicates that the applicability of SRGM corresponding to $b(t)$ function of S type and power function is stronger.

Predictive Performance Analysis. Experiments were conducted on the same data set and the following prediction curves were drawn. The closer the curve is to 0, the better the prediction performance is. The predicted value is greater than 0, indicating a positive prediction, and less than 0, indicating a negative prediction.

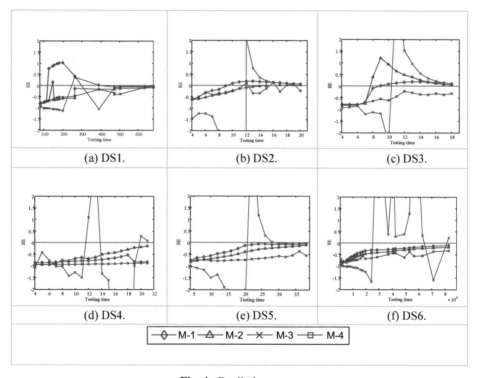

Fig. 4. Prediction curves

The prediction curve of single SRGM, multi-FDS and multi-FDR is drawn in Fig. 4. By analysing the trend of the above curves, it can be found that, on the whole, the prediction curve of most models tends to be stable and close to 0 with the test time, except for the prediction performance deviation of a few models on some data sets. On different data sets, the prediction performance of constant and power function FDR is good, the prediction curve of S-type FDR fluctuates greatly and the prediction performance is poor, and the prediction performance of complex exponential FDR is mediocre. In particular, for most data sets, the prediction curve fluctuates greatly in the early stage and gradually becomes stable in the later stage, indicating that with the growth of the test time, the prediction ability of the model for real data sets gradually improves, indicating that the test software is more skilled in software test environment and test tools.

b(t) **Sequence Analysis.** According to the above fitting and prediction curves, a comprehensive ranking result is given: $b_2(t) > b_3(t) > b_1(t) > b_4(t)$. According to the type of $b(t)$, there is power function > S type > constant type > complex exponential type. The fitting curve of the SRGM model corresponding to power function $b_2(t)$ on most data sets is consistent with the real failure data curve, and the prediction performance is good. The fitting performance of S type $b_3(t)$ is also excellent and performs well on most data sets, but the prediction performance is not stable and fluctuates greatly. The fitting performance of constant type $b_1(t)$ is good, which can basically fit the growth trend of the failure data set and has good prediction performance. The complex exponential $b_4(t)$ cannot fit the data set with concave growth well, and its prediction performance in some data sets is mediocre.

5   Conclusion

This paper focuses on the impact of different FDR models on the performance of SRGM models and performs an empirical analysis. A single SRGM multi-FDS multi-FDR scheme is proposed to derive the partial order sequence of the SRGM model corresponding to the software fault detection rate function and analyse the effect of FDR on the performance of the SRGM model. Four types of fault detection rate, namely, constant, power function, S-type, and exponential, are selected in the experiments, and it is found that the power function type FDR has excellent performance, followed by the S-type FDR, and the exponential type has the worst performance. The research in this paper has some guiding significance for selecting the appropriate fault detection rate to establish the SRGM of distance in the actual software testing process, and provides a reference for testing resource allocation and optimal release.

Based on the assumption of perfect fault exclusion, this paper assumed that no new faults are introduced in the fault repair process, so it is deficient in imperfect fault exclusion. In future research, more forms of software total fault count functions, more quantitative performance metrics, and more realistic underlying assumptions will be combined for analysis to broadly and comprehensively explore the impact of fault detection rates on SRGM.

References

1. Zhang, C., Meng, F.C., Kao, Y.G., et al.: Survey of software reliability growth model. J. Softw. **28**(9), 2402–2430 (2017). (in Chinese)
2. Zhang, C., Liu, H.W., Bai, R., et al.: Review on fault detection rate in reliability model. J. Softw. **31**(9), 2802–2825 (2020). (in Chinese)

3. Ahmad, N., Khan, M.G.M., Rafi, L.S.: A study of testing-effort dependent inflection s-shaped software reliability growth models with imperfect debugging. Int. J. Qual. Reliab. Manag. **27**(1), 89–110 (2010)
4. Huang, C.Y., Kuo, S.Y., Lyu, M.R.: An assessment of testing-effort dependent software reliability growth models. IEEE Trans. Reliab. **56**(2), 198–211 (2007)
5. Kapur, P.K., Pham, H., Anand, S., et al.: A Unified approach for developing software reliability growth models in the presence of imperfect debugging and error generation. Reliab IEEE Trans. **60**(1), 331–340 (2011)
6. Goel, L., Okumoto, K.: Time-Dependent error-detection rate model for software reliability and other performance measures. IEEE Trans. Reliab, R-**28**(3): 206–211 (1979)
7. Yamada, S., Ohba, M., Osaki, S.: S-shaped reliability growth modeling for software error detection. IEEE Trans. Reliab. **32**(5), 475–484 (1983)
8. Yamada, S., Ohtera, H., Narihisa, H.: Software reliability growth models with testing-effort. IEEE Trans. Reliab. **35**(1), 19–23 (1986)
9. Huang, C.Y., Lyu, M.R., Kuo, S.Y.: A unified scheme of some nonhomogenous poisson process models for software reliability estimation. IEEE Trans. Software Eng. **29**(3), 261–269 (2003)
10. Chiu, K.C., Huang, Y.S., Lee, T.Z.: A study of software reliability growth from thr perspective of learning effects. Reliab. Eng. Syst. Saf. **93**(10), 1410–1421 (2008)
11. Pham, H., Nordmann, L., Zhang, X.: A general imperfect-software-debugging model with S-shaped fault-detection rate. IEEE Trans. Reliab. **48**(2), 169–175 (1999)

Research on Vibration Index of IRI Detection Based on Smart Phone

Jingxiang Zeng[1], Jinxi Zhang[1,2(✉)], Qianqian Cao[1], and Wangda Guo[1]

[1] Institute of Transportation, Beijing University of Technology, Beijing 100124, China
zhangjinxi@bjut.edu.cn
[2] Beijing Engineering Research Center of Urban Transport Operation Guarantee, Beijing 100124, China

Abstract. With the development of science and technology, intelligent pavement smoothness detection becomes possible. Intelligent IRI (International Roughness Index) detection is one of the important development directions of pavement performance detection. Different from traditional IRI detection, intelligent IRI detection uses smart phones to collect traffic vibration data. There are many vibration indexes in IRI evaluation unit of driving vibration data, and IRI evaluation can be realized by extracting vibration indexes. In this study, the corresponding relationship between pavement vibration data and IRI is preliminarily proved by driving test. The synthetic vibration acceleration index can reflect the change of IRI. The length of IRI evaluation unit reflects different significance of pavement performance, and the evaluation vibration index extracted is different. When the evaluation unit is short, IRI reflects the local pavement performance of the evaluation unit, and the correlation between the minimum value of vehicle synthetic vibration acceleration and IRI is the best. When the evaluation unit is long, IRI reflects the overall pavement performance of the evaluation unit, and the correlation between the average value of the absolute value of the vehicle synthetic vibration acceleration and IRI is the best.

Keywords: Vibration index · IRI · Detection · Smart phone

1 Introduction

With the development of road transportation, countries all over the world including China have built huge road transportation networks. In China, for example, the total length of roads reached 5.02 million kilometers by the end of 2019 [1]. Among them, expressways reach 150,000 km [2]. The construction of a large number of transportation infrastructure provides convenient ways for people to travel and promotes the rapid development of society and economy [3]. On the other hand, the rapid construction and huge stock of road facilities make road workers face two problems. First, how to carry out the maintenance of existing road facilities, so that road facilities are in a good technical state, to provide safe and comfortable services for road users. The second is how to carry out real-time monitoring of road facilities, so as to discover the existing problems in time and make scientific maintenance decisions. It is an important work to

Z. Qian et al. (Eds.): WCNA 2021, LNEE 942, pp. 1067–1076, 2022.
https://doi.org/10.1007/978-981-19-2456-9_107

realize the real-time and accurate evaluation and monitoring of the technical status of pavement facilities [4].

At present, different countries have established different pavement performance evaluation systems [5]. Pavement performance in China includes seven indexes, such as flatness, damage, rutting, bearing capacity, skid resistance, jumping and abrasion [6]. Different indexes reflect the technical performance of different aspects of pavement surface. The roughness of road surface is usually represented by IRI [7]. As one of the most important pavement performance evaluation indexes, road workers around the world have conducted long-term research on IRI detection methods, and put forward different detection methods such as manual three-meter ruler method, accumulative bumpy instrument and laser flatness detector [8]. The IRI detection is transformed from manual detection to automatic detection [9]. It promotes the speediness and standardization of pavement performance test and promotes the development of road transportation [10]. The current detection methods will have a certain impact on the road traffic operation, which requires a lot of detection costs. How to realize the intelligent evaluation of IRI is an important research topic facing pavement engineers [11].

With the development of science and technology, the functions of smart phones are becoming increasingly powerful [12]. The nine-axis vibration sensor and GPS positioning sensor carried by smart phones provide the possibility for IRI evaluation [13]. In People's Daily driving process, vehicle vibration caused by IRI can be collected by smart phones, and IRI can be pre-detected by analyzing and processing driving vibration data [14]. IRI detection needs to be conducted according to the evaluation unit, and the common evaluation unit includes 10 m, 20 m, 50 m, and 100 m [15]. Different detection units have different evaluation angles for IRI. How to use the driving vibration data in each evaluation unit to calculate the effective vibration index in the time domain of the vibration data is of great significance to further realize the use of smart phones to detect IRI [16]. In this paper, by carrying out driving test, the possibility of reflecting IRI through driving vibration data is verified, the correlation of vibration indicators of driving vibration data in different IRI evaluation units is compared, and a new method is proposed to detect IRI in different evaluation units by using different vibration indicators. The method of IRI detection using traffic vibration data is of great significance in the aspects of detection cost, detection frequency and environmental protection.

2   Test

In order to establish the relationship model between driving vibration data and IRI, this paper carried out speed bump test, urban road test and test site test by using a special smartphone App and cars.

2.1   Test Equipment

In order to collect users' driving data using smart phones, the author has developed a special smart phone App for driving data collection [16]. The App interface is shown in Fig. 1. Data collected by App mainly include triaxial vibration acceleration data, GPS geographic location data and time data. With the map as the background, the App can show users the detection route (Fig. 1 (a)) and see real-time data changes (Fig. 1 (b)). App has the function of viewing historical data to provide users with corresponding services (Fig. 1 (c)). The collected user data can be transmitted to Pavement Condition Map using mobile network or wireless signal. On the history page, you can view all collected data and upload or download data again. As an experimental product, currently users need to register to use the App. Data collection should be agreed by mobile phone users and comply with relevant laws and regulations.

a   b   c

Fig. 1. App interface

The three smart phones used in this paper are common smart phones in the market, HuaWei, MINI and OPPO. All three phones are equipped with sensors that collect triaxial vibration acceleration data, GPS data and time data. In this study, vibration acceleration data were collected at a frequency of 10 Hz and GPS data at a frequency of 1 Hz. According to the public's mobile phone placement habits and test needs, mobile phones will be placed in three postures. The first attitude is horizontal, that is, the coordinate system of mobile phone is consistent with the coordinate system of vehicle. The mobile phone is tightly fixed in the middle of the vehicle with adhesive tape, so that the standard posture mobile phone is closely attached to the vehicle; The second attitude is inclined, that is, the mobile phone bracket is tilted and fixed in the middle of the vehicle; The third pose is a random pose, that is, the mobile phone is placed in the pocket of the driver and the passenger, and the driver and the passenger do not touch the mobile phone artificially in the process of driving. The phone brand and location will be switched after a period of driving. The test vehicles are SUV, car and special test vehicle.

2.2 Speed Bump Test

This paper chooses a newly built road inside the parking lot as a speed bump test road. The section has good IRI and straight line, and the length of the section is about 100 m. There is a relatively new trapezoidal speed belt in the middle of the section, and the size of the speed belt is shown in Fig. 2.

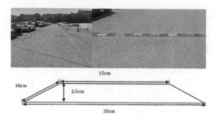

Fig. 2. Speed bumps test pavement conditions

The tester should install the special driving vibration data collection App before the test, open it and install it on the vehicle. The three test mobile phones were fixed on the handrail box in the middle of the vehicle with adhesive tape in a horizontal attitude. The mobile phone bracket is fixed on the air conditioner air outlet in the front of the vehicle in an inclined attitude. Placed in the experimenter's pocket in a random posture. The test vehicle passes through the deceleration belt repeatedly at uniform speed on the test road. When the vehicle passes through the deceleration belt, the tester records the specific time of passing the deceleration belt.

2.3 Urban Road Test

In this paper, a road section with a wide range of IRI indexes was selected to carry out driving test. The test road is a section of 2,044 m, and the IRI of this section varies greatly. IRI has maximum value of 6.24 m/km minimum value of 2.60 m/km average value of 4.0 m/km. The IRI detection unit is 100 m.

Fig. 3. Driving test on urban road

As shown in Fig. 3, The test vehicle and the test phone are consistent with the speed bump test. The test vehicle starts to accelerate before the starting point of the test site, and when it reaches the starting point of the test site, the passing time is recorded and the vehicle keeps driving at a constant speed. Record the passing time when the vehicle passes the end of the test site and repeat it several times.

2.4 Special Road Test

In this paper, the special test site for pavement performance evaluation of the Ministry of Transport is selected to carry out driving test, as shown in Fig. 4. The section of the test site is an annular test site with a length of 4 km. The test site uses special IRI detection equipment to accurately measure IRI, and the IRI detection and evaluation unit is 10 m. The starting position of the test is consistent with the starting and ending position of the section tested by special testing equipment.

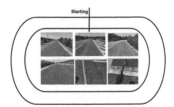

Fig. 4. Driving test on Special road

The test vehicle and the test phone are consistent with the speed bump test. The test vehicle began to accelerate before the starting point, and when it reached the starting point of the test site, the passing time was recorded and the vehicle kept driving at a constant speed. Record the passing time when the vehicle passes the terminal and repeat it several times.

3   The Data Analysis

3.1   Vibration Index of Z-axis Direction under Horizontal Attitude

When analyzing the internal vibration of the vehicle caused by IRI, 1/4 vehicle model can calculate the relationship between IRI and vibration acceleration by means of mechanical calculation. The 1/4 vehicle model simulates the vibration of the vehicle body when IRI changes with a single wheel. The model needs the specific parameters of vehicle suspension system such as body mass and suspension stiffness coefficient. The model proves the influence principle of IRI on vehicle vibration data. According to 1/4 vehicle model, vertical vibration acceleration can reflect IRI. In the speed belt test carried out in this paper, the vibration acceleration in z-axis direction collected by horizontal attitude smart phones is the vibration acceleration data caused by IRI.

As shown in Fig. 5, the data comes from speed belt test, and the black line indicates the vibration acceleration data in z-axis direction collected by a speed belt test. When the vehicle is driving on the test road, the az-axis fluctuates up and down near the gravitational acceleration. When the vehicle passes through the speed belt, the az-axis changes greatly, and the data change time is consistent with the time when the vehicle passes through the speed belt.

Through the speed belt test, it can be seen that there is a corresponding relationship between driving vibration data and IRI, and the method of IRI detection using driving

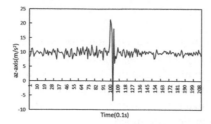

Fig. 5. Vibration acceleration in z-axis direction under horizontal attitude

vibration data is feasible. Although 1/4 of the vehicles can accurately calculate IRI through driving vibration data, relevant parameters of the vehicles need to be accurately calibrated, which is not conducive to the extensive collection of driving vibration data.

3.2 Synthetic Vibration Acceleration Index

Vibration acceleration data in the x-axis, y-axis and z-axis directions collected by the three mobile phone placement methods are shown in Fig. 6, which is part of the data obtained in the test site. When the posture of smart phone is horizontal, the coordinate system of mobile phone is consistent with the coordinate system of vehicle. Therefore, the vibration acceleration data in the x-axis direction can indicate the vibration of the vehicle in the left and right directions, the vibration acceleration data in the y-axis direction can indicate the vibration of the vehicle in the moving direction, and the vibration acceleration data in the z-Axis direction can indicate the vibration of the vehicle in the vertical direction. When the posture of a smartphone is tilted, the vibration acceleration data in the x-axis, y-axis and z-axis directions are the data when the phone is in a fixed posture, but the vibration acceleration data in a single direction has no actual physical significance due to the inconsistency between the vehicle coordinate system and the mobile coordinate system. When the posture of smart phone is random, the vibration acceleration data in the single direction of x-axis, y-axis and z-axis also has no actual physical significance.

In fact, road vibrations were consistent regardless of where the phone was placed. Therefore, in this study, the synthetic acceleration is used as the time-domain effective vibration acceleration index, and the calculation method is shown in formula (1).

$$a_c = \sqrt{a_{X-axis}^2 + a_{Y-axis}^2 + a_{Z-axis}^2} - g \qquad (1)$$

ac is the composite vibration acceleration, ax-axis, ay-axis and az-axis are the vibration acceleration values collected in the x-axis, y-axis and z-axis directions of the mobile coordinate system collected by smart phones respectively, and g is the acceleration of gravity.

3.3 Synthesize Average Value Index of Absolute Vibration Acceleration

IRI divides and evaluates sections according to evaluation units. Due to different driving speeds, each evaluation unit contains different numbers of driving vibration data. The

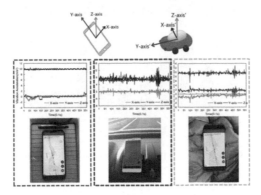

Fig. 6. Data of different mobile phone status

evaluation unit is small for local evaluation of IRI, while the evaluation unit is long for overall evaluation of IRI. The time-domain vibration indexes in the evaluation unit include 9 vibration indexes: maximum value, minimum value, average value, standard deviation, average value of absolute value, maximum value of absolute value, median of absolute value and standard deviation of absolute value. In this paper, the correlation coefficient is used to select the optimal time domain vibration index to detect IRI.

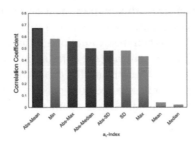

Fig. 7. Correlation of vibration index of 100 m evaluation unit

As shown in Fig. 7, red indicates the related indexes of ac absolute value, green indicates the related indexes of ac, Mean indicates the average index, Min indicates the minimum index, Max indicates the maximum index, SD indicates the standard deviation index, and Median indicates the Median index. The data comes from urban road driving test, and the average value of absolute value per 100 m has the greatest correlation with IRI data. In conclusion, when the evaluation unit is 100 m, the average value of absolute value of synthetic vibration acceleration can best reflect the changes of IRI.

3.4 Synthesize Minimum Vibration Acceleration Index

Compared with the road surface evaluation unit of 100 m, the IRI evaluation unit of the special road test is 10 m. The small evaluation unit is more prominent in the local characteristics of the evaluation unit.

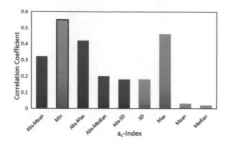

Fig. 8. Correlation of vibration index of 10 m evaluation unit

As shown in Fig. 8, red indicates the related indexes of ac absolute value, green indicates the related indexes of ac, Mean indicates the average index, Min indicates the minimum index, Max indicates the maximum index, SD indicates the standard deviation index, and Median indicates the Median index. The data are from the special road test, and the minimum value per 10 m has the highest correlation with IRI data. In conclusion, when the evaluation unit is 10 m, the minimum value of composite vibration acceleration can best reflect the variation of IRI.

3.5 Conclusion

Different IRI evaluation units have different pavement evaluation purposes, so it is necessary to use different vibration indicators to establish the relationship model between driving vibration data and IRI. When the evaluation unit is short, IRI reflects the local road performance of the evaluation unit, and the correlation between the minimum value of vehicle synthetic vibration acceleration and IRI data is the best. When the evaluation unit is long, IRI reflects the overall road performance of the evaluation unit, and the average value of the absolute value of vehicle synthetic vibration acceleration has the best correlation with IRI data.

4 Conclusion and Prospect

In this paper, App and test vehicles were used to carry out driving test, and IRI was evaluated by collecting driving vibration data. IRI evaluation units are different. By studying the relationship between the length of different evaluation units and vibration indicators, the following conclusions are drawn in this paper: (1) there is a corresponding relationship between driving vibration data and IRI, and it is feasible to detect IRI using driving vibration data. (2) Synthetic vibration acceleration index can reflect IRI changes. (3) When the evaluation unit is short, IRI reflects the local road performance of the evaluation unit, and the correlation between the minimum value of vehicle synthetic vibration acceleration and IRI data is the best. (4) When the evaluation unit is long, IRI reflects the overall road performance of the evaluation unit, and the correlation between the average value of the absolute value of vehicle synthetic vibration acceleration and IRI data is the best. The research on IRI detection using vehicle vibration data is still in its infancy. This paper mainly studies the selection of vibration indicators of different evaluation units. More driving tests will be carried out in the future, and data fusion and big data processing methods will be applied to this study to continuously improve the accuracy of IRI detection.

Acknowledgment. This work was supported by the National Natural Science Foundation of China under Grant No. 51778027 and National Key R&D Program of China No. 2018YFB1600300.

References

1. Zhang, J.: Special Topics on Road Engineering, 2nd edn. Beijing Science press, Beijing (2019)
2. Celaya-Padilla, J.M., Galván-Tejada, C.E., et al.: Speed bump detection using accelerometric features: a genetic algorithm approach, Sensors (2018)
3. Alhasan, A., White, D.J.: Continuous wavelet analysis of pavement profifiles, Autom. Constr. 134–143 (2016)
4. Ozer, E., Maria, Q.F., Feng, D.: Citizen sensors for SHM: towards a crowdsourcing platform. Sensors **15**(6), 14591–14614 (2015)
5. Souza, V.M.A.: Asphalt pavement classification using smartphone accelerometer and complexity invariant distance. Eng. Appl. Artif. Intell. **74**, 198–211 (2018)
6. Li, X., Goldberg, D.W.: Toward a mobile crowdsensing system for road surface assessment. Comput. Environ. Urban Syst. **69**, 51–62 (2018)
7. Wang, S., Zhang, J., Yang, Z.: Experiment on asphalt pavement roughness evaluation based on passengers' physiological and psychological reaction. In: ICCTP 2010: Integrated Transportation Systems: Green, Intelligent, Reliable - Proceedings of the 10th International Conference of Chinese Transportation Professionals, vol. 382, pp. 3852–3863 (2010)
8. Souza, V.M.A., Cherman, E.A., Rossi, R.G., Souza, R.A., Towards automatic evaluation of asphalt irregularity using smartphones sensors. In: International Symposium on Intelligent Data Analysis, pp. 322–333 (2017)
9. Aleadelat, W., Ksaibati, K.: Estimation of pavement serviceability index through android-based smartphone application for local roads. Transp. Res. Rec. **2639**, 129–135 (2017)
10. Allouch, A., Koubaa, A., Abbes, T., Ammar, A.: RoadSense: smartphone application to estimate road conditions using accelerometer and gyroscope. IEEE Sens. J. **17**(13), 4231–4238 (2017)
11. Zhang, J., Zhou, T.: Comprehensive evaluation method of pavement comfort based on D-S evidence method. J. South China Univ. Technol. **47**(02), 106–112 (2019)
12. Zhang, J., Du, Y.: Evaluation method of asphalt pavement roughness based on passenger experience. J. Beijing Univ. Technol. **39**(2), 257–262 (2013)
13. Zhou, T.: Research on Evaluation Method of Pavement Comfort Based on a Variety of Traffic Data (Ed.). Beijing University of Technology (2019)
14. Jiao, J., et al.: A densely connected end-to-end neural network for multiscale and multiscene SAR ship detection. IEEE Access **6**, 20881–20892 (2018)
15. Jingxiang, Z.: Detection and treatment of inter-harmonic (Ed.). Jinan university (2016)
16. Jingxiang, Z., Jinxi, Z., Dandan, C.: Preliminary study on pavement performance characteristics based on intelligent method. In: 2020 IEEE 9th Joint International Information Technology and Artificial Intelligence Conference (ITAIC), pp.1484–1490 (2021)

Certificateless Identity Management and Authentication Scheme Based on Blockchain Technology

Chao Han and Fengtong Wen[✉]

School of Mathematical Sciences, University of Jinan, Jinan 250022, China
wftwq@163.com

Abstract. Identity management and authentication in cyberspace is crucial for all forms of remote communication. The traditional authentication technology has great security risks due to its central third-party structure, such as single point of failure, malicious server attacks and so on. The emergence of blockchain technology provides a new way of thinking to solve this problem. This paper focuses on the identity management and authentication scheme based on blockchain technology. Using the decentralized, open and transparent characteristics of blockchain to make up for the shortcomings of traditional identity management and authentication mechanisms. In this paper, we analyze the BIDaaS [1] identity management and authentication scheme proposed by Jong-Hyouk and point out the obvious shortcomings of the scheme, such as suffer impersonating attack simply, virtual identities are not unique. We combine the specificity of biological characteristics to implement a unique virtual identity on the chain and improve the off-chain identity authentication process using a certificateless scheme to build a reasonable and secure identity management and authentication scheme, which realizes two-way authentication and session key agreement. The analysis shows that the scheme has a high level of safety.

Keywords: Blockchain · Identity authentication · Biometric · Certificateless

1 Introduction

Identity management and authentication is the key technology of information security. With the development of society and the continuous progress of technology, the world has entered the era of informatization, and the Internet communication interaction has been increasing, involving all aspects of people's lives, and a lot of personal information and important information of enterprises and governments are disseminated in the network, and once important information is intercepted or leaked, there will be great security risks. In such an information security context, it is increasingly important to securely manage identity and achieve mutual authentication. Public Key Infrastructure (PKI) [2] is one of the typical representatives of security solutions on the Internet, which is mainly used to provide authentication services and enable users to complete a series of operations such as authentication, access and communication in an environment where they do not trust

© The Author(s) 2022
Z. Qian et al. (Eds.): WCNA 2021, LNEE 942, pp. 1077–1088, 2022.
https://doi.org/10.1007/978-981-19-2456-9_108

each other. At present, most of the websites' public key certificates are often provided by some CA certification service companies or organizations, and the premise of our certification is to recognize the legitimacy of the certificate, but with the improvement of computing level, the risk of attacks on databases in these traditional centralized structures is increasing, and users cannot grasp the initiative of their personal data, so it is easy to have problems such as privacy leakage [3].

Along with the rapid development of Bitcoin, blockchain [4, 5], the underlying support technology for cryptographic digital currencies, has gradually attracted attention. The decentralized, secure and traceable, anonymous and tamper-proof nature of block-chain provides a new idea to solve traditional identity management and authentication, which does not rely on specific central nodes to process and store data, and thus can avoid the risk of centralized server single-point collapse and data leakage. However, there are significant differences between blockchain technology and traditional identity authentication architecture, and many traditional solutions are not applicable in block-chain applications. Coupled with the fact that in blockchain technology, data are stored in scattered nodes without a unified manager, and the performance and security capabilities of nodes vary, it is easy for attackers to compromise some of them, and attackers can even masquerade as legitimate nodes. All kinds of problems will pose a great threat to the identity authentication and privacy protection under blockchain technology. How to build a reasonable identity management and authentication scheme based on block-chain technology is crux.

2 Related Work

Blockchain technology [6, 7] is an integrated application of distributed storage, P2P networks, consensus algorithms, cryptographic mechanisms and other technologies. Its features such as decentralized and non-tamperable bring a new direction to solve the problems of single point of failure, centralization, and key management in traditional identity authentication scheme.

The concept of Bitcoin [8] was first introduced by a man named Satoshi Nakamoto in 2008, and its underlying technology blockchain quickly attracted attention. Because of its distributed and decentralized characteristics, its research in identity authentication is slowly becoming an important research direction. China's Ministry of Industry and Information Technology has even proposed through a white paper [9, 10] that blockchain has a significant role in the application of digital certificates. The design and development of blockchain technology in identity identification and authentication is firstly reflected in the use of this decentralized structure to establish PKI. 2014, MIT scholar Conner proposed the first distributed PKI scheme based on blockchain technology Certcion [11, 12], using blockchain as the core of the technology replacing the traditional CA authentication mechanism with Certcion, through Certcion replaces the traditional CA authentication mechanism by using Certcion to chain certificate information through transactions and directly bind the user's identity to the certificate public key. However, Certcoin directly binds user identity with certificate public key and does not do privacy protection processing will lead to user identity leakage and cannot prevent attackers from illegally occupying the identity of legitimate users. In addition, the calculation cost of this scheme is relatively large. Shocard [13] was an early experiment in blockchain identity management and has evolved to date, with a representative authentication and registration process that forms a consensus on the technical idea that user endpoints store personal data and the blockchain acts as a decentralised exchange commitment to ensure the validity and integrity of the information, paving the way for the creation of subsequent solutions. Blockstack [14] is a decentralised PKI system built on top of the Namecoin blockchain proposed by Muneeb Ali et al. It uses Bitcoin's proof-of-work consensus mechanism to maintain the system's state consistency, and there is no central authority or trusted third party in the system. Authcoin [15] is a decentralized PKI scheme. The protocol uses the decentralized, fault-tolerant, and hard-to-tamper features of blockchain to store data securely, eliminating the reliance on trusted third parties. There have been many subsequent attempts in the West to combine blockchain with identity management. For example, PKIoverheid and Idensys projects [16, 17] in the Netherlands, e-Residents [18] in Estonia, etc. IDHub [19] is the first blockchain-based de-centralized digital identity platform from CenturyLink in China, which is used for iden-tity authentication related to civil rights of new login methods in the network. But the drawbacks of these early attempts are also obvious, most of them use the Bitcoin block-chain, which has a distributed ledger of thousands of nodes. That is time-consuming and extremely inefficient for user authentication. And the bitcoin platform is open to all and the third-party correlation analysis of user behavior also leaks privacy to some extent. Later emerged blockchain technology identity authentication based on identity attributes, construct the KGC in the traditional identity authentication protocol through a decentralized structure, such as the protocol proposed by Wang [20] et al. and certificate-less based blockchain technology identity, such as the protocol proposed by Gervais Mwitende [21] where the blockchain identity manager holds part of the private key and attenuates the authority of third parties. However, the performance cost increases as the number of interactive communication steps increases. Muftic propose a BIX protocol [22], which aims to distribute the role of CAs while retaining security features, but the protocol is still incomplete and lacks steps to revoke and renew certificates.

With the improvement of technology, biometric identification technology [23, 24] is widely used and have become the mainstream technical used for identification in various industries because of its advantages such as being difficult to tamper, uniquenes, stability, convenient and efficient access. Some authentication based on biometric and blockchain technology have been proposed one after another, for example, in 2018 Zhou [25] proposed a two-factor authentication scheme based on blockchain technology for biometric features and password, which uses Hash algorithm and elliptic curve algorithm for authentication of biometric, which reduces the number of signatures and verification by public key algorithm, but the biometric and password need extensive use of cryptographic techniques for encryption and decryption operations, which has defects such as low efficiency and poor timeliness.

The above research results, as blockchain technology is still in its infancy, but its unique features combined with identity authentication will become the main form of authentication in the future, with great development prospects.

3   Relevant Knowledge

3.1   Computational Difficulties

(1) Elliptic curve discrete logarithm problem (ECDLP): it is known that E_p is defined in a finite field F_p on an elliptic curve of the form $E_p : y^2 = x^3 + ax + b(\mathrm{mod}p)$ of an elliptic curve, where p is a prime number $a, b \in F_p, 4a^3 + 27b^2 \neq 0modp$. Given a point on the elliptic curve $P \in E_p$, and a positive integer s, sP denotes the product of s and P, given $P, Q \in E_p$, it is impossible to compute s in polynomial time such that $Q = sP$.

(2) Elliptic Curve Computation Diffie-Hellman Problem (ECCDH): Given three points on an elliptic curve $P, sP, mP \in E_p$, it is impossible to compute in polynomial time $smP \in E_p$.

3.2   Bilinear Pairs

G_1 is an additive group of order q and G_2 is a multiplicative group. The bilinear map $e : G_1 \times G_1 \to G_2$ Satisfies the following properties.

(1) Bilinear: for any $P, Q \in G_1, a, b \in Z_q^*, e(aP, bQ) = e(P, Q)^{ab}$.
(2) Non-degeneracy: The existence of $P, Q \in G_1$ that makes $e(P, Q) \neq 1_{G_2}$, where 1_{G_2} denotes the group G_2 of unit elements.
(3) Computability: there exist efficient algorithms for any $P, Q \in G_1$, we can calculate $e(P, Q)$.

3.3   Fuzzy Extractor

Let the extracted biometric feature value be *BIO* and the fuzzy extractor [26] be a pair of functions$\{Gen(\cdot)Rep(\cdot, \cdot)\}$. The first time the biometric feature value is collected, the random generating function $Gen(\cdot)$ is used to find $(\sigma, \vartheta) = Gen(BIO)$, where σ is a random value instead of*BIO*, and ϑ while is the auxiliary string, which is academically used to recover the error correction code of *BIO*. The deterministic recovery function $Rep(\cdot, \cdot)$ is used when re-extracting to check the biometric eigenvalues, and $\sigma = Rep(\mathrm{BIO}^*, \vartheta)$ is computed for the re-extracted eigenvalues BIO*, using the error correction code ϑ as described above. Thus, σ is recovered with a specific error allowed.

3.4   Blockchain Data Structure

A blockchain is generally considered to be a decentralized, de-trusted, distributed shared ledger system in which blocks of data are assembled in chains in chronological order to form a specific data structure and are cryptographically guaranteed to be tamper-evident and unforgeable. Structurally the blockchain is composed of blocks and chain structure, where each block generally includes two parts: the block header (Header) and the block body (Body), where the block header includes version information, the hash value of the previous block, the timestamp, the target hash of the current block, the random number and the Merkle tree root (Merkle); the block body contains information about all transactions over an interval of time.

4   Scheme Analysis and Improvement

4.1   BIDaaS Review

A new identity service system BIDaaS based on blockchain technology is proposed. The solution involves three parties, user U, BIDaaS provider, and partners of the BIDaaS provider. The system aims to establish mutual authentication between users and partners without sharing any information or security credentials in advance. All three parties are blockchain nodes and have access to the blockchain.

(1)   Virtual identity creation: the user creates a pair of private key k_{pri}^{user} and public key k_{pub}^{user}. The user can securely store k_{pri}^{user}. Then a virtual identity is created using k_{pub}^{user} to create a virtual identity ID_{user}.

(2)   Identity on the chain k_{pub}^{user} and the generated virtual identities ID_{user} from the user to the BIDaaS provider through a secure channel. The BIDaaS provider use its own private key k_{pri}^{pro} will k_{pub}^{user} and ID_{user} are digitally signed $Sig_{k_{pri}^{pro}}\left(k_{pub}^{user}, ID_{user}\right)$. The BIDaaS provider then transfers the k_{pub}^{usr}, ID_{user}, the created signatures $Sig_{k_{pri}^{pro}}\left(k_{pub}^{user}, ID_{user}\right)$ placed in the blockchain. This registration is executed as a blockchain transaction and broadcast to the BIDaaS blockchain node. The registration information is then stored on the BIDaaS blockchain.

(3)   Authentication: When a user wants to access the services provided by a partner, the user simply sends a message $M_1 = (ID_{user}, r, Sig_{k_{pri}^{user}}(ID_{user}, r))$ to the partner provide ID_{user}, r. When the partner receives a service access request from the user, it first accesses the BIDaaS blockchain to check ID_{user} whether it exists on the record of the BIDaaS blockchain. If it exists, the partner obtains the relevant information, such as k_{pub}^{user}. If the verification passes, the partner sends a message $M_2 = (ID_{user}, r + 1, E_{k_{pub}^{user}}\left(ID_{user}, r + 1, k_{pub}^{ptn}\right))$. After receiving the M_2 message, the user uses k_{pri}^{user} decrypts the message and verifies it with $r + 1$. On success, the user obtains k_{pub}^{ptn} through M_2, the user sends a message $M_3 = (ID_{user}, r + 2, E_{k_{pub}^{ptn}}(ID_{user}, r + 2))$. When the partner receives M_3, it uses k_{pri}^{ptn} decrypt the message and verifies the message with $r + 2$. With the BIDaaS blockchain, authentication between the user and the partner is established.

(4)   Additional information requests: the partner may provide the BIDaaS provider with some additional information needed to provide the service to the user. The partner requests the information required by the user through a separate secure channel established with the BIDaaS provider.

4.2   Scheme Analysis

(1)   Impersonation attack: After a legitimate user completes authentication with a particular service provider to obtain a service, the user can masquerade as other users in the chain to spoof the same service provider in the next session. Suppose the attacker is denoted as A, the legitimate user is B, and the service provider is S. A completes the

session for the first time by $M_2 = (ID_A, r+1, E_{k_{pub}^A}(ID_A, r+1, k_{pub}^{ptn}))$ Get the public key of that service provider S. Next A obtains the virtual identity and passphrase of B, and a certain message issued by it, through the on-chain message $M_1 = (ID_B, r, Sig_{k_{pri}^B}(ID_B, r))$. Thereafter A can masquerade as B and use M_1 initiate a session to S. After receiving $M_2 = (ID_B, r+1, E_{k_{pub}^B}(ID_B, r+1, k_{pub}^S))$, the attacker A can still obtain public key of S based on the previous session even if he does not know user private key of B, and thus send $M_3 = (ID_B, r+2, E_{k_{pub}^S}(ID_B, r+2))$.

(2) Server spoofing attack (man-in-the-middle attack): the session does not achieve two-way authentication, attacker A intercepts the message sent by B to M_1, compute $M_2 = (ID_B, r+1, E_{k_{pub}^B}(ID_B, r+1, k_{pub}^A))$ sent to B, through the user's authentication, to achieve server spoofing attack.

(3) De-time synchronization attack: the random number r is designed such that obtaining $M_1 M_2 M_3$ the value of any random number in can be inferred from the other two message values, and there is no guarantee of a de-time synchronization attack. If the random number is replaced with a timestamp, the same impersonation attack exists.

(4) No two-way authentication: This scheme can only achieve service provider to user authentication by M_3. The user can't determine whether the service provider has received the message sent by the user and whether the service provider has decrypted it correctly.

(5) 51% attack: a user independently selects a public-private key pair and the public key for virtual identity creation. A user can select an uncountable number of public-private key pairs, then he can have an infinite number of virtual identities. Since the operation mechanism of blockchain is using consensus, then in a private chain with a limited number of nodes, a user can always create virtual identities of more than half of the number of nodes in that chain, which means he will occupy more than half of the nodes and control this chain, bringing great losses.

4.3 New Scheme

This section proposes a certificate free unique virtual identity management and authentication scheme based on blockchain technology. Users use biometrics to create and upload their identities on smart devices, the chain operator broadcasts node information to the chain, and the chain nodes use certificateless with bilinear pairs to authenticate each other and achieve key agreement.

The scheme is divided into 3 phases: virtual identity creation phase, identity on-chain phase, and identity authentication and key agreement phase. The whole process is carried out using on-chain information storage and off-chain identity authentication, and the used some of the parameters are shown in Table 1.

Table 1. Symbol description.

Symbolic	Connotation
A/B	User/server
ISP	Blockchain operators
G_1/G_2	q-order addition group/q-step multiplicative group
BIO	biological feature
s/P_0	operator master key $\in Z_q^*$/operator's public key $\in G_1^*$
e	bilinear mapping, $G_1 \times G_1 \to G_2$
h	Hash function, $\{0, 1\}^* \to G_1^*$
H	Hash function, $\{0, 1\}^* \times G_2 \to Z_q^*$
H_1	Hash function, $\{0, 1\}^* \to Z_q^*$
Q	Virtual identity on a chain
D/S	partial private key of node $\in G_1^*$/ complete private key of node $\in G_1^*$
x	random numbers. $\in Z_q^*$
k_{pri}, k_{pub}	Private key, public key
sk_{ij}, sk_{ji}	Session key

Both the user and the server act as blockchain nodes and need to be on the chain for transaction processing, session with other nodes and providing and obtaining services, then they need to perform virtual identity creation and identity on the chain. In this scheme, the user or server both act as blockchain nodes with the same attributes.

(1) Virtual identity creation: taking a user as an example, user A enters the unique identity feature BIO through a smart device with a fuzzy extractor, and the smart device obtains derived σ through generating the algorithm $Gen(BIO)$ and calculates the unique identity $ID_A = H_1(\sigma)$. To ensure the virtual nature of the user's identity in the chain, A computes $Q_A = h(ID_A)$. Q_A is the virtual identity of the corresponding node on user A's chain. and sends Q_A to the chain manager ISP. ISP receives Q_A, computes the partial private key $D_A = s \cdot Q_A$, the private key and public key of ISP is s and $P_0 = s \cdot P$; and sends D_A back to user$A$.

(2) Identity on the chain: A sends Q_A to the chain managerISP. ISP receives Q_A, calculates the partial private key $D_A = s \cdot Q_A$, and sends D_A through a secure channel. A randomly selects x_A, calculates the complete private key $S_A = x_A \cdot D_A$, and the public key $< X_A, Y_A >$, where $X_A = x_A \cdot P, Y_A = x_A \cdot s \cdot P = x_A \cdot P_0$ send an uplink request to $ISP(Q_A, Y_A)$. ISP broadcasts the upload request message to the chain and generates the corresponding block.

Similarly, Server B performs the above operations to implement the virtual identity creation and identity on the chain process.

(3) Authentication and Key Agreement: The answering process between the user and the server is completed by the following steps.

Step 1: When user A needs to request a service from server B, A picks a random number $r_i \in Z_q^*$, calculate $R_i = r_i \cdot P$. Random $a \in Z_q^*$. Calculate $w_1 = e(P,P)^a$, $U_1 = v_1 \cdot S_A + a \cdot P$, $v_1 = H(R_i, w_1)$. Send the request message $M_1 = (Q_A, R_i, X_A, U_1, v_1)$

Step 2: Server B receives a request message M_1 from user A, and based on the Q_A search the chain information, find the corresponding block and Y_A. First verify whether the public key matches by checking. $e(X_A, P_0) = e(Y_A, P)$ whether it is validated to verify whether the public key matches, determine the identity of the user, if not, abort the session to deny service, otherwise continue the following verification. Server B obtains the public key of user A $<X_A, Y_A>$, calculate $w_1 = e(U_1, P) \cdot e(Q_A, -Y_A)^{v_1}$ check $v_1 = H(R_i, w_1)$ whether it is validated. If it holds, the verification passes, otherwise the session is aborted. B choose $r_j \in Z_q^*$, calculate $R_j = r_j \cdot P$, $k_{ji} = r_j \cdot R_i$, $Auth_{ji} = h(Q_A \| Q_B \| k_{ji} \| R_j \| T)$ where T is the time stamp. Random $b \in Z_q^*$, compute $w_2 = e(P,P)^b$, $U_2 = v_2 \cdot S_B + b \cdot P$, $v_2 = H(R_j, w_2)$. Send the response message $M_2 = (T, Q_B, R_j, X_B, Auth_{ji}, U_2, v_2)$.

Step 3: User A receives a response message M_2 from Server , first checking the time T to determine ΔT whether it is within a reasonable range. Then according to the Q_B search the information on the chain, find the corresponding block and Y_B, first verify whether the public key matches by checking $e(X_B, P_0) = e(Y_B, P)$ whether it is validated to verify whether the public key matches, determine the identity of the server, if not, abort the session to deny service, otherwise continue the following verification, user A obtains the public key of server $B\langle X_B, Y_B \rangle$, calculate $w_2 = e(U_2, P) \cdot e(Q_B, -Y_B)v_2$, check $v_2 = H(R_j, w_2)$ whether it is validated. If it holds, the verification passes, otherwise the session is aborted. calculate $k_{ij} = r_i \cdot R_j$ If it is valid, then the session is aborted $Auth_{ji} = h(Q_A \| Q_B \| k_{ij} \| R_j \| T)$ and if it holds, then authentication is achieved. Compute the session key $sk_{ij} = h(Q_A \| Q_B \| k_{ij})$, which further hides the session key, computes $M_3 = sk_{ij} \oplus w_1 \oplus w_2$, $h(T)$, send M_3 to server B.

Step 4: The server receives the $M_3 = sk_{ij} \oplus w_1 \oplus w_2$, $h(T)$, first check the time T. Since T is self-selected, it can effectively avoid denial of service etc. caused by time synchronization attacks. By w_1, w_2 obtain sk_{ij}, calculate $sk_{ji} = h(Q_A \| Q_A \| k_{ji})$ and verify that. $sk_{ij} = M_3 \oplus w_1 \oplus w_2$, whether it holds. If it holds, the two parties complete mutual authentication and establish the session key.

4.4 Security Analysis

(1) Avoid single point of failure: the identity management and authentication scheme of this paper built based on the decentralized characteristics of blockchain can effectively avoid the single point of failure problem under traditional identity authentication; at the same time, in order to avoid the possible security problems caused by the existence of blockchain operators in this scheme, we adopt the design of partial private key and complete private key to realize the autonomy of user keys and public key self-certification.

(2) Resistant DOS attack: the server node itself picks the timestamp T and verifies the timeliness of T by itself, and the user node does not need to pick the parameters, after the server node completes the parameter update, there is no need to worry about clients failing to update the parameters successfully for some reason, causing obstacles to further communication.

(3) Unique virtual identity: Unlike the traditional way of password and smart card, biometric features are unique, lifelong and stable. In our scheme, users or servers need to collect biometric features through smart devices and correspond to unique virtual identity through fuzzy extractor and specific operation, then unique users or servers can only have unique nodes, avoid 51% attacks generated by the consensus mechanism in the blockchain.

(4) Resistant to replay attacks: the authentication process incorporates elements such as timestamp T to avoid replay attacks, and we use certificateless scheme to ensure that the information is not altered. On the other hand, we use a certificate with the user's private key to further ensure that the message will not be tampered with, the receiver will verify the message by the public key of the sender, thus resisting replay attacks.

(5) Resistant impersonation attack: Although any node can obtain the virtual identity and Y of other users on the chain then can also intercept the node to send information M to obtain X, thus obtain the user's virtual identity and public key, but we use the certificateless scheme, the attacker can't obtain the private key. Suppose the attacker is denoted as A, the legitimate user is B:

Here we consider the following cases: our security is based on blockchain technology, which is achieved by calculating W and verifying that V is equal, U achieves the hiding of the private key, V ensures that the information is not modified, and blockchain technology ensures that the user matches the public key.

1:A changes the R_i sent by user B for session key acquisition but without changing information such as $w1$, U_1, v_1. Then the receiver checks $v_1 = H(R_i', w_1)$ the equation does not hold and can't be verified. 2:A changes the R_i, w_1 sent by user B but without changing information such as U_1, v_1. Then the receiver checks $v_1 = H(R_i', w_1')$ the equation does not hold and can't be verified.3: A changes the R_i, w_1, v_1 sent by user B but without changing the U_1. The receiver can't calculate w_1 correctly. Assuming that the receiver calculates a new $w_1' = e(U_1, P) \cdot e(Q_A, -Y_A)v_1' = e(P, P)a \cdot e(v_1 x_A s Q_A, P) \cdot e(Q_A, -x_A s P)v_1'$, $v_1' = H(R_i', w_1')$ the equation does not hold and can't be verified. 4: A changes the UR_A, w_1, v_1, U_1 sent by user B. A calculate a new $1' = v_1' \cdot S_A' + a' \cdot P$ .The receiver can't calculate $w1$ correctly. Assuming that the receiver calculates a new $w_1' = e(U_1', P) \cdot e(Q_A, -Y_A)v_1'$
$= e(P, P)a \cdot e(v_1 S_A' Q_A, P) \cdot e(Q_A, -S_A P)v_1'$, $v_1' = H(R_i', w_1')$ the equation does not hold and can't be verified.

It means that the attacker who does not hold the node private key cannot complete the disguise, and the node private key is determined by a random number chosen by the node itself.

(6) Resistant internal attacks: Blockchain operators provide part of private keys for nodes, which ensures that node keys are generated by themselves and no identical information between different nodes, then internal nodes, whether other users or other servers and operators, cannot carry out internal attacks. The decentralization and consensus mechanism of blockchain also guarantee the scheme resistance to internal attacks.

4.5 Efficiency Analysis

(1) No need to store authentication table: Unlike traditional solutions, we use biometric features combined with fuzzy extractor to eliminate the process of storing authentication table to verify whether a user is a scheme user by the management center in the past, saving a lot of storage space. And the user calculates and manages the public and private keys independently, no certificate is required.

(2) Two-way multiple authentication: in our authentication process, nodes first verify each other's identity by X and Y, and then use certificateless for another authentication, in addition to adding the authentication information Auth for further authentication, the authentication process is more robust.

(3) Operational complexity: Firstly, the authentication process in our scheme has only three message passes, which completes the two-way authentication and achieves session key agreement by the minimum number of times. Secondly, only two iso-or, nine hashes, five bilinear pairs, and three signature operations are applied in our scheme.

5 Summary

In this paper, we propose a certificate free unique virtual identity management and authentication scheme based on blockchain technology for identity management and mutual authentication with the help of blockchain technology. We use biometric features to ensure the uniqueness of the virtual identity of the node from the user or server, use certificateless to ensure privacy and secure the information, and achieve mutual authentication and key agreement between the two parties with the help of decentralization, immutability and openness and transparency of blockchain technology. Through analysis our solution has high security and efficiency.

The identity authentication based on blockchain technology also has the function of cross-domain authentication, how to improve the scheme in this paper so as realize the cross-domain authentication between different private chains or federated chains to make the identity management and authentication in cyberspace more convenient and secure is the direction we want to study.

References

1. Lee, J.H.: BIDaaS: blockchain based ID as a service. IEEE Access **6**, 2274–2278 (2018)
2. Maurer, U.: Modelling a public-key infrastructure. In: Bertino, E., Kurth, H., Martella, G., Montolivo, E. (eds.) ESORICS 1996. LNCS, vol. 1146, pp. 325–350. Springer, Heidelberg (1996). https://doi.org/10.1007/3-540-61770-1_45
3. Fromknecht C., Velicanu D.: A decentralized public key infrastructure with identity retention. Technical Report, Massachusetts Institute of Technology **803** (2014)
4. Yuan, Y., Wang, F.: Development status and prospects of blockchain technology. Acta Autom. Sin. **42**(04), 481–494 (2016)
5. Shao, Q., Jin, C., Zhang, Z., Qian, W.: Blockchain technology: architecture and progress. Chin. J. Comput. **41**(05), 969–988 (2018)

6. Xin, S., Qingqi, P., Xuefeng, L.: Summary of blockchain technology. J. Netw. Inf. Secur. **2**(11), 11–20 (2016)
7. Pan, W., Huang, X.: Identity management and authentication model based on smart contract. Comput. Eng. Des. **41**(4), 915–919 (2020)
8. Nakamoto, S.: Bitcoin:A peer-to-peer electronic cash sytem[EB/OL]. https://bitcoin.org/bit coin.pdf. Accessed 2008
9. Ministry of Industry and Information Technology. White paper for Chinese blockchain technology and application development. Beijing: Ministry Ind. Inf. Technol. **23** (2016)
10. Blockchain White Paper. Beijing: China Academy of Information and Communications Technology (2019)
11. Fromknecht, C., Velicanu, D.: CertCoin: A NameCoin Based Decentralized Authentication System. Technical Report, 6.857 Project, Massachusetts Institute of Technology (2014)
12. Fromknecht, C., Velicanu, D.: A decentralized public key infrastructure with identity retention. Technical Report, 803, Massachusetts Institute of Technology (2014)
13. Travel Identity of the future [EB/OL], https://shocard.com. Accessed 06 July 2019
14. Ali, M., Nelson, J., Shea, R., Freedman, M.J.: Blockstack: a global naming and storage system secured by blockchains. In: 2016 USENIX Annual Technical Conference (USENIX ATC 16), Denver, CO, pp. 181–194 (2016)
15. Leiding, B., Cap, C.H., Mundt, T., et al.: Authcoin: validation and authentication in decentralized networks. arXiv preprint arXiv:1609.04955 (2016)
16. PKI overheid-Logius [EB/OL], https://www.logius.nl/diensten/pkioverheid/. Accessed 14 Mar 2018
17. GDI[EB/OL].   https://www.digitaleoverheid.nl/digitaal-2017/digitalisering-aanbod/gdi. Accessed 14 Mar 2018
18. Estonia's new e-residents are surpassing the country's birthrate [EB/OL]. https://thenex tweb.com/eu/2017/07/25/estonias-new-e-residents-surpassing-countrysbirth-rate/. Accessed 14 Mar 2018
19. IDHu Digital Identity White Paper [Z] (2017)
20. Wang., Z., et al.: ID authentication scheme based on PTPM and certificateless public key cryptography in cloud environment. Softw. **27**(6), 1523–1537 (2016)
21. Gervais, M., Sun, L., Wang, K., Li, F.: Certificateless authenticated key agreement for decentralized WBANs. In: International Conference on Frontiers in Cyber Security, pp. 268–290. Springer (2019)
22. Muftic, S.: Bix certificates: cryptographic tokens for anonymous transactions based on certificates publicledger. Ledger **1**, 19–37 (2016)
23. Li, S.Z.: Encyclopedia of Biometrics, pp. 1–22. Springer, NJ, USA (2009)
24. Jain, A.K.: Biometric recognition: Q&A. Nature **449**, 38–40 (2007)
25. Zhicheng, Z., Lixin, L.: Biometric and password two-factor cross domain authentication scheme based on blockchain technology. J. Comput. Appl. **38**(6),1620–1627 (2018)
26. Wang, D., Li, W.T., Wang, P.: Cryptanalysis of three anonymous authentication schemes for multi-server environment. J. Softw. **29**(7), 1937–1952 (2018)

Simulations of Fuzzy PID Temperature Control System for Plant Factory

Hongmei Xie[1($\boxtimes$)], Yuxiao Yan[1], and Tianzi Zeng[2]

[1] Northwestern Polytechnical University, Xi'an 710072, Shaanxi, China
xiehm@nwpu.edu.cn
[2] Northwestern Polytechnical University, Shaanxi 710000, Chang'an, China

Abstract. The five key factors that affect plant growth are temperature, humidity, CO_2 gas density, nutritious liquid density and light intensity. The monitoring and controlling of these factors are vital. Fuzzy PID controller technology for plant factory environment parameter controlling was proposed and temperature controlling using three different methods were given out. The physical and mathematical models of ordinary differential equation used in temperature subsystem in plant factory was established, traditional PID controller was discussed and specifically the fuzzification interface, membership function, fuzzy inference rule and the defuzzification procedure were designed for mere fuzzy and fuzzy PID controllers. Simulations for temperature controlling using pure PID, mere fuzzy and fuzzy PID control algorithm were performed respectively. The experimental results show that the performance of the novel fuzzy PID controller is best since it outperforms the other controllers in terms of stable error, overshooting and stabling time. The stable error, overshooting and time to stable for fuzzy PID are 0, 0.1% and 170 s respectively, all are the minimum among the three controllers.

Keywords: Internet of Things · Plant factory · Mere fuzzy controller · Fuzzy PID controller · Performance simulation

1 Introduction

The plant factory (PF) can stably cultivate high-quality vegetables in any environment by manually controlling the plant growth environment. Nowadays, with the increasing of population, reduction of arable land and degradation of the environment, there is an urgent need for artificial plant factory to grow vegetables or cultivate seeds under severe conditions like space-station or scientific investigation sites in Antarctica [1]. Meanwhile the requirements for high quantity and quality of food have continued to increase, therefore, plant factory was proposed all around the world to meet these urgent demands [2–5]. Based on the urgent needs and current technology, we designed a prototype control system [6] using ARM and wireless communication techniques like Zigbee for a plant factory for green-leaf vegetable growing. Nowadays, the intelligent fuzzy theory-based environment parameter controlling and the corresponding mini realization is a trend in research field, so PF temperature adjustment algorithms using advanced theory need to be investigated thoroughly.

© The Author(s) 2022
Z. Qian et al. (Eds.): WCNA 2021, LNEE 942, pp. 1089–1099, 2022.
https://doi.org/10.1007/978-981-19-2456-9_109

The plant factory is divided into two parts: a set of wireless sensor networks and an embedded human-machine controlling platform. The system has a clear structure, and strong versatility, which provides a broad application prospect for agricultural development.

Temperature plays a key role in plant growth, so researchers have proposed various controlling methods [7–10] for temperature controlling. This study took the temperature controlling of a plant factory as the research object. We established the controlled object model, analyzed the classical proportion-integration-differential controlling (PID-C) and the mere fuzzy controlling (FC) methods. After that we presented a novel fuzzy proportion-integration-differential controlling (F-PID-C) strategy, implemented and tested it in terms of some objective controlling metrics.

Fuzzy controlling is a method to mimic human's experience and knowledge to control a system. This research aims to take advantage of the capability of fuzzy controlling system and apply it to plant factory. Wang H.Q. et.al. [10] compared the pure PID controller and fuzzy PID controller for plant temperature. In [9,10], the authors proposed a fuzzy logic controller for robots to control the wheels' speed and moving direction. And some other embedded systems based on fuzzy controlling were discussed in [11, 12]. This paper designed, coded using higher and lower-level programming languages using the developed hardware prototype and fuzzy control theory.

The following organization of this paper is as below. Section 2 gives out the mathematical modelling and various methods which including PID-C, FC and the proposed F-PID-C for temperature. Simulations and experimental results are shown in Sect. 3. Discussions, summary and conclusion are given in the last section.

2   Modeling and Algorithms for Temperature Controlling

2.1   Mathematical Modeling of Temperature Controlling System

The temperature is adjusted by heating and cooling controllers. Here we took the heating process as our T, after theory and experimental analysis, we found that the dynamic behavior of the plant factory can be modeled as ideal 1-order inertia time-delay model as "Eq. (1)".

$$G(s) = \frac{Ke^{-\tau s}}{Ts + 1} \tag{1}$$

Here K is the static gain, T is the time constant and τ is the pure delay time of the object. Here we analyze the 3 types of controlling strategies as following: PID controlling (PID-C) has simple structure, reliable performance and it can eliminate the stable error in most cases. Fuzzy controlling (FC) has short response time and small overshoot and it can simulate human reasoning and decision-making based on prior knowledge and expert experience. Fuzzy PID controlling (F-PID-C) has fast response speed and it integrated the intelligent fuzzy controlling with the basic PID structure, which is stronger and more accurate.

2.2 Design and Analysis of Different Controlling Methods

PID Controlling (PID-C). PID-C has proportion, integration and differential components connected in parallel. Controlling bias is the required value minus the output value.
The relationship between input and output is as "Eq. (2)".

$$u(t) = K_p e(t) + K_i \int_0^t e(\tau)d\tau + K_d \frac{d}{dt} e(t) \tag{2}$$

where $u(t)$ is the output, $e(t)$ is the input, Kp is the proportional coefficient, K_i is the
integration coefficient, K_d is the differential time coefficient respectively.

PID controller is implemented by PID controlling algorithms program. The input
signal is analog and it must be converted to digital signals via sampling/holding and
quantization. To simplify the writing, $e(kT)$ is denoted as $e(k)$. Transformed equation
is as "Eq. (3)". The controlled parameter's increasing value is as "Eq. (4)"

$$u(k) = K_p e(t) + K_i \sum_{j=0}^{k} e(j) + K_d[e(k) - e(k-1)] \tag{3}$$

$$\Delta u(k) = K_p \Delta e(k) + K_i e(k) + K_d[\Delta e(k) - \Delta e(k-1)] \tag{4}$$

The controlled parameter's increasing value $\Delta u(k)$ can be get using the former three
measured bias values since general control system using constant sampling period T.
Note that we adopted 4-points center difference methods to merge the difference terms
for PID controlling design. The difference terms is as "Eq. (5)". By weighted summation,
the approximated differential term are as "Eq. (6)".

$$\bar{e}(k) = \frac{[e(k) + e(k-1) + e(k-2) + e(k-3)]}{4} \tag{5}$$

$$\frac{\Delta\bar{e}(k)}{T} = \frac{1}{6T}[e(k) + 3e(k-1) - 3e(k-2) - e(k-3)] \tag{6}$$

Fuzzy Controlling (FC). FC is a kind of computer digital control based on fuzzy set,
fuzzy language variables and fuzzy logic inference system [13]. FC technology mim-
ics human's thinking and accepts inaccurate and incomplete information for logical
reasoning. The structure diagram of FC is shown as Fig. 1.

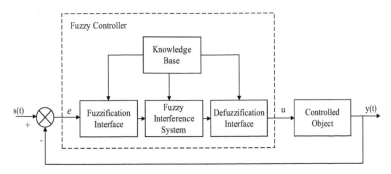

Fig. 1. Block diagram of a FC system.

In Fig. 1, $y(t)$ is the output of the controlled object, u is the input of the controlled object, $s(t)$ is the reference/required input, e is the error.

In real application, FC can be composed by two ways. One is to use fuzzy logic chip and this manner has characteristics like fast speed but the corresponding I/O and controlling rule are limited. Another way is to use MCU to realize FC. In plant factory, the FC is realized by the latter way.

The fuzzy controller is mainly composed of the following four parts:

Fuzzification Interface. The input of fuzzy part is not only the error e but also the changing rate of error Δe. We convert e and Δe into ambiguous variable by membership function. The commonly used triangular membership function is shown as Fig. 2.

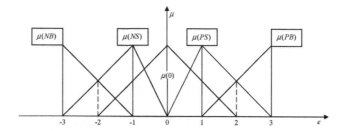

Fig. 2. Triangular membership function.

Knowledge Base (KB). The knowledge base (KB), as the name implies, stores all the knowledge about the fuzzy controller. Input and output refer to the fuzzy controlling rules table. The inputs of E and EC together determine the output. The input values and output value are expressed in fuzzy language as Negative Big (NB), Negative Medium (NM), Negative Small (NS), Zero (ZO), Positive Small (PS), Positive Medium (PM), Positive Big (PB).

Fuzzy Inference system. The input quantities are E and EC, which is updated at each sampling time. E stands for the vector $\mathbf{A}'$, and EC corresponds to the B' and then the reasoning result vector C is shown as "Eq. (7)".

$$C' = (A'XB') \circ R \tag{7}$$

Defuzzification Interface. Using defuzzification algorithm like maximum DoMF, gravity center or median methods, the controlling parameter μ can be obtained. Readers can refer to [10] for the detail information of the three defuzzification methods. Here, weighted averaging is adopted. It can be expressed as "Eq. (8)".

$$C(k) = \frac{\sum_{i=1}^{n} k_i c_i}{\sum_{i=1}^{n} k_i} \tag{8}$$

Where the coefficient k_i can be selected accordingly. The weighted averaging method is very flexible. Finally, the actual output is obtained by inverse domain transformation.

Fuzzy PID Controlling (F-PID-C). F-PID-C is a combination of PID and fuzzy control algorithms. It realized on-line self-tuning of three PID parameters through the control of fuzzy system. The input of the fuzzy system are deviation E and the change rate EC, and the change values of the three PID parameters, and are used as outputs, F-PID-c takes into account the advantages of PID control system, such as simple principle, convenient use, strong robustness, etc. and makes the controlled system have good performance in both static and dynamic environments, which makes it easy to implement with a single-chip microcomputer. Here we majorly solve the temperature controlling problem. Based on the influence of parameters K_d, K_i and K_p, at different E and EC, the requirements for the parameters are as following:

(1) When the value of E is large, K_p should be increased and K_d should be reduced to fasten the response speed and the integral effect should be removed (i.e. $K_i = 0$) to prevent saturation of the integral and avoid large overshoot in the system response.
(2) When the value of E and EC are of medium value, the three parameters should be increased. We should reduce K_p values slightly, and keep K_i and K_d moderate to ensure the system's responding speed.
(3) When the value of E is small, the value of K_p and K_i should be increased to make the system have good performance in stability. Meanwhile, considering the oscillation amplitude and anti-interference ability of the system. The setting principle of K_d is: when EC is small, K_d can be increased, usually a medium value; when EC is large, K_d should be reduced. The adjusting equation for K_p, K_i and K_d as "Eq. (9)".

$$K_P = K'_P + \Delta K_P$$
$$K_i = K'_i + \Delta K_i \tag{9}$$
$$K_d = K'_d + \Delta K_d$$

The initial values of K'_p, K'_i and K'_d are obtained by conventional methods. During system operation, the three parameters are optimally tuned by means of a fuzzy controller. The specific steps are as following:

(1) The first fuzzy controller is established according to the fuzzy control rules of the proportional section, and the input amount (E and EC) and output amount of the first fuzzy controller are fuzzy variables {NB, NM, NS, ZO, PS, PM, PB}. Proportional partial self-tuning is achieved with the first fuzzy controller.
(2) The second fuzzy controller is established according to the fuzzy control rules of the integration section, and the input amount (E and EC) and output amount of the second fuzzy controller are fuzzy variables {NB, NM, NS, ZO, PS, PM, PB}. Integration partial self-tuning is achieved with the second fuzzy controller.
(3) The third fuzzy controller is established according to the fuzzy control rules of the differential components section, and the input amount (E and EC) and output amount of the third fuzzy controller are fuzzy variables {NB, NM, NS, ZO, PS, PM, PB}. Differential components partial self-tuning is achieved with the third fuzzy controller.

3   Simulation Experiments

The real developed plant factory prototype is shown in Fig. 3 and we established simulation models based on the mathematical model of the real system.

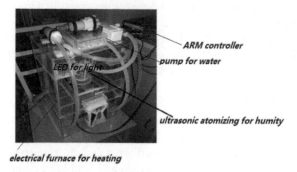

ARM controller

pump for water

LED for light

ultrasonic atomizing for humity

electrical furnace for heating

Fig. 3. The prototype of our developed plant factory.

As shown in Fig. 3, the designed plant factory has the following environmental control facilities: electric furnace for heating, semiconductor cooling circuits for cooling, ultrasonic atomizing chips for humidification, LED lights for illumination, and the corresponding supporting equipment. And by using the ARM and ZIGBEE development platform, we designed a set of data acquisition and control systems.

Then the authors simulated the system environment, analyzed, calculated, and studied on PC by programming/coding and fuzzy GUI toolbox interface to obtain the true quantitative relationship.

3.1   Modeling of the Plant Factory Temperature Controlling System

Experimental environment and initial condition setup: the temperature changing from 12 °C−28 °C. The nominal voltage and power of electric furnace are 220 V and 250 W and the test voltage is 45 V. The step response curve (also known as rising curve) can be obtained by experimental test. Using the method of flying curve measuring, we can obtain the mathematical model of the control object. Here $t_{0.284}$ and $t_{0.632}$ are the corresponding time when the rising curve reaches 28.4% and 63.2%, of the steady-state value respectively. Then the transfer function for the plant factory temperature controlling system is a one-order ODE system. Gain K was determined according to "Eq. (1)". Then the other two parameters τ and T are obtained by approximately calculating method, which are shown as "Eq. (10) and (11)"

$$\tau = 1.5(t_{0.284} - \frac{1}{3}t_{0.632}) \tag{10}$$

$$T = 1.5(T_{0.632} - T_{0.284}) \tag{11}$$

3.2 Experimental Simulation Results

Parameter Determination for PID Controller. The following is an introduction to common methods. Empirical data method could provide data range according to long-term practical experience.

Table 1. Ziegler-Nichols empirical formula.

	K_p	T_n	T_v	K_i	K_d
P	$0.5K_{pcrit}$	–	–	–	–
PD	$0.8K_{pcrit}$	–	$0.12T_{crit}$	–	K_pT_v
PI	$0.45K_{pcrit}$	$0.85T_{crit}$	–	$\frac{K_p}{T_n}$	–
PID	$0.6K_{pcrit}$	$0.5T_{crit}$	$0.12T_{crit}$	$\frac{K_p}{T_n}$	K_pT_v

The optimal value of parameter will change with the change of controlled object. Ziegler-Nichols regularizing can calculate parameter values quickly and accurately. Here we obtained parameters according to the Ziegler-Nichols empirical formula (as shown in Table 1).

The stability limit is determined by the proportion part. This limit will be reached when steady state oscillation occurs, thereby determining the values of K_{pcrit} and T_{crit} .Where $K_{pcrit} = 0.19$ and $T_{crit} = 125$. And when the desired value is 20 °C, the simulated response curve is shown in Fig. 6.

Software Composition of Fuzzy Controlling. MATLAB has fuzzy control toolbox and Simulink simulation platform. We use MATLAB and Simulink platform to build the entire fuzzy control system and conduct simulation research.

First, we constructed the following Mamdani-type fuzzy controller as shown in Fig. 4, and created a FIS-type file named fuzzy. FIS which inputs the relationship of fuzzy controller input variables$E$andEC given at Table 1 and the output of controller is shown by oscillator.

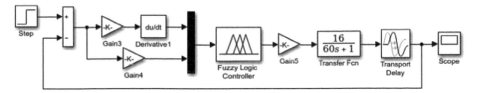

Fig. 4. Structure of FC in simulation.

When the given value is 20 °C, the fuzzy controller controls the electrical heating temperature control system and the simulation response curve is shown in Fig. 7.

Structure of Fuzzy PID. The structure of PID FC is shown in Fig. 5. And similar steps were taken here as the fuzzy controller, the only difference is that when creating the FIS file, we used the commonly referred three tables in ref[12]. The simulation result for PID FC is shown in Fig. 8.

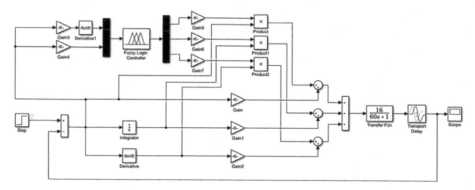

Fig. 5. Structure of fuzzy PID controller in simulation.

Simulation Results and Analysis. The modeled plant factory temperature system is $G(s) = \frac{16}{60s+1}e^{-40s}$, it is a one-order ODE. The controlling performance is evaluated by stable error, stabling time and overshoot, which is defined as following: Stable error is the difference between true value and ideal value. Stabling time is the interval from the beginning point to 90% of the stable value. Overshoot is the maximum deviation of the adjusted parameter from the given value.

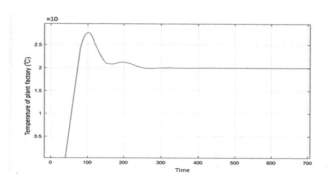

Fig. 6. PID controller ($P = 0.114, I = 0.001824$)

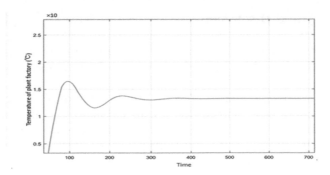

Fig. 7. Fuzzy controller ($K_{ec} = 3$, $K_e = 0.49$, $K = 0.2$)

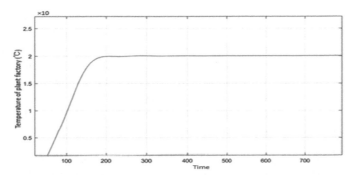

Fig. 8. Fuzzy   PID   controller($K_{ec} = 3.3$, $K_e = 1$, $P = 0.114$, $I = 0.0006$, $D = 2.0387$, $K_p = 0.045$, $K_i = 0.00006$, $K_d = 0.081$)

Comparison of simulation results of three different controllers is show in Table 2.

Table 2. Simulation results.

	Overshoot	Stabling time	Stabling error
PID	37%	350 s	0
Pure Fuzzy	24%	450 s	0.681
Fuzzy PID	0.1%	320 s	0

From Table 6, the experimental results show that the fuzzy PID controller is the best controller in terms of all the three metrics.

4 Summary and Discussion

The temperature adjustment facility now used is a heating oven in the developed prototype and later it can be replaced by some more advanced devices such as semi-conductor circuits.

This paper majorly discussed the fast and accurate control of temperature and proposed a novel fuzzy PID controller and test its advantages for the commonly used one-order dynamic system. Later work can continue to develop fuzzy-based control subsystem for other environment as water pumping in or out, humidity or carbon oxide adjustment. And if higher order system is involved, the corresponding two-order system controlling and control effect evaluation should be done.

The focus and main contribute of the work is that a fuzzy PID controlling method is presented and tested based on the math model of the temperature control system for the plant factory. The fuzzy PID controlling method combines the advantages of the other two controlling methods, and achieves the ideal performance of shorter system adjustment time, smaller overshoot, and smaller steady state error. And the conclusion that the best controlling strategy is the fuzzy PID controlling in terms of control stability, adjust time and speed. Therefore, the fuzzy PID should be given priority in the temperature of the plant factory instead of pure PID or mere fuzzy controller.

To make the method more useful, future work can also focus on the embedding system implementation of the F-PID-C into the plant factory application field.

Acknowledgments. This research was funded by National Nature Science Foundation of China, grant number: 617024419. The authors would give their sincere thanks to the anonymous reviewers for their kind reviewing and valuable remarks.

References

1. Kozai, T., Niu, G., Takagaki, M.: Plant Factory: An Indoor Vertical Farming System for Efficient Quality Food Production. 2nd edn. Academic Press (2015)
2. Mamdani, E.H.: Application of fuzzy algorithm for control of simple dynamic plant. Proc. IEEE **121**, 1585–1588 (1974)
3. Seo, K.A.: Simulation study on an artificial neural network based automatic control system of a plant factory. Int. J. Control Autom. **5**(6), 127–136 (2013)
4. Song, S.J.; Liu, Y.F.; Zhao, W.Q.: Fuzzy control of microwave dryer for drying Chinese Jujube. In: Proceedings of FSDM, pp. 181–190 (2019)
5. Cui, S., Chen, M., Zhang, Y., He, L.: Study on decoupling control system of temperature and humidity in intelligent plant factory. In: 2020 IEEE 9th Joint International Information Technology and Artificial Intelligence Conference (ITAIC), pp. 2160–2163 (2020). https://doi.org/10.1109/ITAIC49862.2020.9339036
6. Shi, J.F.: Design of the Environment Control for Plant Factory Base on ARM and ZigBee. Northwestern Polytechnical University, Xi'an, China (2016)
7. Hou, S.W., Tong, S.R.: Fuzzy logic based assignable caused ranking system for control chart abnormity diagnosis. IEEE Int. Conf. Fuzzy Syst. (FUZZ) **1–6**, 49–53 (2008)

8. Millington, I., et.al.: Artificial Intelligence for Games. Morgan Kaufmann Publisher. 2nd edn (2009)
9. Ohyama, K.: Coefficient of performance for cooling a home-use air conditioner installed in a closed-type transplant production system. J. High Technol. Agri. **14**, 141–146 (2002)
10. Wang, H.Q., Ji, C.Y., et al.: Modeling and simulation of fuzzy self-tuning PID temperature control system. Comput. Eng. **4**(38), 233–239 (2012)
11. Gao, S.Y., Xu, F.Z., Zhang, H.X.: A fuzzy logic cross-coupling controller for mobile robots. WIT Trans. Inf. Commun. Technol. **67**, 59–68 (2014)
12. Li, S.Y.: Fuzzy Control. Haerbin Institute Technology PublishingHouse (2011)
13. O'Dwyer, A.: Handbook of PI and PID Controller Tuning Rules. World Scientific (2006)
14. Shi, Z., Wang, C.: Application of fuzzy control in temperature control systems. In: Proceedings of 2011 International Conference on Electronic & Mechanical Engineering and Information Technology, pp. 451–453 (2011)
15. El Maidah, N., Putra, A.E., Pulungan, R.: A fuzzy control system for temperature and humidity warehouse control. Inform. J. **1**(2), 7277–7283 (2018)

An Effective GAN-Based Multi-classification Approach for Financial Time Series

Lei Liu, Zheng Pei, Peng Chen$^{(\boxtimes)}$, Zhisheng Gao, Zhihao Gan, and Kang Feng

School of Computer and Software Engineering, Xihua University, Chengdu, China
chenpeng@mail.xhu.edu.cn

Abstract. Deep learning has achieved significant success in various applications due to its powerful feature representations of complex data. Financial time series forecasting is no exception. In this work we leverage Generative Adversarial Nets (GAN), which has been extensively studied recently, for the end-to-end multi-classification of financial time series. An improved generative model based on Convolutional Long Short-Term Memory (ConvLSTM) and Multi-Layer Perceptron (MLP) is proposed to effectively capture temporal features and mine the data distribution of volatility trends (short, neutral, and long) from given financial time series data. We empirically compare the proposed approach with state-of-the-art multi-classification methods on real-world stock dataset. The results show that the proposed GAN-based method outperforms its competitors in precision and F1 score.

Keywords: Financial time series · GAN · Convolutional LSTM · Classification

1 Introduction

In the past two decades, people have become more and more interested in the classification of time series, and more and more scholars at home and abroad have joined the research. Moreover, with the advent of the 5G era, big data is closely related to our lives. Time series data is everywhere, especially in the medical industry, industrial industry, and meteorology [1–3].

Time series classification is a critical issue in the research of time series data mining. Time series classification (TSC) accurately classifies a series of unknown time series according to the known "category" labels in the time series, and TSC can be regarded as a "supervised" learning mode. TSC has always been regarded as one of the most challenging problems in data mining, and it is more challenging than traditional classification methods [4]. First of all, time series classification needs to consider the numerical relationship between different attributes and the order relationship of all time series points. In addition, the financial time series has complex, highly noisy, dynamic, non-linear, non-parameters and chaos characteristics, so how the model can learn the characteristics of the sequence to have a better performance in classification performance will be very challenging. Since 2015, hundreds of TSC algorithms have been proposed [5]. Traditional time series classification methods based on sequence distance have proven to achieve

© The Author(s) 2022
Z. Qian et al. (Eds.): WCNA 2021, LNEE 942, pp. 1100–1107, 2022.
https://doi.org/10.1007/978-981-19-2456-9_110

the best classification performance in most fields. In addition, there are feature-based classification methods that have excellent classification performance based on existing good features. However, it is challenging to design good features when faced with financial time series to capture some inherent properties. Although the methods based on distance or feature are used in many research works, these two methods have caused too much calculation for many practical applications [6]. As many researchers apply deep learning methods to TSC, more and more TSC methods are proposed, especially with new deep structures such as residual neural networks and convolutional neural networks. These methods are applied in image, text, and audio areas to process time series data and related analysis. Such as Fazle et al. proposed a multivariate LSTM-FCNs for time series classification, which further improved the model's classification accuracy by improving the structure of the full convolution block [7].

Inspired by the classification application of deep learning in the image field, such as GAN, which has achieved remarkable success in generating high-quality images in computer vision, we explore a deep learning framework for multivariate financial time series classification. The model uses ConvLSTM as the generator to learn the distribution characteristics of the data and MLP as the discriminator to discriminate whether the output data of the generator is true or false. We evaluated the performance of our model on publicly available stock datasets and selected several classic comparison methods. The experimental results show that the classification performance of the GAN on the MSFT is significantly improved compared to other models and less pre-processing. We summarize our contributions as follows:

- We propose an effective GAN-based volatility trends multi-classification model for multivariate financial time series based on stock data with multiple technical indicators.
- We improved the generator of GAN by adopting ConvLSTM to capture temporal dependencies and classify various volatility trends efficiently.

The organizational structure of this paper is as follows: Sect. 2 reviews relevant research work. Section 3 introduces the proposed improved model. The Sect. 4 presents the experiments done. Finally, we draw our conclusions in Sect. 5.

2 Related Work

In the classification research of time series, many deep learning methods have been applied. For example, Michael [8] and others took the lead in applying recurrent neural networks (RNN) to time series classification. Recently, Yi et al. [9] have proposed multi-channels deep convolutional neural networks (MC-DCNN) by improving convolutional neural networks (CNN). This model automatically learns the features of a single variable time series in each channel [10] has achieved great success in computer vision, especially in graphic recognition tasks, such as GAN has been achieved remarkable success in computer vision high-quality image generation. The application scenarios of GAN have been rapidly developed, covering images, texts, time series. With the continuous investment of researchers, GAN has been researching more and more in data generation, anomaly detection, time series prediction, classification. Ian Goodfellow and others first proposed the GAN to generate high-quality pictures [11]. Later, Xu, Zhan, and others [12] used improved GAN and LSTM to predict satellite images, thereby obtaining important resources for weather forecasting. In recent years, there have been more and more researches using generative confrontation networks on financial time series, and the research on the price trend fluctuation prediction is of great practical value. Zhang et al. [13] applied GAN to stock price prediction, tried to use GAN to capture the distribution of actual stock data, and achieved good results compared with existing deep learning methods. Feng [14] and others proposed a method based on adversarial training to improve the generalization of neural network prediction models. The results show that their model performs better than the existing methods. According to the characteristics of financial time series, we know that the challenge of this research is how to let GAN learn the price data trend distribution of the original data to have a better performance in the end-to-end classification. Meanwhile, the three-classification research on the financial time series price trend is more challenging than binary classification. However, it has an outstanding good reference value for stock trading.

3   Methodology

We propose a new GAN architecture for end-to-end three-classification of stock closing price trends based on this principle. Based on the improvement on GAN. We will show the detailed structure description in Fig. 1. It shows that the model's input is $X = \{ x_1, x_2, \cdots, x_t \}$ composed of daily stock data for t days. Both X_{fake} and X_{real} are a probability matrix with one row and three columns of the discriminator's output. In the GAN, both the generator and the discriminator try to optimize a value function, and eventually, they reach an equilibrium point called Nash equilibrium. Therefore, we can define our value function $V(G, D)$ as:

$$\min_{G} \max_{D} V(G, D) = E[\log D(X_{real})] + E[\log(1 - D(X_{fake})] \tag{1}$$

When calculating the error of the probability matrix one-hot encoding, we usually use the cross- entropy loss function. Given two probability distributions p and q, the cross-entropy of p expressed by q is defined as follows:

$$H(p, q) = - \sum p(x) \log q(x) \tag{2}$$

where p represents the actual label and q represents the predicted label. We get the probability matrix $\hat{C}_{t+1}$ and calculate the cross-entropy loss with the actual probability matrix C_{t+1} at that moment.

$$D_{loss} = \frac{1}{m} \sum_{i=1}^{m} H(D(X_{real}), D(X_{fake})) \tag{3}$$

$$G_{loss} = \frac{1}{m} \sum_{i=1}^{m} H(C_i, \hat{C}_i) \tag{4}$$

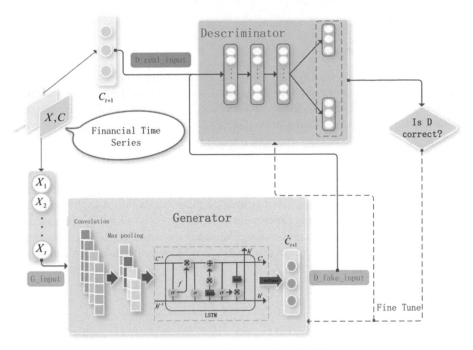

Fig. 1. The architecture of our GAN.

The eleven technical indicators are: 'Close', 'High', 'Low', 'Open', 'RSI', 'ADX', 'CCI', 'FASTD', 'SLOWD', 'WILLER', 'SMA' [15]. Each input X is a vector composed of the above eleven features. Based on the generator, we extract the output of ConvLSTM and put it into a fully connected layer to generate three types of probability matrices of short, neutral, and long through the softmax activation function, which is defined as follows:

$$C_{t+1} = [\alpha, \beta, \gamma], (\alpha + \beta + \gamma = 1) \tag{5}$$

The goal is to let $\hat{C}_{t+1}$ approach C_{t+1}, and we can get $\hat{x}_{t+1,C}$ from $\hat{x}_{t+1}$ so that we can get the probability matrices. The output of generator $G(X)$ defined as follows.

$$h_t = g(x) \tag{6}$$

$$G(X) = \hat{C}_{t+1} = \delta(W_h^T h_t + b_h) \tag{7}$$

Where $g(\cdot)$ denotes the output of ConvLSTM and h_t is the output of the ConvLSTM with $X = \{ x_1, x_2, \cdots, x_t \}$ as the input δ stands for the softmax activate function. W_h and b_h denote the weight and bias in the fully connected layer. We also use dropout as a regularization method to avoid overfitting. In addition, we can use the idea of a sliding window to predict $\hat{C}_{t+2}$ by $\hat{C}_{t+1}$ and X.

4 Experiments

4.1 Dataset Descriptions

We selected actual stock trading data from the Yahoo Finance website (https://finance. yahoo.com/) to evaluate our model and selected several classic deep learning methods as baseline methods. These stock data is Microsoft Corporation (MSFT). We construct our label data through the closing price (Close) and define $x_{Close,i}-x_{Close,i+1} > \mu$ as short, $x_{Close,i+1}-x_{Close,i} < \theta$ as long, and $x_{Close,i+1}-x_{Close,i} = \lambda$ as neutral $(0 < i < n)$, where $\mu, \theta > 0$, $\lambda = 0$ is the parameter we set according to the corresponding stock. We first normalize the data with Z-score to eliminate the influence of dimensions between different variables. Our goal is to predict the trend of the stock's closing price on the next day and get the trend of the closing price on the $t + 1$ day through the input X_t of the past t days. Through repeated experiments, we set t to be 30. Our data is divided into three parts: training, validation and testing. We select the first 85%–90% of the data on each stock as the training set and the rest (10%–15%) part as the validation and test set. We will give the trend chart in Fig. 2.

Fig. 2. The trend image of MSFT.

From Fig. 2, we can intuitively see that the MSFT data's price trends fluctuate from the beginning. When it rose to 2000, it began to decline in an oscillating trend and then remained in a long-term turbulence "stable" until it began to rise in 2012. As a result, it can be seen that MSFT can better test the robustness of different models. The MSFT data set started from 1999/1/4 to 2018/12/31, the length is 5031, the training set length is 5031, the validation set length is 252, and the test set length is 503.

4.2 Experiment Setting

In our model, the ConvLSTM's filters in the convolutional layer set to 256, 128, the size of the convolution kernel is 2. After the convolutional layer, we add a pooling layer of size 2, followed by the convolutional layer is connected to the LSTM layer, the number of cells is 100, 100. Then a fully connected layer is output with the softmax activation function. We also use the generator parameter settings in the ConvLSTM benchmark method. The cells in the four layers of the discriminator set to 256, 128, 100, 3, and the softmax activation function is used in the last fully connected layer. The training epochs are usually kept at 1000, and we set the initial batch size to 60. We add a dropout layer with a value of 0.2 after the CNN and LSTM layers to prevent overfitting. The learning rate of the generator is 1e−3, the final learning rate is 1e−4. Every 50 epoch, if the recall index on the validation set does not improve, the learning rate will decrease by 2e−5 until the final learning rate reaches. All model training is performed with the Keras version 2.3.1 library of TensorFlow version 2.0 background. The experimental operating system is Ubuntu 16.04 and using NVIDIA GeForce GTX 1080Ti GPU. Some third-party libraries, such as the use of Talib to calculate technical indicators.

4.3 Experiment Results

We conducted a detailed experimental analysis on the MSFT based on several different comparison methods. First, we selected Macro and Weighted based on the multi-classification indicators. Among them, the macro and weighted include the corresponding precision, recall, and f1-score indicators. For ease of description, the bold font in

Indicator	LSTM	GRU	CNN	ConvLSTM	Proposed Method
Weighted-precision	0.3670	0.3407	<u>0.3690</u>	0.3588	**0.3732**
Weighted-recall	0.3664	**0.4040**	0.3597	0.3450	<u>0.3705</u>
Weighted-f1-score	0.3299	0.3414	<u>0.3575</u>	0.3490	**0.3607**
Macro-precision	0.3506	0.3179	<u>0.3575</u>	0.3438	**0.3609**
Macro-recall	0.3528	0.3450	**0.3585**	0.3425	<u>0.3536</u>
Macro-f1-score	0.3134	0.3011	<u>0.3519</u>	0.3400	**0.3563**

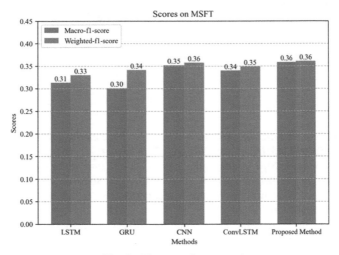

Fig. 3. The experiment results

our table represents the best value in the comparison method, and the underlined data indicates the secondary. At the same time, the Macro-f1-score and Weighted-f1-score indicators of different methods on the MSFT are shown in Fig. 3.

From experimental results, we can see that the proposed method performed better than the contrasted deep learning methods on four indicators, primarily the Weighted-precision indicator reached 0.3732. Compared with the highest 0.3690 in the comparison method, it is improved by 0.0042. As shown in Fig. 3, compared to others, the proposed method has slightly improved in average Macro. It should be noted that we select the best performance among other methods to compare with our method. Moreover, it can be seen that ConvLSTM is added as a generator to the generative confrontation network, and the classification performance is improved compared to the end-to-end ConvLSTM on the indicators.

5 Conclusion

In the research on the movement trend classification of financial time series prices, an improved generative model based on ConvLSTM and MLP is proposed to capture temporal features effectively and mine the data distribution of volatility trends from given financial time series data. The experimental results show that the proposed method has been further optimized under the above circumstances. Our model improves the overall classification performance and guides actual transactions. Moreover, our model outperforms the baseline methods on the datasets with complicated distribution characteristics. However, the limitation of the experiments is that the eleven technical indicators we selected in this experiment may not be the best. Different indicator combinations may have different effects on the performance of the model. Therefore, detailed experimental comparisons of the impact of different indicator selections on model performance are also follow-up work arrangements.

Acknowledgement. This work is partially supported by China Scholarship Council, Science and Technology Program of Sichuan Province under Grant 2020JDRC0067 and 2020YFG0326, and Talent Program of Xihua University under Grant Z202047.

References

1. Maleki, M., et al.: Time series modelling to forecast the confirmed and recovered cases of COVID-19. Travel Med. Infect. Dis. **37**, 101742 (2020)
2. Sezer, O.B., Gudelek, M.U., Ozbayoglu, A.M.: Financial time series forecasting with deep learning: a systematic literature review: 2005–2019. Appl. Soft Comput. **90**, 106181 (2020)
3. Gao, Z.-K., Small, M., Kurths, J.: Complex network analysis of time series. EPL (Europhys. Lett.) **116**(5), 50001 (2017)
4. Yang, Q., Xindong, W.: 10 challenging problems in data mining research. Int. J. Inf. Technol. Decis. Mak. **5**(04), 597–604 (2006)
5. Bagnall, A., et al.: The great time series classification bake off: a review and experimental evaluation of recent algorithmic advances. Data Min. Knowl. Discov. **31**(3), 606–660 (2017)

6. Xiaopeng, X., et al.: Fast time series classification using numerosity reduction. In: Proceedings of the 23rd International Conference on Machine Learning (2006)
7. Karim, F., et al.: LSTM fully convolutional networks for time series classification. IEEE Access **6**, 1662–1669 (2017)
8. Hüsken, M., Stagge, P.: Recurrent neural networks for time series classification. Neurocomputing **50**, 223–235 (2003)
9. Zheng, Y., Liu, Q., Chen, E., Ge, Y., Zhao, J.L.: Time series classification using multi-channels deep convolutional neural networks. In: Li, F., Li, G., Hwang, Sw., Yao, B., Zhang, Z. (eds.) Web-Age Information Management. WAIM 2014. LNCS, vol. 8485. Springer, Cham. https://doi.org/10.1007/978-3-319-08010-9_33
10. Krizhevsky, A., Sutskever, I., Hinton, G.E.: Imagenet classification with deep convolutional neural networks. Adv. Neural. Inf. Process. Syst. **25**, 1097–1105 (2012)
11. Goodfellow, I., et al.: Generative adversarial nets. Adv. Neural Inf. Process. Syst. **27** (2014)
12. Xu, Z., et al.: Satellite image prediction relying on GAN and LSTM neural networks. In: ICC 2019–2019 IEEE International Conference on Communications (ICC). IEEE (2019)
13. Zhang, K., et al.: Stock market prediction based on generative adversarial network. Procedia Comput. Sci. **147**, 400–406 (2019)
14. Feng, F., et al.: Enhancing stock movement prediction with adversarial training. arXiv preprint arXiv:1810.09936 (2018)
15. Patel, J., et al.: Predicting stock and stock price index movement using trend deterministic data preparation and machine learning techniques. Expert Syst. Appl. **42**(1), 259–268 (2015)

The Ground-State Potential Energy Surface
of F-Li$_2$ Polymer

Yue Wang[✉], Qingling Li, Guoqing Liu, Wenhao Gong, Shijun Yu, Yu Liu,
Xiaozhou Dong, Shiwen Chen, and Chengwen Zhang

Department of Electrical Engineering, Tongling University, Tongling 244000, Anhui,
People's Republic of China
wangyue8001@qq.com

Abstract. The first three-dimensional potential energy surface (PES) for the
ground-state of F-Li$_2$ polymer by CCSD(T) method were present. Two Jacobi
coordinates, R and θ and the frozen molecular equilibrium geometries were used.
We mixed basis sets aug-cc-pCVQZ for the Li atom and aug-cc-pCVDZ for the
F atom, with an additional (3s3p2d) set of midbond functions. The total of about
365 points were generated for the PES. Our ab initio calculations were consistent
with the experimental data very well.

Keywords: Ab initio calculation · PES · F-Li$_2$ polymer

1 Introduction

In recent years, Lithium is found to be form stoichiometric polymer with various elements. On the other hand, There are a lot of practical application of fluoride, such as
the six lithium fluoride phosphate is the core of the electrolyte materials, and is one of
the key materials necessary for the lithium battery electrolyte; LiF and other electronic
injection material introduction of organic optoelectronic devices have become a good
luminescent material [1–4]. F-Li$_2$ Polymer belongs to super valence compounds containing odd electronic, it has good nonlinear optical properties, so the scientists study
on super molecular structure of alkali metal fluoride has always maintained a strong
interesting in F-Li$_2$[5–7].

When we study reaction kinetics characteristics, the first thing is to build precise PES.
In the past ten years, some studies polarization molecular science of the system offers F-
Li$_2$ polymer structure and the dynamic response process [8–11]. Through investigation
we learned that most of the potential energy surface of F-Li$_2$ polymer before, is the
method by semi-empirical fitting.

Our calculations are covered a wide range of interaction energy of the potential
energy surface. First, considering vibrational weakly bound van der Waals complexes
and the good performance on similar optimization, we used the CCSD (T) calculation
method for single point of interaction energy. And then we described the features of the
F-Li$_2$ PES. At last we focus our attention on the ground state energy of this system.

© The Author(s) 2022
Z. Qian et al. (Eds.): WCNA 2021, LNEE 942, pp. 1108–1113, 2022.
https://doi.org/10.1007/978-981-19-2456-9_111

2 Ab Initio Calculations

When we do some calculation for alkali metal diatomic molecules the electronic related functions must be considered. The basis sets used for frequency calculations consist of aug-cc-pCVQZ for the Li atom and aug-cc-pCVDZ for the F atom. At the same time, we added with an additional (3s3p2d) set of midbond functions. In order to improve the convergence of basis set, we joined Midbond functions (mf) at the midpoint of R. We used quantum analysis framework in the process of computing the Jacobi coordinates system (r, R, θ). As shown in Fig. 1. The r is the distance of Li-Li, the R is the length of the vector connecting the Li-Li center of mass and the F atom, and θ is the angle between R and the x axis. For a given value of R, the angle θ changes from 0° to 90° in steps of 10°. We calculated 365 geometries for the whole interaction energy.and the ground state of the spacing is $r_{eq} = 2.696 \, a_0$ [12].

To ensure that the basis permits polarization by Li, we added diffuse augmentation functions. In the well range (the short range) $(0a_0 \leq R \leq 4a_0)$, while $\theta{=}0°$ and $\theta{=}90°$,we used the interval equal step way $\Delta R = 0.1a_0$. In the long range $(4a_0 \leq R \leq 11a_0)$, with $\Delta R = 1a_0$.

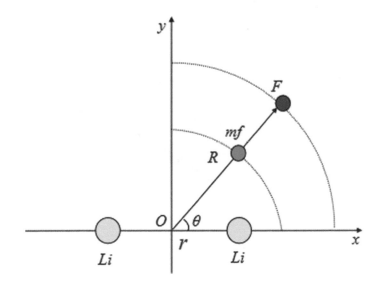

Fig. 1. Jacobi coordinates system

The *ab initio* calculations have been calculated with Gaussian 09W perform packet [13]. We considered all electronic correlation calculation process. The method of supramolecular was used when we calculated the interaction between Alkali metal pairs to the atom fluoride.

3 Results and Discussion

We show the behavior of the potential energy surface from ten different anglers as we can see In Fig. 2(a). When $R < 2a_0$, with the increase of R ten different points of view of potential energy are gradually increase. After reaching different peaks the potential energy reducing with R increasing. In the scope of $R > 5a_0$ the potential energy changes flatten. In Fig. 2(b) We can clearly see that an obvious potential barrier appears at $\theta = 30°$ and at $\theta = 90°$ a shallow potential well appears about the range ($1.8a_0 \leq R \leq 2.2a_0$).

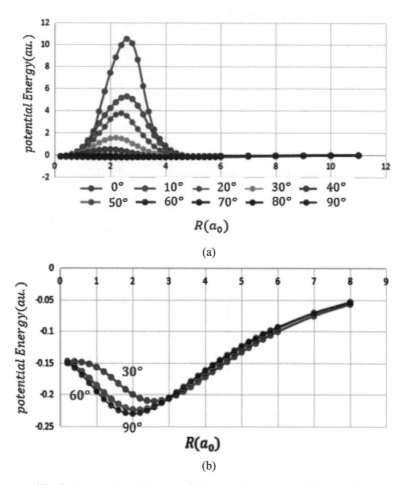

Fig. 2. Orientational features of the potential energy surface of F-Li$_2$.

In Fig. 3 we can see clearly that as the R increasing in the large area of the long-range the interaction converge to the same asymptotic value. The shape of a "T" backwards Li–F–Li is the lowest energy configuration (-3.87eV(-1.763e^{-5}Hartree) at $R = 2a_0$).

In Fig. 4 we show the 3D-PES for angles $\theta = 0°-360°$. The figure shows that the potential energy changes present strong anisotropy. The saddle point is located at R $= 2.6$Å and $\theta = 0°$. Clearly we can see that a shallow well appears at $\theta = 90°$. The absolute dissociation energy we can get is -3.87eV(-1.763e^{-5}Hartree), which is close to that obtained from the experiment [14]. This result reflected the potential energy changes in large angle is anisotropic.

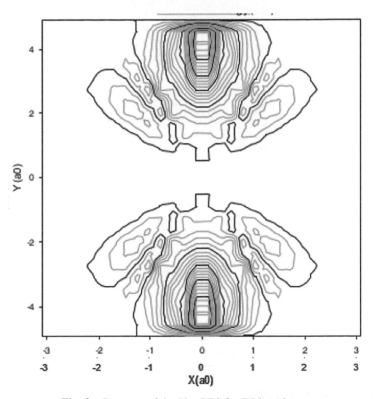

Fig. 3. Contours of the V$_{00}$ PES for F-Li$_2$ polymer

In Fig. 4, there are two obvious peaks on the ground state potential energy surface. Peak corresponds to the left is F + Li$_2$ and the right peak corresponds to the Li - F - Li reactants. We can easily see the whole potential energy is anisotropic.

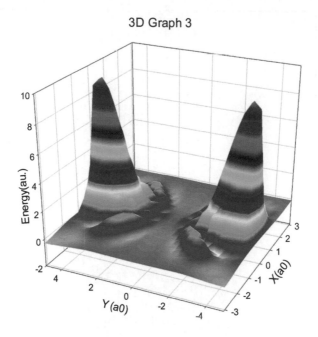

Fig. 4. PES for the Li-Li-F (angle $\theta = 0°-360°$)

4 Concluding Remarks

We adopted ab initio calculation method to calculate the ground state potential energy of F-Li$_2$ polymer. By the continental scientific drilling (CCSD (T) method and aug-cc-pCVQZ /aug-cc-pCVDZ + 332 basis set, we draw out the potential energy surface in the whole process of the three dimensional space. Compared with previous two-dimensional potentials with fixed $r_e = 2.696a_0$, Our theoretical results agree well with the experimental data.

Acknowledgments. This work is Supported by the Key projects of science research in University of Anhui Province (Grant: KJ2020A0695, KJ2020A0699), the teaching demonstration class project in Anhui Province (Grant: 2020SJJXSFK2400), the Innovation Project of Excellent Talents Training in Anhui Province(Grant: 2020zyrc153), the College Students' innovative training program (Grant: tlxy202010381316, tlxy202010381380, tlxy202010381574).

References

1. Fernandez-Lima, F.A., Henkes, A.V., da Silveira, E.F., Chaer Nascimento, M.A.: Alkali halide nanotubes: structure and stability. J. Phys. Chem. C. **116**, 4965-4969 (2012)
2. Senturk, S., Naturforsch, Z.: Phys. Sci. A. **66**, 372 (2011)

3. Bhowmick, S., Hagebaum-Reignier, D., Jeung, G.-H.: Potential energy surfaces of the electronic states of Li2F and Li2F-. J. Chem. Phys. **145**, 034306 (2016)
4. Srivastava, A.K., Misra, N.: M2X (M= Li, Na; X= F, Cl): the smallest superalkali clusters with significant NLO responses and electride characteristics. Molecular Simul. **42**, 981 (2016)
5. Wang, K., Liu, Z., Wang, X., Cui, X.: Enhancement of hydrogen binding affinity with low ionization energy Li2F coating on C60 to improve hydrogen storage capacity. Int. J. Hydrogen. Energ. **39**(28), 15639 (2014)
6. Srivastava, A.K., Misra, N.: Unusual properties of novel Li 3 F 3 ring: (LiF 2--Li 2 F) superatomic cluster or lithium fluoride trimer, (LiF) 3? Rsc. Adv. **4**(78), 41260 (2014)
7. Srivastava, A.K., Misra, N.: Can Li2F2 cluster be formed by LiF2/Li2F--Li/F interactions? an ab initio investigation. Mol. Simul. **41**(15), 1278 (2014)
8. Yokoyama, K., Haketa, N., et al.: Ionization energies of hyperlithiated Li2F molecule and LinFn− 1 (n= 3, 4) clusters. Chem. Phys. Lett. **330**, 339 (2000)
9. Olivera, M., Miomir, V., et al.: Rapid. Commun. Mass. Sp. **17**, 212 (2003)
10. Veličković, R.S., Vasil Koteski, J., et al.: Chem. Phys. Lett. **14**, 151 (2007)
11. Jasmina, D., Suzana, V., et al.: Dig. J. Nanomater. Bios. **8**(1), 359 (2013)
12. Colbert, D.T., Miller, W.H.: A novel discrete variable representation for quantum mechanical reactive scattering via the S-matrix Kohn method. J. Chem. Phys. **96**, 1982 (1992)
13. Gaussian 09W is a package of ab initio programs written by M J Frisch, G W Trucks with contributions from others; for more information, see <http://gaussian.com/glossary/g09/>
14. Chan, K.W., Power, T.D.: An ab initio study of He--F 2, Ne--F 2, and Ar--F 2 van der Waals complexes. J. Chem. Phys. **110**, 860 (1999)

New Tendencies in Regulating Personal Data. Can We Achieve Balance?

Mikhail Bundin[1]([⊠]) ⓘ, Aleksei Martynov[1] ⓘ, and Lyudmila Tereschenko[2]

[1] Lobachevsky State University of Nizhny Novgorod (UNN), Nizhny Novgorod 603950, Russia
mbundin@mail.ru, avm@unn.ru
[2] The Institute of Legislation and Comparative Law under the Government of the Russian Federation, Moscow 117218, Russia

Abstract. The article seeks to emphasize existing tendencies in regulation personal data in Russia and in foreign countries. The wide use of modern technologies of data processing "big data", "artificial intelligence", "internet of things" does not only open new opportunities for business and people but also makes more evident the gap between individual's interests for control of his data processing and thus protecting its privacy and commercial use of data by Internet companies. The state, on the other hand, seeks to get a more wide and exclusive access to the data collected by business entities, trying to apply a renewed concept of data sovereignty using its citizens' personal data protection as a legal ground. The author notes the growing desire from both the state and business entities to undermine individual's right to control his data processing as an inherent right of a data subject in order to facilitate the access to them and guarantee their interests. Awareness by the state and business of the new opportunities given by processing metadata including personal data, as a fundamental resource for the digital economy development can potentially lead to the situation where an individual will no longer be able to participate in determining the key parameters of their use. Most recent changes in Russian legislation on open access personal data that are to come into force in 2021 also leave much ground for uncertainty. In fact, they can shift the balance even more towards the interests of big business and the state.

Keywords: Personal data · Big data · Digital economy · Data sovereignty · Human rights

1 Introduction

Over the past few decades, the issue of personal data protection has been addressed a great number of times and in a variety of aspects. It makes to suggest that this issue has long been exhaustively studied and discussed. However, this is not so, there are many reasons to address again and again at the problematics of personal data protection from new angles and with new approaches. Let's name some of them that seem to be on the top now.

Firstly, personal data are closely related to the individual and his rights and freedoms. Human right theory and individual's legal status is some kind of a "living matter" that

© The Author(s) 2022
Z. Qian et al. (Eds.): WCNA 2021, LNEE 942, pp. 1114–1124, 2022.
https://doi.org/10.1007/978-981-19-2456-9_112

is evolving towards empowerment a human being with news rights and freedoms as necessary remedies versus social and technological evolution challenges. So, the more new technologies penetrates into our life, the greater will be the value of the human rights and freedoms associated with the information and data processing.

Secondly, the emergence of new information technologies for data processing may not only give new and unprecedented before opportunities for modern economy, but also create a threat of uncontrolled use of data and, therefore, undermine the humans rights and freedoms. Among these technologies are increasingly called "big data", "cloud computing", "artificial intelligence", "internet of things", etc. However, they are often used together in different sorts of combinations to collect and process data and a refusal to use them would mean a serious technological lag behind competitors [28].

Russia is not an exception in this respect. New program and policy documents seriously pay attention to the problem of personal data protection as a priority principle [6]. The current Doctrine of Information Security [9] puts the problem of personal protection in the information sphere on one of the first places, including the problem of ensuring privacy in the use of information technology. The Strategy for the Development of the Information Society for 2017–2030 [27], responding to the challenges of the modern technological revolution, in particular, "big data", says about the need to preserve and ensure the balance of interests of the individual and his right to personal and family secrets and the introduction of new technologies ("big data") for information processing. This is expected to be achieved through their storage on Russian territory and transmission only through Russian operators, as well as by preventing the illegal collecting of data on Russian citizens. The state program "Digital economy of the Russian Federation" [7, 12] also contains in its Roadmap a number of measures aimed at ensuring the protection of the individuals' rights and legitimate interests in the circumstances of digital economy, especially when processing big users' data in social networks and other means of social communication.

The international community also does not remain indifferent to this issue. The new General Data Protection Regulation [13] in the EU notes the need to strengthen, harmonize and develop measures for the protection of personal data in the context of new technological challenges that have arisen after the adoption of the well-known Directive 95/46/EC [8].

Despite such an abundance of normative and policy documents that seek to consolidate and establish individual's rights on personal information as something inherent to a human being in the information society, in fact, there can be seen other contradictory tendencies that eventually may undermine the existing concept of data protection as well as the right of an individual to control his data processing. Moreover, these trends are common not only for Russia but also for other countries [14, 23, 25, 29].

2 Methodology

The authors used quantitative and qualitative analysis of existing Russian and foreign publications in open sources in international and Russian science-citation databases.

Considering the topic of the research, the main emphasis was made on publications indexed in the Russian scientific citation database (E-library[1]), Scopus and ScienceDirect.

In addition to analyzing the state of modern scientific research, the authors used for qualitative analysis statistical data on digital economy in Russia and in the world, for comparative analysis existing program and policy documents on digital economy, information security, information society as well as existing legal texts and bills envisaged for adoption in the nearest future.

3 Literature Review

The problem of personal data protection has long been of serious interest to Russian and foreign scholars [2, 4, 17–19, 25, 30, 32, 33]. The direct correlation between the data protection and human rights and freedoms makes this topic far from being exhausted.

At the same time, for the scope and aim of this study, the most relevant and significant studies are studies of the legal nature and considerations of personal data as an object of ownership [3, 15, 19, 22, 24]. The problematics of 'propertisation' of personal data has been long studied by scholars and for now has no a universal solution, especially in the frame of diversity in understanding of 'ownership' by different legal systems and national peculiarities.

Another important component of the study is the consideration of the problems of commercialization of personal data as a product or a service, as well as various proposals to simplify the procedure for obtaining consent and to protect the rights of operators on the databases created by them [14, 25].

To some extent, new and interesting for the purposes of the study is the concept of digital or data sovereignty [10] describing the desire of the state to control data processing and information flows and to have access to personal data accumulated by metadata operators, international, transnational and national Internet companies including social networks [16, 18, 23, 26].

4 Personal Data in Digital Economy Environment

Firstly, for a modern economy based on knowledge and data, where the data itself, including personal data, are a crucial element-source without which the digital economy simply cannot function. The issue of strong contradiction between the concept of "big data" and "personal data" has been repeatedly addressed and is increasingly finding its supporters [20, 25]. It seems to be clear that the principles of data protection could hardly be compatible with the three 'V' concept of big data processing and here lies the most important contradiction and awareness of large Internet business entities. The fact that data controller is dependable on the consent of an individual who has an absolute right to withdraw from data processing constitutes a serious risk accompanied with more complex issue of necessity to comply not only with one national jurisdiction rules but

---

[1] https://elibrary.ru.

to face other jurisdictions' requirements that potentially may contradict each other and lead to possible sanctions.

All this makes data processing a wary ground and explains from one point the strong intention of data controllers in minimization of possible risks by simplifying the consent obtaining from an individual or by establishing their own concept of propertisation or commercializing of data, including personal data, to defend their interest through long and well-known concept of "ownership" [14, 19].

This intention is supported by day to day practice and sometimes by neglection of a large part of users to their privacy protection [23]. It is commonly well-known that even in case of adoption and publishing of privacy policies by data controllers on their web sites as well as the announcement to admit them in order to obtain web services or get other benefit from an Internet company users mostly accept them and without a real possibility to properly read and understand their content because of its complexity and a lack of any professional skills. The problematics of complexity of user's agreements was addressed several times and always with no coherent solution. The existing trend on making law provisions more robust and detailed in data protection make them generally even more complicated and harder for understanding and thus practically useless for the purpose of giving a coherent and clear user's consent on his data processing.

At the same time, the so-called profiling of online users (web profiling) is becoming a usual practice in return for better (users' oriented) services, which presumes tracking their online activities on the Internet, preferences and interests. Profiling is used in a variety of areas, primarily in Commerce, in the use of contextual advertising allowing to provide targeted advertising and, ultimately, to optimize selling, production and increase profits [31].

This makes Internet companies seek for more benefit from data processing by share them with third parties or even to sell them. The existence of a whole market of "personal data", sometimes latent, is no longer to be something outstanding or unpredictable that is justified by several major revelations over the past few years. Hence the understandable intention to legalize the already existing practices of processing and transmission of metadata and reduce the risks associated with legislative barriers, which they consider, apparently, as annoying obstacles [14].

This explains the proposal for the monetization of obtaining the individual's consent or the creation of a unified database or a sort of 'individuals' consent database. The latter is supposed to be a single register of individuals' consent on their data processing. The consent includes the description of datasets that an individual gives permission for collecting and processing by any data controller. This system could make possible for data controllers to start data processing without directly contacting data subject for consent.

The problem of monetization or use of the category of ownership for personal data or propertization has repeatedly become an issue for a number of studies but unfortunately with no clear answer to this complex question [15, 24]. The concept of ownership could be possibly applied (that is also under question) but only to some extend and for sure not to all the categories of personal data. Some data as DNA are unique to an individual and couldn't be transferred to any one as property or somehow [15]. At the same time, the idea to use ownership for personal data, used in USA, can be considered as the most adequate response to the diversity of states' legal systems with no clear provisions on federal level [24]. In these circumstances, the ownership of personal data could be a universal remedy for human rights protection.

On the contrary in the European tradition, personal data are regarded as a mean to protect human rights and freedoms – a sort of inherent right of an individual to control his data processing as part of his individuality. In this sense, the role of an individual as the 'data subject' is in determining the key parameters of any data processing including the right to withdraw from it [29].

At the same time, it is impossible not to recognize a significant interest of data controllers (Internet companies) companies including the commercial value of data sets in obtaining and protecting their rights to personal data. In fact, it is even difficult to assume the cost and real value of investments of data controllers [14] i.e., the owners of social networking services or e-commerce projects in the processing of personal data. It seems to be logical recognize not only the existence of such an interest, but also the fairness of such claims for investment protection and stability of digital economy functioning, as well as their dependence from personal data protection regulation changes.

5 Personal Data and Digital Sovereignty

It should be of no more a secret that the state also seeks to learn more about an individual, his personal or private life, intending in some cases to get full exclusive access to his data. We could observe now a clear and unambiguous tendency to expand the powers of the state as the operator of personal data and the reducing number of cases where an individual may interfere as data subject and influence or control his data processing. Surely that may have an explanation and legal ground as we cannot deny the increased presence of terrorist and extremist organizations, as well as simply illegal content on social networks and the Internet in general. On the other hand, those restrictions aimed to control online activities of users and collecting data on them do not leave any coherent guaranties of how this information is used exactly and whether there is an abuse control system operated by competent authority.

The only thing to admit here that the state is already aware of the benefits of "big data", "artificial intelligence", "Internet of things" technologies and has long been one of the major data controllers. It remains only to take a few steps to erase the barriers between different state information processing systems to process metadata and to adopt another exclusion in the Data protection legislation for that reason. The state is clearly understanding the value of metadata on online users' activities accumulated by third parties – private entities and large Internet companies. Those data were long kept a secret from state authorities. But the situation has greatly changed since. It is not necessary to blame only Russia or regard it as a unique case - other countries also seek to openly or covertly use big data technologies and get wide access to third parties' datasets with more or less success, pursuing a variety of goals [16, 32]. It becomes no sensation publicly revealed facts of leaking metadata from social networks to state intelligent services or other investigative authorities.

In many ways, this contributes also to the rooting and active promotion of the concept of 'information/digital sovereignty' or data sovereignty [1]. Perhaps only this can logically explain the recent steps of the Russian state.

This concept was very convenient for the protection of the interests of the state in the information sphere and is now actively used by some countries. In fact, the state is looking for control over the flow of data that has any connection with it, as well as the technological infrastructure on its territory. By adopting in 2015 legislative provisions on the mandatory storage of at least a copy of data on Russian citizens on Russian territory, the state made another step to establish control over the data accumulated by Internet companies providing e-services to Russian citizens. The second important step was to establish a requirement to disclose the source code for encryption used for a secure connection when using network services [21].

Later, all this was supplemented by the requirement to store all information about the connection and the content received by the user from the internet and telecommunication service provider for 6 months. Those decisions are well-known as 'Yarovaya' Bill. All this clearly underlines the state's desire to control its information space and often use data on citizens (the need to protect personal data or individual's information security) as a reason to control data flows and get access to them [21].

6 Current Legislative Initiatives and Data Regulation Perspectives

Recently, the Russian legislator has increasingly addressed to the topic of personal data protection. Undoubtedly, the pandemic period has further strengthened the above-mentioned trends and is likely to be a subject for discussion and the time for more thorough analysis will come, including in terms of the protection of human rights and personal data. Large-scale leaks of personal data of patients who have had COVID-19 cause serious concern to the Russian society and can hardly be ignored [34]. One of the consequences became serious tightening of liability for violations of the legislation on the protection of personal data. In many cases, the amount of fines was almost doubled, simultaneously with the replacement of the 'warning' with real punishment.

However, the most recent attempt to resolve the issue of the legal regime of publicly available data is of particular interest. For a long time, Russian lawmakers have explicitly used such a concept as "publicly available (open access) personal data", which became such in the case of a law on disclosure of information (for example, the income of high-ranking civil servants), or if the data subject himself made them so. Under this concept, personal data actively posted by users of social networks became open, and their processing by third parties did not seem to require special consent for processing. Later, this concept was abandoned, it was presumed that only the data published on publicly available resources of personal data under data subject's direct and explicit consent for their openness could be processed freely.

Nevertheless, Russian legislation and practice in this case demonstrated the ambiguity of this position. The starting point here was the well-known case of Vkontakte v. Double [5]. As a matter of fact, the main issue in this case was the question of the legality of the use of open data of users of the social network by third-party services that process such information. After many twists and turns, the court concluded that personal data becomes publicly available only if it is provided by the subject himself and is available to an indefinite circle of persons. The court did not recognize the social network as an open access source of personal data, primarily due to the lack of consent of the subject to post them on social networks. This position was actively expressed by the Russian Data Protection Authority (Roskomnadzor) supporting the need for the consent of the personal data subject to the collection and processing of personal data posted by users in open access on social networks. However, the latest decision in this case quite clearly indicated that there were no violations of the law on personal data, if the online service carried out indexing and caching of the data of social network pages similar to a search engine and if the users using the tools of the social network itself gave their consent to the indexing of their pages by search engines.

In parallel with this decision, the Russian IT community was puzzled by a new legislative initiative, which comes into force on April 1, 2021 [11], regarding the appearance of a new category - "personal data allowed by the subject of personal data for distribution". As a matter of fact, these are the personal data in respect of which the user has unequivocally, and in a special form, agreed to unlimited(open) access to them by third parties. In other words, third parties can freely process such data, and the operator can transfer, distribute or allow access to it. At the same time, the subject has the right to stipulate certain conditions or set exceptions for the transfer of data to certain persons. The consent must name specific categories of data for which such a regime is established, and can be withdrawn at any time by the subject without giving reasons. It is extremely specific that such consent can be provided by the subject directly to the operator or through a special information system, operated by Roskomnadzor.

It is obvious that these changes have raised a lot of questions, including quite practical ones, from the point of view of the functioning of the Roskomnadzor consent register, as well as the need to bring the existing practice of social networks and many other online services in accordance with these provisions and the legal formalization of user consent, which have yet to be resolved in the nearest future.

7 Conclusion

Currently, we can say that we live in a time of changing the paradigm of views on the problem of personal data protection. In fact, the well-known concept of personal data protection as an inalienable right of any person with a large number of internal elements-the rights of the data subject to control and determine the key parameters of data processing-no longer seems so indisputable. The realities of the data economy force data controllers to challenge the existing principles of data protection regulation, which obviously hinder the further development of the digital economy. It's no secret that many multinational Internet companies are now seeking 'better' jurisdiction to avoid national legal barriers to the use of big data and other modern technologies to process personal metadata or host technological infrastructure. They are trying to lobby for a new legal framework for the protection of personal data, actively supporting the "propretization" and "commercialization" of personal data, turning it into a kind of commodity for free circulation with less risk of being held accountable. Of course, we need to talk about the beginning of this initiative, but the trend is clearly visible.

However, Russia is hardly one of the countries with an established tradition of respect for personal data. In fact, the legislation on personal data itself has been in full force for about 15 years, and some drastic changes in the legal consciousness of citizens in this regard can hardly be expected.

On the other hand, we can assume the emergence of another very interesting trend, which reflects the interest of the state not only to accumulate large personal data in state or "affiliated" information systems, but also to have access to or at least control large data accumulated by private economic entities. It is still difficult to say with certainty what awaits the concept of personal data soon, but much is already becoming obvious. The international community and international organizations would probably play a more important role in addressing these issues. There is no doubt that significant changes in the legislation on personal data in one group of States can have significant consequences for others in the context of the globalization of the digital economy. The most striking example of this is the numerous changes in the privacy policies of the largest social network operators as a result of the adoption of the General Data Protection Regulation in the EU.

In any case, the necessary balance between restricting access to personal data, on the one hand, and freedom of business, on the other, has yet to be found.

The biggest regret here can only be that all these trends are surprisingly common in the matter of depriving a person of his rights to control the processing of his personal data. This is an awful prospect, and none of us should forget about the purpose of personal data as part of the human rights protection system and, in most cases, the only means of providing it. Recent legislative decisions in Russia, which, undoubtedly, were initially aimed at significantly expanding the tools of the data subject in determining the regime of his open access data, are unlikely to change the situation. Despite a number of positive aspects and the emergence of transparency in relations between the controller, third parties and the data subject, it is still worth noting that it will be more likely to benefit the state and the IT-business. In fact, there is at list three reasons to be thoroughly addressed in this case:

1. As a rule, such consent to public availability (open access) will be conditioned on the provision of digital services "necessary/indispensable" for the user – the refusal of which may block their use.
2. Considering today huge arsenal of big data solutions, strong artificial intelligence, capable of self-learning, even an experienced user will find it increasingly difficult to assess and assume the possible consequences of his consent and recognize threats. Ultimately, this solution will certainly allow to legalize the work of many network services, which will use personal data even more freely.
3. New technologies should be considered as a mean not only for personal data collecting or processing but also as a powerful tool for data breaches detecting. Russian Data Protection Authority – Roskomnadzor is seeking to create an internet platform capable to detect unlawful personal data collecting in the Internet.

Funding. The reported study was funded by RFBR, project number 20–011-00584.

References

1. Adonis, A.A.: International law on cyber security in the age of digital sovereignty, E-International Relations, 14 Mars 2020, https://www.e-ir.info/2020/03/14/international-law-on-cyber-security-in-the-age-of-digital-sovereignty/
2. Arkhipov, V., Naumov, V.: The legal definition of personal data in the regulatory environment of the Russian Federation: between formal certainty and technological development. Comput. Law Secur. Rev. 32(6), 868–887 (2016). https://doi.org/10.1016/j.clsr.2015.08.006
3. Bataineh, A.S., Mizouni, R., El Barachi, M., Bentahar, J.: Monetizing personal data: a two-sided market approach. Procedia Comput. Sci. 83, 472–479 (2016). https://doi.org/10.1016/j.procs.2016.04.211
4. de Terwangne, C.: Council of Europe convention 108+: a modernised international treaty for the protection of personal data. Comput. Law Secur. Rev. 40, 105497 (2021). https://doi.org/10.1016/j.clsr.2020.105497
5. Decision of the Court of Arbitration of the City of Moscow # A40–18827/17–110–180, https://kad.arbitr.ru/Document/Pdf/1f33e071-4a16-4bf9-ab17-4df80f6c1556/5f0df387-8b34-426d-9fd7-58facdb8a367/A40-18827-2017_20210322_Reshenija_i_postanovlenija.pdf?isAddStamp=True
6. Decision of the Supreme Eurasian economic Council of 11.10.2017 № 12 "On the Main directions of the digital agenda of the Eurasian Economic Union until 2025". https://docs.eaeunion.org/docs/ru-ru/01415258/scd_10112017_12
7. Digital Economy Regulation – Skolkovo Community. http://sk.ru/foundation/legal/. Accessed 23 July 2021
8. Directive 95/46/EC of the European Parliament and of the Council of 24 October 1995 on the protection of individuals with regard to the processing of personal data and on the free movement of such data. https://eur-lex.europa.eu/eli/dir/1995/46/oj. Accessed 23 July 2021
9. Doctrine of Information Security of Russian Federation. President of Russian Federation (2016). http://www.kremlin.ru/acts/bank/41460. Accessed 23 July 2021
10. Efremov, A.: Forming the concept of state information sovereignty, Law. J. High. School Econ. 1, 201–215 (2017). https://doi.org/10.17323/2072-8166.2017.1.201.215
11. Federal Law No. 519-FZ of 30.12.2020 " On Amendments to the Federal Law "On Personal Data", http://publication.pravo.gov.ru/Document/View/0001202012300044. Accessed 23 July 2021
12. Federal Program "Digital Economy in Russian Federation". Federal Government of Russian Federation (2017). http://static.government.ru/media/files/9gFM4FHj4PsB79I5v7yLVuPgu4bvR7M0.pdf. Accessed 23 July 2021
13. General Data Protection Regulation (GDPR). https://gdpr.eu/tag/gdpr/. Accessed 23 July 2021
14. Hare, S.: For your eyes only: U.S. technology companies, sovereign states, and the battle over data protection. Bus. Horizons, 59(5), 549–561 (2016). https://doi.org/10.1016/j.bushor.2016.04.002
15. Janeček, V.: Ownership of personal data in the Internet of Things. Comput. Law Secur. Rev. 34(5), 1039–1052 (2018). https://doi.org/10.1016/j.clsr.2018.04.007
16. Jasserand, C.: Law enforcement access to personal data originally collected by private parties: missing data subjects' safeguards in directive 2016/680. Comput. Law Secur. Rev. 34(1), 154–165 (2018). https://doi.org/10.1016/j.clsr.2017.08.002
17. Lloyd, I.: From ugly duckling to Swan. The rise of data protection and its limits. Comput. Law Secur. Rev. 34(4), 779–783 (2018). https://doi.org/10.1016/j.clsr.2018.05.007
18. Malatras, A., Sanchez, I., Beslay, L., et al.: Pan-European personal data breaches: mapping of current practices and recommendations to facilitate cooperation among Data Protection Authorities. Comput. Law Secur. Rev. 33(4), 458–469 (2017).https://doi.org/10.1016/j.clsr.2017.03.013

19. Malgieri, G., Custers, B.: Pricing privacy – the right to know the value of your personal data. Comput. Law Secur. Rev. **34**(2), 289–303 (2018). https://doi.org/10.1016/j.clsr.2017.08.006
20. Mantelero, A.: AI and Big Data: a blueprint for a human rights, social and ethical impact assessment. Comput. Law Secur. Rev. **34**(4), 754–772 (2018). https://doi.org/10.1016/j.clsr.2018.05.017
21. Moyakine, E., Tabachnik, A.: Struggling to strike the right balance between interests at stake: the 'Yarovaya', 'Fake news' and 'Disrespect' laws as examples of ill-conceived legislation in the age of modern technology. Comput. Law Secur. Rev. **40**, 105512 (2021), ISSN 0267-3649. https://doi.org/10.1016/j.clsr.2020.105512
22. van de Waerdt, P.J.: Information asymmetries: recognizing the limits of the GDPR on the data-driven market. Comput. Law Secur. Rev. **38**, 105436 (2020), ISSN 0267-3649, https://doi.org/10.1016/j.clsr.2020.105436
23. Prince, C.: Do consumers want to control their personal data? Empirical Evidence Int. J. Hum. Comput. Stud. **110**, 21–32 (2018). https://doi.org/10.1016/j.ijhcs.2017.10.003
24. Purtova, N.: Property rights in personal data: learning from the American discourse. Comput. Law Secur. Rev. **25**(6), 507–521 (2009). https://doi.org/10.1016/j.clsr.2009.09.004
25. Savelyev, I.: Problems of application of the legislation on personal data in the era of "Big data" (Big Data), Law. J. High. School Econ. **1**, 43–66 (2015), https://law-journal.hse.ru/data/2015/04/20/1095377106/Savelyev.pdf
26. Steppe, R.: Online price discrimination and personal data: a general data protection regulation perspective. Comput. Law Secur. Rev. **33**(6), 768–785 (2017). https://doi.org/10.1016/j.clsr.2017.05.008
27. Strategy for Information Society Development 2017–2030. Adopted by the Russian President's Decree on 9 May 2017. http://www.kremlin.ru/acts/bank/41919. Accessed 22 July 2021
28. Tankard, C.: What the GDPR means for businesses. Netw. Secur. **2016**(6), 5–8 (2016). https://doi.org/10.1016/S1353-4858(16)30056-3
29. Tikkinen-Piri, C., Rohunen, A., Markkula, J.: EU general data protection regulation: changes and implications for personal data collecting companies. Comput. Law Secur. Rev. **34**(1), 134–153 (2018). https://doi.org/10.1016/j.clsr.2017.05.015
30. Wang, Z., Yu, Q.: Privacy trust crisis of personal data in China in the era of Big Data: the survey and countermeasures. Comput. Law Secur. Rev. **31**(6), 782–792 (2015). https://doi.org/10.1016/j.clsr.2015.08.006
31. Wu, Y.: Protecting personal data in E-government: a cross-country study. Gov. Inf. Q. **31**(1), 150–159 (2014). https://doi.org/10.1016/j.giq.2013.07.003
32. Wu, Y., Lau, T., Atkin, D.J., Lin, C.A.: A comparative study of online privacy regulations in the U.S. and China. Telecommun. Policy, **35**(7), 603–616 (2011). https://doi.org/10.1016/j.telpol.2011.05.002
33. Xue, H.: Privacy and personal data protection in China: an update for the year end 2009. Comput. Law Secur. Rev. **26**(3), 284–289 (2010). https://doi.org/10.1016/j.clsr.2015.08.006
34. Zhukova, K.: "The situation is critical": what threatens the largest data leak of coronavirus patients, Forbes Russia, https://www.forbes.ru/tehnologii/415857-situaciya-vesma-kritichna-chem-grozit-krupneyshaya-utechka-dannyh-zabolevshih

An Insight into Load Balancing in Cloud Computing

Rayeesa Tasneem[(⊠)] and M. A. Jabbar

Vardhaman College of Engineering, Hyderabad, India
rayeesa.tasneem3@gmail.com, jabbar.meerja@gmail.com

Abstract. Cloud Computing has emerged as a High-performance computing model providing on-demand computing resources as services via the Internet. Services include applications, storage, processing power, allocation of resources and many more. It is a pay-per-use model. Despite of providing various services, it is also experiencing numerous challenges like data security, optimized resource utilization, performance management, cost management, Cloud migration and many more. Among all, Load Balancing is another key challenge faced by Cloud. Effective load balancing mechanism will optimize the utilization of resources and improve the cloud performance. Load balancing is a mechanism to identify the overloaded and under loaded nodes and then balance the load by uniformly distributing the workload among the nodes. Various load balancing mechanisms are proposed by various researchers by taking different performance metrics. However existing load balancing algorithms are suffering from various drawbacks. This paper emphasizes the comparative review of various algorithms on Load Balancing along with their advantages, shortcomings and mathematical models.

Keywords: Cloud Computing · Challenges · Load balancing · Static load balancing · Dynamic load balancing · Scalability · Fault-tolerance · Performance metrics

1 Introduction

In this computer world, Cloud Computing is the biggest buzz these days. The term Cloud is obtained like a metaphor for the Internet. Generally, in the diagrams related to the network, the Internet is figured as a Cloud, which means that the area is not of user concerned. So in this idea, it is most relevant to the notion of Cloud Computing. It is a subscription-based service where a user can acquire computer resources and networked storage space [10]. It is a type of computing wherein, resource sharing is done rather than ownership. Users just had to pay for the resources they use. After usage of these resources, they are released.

The beauty of Cloud Computing is that users need not worry about software installations, upgrades and maintenance. It is the service provider's responsibility to keep updated otherwise they lose customers. Amazon was the first company to offer cloud services to the public [2]. Many more companies including Google, Microsoft, and others also came forward to provide services.

© The Author(s) 2022
Z. Qian et al. (Eds.): WCNA 2021, LNEE 942, pp. 1125–1140, 2022.
https://doi.org/10.1007/978-981-19-2456-9_113

As there is a huge increase in demand for Cloud Computing technology, the demand for services is also increased. Thereby, the workload on the servers needs to be balanced. This balancing of workload is done by Load Balancers. There exist different types of load in Cloud Computing namely, network load, CPU load, memory load etc. Load Balancing has a very significant role in the field of Cloud Computing environment. It is a method of distributing the workload uniformly among all the servers. For balancing the load efficiently different load balancing algorithms are discussed in this paper. Furthermore, these algorithms aim to minimize response time, increase the throughput, maximize resource utilization and enhance the performance of the system.

This research study mainly emphasizes on the analysis of different static and dynamic load balancing algorithms in Cloud Computing. The comparison of these discussed algorithms is done based on the performance parameters of load balancing algorithms as shown in Table 1.

2 Load Balancing

To carry out the distribution of load properly, a Load Balancer is used which receives jobs from various locations and distributes them to the data center. It is a device that works like a reverse substitute and distributes application network load over various servers [4]. The goal of Load Balancing is to enhance the performance, sustain stability and scalability for accommodation if there is an increase in large-scale computing, the backup plan is necessary at the time of system crash and decrease the associated costs [4].

Load Balancing is extremely important in Cloud Computing as it reduces response time, execution time, waiting time of users and so on [3]. The load balancer maintains the load in such a way that, if it finds overloaded nodes, then it transfers some of the jobs of overloaded nodes to underloaded nodes to carry out the faster execution and also the user's waiting time is reduced. The ultimate purpose of Load Balancing is to utilize the processors efficiently by keeping them busy. The processor should not remain idle otherwise; the overall performance of the system is affected. Distributed systems contain many processors working together or independently either linked to each other or not [3]. The work on each processor is distributed based on its processing speed and processing capacity to minimize the waiting and execution time of users.

Some of the major functions of a load balancer are [11]:

- The client requests are distributed efficiently among several servers.
- It guarantees high reliability and scalability by transmitting requests only to those servers which are online.
- It offers flexibility to append or remove servers on demand.

Based on the load balancing algorithms supported by load balancers, the load balancers can figure out whether a particular server (or the set of servers) is prone to get heavily-loaded or not, and if it is, then the load balancer forwards the workload to the nodes which are with minimum load [12].

2.1 Load Balancing Types

Based on the initiation of a process, Load Balancing algorithms are categorized into three types as stated in [1].

- Sender initiated:
  The sender finds that there are many tasks to be executed, so the sender node takes the initiative to transmit the request messages until it discovers a receiver node that can share its workload.
- Receiver initiated:
  Here, the algorithm is initiated by the receiver node sending a message request to get a job from a sender (heavily loaded server).
- Symmetric:
  It is the combination of both types of algorithms i.e., sender initiated algorithm and the receiver-initiated algorithm.

Based on the system's current state, load balancing algorithms are classified into two categories:

- Static Algorithm:
  It is independent of the system's current state. Prior information regarding the system requirements (server capacity, memory, computation power, network performance) and all the requirements of users are known earlier before execution. Once the execution starts, the user requirements are not changed and also the load remains constant.
- Dynamic Algorithm:
  Unlike static algorithms, dynamic algorithms consider the system's current state while taking decisions. Information regarding user or system requirements is not known in advance. The Dynamic algorithms work in such a way that the jobs are assigned at runtime upon the request from the users. Depending on the situation, jobs are transferred from overloaded nodes to underloaded nodes; so consequently, these algorithms have a significant improvement in the performance over static algorithms. The only drawback is that it is a little difficult to implement but the load is balanced effectively.

3 Existing Load Balancing Algorithms

There exist various types of static load balancing algorithms. A few of the algorithms are briefly described below.

3.1 Round Robin Load Balancing Algorithm [1, 5]

This is a static load balancing algorithm and its implementation is the simplest of all algorithms. In these algorithms, the allocation of jobs to processors is done circularly. Initially, it selects any random node and allocates a job to it, then it moves to other nodes to allocate in a round-robin approach, without showing any priority. Here, each node is assigned with some time quantum in which it has to execute the job, if the job is not finished it has to wait for the next slot to resume its execution.

Advantages:

- The main advantage is that the fastest response time of the processes.
- It doesn't lead to starvation. The process need not wait for a long time to execute its job.

Shortcomings:

- Due to the uneven distribution of workload, some of the nodes get overloaded and underloaded as the execution time of the process is not determined earlier.

Mathematical model:

This mathematical model is provided by [13]. It is proposed to optimize the value of Time Quantum (TQ) and also to reduce the waiting time of jobs as shown below. There are certain assumptions regarding this mathematical model. It is considered that there exist a total 'n' number of processes that are waiting in a ready queue and they are dispatched circularly. Each process has a Burst Time which is well-known in advance and is available [13].

The parameters considered in this model are stated below as shown in [13]:

n: Overall number of ready processes initially.
S_i: Burst of the ith process.
TAT_i: Turn Around Time of the ith process.
W_i: Waiting time of the ith process.
R_i: Overall number of the times the processor is utilized by the ith process.
Lq_{im}: The final time quantum used by the ith process.
PP_{ij}: Overall burst time of the processes that are similar to j, which are waiting in the ready queue before the execution of the ith process.
PS_{ij}: Overall burst time of the processes that are similar to j, which are waiting in the ready queue after the execution of the ith process.
CT: Time required for context switching.
q: The time quantum required for the execution of the process.

$$MinW = \frac{\sum_{i=1}^{n} W_i}{n} \tag{1}$$

$$TAT_i = (R_i - 1)(q + CT) + Lq_i + \sum_{j<i} PP_{ij} + \sum_{i<j} PS_{ij} \tag{2}$$

$$W_i = TAT_i - S_i \tag{3}$$

$$R_i = \left\lceil \frac{S_i}{q} \right\rceil \tag{4}$$

$$Lq_i = S_i - \left\lceil \frac{S_i}{q} \right\rceil . q \tag{5}$$

$$PP_{ij} = \begin{cases} R * (q + CT) \text{ if } R_i < R_j \\ (R_j - 1) * (q + CT) + Lq_j + CT \text{ otherwise } \forall j < i \end{cases} \tag{6}$$

$$PS_{ij} = \begin{cases} (R_j - 1) * (q + CT) \text{ if } R_i \leq R_j \\ (R_j - 1) * (q + CT) + Lq_j + CT \text{ otherwise } \forall j > i \end{cases} \tag{7}$$

$$Q : integer > 0, \tag{8}$$

where,
Equation (1) shows the average waiting time of the process which is to be minimized as far as possible. Equation (2) computes the total turnaround time of the process which includes the number of times the process acquires the complete quantum from the processor, context switching time, plus the amount of last time quantum, plus the total sum of execution times of the predecessor and successor processes of the ith process [13]. Equation (3) computes the waiting time of the ith process. Equations (4) and (5) computes the total number of times i^{th} process acquires the processor and the amount of the last required time quantum respectively. Equations (6) and (7) compute the total execution times of the predecessor and successor processes respectively [13]. Equation (8) indicates the condition that the time quantum 'q' should be an integer value [13].

3.2 Opportunistic Load Balancing Algorithm

The primary goal of the OLB algorithm is to keep every node busy [5]. The present (current) workload of the virtual machine is not considered. OLB takes an unexecuted job from the ready queue and allocates it to the node which is available currently in a random approach irrespective of the current state of the virtual machine (node's current workload) [5]. As the node's execution time is not computed, the processing of the job is done very slowly [5].

Advantages:

- Virtual machines are kept busy all the time.
- Unexecuted tasks are handled quickly by assigning them to nodes randomly.

Shortcomings:

- Processes are executed slowly as the node's execution time is not computed.

Mathematical model:
Let us suppose there exist a total of three VMs, VM1, VM2 and VM3 having various loads, for instance, VM1 has 10 s, VM2 has 80 s and VM3 has 30 s [14]. Let J1 is the new job that has arrived for execution, then the scheduler ought to choose one virtual machine from the three VM1, VM2 and VM3 and assign a job to it. The scheduler chooses the virtual machine which has a minimum load i.e., 10 s. Here the significance of load is referred to the level of a preoccupation of virtual machines with current jobs

[14]. VM1 will accomplish the allocated jobs after 10 s, similarly VM2 in 80 s and VM3 within 30 s. Therefore, the scheduler chooses VM as it is least loaded [14]. The working of this algorithm is shown in Fig. 1.

$$index \leftarrow Min\{v.getready()|\forall VML\} \qquad (9)$$

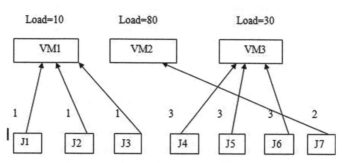

Fig. 1. Working approach of OLB

3.3   Min-Min Load Balancing Algorithm

In this algorithm, firstly for all the tasks, minimum completion time is calculated then among them, the task that has minimum completion gets assigned to that corresponding machine/node which has minimum completion time (fastest response) [5]. Then, all the remaining tasks are updated on that machine. The allocated task is deleted from the record. Similarly, all the remaining tasks are allocated with a resource. Performance of this algorithm is enhanced when smaller tasks (small execution time) are more in number compared to larger tasks (large execution time) otherwise this approach can lead to starvation [5].

Advantages:

- Performance is better in the case when the smaller tasks (execution time is less) are greater in number than larger tasks (execution time is large) [1].

Shortcomings:

- This algorithm leads to starvation for larger tasks.

Mathematical Model:
The key motive of this algorithm is to minimize makespan as far as possible. In a task set, for each task, the expected execution times on each machine are computed accurately before the execution [15]. This is done with the help of the Expected Time to Compute (ETC) matrix model which contains ETC(ti,mj) where task ti is performed on machine mj [15].

Let us consider the Metatask T which comprises a set of tasks t1, t2, t3,...., tn in the scheduler.

Let R be the Resource Set which comprises of set of Resources m1, m2, m3,...., mk those are existing at the task arrival.

Then, the makespan for this algorithm is calculated using the formulae shown in Eqs. (10) and (11):

$$makespan = \max\left(CT\left(t_i, m_j\right)\right) \tag{10}$$

$$CT_{ij} = R_j + ET_{ij} \tag{11}$$

where,

$CT \rightarrow$ Completion time of machines

$ET_{ij} \rightarrow$ Expected execution time of job i on resource j

$R_j \rightarrow$ Availability time or Ready time of resource j after the execution of earlier assigned jobs.

3.4 Max-Min Load Balancing Algorithm [5, 6]

The Max-Min algorithm is identical to the above Min-Min algorithm, once the minimum completion time of all the available tasks is computed, then among these, the task which has maximum completion time among all the tasks as assigned to the corresponding node that has minimum completion time. Then all the remaining tasks on that node are updated and that allocated task is deleted from the record. Similarly, all the remaining tasks are allocated with a resource. In this algorithm, smaller jobs (less execution time) are executed simultaneously along with the larger jobs (large execution time), so the makespan (total time taken for executing all the tasks) is reduced and resources are utilized efficiently unlike in the Min-Min algorithm.

Advantages:

- The waiting time of large size jobs is reduced.
- Resources are utilized efficiently and makespan is reduced.

Shortcomings:

- Same as the Min-Min load balancing algorithm, this algorithm is also applicable only to small-scale distributed systems.

Mathematical Model [27]:

The main motive of this algorithm is to reduce the waiting time of the larger jobs (large execution time) as far as possible. Here, smaller tasks are simultaneously executed along with larger tasks, so thereby the makespan is reduced and the resources are utilized properly [6]. The mathematical model of this algorithm is the same as the above Min-Min load balancing algorithm which uses the ETC matrix model to compute the expected execution time of the tasks before execution.

Let us consider the Metatask T which comprises a set of tasks namely t1, t2, t3,...., tn in the scheduler.

Let M be the Machine set which comprises a set of machines namely m1, m2, m3,...., mk those are existing at the task arrival.

Then, the expected completion time for any algorithm can be computed as shown in Eq. (12) [27]:

$$Et(t_i, m_j) = Mch(m_j) + MT(t_i, m_j) \tag{12}$$

where,

$Mch(m_j) \rightarrow$ Idle time of Machine i.e. the time at which machine finishes any earlier assigned jobs.

$MT(i, j) \rightarrow$ Execution time estimated for the task t_i on machine m_j.

$ET_{ij} \rightarrow$ Expected Completion Time of task t_i on machine.

3.5 Throttled Load Balancing Algorithm [7, 8]

According to this algorithm, the total number of VMs are maintained in the form of a table by the load balancer and their states (BUSY/AVAILABLE). Firstly, the user requests the data center controller to obtain a VM to execute the task. Then the datacenter controller requests the load balancer for the distribution of VMs. The load balancer checks the index table of VMs starting from the top till it finds the first available VM. If it finds the VM, then the corresponding VM id is sent to the data center controller then the datacenter controller requests the VM defined by that id to the load balancer, and the task is allocated to a virtual machine. After allocation of a task, the data center controller notifies the load balancer about the new allotment then the load balancer updates the index table. While processing a user request if the corresponding virtual machine is not available then the load balancer replies with '−1' to the datacenter.

Advantages:

- Resources are utilized efficiently and good performance is obtained.

Shortcomings:

- The current workload of VM is not considered.
- VM Index table should be scanned from the top at every arrival of the request due to which response time.

Mathematical Model [17]:

Modified Throttled Load Balancing algorithm proposed by [17] provides flexibility to the client for acquiring services from the service provider of Cloud. This algorithm is discussed in three stages. The foremost stage is the initialization stage. In the initialization stage, the expected response time of each virtual machine is computed. The second stage is to discover the efficient virtual machine. The third and final stage is to return the ID of an efficient virtual machine. The expected response time of VM can be computed by using the following formula shown in Eq. (13) [17]:

$$ResponseTime = Fint - Arrt + TDelay \tag{13}$$

where,

Arrt → Arrival time of user request

Fint → Finish time of user request.

TDelay → Transmission delay which is computed using the below formula shown in Eq. (14).

$$TDelay = Tlatency + Ttransfer \qquad (14)$$

where,

TDelay → Transmission delay

Tlatency → Network latency

Ttransfer → The amount of time required for transmission of the data size of the single request (D) from a source location to destination location which is computed by using the below formula shown in Eq. (15).

$$Ttransfer = D/Bwperuser \qquad (15)$$

where,

Bw → Bandwidth per user is computed using below formula shown in Eq. (16).

$$Bwperuser = Bwtotal/Nr \qquad (16)$$

where,

Bwtotal → Total available bandwidth

Nrl → Number of user requests which are currently in transmission.

By using the above formulae, the response time of the virtual machines is computed and then an efficient virtual machine can be obtained among them.

3.6 Active Clustering Load Balancing Algorithm [8]

Generally, this algorithm is referred to as a self-aggregation algorithm which is according to the concept of grouping identical nodes as one group and working on them. Initially, the process is started by a node which is known as an initiator node and from the neighbor nodes, it selects another node known as a matchmaker node which should be a different type compared to the initiator node. Then this matchmaker node links with one of its neighbor nodes which should satisfy the criteria of the initiator node. Finally, the matchmaker node deletes the link which is connecting between itself and the initiator node. This procedure is continued iteratively till the load is balanced among all the nodes.

Advantages:

- Resources are utilized efficiently as the virtual machines are grouped as a cluster with similar properties.

Shortcomings:

- The system's performance is decreased when the variety of nodes increases.

Mathematical Model [19].

This algorithm proposed by [19] divides the similar capacities of virtual machines into groups which is known as a cluster and this is done by using the K-means clustering method. Euclidean distance formula is selected for allocating virtual machines to clusters. The value of K i.e., the total number of clusters is selected in such a way that it is the greatest prime factor of n where n gives the number of virtual machines. Clustering of n virtual machines is done into K-number of clusters using three types of resources as parameters. They are CPU processing speed, the bandwidth of network and Memory. To compute the distance of VMs with centers of other clusters:

$$EUD(VM_i)(C_i) = sqrt\left[(CPU_i - CPU_j)^2 + (Mem_i - Mem_j)^2 + (BW_i - BW_j)^2\right]$$
(17)

The cluster's new mean when a node is allocated to it is computed by the following formulae:

$$CPU_j = \frac{(CPU_i + CPU_j)}{2}$$
(18)

$$Mem_j = \frac{(Mem_i Mem_j)}{2}$$
(19)

$$BW_j = \frac{(BW_i + BW_j)}{2}$$
(20)

3.7   Ant Colony Optimization Load Balancing Algorithm [4, 9]

The main goal of this load balancing algorithm is to explore an optimal path between the food source and colony of ants according to the behavior of the ant. Its objective is to efficiently distribute the workload among all the nodes. Firstly, when the request is made, the ant begins moving in the direction of the food source from the head node. While moving ahead, ants keep a record of every node they have visited for making future decisions. During their movement ants deposit the pheromones so that it helps further ants to choose the next node. The strength of pheromones depends on the components such as food quality, the distance of food etc. Denser pheromone is attracted by many ants. The pheromones are updated when the jobs are executed.

There are two kinds of pheromones in the Ant Colony Optimization algorithm. One is the Foraging pheromone which is used to find nodes that are overloaded by moving forward while the Trailing pheromone is used for discovering its path to get back to the node which is underloaded. It means if an ant discovers a heavily loaded node it begins moving back to the underloaded node for assigning a job to it. Every single ant develops result set and then it builds to get a complete solution. The ant attempts to update a single result set continuously instead of updating the result set of their own. The solution set is also continuously updated. The node commits suicide once it discovers the target node as a result; the number of ants gets reduced in the network.

Advantages:

- This algorithm overcomes heterogeneity and is adjustable for dynamic environments
- It enhances the performance of the system.
- Scalability is good and has excellent fault tolerance.

Shortcomings:

- Network overhead is increased
- Delay in moving forward and backward [16].

Mathematical Model [28].
In this algorithm, the main objective of ants is the redistribution of work among the nodes. The cloud network is traversed by ants to select nodes for their next step using the classical formula shown below, where the probability P_k of an ant that is presently on node 'r' choosing the neighboring node 's' for traversal is shown in Eq. (21):

$$P_k = (r, s) = \frac{[\tau(r, s)][\eta(r, s)]^\beta}{[\tau(r, u)][\eta(r, u)]^\beta} \tag{21}$$

where,

$r \to$ Current node

$s \to$ Next node

$\tau \to$ Pheromone concentration of the edge

$\eta \to$ The desirability movement of ants (the move is highly desirable if it is from overloaded nodes to underloaded nodes or vice versa.)

$\beta \to$ Depends on the relevancy between the pheromone concentrations with with the distance moved.

The formula for updating the Foraging Pheromone is shown in Eq. (22)

$$FP(t + 1) = (1 - \beta_{eva})FP(t) + \sum_{k=1}^{n} \Delta FP \tag{22}$$

where,

$\beta_{eva} \to$ Evaporation rate of the Pheromone

$FP \to$ Foraging Pheromone of the edge before the move

$FP(t + 1) \to$ Foraging Pheromone of the edge after the move

$\Delta FP \to$ Change in the Foraging Pheromone.

'The formula for updating the Trailing Pheromone is shown in Eq. (23):

$$TP(t + 1) = (1 - \beta_{eva})TP(t) + \sum_{k=1}^{n} \Delta TP \tag{23}$$

where,

$\beta_{eva} \to$ Evaporation rate of the Pheromone

$TP \to$ Trailing Pheromone of the edge before the move

$TP(t + 1) \to$ Trailing Pheromone of the edge after the move

$\Delta TP \to$ Change in the Trailing Pheromone.

4   Research Performance Parameters Used for Different Load Balancing Algorithms

1) Throughput: This parameter helps to compute the overall number of jobs whose execution is accomplished. Throughput should be high for the good performance of the system.
2) Overhead: Overhead involves additional cost required, inter-processor and inter-process communication and migration of tasks while executing a load balancing algorithm [1]. It should be minimized to obtain the efficiency of an algorithm.
3) Fault tolerance: It can be defined as the ability of the system to keep processing without any interruption even when one or more system elements fail to work. For good load balancing, fault tolerance should be high.
4) Migration time: This parameter is defined as the amount of time needed to migrate a task or resources from one node to other nodes. Migration time should be less.
5) Response time: It is defined as the time period between the sender's request and the receiver's response. It must be reduced to enhance the system's performance.
6) Resource Utilization: It helps to check whether the system resources are utilized properly or not. The resource utilization should be optimum.
7) Scalability: It is the ability of a system to increase the number of nodes with the same QOS (Quality Of Service) if the number of users increases.
8) Performance: With the help of this parameter the overall system efficiency is checked. It must be enhanced at an acceptable cost [26–28].

5   Research Findings

The above discussed static and dynamic load balancing algorithms satisfy certain performance metrics which are presented in Table 1.

Table 1. Comparison of Load Balancing algorithms by considering above performance metrics

Static algorithms	Throughput	Resource utilization	Overhead	Scalability	Response time	Migration time	Fault tolerance	Performance
Round robin	Yes	Yes	Yes	No	Yes	No	No	Yes
OLB	Yes	No	No	No	No	No	No	Yes
Min-Min	Yes	Yes	Yes	No	Yes	No	No	Yes
Max- Min	Yes	Yes	Yes	No	Yes	No	No	Yes
Throttled	Yes	Yes	No	Yes	Yes	Yes	Yes	Yes
Active clustering	Yes	Yes	Yes	No	No	Yes	No	No
Ant colony optimization	Yes	Yes	Yes	Yes	No	Yes	Yes	Yes

In view of this, the below Table 2 shows a comprehensive survey of different techniques employed by the researchers for load balancing in the field of Cloud Computing along with their pros and cons.

Table 2. Survey of different techniques used by researchers for Load Balancing in Cloud Computing

S.no	Algorithm	Approach used	Environment	Simulator	Pros/Cons
1	[20]	Used the concept of honey bee for allocating the existing resources to the network to decrease makespan	Heterogeneous	Cloud sim and workflow	Scalable, Fault-tolerant, minimum associated overhead but throughput is less
2	[21]	According to the method of soft computing algorithm and SA is used to resolve the problem of balancing the load dynamically among distinct resources [21]	Heterogeneous	Cloud analyst	Only response time and associated cost is good
3	[22]	Suggested a load balancing algorithm by the combination of two algorithms to minimize the overall processing cost and also processing time	Homogeneous	Cloud Sim	Utilization of resources and job response time is improved but performance is reduced as system diversity increases
4	[23]	Proposed an algorithm based on the data locality using ranging and tuning functions and to solve scheduling problems in Cloud Computing environment	Heterogeneous	Cloud sim	Makespan and cost is reduced and resources are utilized efficiently

(*continued*)

Table 2. (*continued*)

S.no	Algorithm	Approach used	Environment	Simulator	Pros/Cons
5	[24]	Proposed to decrease active physical servers so that the underutilized servers are scheduled to save energy	Both	Cloud sim	Resources are utilized efficiently and power consumption is reduced
6	[25]	Proposed a mathematical model with the help of GT(Group Technology)	Heterogeneous	Grid Sim	Resources are utilized efficiently but the computation time is more
7	[18]	Proposed an algorithm based on the honeybee foraging method to minimize execution time and average response time [18]	Heterogeneous	Cloud Sim	Response time and Execution time is good but the migration process is not efficient
8	[26]	Proposed a load balancing algorithm that is energy efficient with the help of the FIMPSO algorithm [26]	Heterogeneous	MATLAB	CPU utilization is maximum with 98%, average response time is least with13.58 ms, [26]

6 Conclusion

Cloud Computing is a rising trend in the IT industry which has a very large number of requirements such as infrastructure, resources, and storage. Among all the challenges faced by Cloud Computing, Load Balancing is also another key challenge. Load Balancing is the method of uniform distribution of workload among the nodes to improve utilization of resources and enhance the system performance. This paper briefly describes the importance of Load Balancing, its benefits and its types. This research also focuses on the survey of different Load Balancing algorithms proposed by researchers. Algorithms are briefly explained with their advantages, shortcomings and mathematical models. Various performance parameters such as scalability, throughput, performance etc., are considered to compare these load balancing algorithms. The tabularized comparison depicts that in comparison with dynamic load balancing algorithms static load balancing algorithms are more stable. However, dynamic load balancing algorithms are more preferable because of certain parameters such as overhead rejection, resource utilization, reliability, cooperativeness, adaptability, fault tolerance, throughput, and waiting and response time. In future our research will focus on various cloud resource utilization issues.

References

1. Aditya, A., Chatterjee, U., Gupta, S.: A comparative study of different static and dynamic load balancing algorithm in cloud computing with special emphasis on time factor. Int. J. Curr. Eng. Technol. **5**(3), 1898–1907 (2015)
2. Velte, A.T., Velte, T.J., Elsenpeter, R.: Cloud computing: a practical approach, pp. 135–140 (2010)
3. Mukati, L., Upadhyay, A.: A survey on static and dynamic load balancing algorithms in cloud computing. In: Proceedings of Recent Advances in Interdisciplinary Trends in Engineering & Applications (RAITEA) (2019)
4. Kumar, S., Rana, D.S.: Various dynamic load balancing algorithms in cloud environment: a survey. Int. J. Comput. Appl. **129**(6), 16 (2015)
5. Shah, N., Farik, M.: Static load balancing algorithms in cloud computing: challenges & solutions. Int. J. Sci. Technol. Res. **4**(10), 365–367 (2015)
6. Sharma, N., Tyagi, S., Atri, S.: A comparative analysis of min-min and max-min algorithms based on the makespan parameter. Int. J. Adv. Res. Comput. Sci. **8**(3), 1038–1041 (2017)
7. Volkova, V.N., et al.: Load balancing in cloud computing. In: 2018 IEEE Conference of Russian Young Researchers in Electrical and Electronic Engineering (EIConRus). IEEE (2018)
8. Mukundha, C., Venkatesh, N., Akshay, K.: A comprehensive study report on load balancing techniques in cloud computing. Int. J. Eng. Res. Dev. **13**(9), 35–42 (2017)
9. Kashyap, D., Viradiya, J.: A survey of various load balancing algorithms in cloud computing. Int. J. Sci. Technol. Res **3**(11), 115–119 (2014)
10. Liang, J., Bai, J.: Data security technology and scheme design of cloud storage. In: Atiquzzaman, M., Yen, N., Xu, Z. (eds.) 2021 International Conference on Big Data Analytics for Cyber-Physical System in Smart City. Lecture Notes on Data Engineering and Communications Technologies, vol. 103. Springer, Singapore (2022). https://doi.org/10.1007/978-981-16-7469-3_9
11. Rimal, B.P., Choi, E., Lumb, I.: A taxonomy and survey of cloud computing systems. In: 2009 Fifth International Joint Conference on INC, IMS and IDC. IEEE (2009)
12. Kaur, R., Luthra, P.: Load balancing in cloud computing. In: Proceedings of International Conference on Recent Trends in Information, Telecommunication and Computing, ITC (2012)
13. Saeidi, S., Baktash, H.A.: Determining the optimum time quantum value in round robin process scheduling method. IJ Inf. Technol. Comput. Sci. **10**, 67–73 (2012)
14. Mohialdeen, I.A.: Comparative study of scheduling algorithms in cloud computing environment. J. Comput. Sci. **9**(2), 252–263 (2013)
15. Kokilavani, T., Amalarethinam, D.G.: Load balanced min-min algorithm for static meta-task scheduling in grid computing. Int. J. Comput. Appl. **20**(2), 43–49 (2011)
16. Sajjan, R.S., Yashwantrao, B.R.: Load balancing and its algorithms in cloud computing: a survey. Int. J. Comput. Sci. Eng. **5**(1), 95–100 (2017)
17. Shah, M.R., Manan, D., Kariyani, M.A.A., Agrawal, M.D.L.: Allocation of virtual machines in cloud computing using load balancing algorithm. Int. J. Comput. Sci. Inf. Technol. Secur. (IJCSITS) **3**(1), 2249–9555 (2013)

18. Hashem, W., Nashaat, H., Rizk, R.: Honey bee based load balancing in cloud computing. KSII Trans. Internet Inf. Syst. **11**(12), 5694–5711 (2017)
19. Kapoor, S., Dabas, C.: Cluster based load balancing in cloud computing. In: 2015 Eighth International Conference on Contemporary Computing (IC3). IEEE (2015)
20. Vasudevan, S.K., et al.: A novel improved honey bee based load balancing technique in cloud computing environment. Asian J. Inf. Technol. **15**(9), 1425–1430 (2016)
21. Mondal, B., Choudhury, A.: Simulated annealing (SA) based load balancing strategy for cloud computing. Int. J. Comput. Sci. Inf. Technol. **6**(4), 3307–3312 (2015)
22. Ghumman, N.S., Kaur, R.: Dynamic combination of improved max-min and ant colony algorithm for load balancing in cloud system. In: 2015 6th International Conference on Computing, Communication and Networking Technologies (ICCCNT). IEEE (2015)
23. Valarmathi, R., Sheela, T.: Ranging and tuning based particle swarm optimization with bat algorithm for task scheduling in cloud computing. Clust. Comput. **22**(5), 11975–11988 (2017). https://doi.org/10.1007/s10586-017-1534-8
24. Liu, X.-F., et al.: An energy efficient ant colony system for virtual machine placement in cloud computing. IEEE Trans. Evol. Comput. **22**(1), 113–128 (2016)
25. Shahdi-Pashaki, S., Teymourian, E., Tavakkoli-Moghaddam, R.: New approach based on group technology for the consolidation problem in cloud computing-mathematical model and genetic algorithm. Comput. Appl. Math. **37**(1), 693–718 (2016). https://doi.org/10.1007/s40314-016-0362-4
26. Devaraj, A., Saviour, F., et al.: Hybridization of firefly and improved multi-objective particle swarm optimization algorithm for energy efficient load balancing in cloud computing environments. J. Parallel Distrib. Comput. **142**, 36–45 (2020)
27. Suntharam, S.M.S.: Load balancing by max-min algorithm in private cloud environment. Int. J. Sci. Res. (IJSR) **4**, 438 (2013). ISSN (Online): 2319–7064 Index Copernicus Value (2013): 6.14 Impact Factor
28. Nishant, K., et al.: Load balancing of nodes in cloud using ant colony optimization. In: 2012 UKSim 14th International Conference on Computer Modelling and Simulation. IEEE (2012)

Fuzzing-Based Office Software Vulnerability Mining on Android Platform

Yujie Huang, Zhiqiang Wang[⊠], Haiwen Ou, and Yaping Chi

Departmentof Cyberspace Security, Beijing Electronic Science and Technology Institute, No. 7 Fufeng Road, Beijing, Fengtai District, China
wangzq@besti.edu.cn

Abstract. The wide application of mobile terminals that makes the software and hardware of mobile platforms gradually become the important target of malicious attackers. In response to the above problems, this paper proposes a vulnerability mining scheme based on Fuzzing. In this scheme, many methods are used to generate a large number of test cases. After the application receives the corresponding test cases, it analyzes the output results and the exceptions thrown. The experimental results show that the scheme can effectively excavate the vulnerabilities of mobile office software on the Android platform, and has certain reliability.

Keywords: Fuzzing · Mobile office · Memory corruption

1 Introduction

Nowadays, Android has become the mobile phone operating system with the largest market share, and its development boom has also brought about new network security issues [1, 2], such as criminals taking advantage of mobile phone program vulnerabilities to seek benefits, and leaking user privacy. Therefore, vulnerability testing of Android applications is essential before facing users [3].

There are few types of research onvulnerability mining of office software onthe Android platform, and the design of test cases is relatively simple. To better solve the threat of Android memory corruption vulnerability, this paper designs, and implements a Fuzzing-based Android platform domestic office software vulnerability mining system. Under the Android platform, office software constructs special test cases, observes the exceptions thrown and the process crashes to find out the possible vulnerabilities, and ensures the security of the mobile offices.

The main contributions of this paper are as follows:

1. Generate test cases by mutation-based, generation-based, and Char-RNN-based methods to ensure the coverage of test cases and detect applications from multi-plesides.
2. Analyze the operating mechanism of office software applications under the Android platform, and construct a set of effective fuzzing test schemes, which can run successfully under various versions of Android and have a wide range of applications.

© The Author(s) 2022
Z. Qian et al. (Eds.): WCNA 2021, LNEE 942, pp. 1141–1149, 2022.
https://doi.org/10.1007/978-981-19-2456-9_114

3. Design and implement a set of office software vulnerability mining systems based on Fuzzing technology to find possible vulnerabilities [4]. The system is simple and easy to use, displays the process and results intuitively, and reduces the threshold of use. The system adopts a modular design, and each module runs independently to facilitate the subsequent functional debugging and upgrading of the vulnerability mining system [5].

The structure of this paper is as follows: Chapter One gives a brief introduction, Chapter Two designs the overall framework and various modules of the system, Chapter Three implements the system, Chapter Four conducts experiments and evaluations, Chapter Five summarizes and puts forward the improvement direction.

2 System Architecture Design

The system is divided into four modules: visualization platform module, test case generation module, fuzzing module, and automatic analysis module. The visualization platform module constructs the graphic page of the entire system, the test case generation module is responsible for constructing semi-effective test cases, the fuzzing module is responsible for the entire process of test cases from sending to running, and the automatic analysis module is responsible for analyzing the crash information and logs that appear during the test to discover the security vulnerabilities that exists. As shown in Fig. 1:

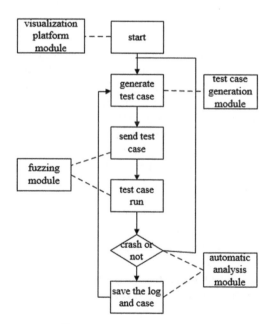

Fig. 1. System module division.

3   Implementation of System Module

3.1   Test Case Generation Module

Mutation-based Method. Mutation-based test case generation requires samples to be obtained in advance, and the steps for generating PDF and HTML are similar. Take the generation of a PDF file as an example, collect a malicious PDF sample set from GitHub as input for subsequent mutation operations. In the program, use the generate_dumb_pdf_sample() method to achieve. By controlling the number of mutations, the input files are mutated to different degrees to ensure the coverage of the generated samples. The specific process is shown in Fig. 2:

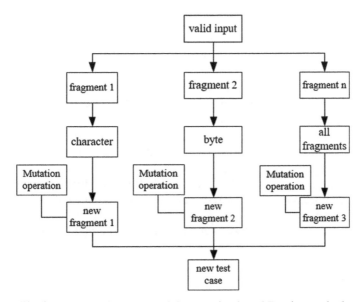

Fig. 2. The specific process of the mutation-based Fuzzing method.

The main steps arc as follows:

(1) Use the choice() function of the random module to randomly select one from the preset sample library as the given valid input;
(2) Obtain the length of the file, use the randrang() function in the random module to randomly select a position "start" as the starting point for subsequent operations;
(3) Determine the text length "len" for mutation, and choose arbitrarily on the premise that it does not exceed the maximum length of the file;
(4) Perform mutation operations based on the values of "start" and "len", such as inserting a random character, deleting a character or flipping a character, etc.;
(5) Write the content obtained after mutation into a new PDF file for subsequent fuzzing.

Generation-based Method. The system made some modifications to the grammar rules of the Google Domato open-source fuzzing test tool to generate PDF files and HTML files for testing. To generate HTML, just call the gen_new_jscript_js() function in Domato. Generate PDF test cases using m PDF (a PHP library) method, the generation steps are as follows:

(1) Call the header() method in mpdf to write the file header of the pdf, where "%PDF-1.1" is used.
(2) Call the indirect object() method in mpdf to write the object.
(3) Call the gen_new_jscript_js() method to randomly select and generate a javaScript script from the modified Domato grammar rule library and write it into the object.
(4) Call the xref And Trailer() method in mpdf to write the cross-reference table and tail of the pdf.

Char-RNN-based Method. The system uses Char-RNN to generate test cases as a supplement to ensure the comprehensiveness of test cases and uses TensorFlow to quickly build the Char-RNN framework. The specific process is as follows:

(1) Read and decode the sample set, and convert it to UTF-8 encoding. Vectorize the sample and establish the mapping relationship between strings and numbers.
(2) The text is divided into text blocks with the growth of $x + 1$. Each input sequence contains x characters in the text, and the corresponding target sequence is moved one character to the right. Rearrange and package the data into batches.
(3) Use tf. keras. Sequential to define the model.
(4) Add optimizer and loss function. Apply the tf. keras. Model. compile method to configure the training steps.Use tf. keras. optimizers. Adam with default parameters and loss function.
(5) Use tf. keras. callbacks. Model Checkpoint to ensure that checkpoints are saved during training.

3.2   Fuzzing Module

Fuzzing is the core part of the entire vulnerability mining system. Before running the system, get the device id of the Android device. After installing adb under windows, use a data cable to connect the Android device to the PC. Set the Android device connection mode to "USB MIDI", and enter the "adb devices" command to get the device id of the currently connected device.Take WPS as the test object for fuzzing. The test process is shown in Fig. 3.

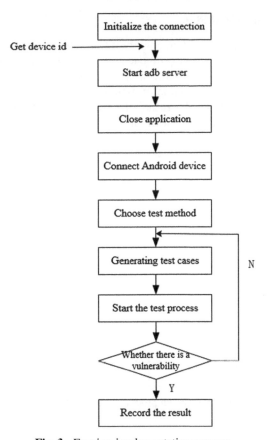

Fig. 3. Fuzzing implementation process.

(1) Call the adb_connection_int() method to initialize the connection. Restart the adb server, connect to the Android device and clear its background according to the WPS package name "cn.wps.moffice_eng" to minimize the interference of other factors in the subsequent testing process.

(2) Enter "http://192.168.189.1:1337/" in any browser to open the visualization page, select the fuzzing test method on this page, and click the "Start" button to start the test.

(3) The background receives the information from the front endand generates the corresponding PDF test case according to the fuzzing method selected by the user. Call the pdf_fuzz() method to start the fuzzing process. Run the WPS application after unlocking the screen of the device, then open the test file and collect all kinds of information feedback from the application during the running process. Execute "adb shell am force-stop cn. wps. moffice_eng" to stop the application.Wait for a while of time before the next fuzzing operation to prevent problems caused by the long-time load operation of the equipment.

3.3   Automatic Analysis Module

The automatic analysis process filters the log information collected during the fuzzing process.Due to the influence of many human factors and uncontrollable factors such as equipment, server, operating environment, etc., Fuzzing technology has the possibility of false alarms, that is, the abnormal information thrown maybe just some bugs, which cannot be called vulnerabilities. Therefore, the automatic analysis function is added to the system. The specific process is shown in Fig. 4.

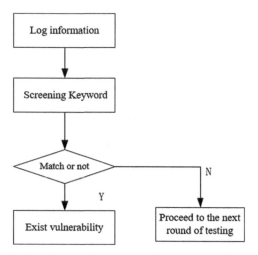

Fig. 4. The implementation process of automatic analysis module.

Use the "adb logcat -d" command to view the corresponding log information, call "subprocess. Popen()" to run the command as a subprocess and get a return value, which is the log information. Use a loop to determine whether there are key signals about *vulnerabilities* predefined in the setting file in the log information, as shown in Table 1. If it exists, save this piece of log information and the test case that caused the log information in the specified folder. Finally, use the adb command "adb logcat -c" to clear the old logs and enter the next test process.

Table 1. Linux abnormal signal comparison table.

Signal	Meaning
SIGTERM	Termination request sent to the program
SIGSEGV	Illegal memory access (segmentation fault)
SIGINT	External interrupt, usually initiated by the user
SIGILL	Illegal program image, such as illegal instruction
SIGABRT	Abnormal termination conditions, such as those initiated by abort()
SIGFPE	Wrong arithmetic operation, such as dividing by zero

4 Experiment and Evaluation

4.1 Experimental Environment

The equipment used in this system includes a PC device and an Android device. The system of the PC device is win10 system, and the IP address is 192.168.189.1. The system of the Android device is Android 4.The mobile office applications tested are WPS Office and UC browser, In addition, Adobe Reader and Chrome browsers are selected as test comparisons.The applications are downloaded from regular channels.

4.2 Experimental Results

Use the system to test different mobile office applications, and the results are shown in Table 2:

Table 2. Mobile office application test results.

Application	Test case type	Number	Time	The number of bug
WPS Office	Based on mutation	1000	14277	0
	Based on generation	1000	15348	0
	Based on Char-RNN	1000	15368	0
UC web	Based on mutation	1000	12731	0
	Based on generation	1000	12468	0
	Based on Char-RNN	1000	15936	0
Adobe reader	Based on mutation	1000	14726	0
	Based on generation	1000	14976	0
	Based on Char-RNN	1000	15324	0
Chrome	Based on mutation	1000	13561	0
	Based on generation	1000	10553	7
	Based on Char-RNN	1000	16008	0

4.3 Evaluation

Among the three test case generation methods, the mutation-based method has the least amount of calculation and the fastest generation speed, while the Char-RNN based method has the largest amount of calculation and the slowest generation speed. On the effectiveness of test cases, the method based on generation is the best, the method based on char RNN is the second, and the method based on variation is the worst. The overall test speed of the same type of application is similar. Compared with the PDF Reader, the browser is more likely to be attacked in DOM parsing [9].

```
D/ADB_SERVICES( 8089):  closing because is_eof=1 r=-1 s->fde.force_eof=0

W/ADB_SERVICES( 8089): create_local_service_socket() name=shell:input keyevent 82

D/ADB_SERVICES( 8089): Calling send_ready local=13594, remote=4644

W/ADB_SERVICES(16679): adb: unable to open /proc/16679/oom_adj

D/AndroidRuntime(16680):

D/AndroidRuntime(16680): >>>>>> AndroidRuntime START com.android.internal.os.RuntimeInit (tool) <<<<<<

D/AndroidRuntime(16680): CheckJNI is OFF
```

Fig. 5. View log files.

Enter the crash folder to view the recorded log file, as shown in Fig. 5.

Check the log files of all the vulnerabilities and find that they all contain the "SIGSEGV" keyword, and all appear "Fatal signal 11 (SIGSEGV) at 0x0000413d (code = -6), thread 16718 (CrRenderer Main)" type of crash, indicating that the problem of null pointer triggers the vulnerability and then causes the application to crash. The backtrace file in the log records the specific information when the application crashes, and the result is shown in Fig. 6. It can be seen from the figure that there is a problem with the so file, that is, an overflow of the static data area of the application.

```
I/DEBUG    (18820): backtrace:

I/DEBUG    (18820):      #00  pc 00a086d8  /data/data/com.android.chrome/lib/libchrome.1985.135.so
```

Fig. 6. View back trace.

5 Conclusion

Currently, the vulnerability of office software under the Android platform has security risks. In response to this problem, this paper designs and implements a domestic office software vulnerability mining system based on Fuzzing technology, analyzes the vulnerabilities that may cause it to crash, generates a large number of test cases, and conducts vulnerability mining through the method of fuzzing. The experimental results show the feasibility of the designed system, which can provide support for developers to improve the application program and improve the completeness of the application program.

The system designed in this paper has certain limitations. It can only detect specific vulnerabilities in specific types of applications, that is, memory vulnerabilities in mobile office software. It is not yet possible to conduct comprehensive vulnerability detection on all Android applications. More in-depth research is needed in the future.

Acknowledgement. This research was financially supported by the National Key RD Program of China (2018YFB1004100), China Postdoctoral Science Foundation-funded project (2019M650606), and the First-class Discipline Construction Project of Beijing Electronic Science and Technology Institute (3201012).

References

1. Enck, W., Ongtang, M., Mc Daniel, P.: Understanding android security. IEEE Secur. Priv. Mag. (Pennsylvania: Berlin), **7**, 50–57 (2009)
2. Ding, L.P.: Security analysis of Android operating system NETINFO SECURITY 28–31 (2012)
3. Feng, S.Q.: Android Software Security and Reverse Analysis ed Chen B and Fu Z, pp. 236–64 (2013)
4. Li, T., Huang, X., Liu, H.Y., Huang, R.: Software vulnerability mining technology based on fuzzing. J. Val. Eng. **3**, 197–199 (2014)
5. Yan, L.K., Yin, H.: DroidScope: seamlessly reconstructing the OS and Dalvik semantic views for dynamic Android malware analysis USENIX Association (New York: Berkeley), p. 29 (2012)
6. Papaevripides, M., Athanasopoulos, E.: Exploiting mixed binaries. ACM Trans. Priv. Secur. (Cyprus: New York), **24**, 1–29 (2021)
7. Palit, T., Monrose, F., Polychronakis, M.: Mitigating data-only attacks by protecting memory-resident sensitive data digital threats: research and practice. Dig. Threats Res. Practice, 1–26 (2020)
8. Schmeelk, S., Yang, J., Aho, A.: Android malware static Analysis Techniques. In: Proceeding of the 10th Annual Cyber and Information Security Research ConfACMVol 6 (Bellevue: New York), pp. 569–584 (2015)
9. Mulliner, C., Miller, C.: Fuzzing the phone in your phone Black Hat USA (LAS VEGAS) (2009)

A Multi-modal Time Series Intelligent Prediction Model

Qingyu Xian[✉] and Wenxuan Liang

International College, Beijing University of Posts and Telecommunications, Beijing, China
xianqingy321@163.com

Abstract. The power load prediction can ensure the power supply and dispatch, which will be useful for market participants to plan and make strategic decisions to enhance reliability, save operation and maintenance costs. Short-term load data series have obvious approximate periodicity, while long-term load data series show variability and dynamic features. In addition, time series data of various modalities, such as market reports and production management data, could play a role in load prediction. One kind of multi-modal CNN-BiLSTM architecture is proposed to predict short-term and long-term load data, which have an improved shared parameter convolutional network to learn feature representation and an improved attention-based BiLSTM mechanism, which could model the dynamic features of multimodal on time series data. Experimental results on multimodal dataset show that, compared with other baseline systems, this model has some advantages in the prediction accuracy.

Keywords: Power load prediction · Multi-modal · CNN-BiLSTM · Attention-based BiLSTM

1 Introduction

Load prediction is a key link in power supply planning, as well as a basic feature and important calculation basis for intelligent power supply planning. In addition to traditional machine learning models, deep neural networks, as the most popular intelligent research framework at present, have been widely implied by researchers in the active distribution network load prediction research. Active distribution network load prediction data can be regarded as time series data, which means it could be classified by chronological order. Time series analysis method describes and interprets phenomena that change over time to derive various predictive decisions. Deep learning neural networks can automatically learn arbitrarily complex mapping from input to output, and support multiple inputs and outputs [1]. It provides many ways for time series prediction tasks, such as automatic learning of time dependence or trends and seasonality automatic processing of data based on time structure.

Although deep neural networks can approximate any complex function arbitrarily and perform good non-linear modelling of a variety of data, in the historical data used in the active distribution network load prediction, the short-term load data sequence has

© The Author(s) 2022
Z. Qian et al. (Eds.): WCNA 2021, LNEE 942, pp. 1150–1157, 2022.
https://doi.org/10.1007/978-981-19-2456-9_115

obvious approximate period characteristics, and the long-term load data sequence shows the variability and rich dynamic characteristics. Besides, with the development of the Internet and big data technology, it will improve the performance of active distribution network load prediction by importing some kinds of time series data, such as market reports and production management data and other modalities. LSTM (Long Short-Term Memory) and other RNN (recurrent neural network) structures could not effective in predicting the difference between peak hours and minimum power consumption times, and usually requires higher computational cost.

This paper proposes a multi-modal CNN-BiLSTM (Convolutional Neural Network-Bidirectional Long Short-Term Memory) architecture, which have an improved shared parameter parallel convolutional network to learn feature representations for short-term load data sequences, and an improved bidirectional attention LSTM network. The model presents the dynamic changing characteristics of data affected by some disturbances with the text features, such as temperature and holidays. On the 24 months of load and market report data set, the method is compared with the convolutional neural network and the bidirectional long short-term memory neural network. The experimental results show that the model has some advantages on the computational speed and accuracy.

The rest of this paper includes: The part II introduces the characteristics of the load sequence data and the variables that may affect the prediction. The third part introduces the multi-modal deep learning. The fourth part details the structure of the proposed multi-modal. The experimental and evaluation results are given in the fifth part and the last one is the summary.

2   Load Feature Extraction and Prediction

2.1   Load Feature Extraction

The load types can be distinguished according to the reaction guidance mechanism and the non-reaction guidance mechanism, which are respectively controllable load and uncontrollable load. The load type is divided into friendly load and non-friendly load. The load prediction model can be constructed by analysing the active load characteristics and energy storage characteristics including friendly load and according to the constraint conditions [2]. Another method is to use the bottom-up prediction method [3], in the small area divided according to certain properties, first perform load prediction, and finally superimpose the obtained load demand curve to obtain a complete load prediction result.

For example, a large amount of data can be processed in parallel through the cloud computing platform, the maximum entropy algorithm can be used to classify the data, the abnormal data and the available data can be distinguished, and the local weighted linear regression model can be combined with the Map-Reduce model framework to realize the active configuration of cloud computing [4].

The Spark platform is used to divide all the obtained data and compute them in parallel to speed up the processing of big data. First, the data is pre-processed through feature extraction, and the input that meets the requirements of the model is obtained, which input into the multivariate L2-Boosting for training and learning and get the final regression model [5]. The grey prediction method is also a common method of load prediction, which added secondary smoothing processing through historical data to eliminate the interference factors of historical data with Markov chain and grey theory to predict the residual sequence and the sign of the future residual together to revise the results [6].

2.2 Load Feature Prediction

As a type of time series data, load prediction can also be implemented using neural network technology. In monthly and quarterly time series, time series prediction based on neural network has more obvious advantages than traditional statistical methods and artificial judgment methods compared with traditional statistical time series methods [7]. Mbamalu et al. believe that load prediction is an autoregressive process, and use iterative re-weighted least squares to estimate model parameters [8]. Based on the combination prediction model of neural network, by learning the weights of different prediction models in the combination, the variable weight coefficient combination prediction model is shown in Eq. 1.

$$y_{ij} = \sum_{t=1}^{K} w_t(i,j)\left(f_{tij} + e_{tij}\right) \tag{1}$$

Where y_{ij} is the actual load of month i in year j, f_{tij} is the predicted value of month i in year j of the first method, $e_{tij} = y_{ij} - f_{tij}$ and $w = \text{Min} \sum_{i=1}^{n} \sum_{j=1}^{12} \left[y_{ij} - g\left(f_{1ij}, f_{2ij}, \ldots, f_{Kij}\right)\right]^2$.

Since there is a relatively complicated non-linear relationship between the actual prediction input and the final output, a three-layer forward neural network is used to fit an arbitrary function. Through the continuous iteration of the network and the update of the gradient back propagation, the final reasonable parameters are obtained. And by these parameters, the combined predicted value of any predicted input value is realized. The load forecasting results by Autoregressive Integrated Moving Average and Seasonal Autoregressive Integrated Moving Average showed that obtained 9.13% and 4.36% mean absolute percentage error respectively. With deep learning Long Short-Term Memory model, it will reduce to 2% [9].

3   Multi-modal Deep Learning

Deep neural networks have been widely used on single modal data such as text, images or audio, which included a variety of supervised and unsupervised deep feature learning model architectures [10]. Multi-modal deep learning refers to training new deep network applications to learn the features of multiple modes. For example, in emotion recognition technology, the voice and text information fusion can improve the effect of emotion recognition [3]. Establishing a private domain network (for visual information and audio information in short videos to extract individual features) and a public domain network (for acquiring joint features) could solve the problem of short video classification [8].

The principle of multi-modal feature learning is, if there are multiple modalities at the same time, one of the modalities can be learned better than a single modal in-depth feature. It can also be learned by sharing representations between multiple modalities to further improve the accuracy index on specific tasks. Researchers have begun to carry out research in various fields for multi-modal model, such as multi-modal model based on fuzzy cognitive maps [5], which first extract a subset from the complete data and trained separately on each subset, then used fuzzy cognitive maps for modelling and prediction, and finally the output was fused from each subset by the information granulation.

The time series data is widely available, such as holidays, weather and other data, which can be used to jointly predict the city's traffic conditions [6]. Firstly, the holiday and weather feature information were extracted, and the Prophet algorithm is selected to predict the traffic flow characteristics during the holidays with one DCRNN network to predict the traffic flow on the combination of road network structure data and flow data. Besides, image and time series data are indispensable in the automatic driving system. The time series refers to the speed series and steering wheel angle series. The multi-modal network serving the autonomous driving system includes CNN, RNN, horizontal control network and vertical control network. The time series data is input into the RNN network for processing, and the image data is input into the CNN network for feature extraction. The extracted features are input into the horizontal and vertical control network respectively. Finally, the predicted value of the steering wheel and speed is obtained to guide the steering wheel angle and the speed.

4   An Improved Multi-modal CNN-LSTM Prediction Model

Although classic time series prediction algorithms can be used for load prediction, the fluctuation of load does not only depend on historical time series data. Due to the diversification of intelligent load management requirements, it is manifested as a multi-modal data form in time series.

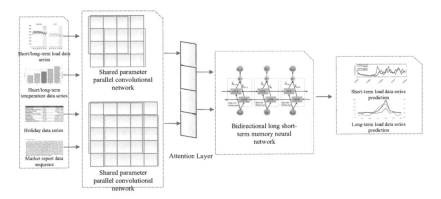

Fig. 1. Multi-modal CNN-BiLSTM network structure.

This paper proposes a multi-modal convolutional neural network-long short term memory neural network prediction method on load data and its primary structure is shown in Fig. 1. For short-term load data series, introduce data such as temperature and holidays, and use an improved shared parameter parallel convolutional network to learn feature representation; and use an improved two-way attention mechanism long and short-term memory neural network, combined with medium and long-term load sequences and effects. The relevant text data is introduced in this model for its dynamic change features.

In the multi-modal convolutional neural network-bidirectional long and short term memory neural network structure in Fig. 1, two parallel convolutional neural networks are used to extract features from the original historical load and other modal data sequences such as temperature and text. These convolutional neural networks share parameters. The first convolutional layer includes two convolution kernels with sizes 4*4 and 5*5. The number of convolution kernels is 64, and then a shared connection is used. The structure is to extract some of the convolution kernels from the previous layer of convolution kernels to form the current layer of convolution kernels. The fully connected output needs to be sent to the attention layer, trained according to the attention mechanism, and output to the BiLSTM network. The size of the hidden state is 64. The final output is the short-term load data sequence and the long-term load data sequence.

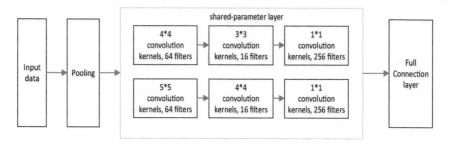

Fig. 2. Shared-parameter convolutional neural network structure.

5   Experiments and Results

In this section, we introduce the experimental evaluation methods and results of the baseline system and the above-mentioned improved methods on existing data sets. The data set contains unit hour load data of a city in North China for about 2 years, local daily maximum temperature, minimum temperature, average temperature and precipitation data, local public holiday date data, and local quarterly market operation information report data within 2 years. The entities and their types in the maximum and minimum temperatures, holiday information, and text are represented as vectors of length 128. The load value is divided into short-term load data series and long-term load data series according to the time period. The former contains the load data series within a quarter, and the latter contains the load data series greater than one quarter. Use these data to predict the unit hour load value on a specified time series period.

The evaluation index is the mean absolute percentage error (MAPE) based on the short-term load data series and the long-term load data series prediction and its calculation method is shown in Eq. 2.

$$\text{MAPE} = \frac{1}{N} \sum_{k=1}^{N} \left| \frac{\hat{v}(k) - v(k)}{v(k)} \right| \times 100\% \tag{2}$$

Where N represents the total number of samples in the test set, $v(k)$ represents the actual value, and $\hat{v}(k)$ represents the predicted value.

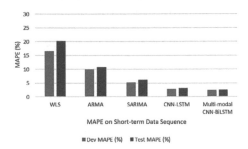

Fig. 3. MAPE results of short-term load prediction.

The baseline system adopts weighted least squares method WLS, autoregressive moving average ARMA, seasonal autoregressive integrated moving average SARIMA and CNN-LSTM architectures, and divides a total of 731 days * 24 h of data into training data and verification data in chronological order And the test data, the ratio is 4:2:4. Under the four baseline systems and the multi-modal CNN-BiLSTM model, the average absolute percentage error MAPE results and the average error MAE results of short-term load data series prediction and long-term load data series prediction are obtained, as shown in Fig. 3 and Fig. 4, respectively. The figure shows that the multi-modal CNN-BiLSTM method has certain advantages for short-term load data sequence prediction and long-term load data sequence prediction on the training set and testing dataset. Compared with the CNN-LSTM architecture, it has a certain error reduction. Especially in the long-term load data series prediction, it has higher prediction accuracy than the short-term load data series.

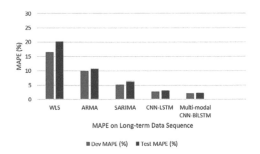

Fig. 4. MAPE results of long-term load prediction.

6 Conclustion

Load prediction has the characteristics of time trend. There are obvious differences in load in different seasons. Precise prediction is helpful for efficient decision-making and reasonable planning. This paper proposes a multi-modal convolutional neural network-bidirectional long and short-term memory neural network architecture, which uses a parallel convolutional network with shared parameters and a bidirectional attention mechanism. The long-term and short-term memory neural network processes load data, temperature data and text data. The multi-modal data sequence, etc., can predict the short-term load data sequence and the long-term load data sequence. The experimental results verify that the network structure can achieve a certain improvement in prediction accuracy compared with other baseline systems.

References

1. Brownlee, J.: Deep learning for time series forecasting: predict the future with MLPs, CNNs and LSTMs in Python. Machine Learning Mastery (2018)
2. Zhuang, H., Zhang, J.: Coordinated voltage control based on model prediction in active distribution networks. Electric Power, **12** (2016)
3. Kuzle, I., Bosnjak, D., Pandzic, H.: Comparison of load growth prediction methods in distribution network planning. In: CIRED 2009–20th International Conference and Exhibition on Electricity Distribution-Part 1, pp. 1–4. IET (2009)
4. Zhang, S., Liu, J., Zhao, B., Cao, J.: Cloud computing-based analysis on residential electricity consumption behavior. Power Syst. Technol. **37**(6), 1542–1546 (2013)
5. Du, D., Xie, J., Fu, Z.: Short-term power load forecasting based on spark platform and improved parallel ridge regression algorithm. In: 2018 37th Chinese Control Conference (CCC), pp. 8951–8956. IEEE (2018)
6. Niu, Y., Wang, Z.Y., Wang, H.J., Sun, Y., Li, X.: Application of improved grey model for mid and long-term power demand forecasting. J. Northeast Dianli Univ. (Nat. Sci. Ed.), **2** (2009)
7. Nelson, M., Hill, T., Remus, W., O'Connor, M.: Time series forecasting using neural networks: should the data be deseasonalized first? J. Forecast. **18**(5), 359–367 (1999)
8. Mbamalu, G.A.N., El-Hawary, M.E.: Load forecasting via suboptimal seasonal autoregressive models and iteratively reweighted least squares estimation. IEEE Trans. Power Syst. **8**(1), 343–348 (1993)
9. Nguyen, H., Hansen, C.K.: Short-term electricity load forecasting with Time Series Analysis. In: 2017 IEEE International Conference on Prognostics and Health Management (ICPHM), pp. 214–221. IEEE (2017)
10. Ngiam, J., Khosla, A., Kim, M., Nam, J., Lee, H., Ng, A.Y.: Multimodal deep learning. In: ICML (2011)

Research on the Deployment Strategy of Enterprise-Level JCOS Cloud Platform

Jianfeng Jiang[1(✉)] and Shumei An[2]

[1] SuzhouIndustrial Park Institute of Services Outsourcing, No.99 Ruoshui Road, Suzhou Industrial Park 215123, China
alaneroson@126.com
[2] RuijieNetworks Co., Ltd., Fuzhou 350028, China
ansm@ruijie.com.cn

Abstract. Ruijie JCOS cloud management platform is the first cloud management platform based on OpenStack principle in China. It has the advantages of stable operation, fast deployment, wide compatibility and high performance. Taking the basic technology of cloud platform management as the core, this paper gives a general description of the deploy of the whole cloud platform, from which we can understand and analyze the shortcomings of building traditional data center, and then illustrate the general process of integrating resources and reducing costs by virtualization technology in combination with real application practice.

Keywords: JCOS (Jie Cloud Operating System) · Virtualization · Clouding platform

1 Introduction

Cloud computing is a technology developed on the basis of distributed computers, parallel computing and network computing, and it is an emerging business model. Cloud computing has had a huge impact on the development of society in just a few years. Currently, cloud computing has swept various IT industry fields.

The full name of JCOS is Jie Cloud Operating System which is an enterprise-level openstatck management platform. It is a SaaS cloud computing management platform for enterprise-level users to uniformly manage multiple cloud resources. Through the comprehensive application of technologies such as hyper-convergence, software-defined networking, containers, and automated operation and maintenance, enterprises can quickly realize the "cloudification" of IT infrastructure with the smallest initial cost. At the same time, the product can achieve "building block stacking" flexible expansion and upgrade on demand with the expansion of the scale of the enterprise and the growth of its own business.

1.1 Structure System of JCOS Cloud Platform

JCOS is a mature cloud computing product. It is a professional cloud computing management platform developed in accordance with the OpenStack open source architecture. By

© The Author(s) 2022
Z. Qian et al. (Eds.): WCNA 2021, LNEE 942, pp. 1158–1166, 2022.
https://doi.org/10.1007/978-981-19-2456-9_116

deploying the JCOS platform, you can experience convenient, safe, and reliable cloud computing services. It integrates management, computing, network, storage and other services into one, and ultra-convenient cloud services that can be clouded out of the box can be realized through UDS all-in-one. The architecture of Jieyun is shown in Fig. 1 below.

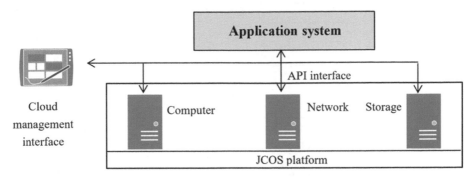

Fig. 1. JCOS architecture diagram

There are four core units in the JCOS architecture. The four major units can provide a powerful cloud computing service experience, which are computing unit, network unit, storage unit, and management unit.

2   Enterprise-level JCOS Cloud Platform Design

In order to improve deployment efficiency and reduce errors caused by manual configuration, this solution JCOS uses the open-source openstack deployment tool fuel. The fuel is a customized JCOS deployment end. JCOS uses fuel for automated deployment, which can improve deployment efficiency and reduce possible errors caused by manual configuration. Therefore, the controller fuel master needs to be prepared before deployment. Fuel master can be deployed on a physical machine or a virtual machine. Generally, it can be deployed on a virtual machine.

The basic deployment process of the JCOS platform is shown in Fig. 2.

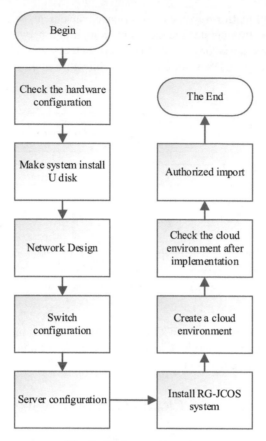

Fig. 2. Deployment process

2.1 Server Configuration

Virtualization Settings

Since UDS nodes will be used as computing nodes, each computing node needs to enable the virtualization setting support as shown in Fig. 3.

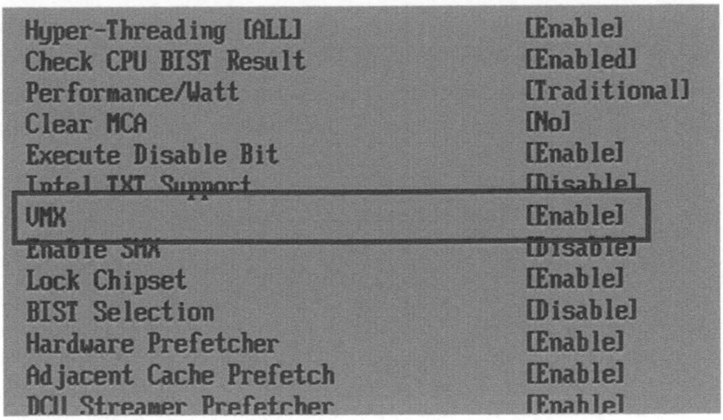

Hyper-Threading [ALL]	[Enable]
Check CPU BIST Result	[Enabled]
Performance/Watt	[Traditional]
Clear MCA	[No]
Execute Disable Bit	[Enable]
Intel TXT Support	[Disable]
VMX	[Enable]
Enable SMX	[Disable]
Lock Chipset	[Enable]
BIST Selection	[Disable]
Hardware Prefetcher	[Enable]
Adjacent Cache Prefetch	[Enable]
DCU Streamer Prefetcher	[Enable]

Fig. 3. VT enable

Server Startup Sequence

After the server virtualization is set up, you need to set the server's startup sequence to the hard disk in the first startup sequence and the network in the second startup sequence. If there are both UEFI and Leagacy boot modes, select Leagacy.

Configure Node IPMI Address

The node IPMI address can be set in the BIOS, or you can open the server management interface to modify the IPMI address through the browser using the default IPMI address. If it is set in the BIOS, go to the BMC network configuration under Server Mgmt to configure it.

Hard Disk RAID Settings

According to the system prompt when the server starts, you can make the corresponding RAID configuration. Press Ctrl + R during startup to enter the RAID card setting interface, you can set RAID5 to improve the reliability of data storage.

2.2 Cloud Computing Network Planning

The servers participating in the deployment of JCOS are called nodes, and the interconnection needs to be through an external switch, and the vlan or port on the external switch is isolated to form a network. Among them, there are 6 JCOS platform deployment networks, which are shown in the following Table 1.

Because of the large number of server network interfaces, these networks are isolated directly through ports. At this time, it is only necessary to determine the corresponding

Table 1. Deployment networks

Network name	Network details
External network/floating IP network	The external network is the only network that the OpenStack cluster connects to the outside world, that is, the actual network of the customer. The other JCOS networks are actually private networks inside the cluster, which are not visible to the outside world
Business private network	A business private network is a virtual network created by OpenStack tenants and a real network assigned to virtual machines by JCOS, so it is generally called a tenant network. The virtual machines communicate through the business private network. The all-in-one machine supports two kinds of virtual networks, VLAN and GRE tunnel. The GRE tunnel is point-to-point and does not require redundant configuration. In VLAN mode, you need to configure a trunk port on the corresponding switch interface to ensure communication between VLANs
Management Network	The management network is the network used for communication between various components of OpenStack cloud computing. The network carries the heartbeat and voting of high-availability clusters, databases, message queues, API calls between components, and virtual machine migration. It is recommended to use 10Gb or better Ethernet
Storage network	Storage network is a network used by computing nodes to access distributed storage. It is recommended to use 10Gb or better Ethernet. Redundant replication of data within distributed storage nodes also requires the use of this network
Deployment/PXE network	The deployment network is used to implement the deployment of the cloud platform. The server with the deployment end and the server to be deployed are connected in the same network, and the server to be deployed is automatically deployed through PXE. The server installs the operating system and completes the installation and configuration of various components through PXE

(continued)

Table 1. (*continued*)

Network name	Network details
IPMI network	The IPMI network is a network for remote management of physical machines. The physical opportunity provides a separate IPMI network port. The high-availability function of the computing node of the all-in-one machine requires the use of the network to shut down and restart the physical machine

relationship between different networks and different network interfaces. The connection topology diagram of the UDS server is shown in Fig. 4.

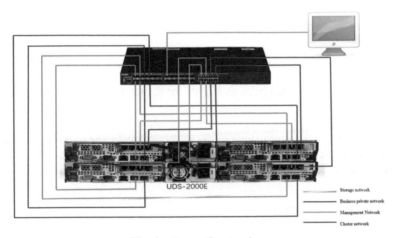

Fig. 4. Connection topology

Isolating the network through ports eliminates the need to plan VLANs. We generally only need to plan the IP addresses of the external network and the floating network, and use the default IP addresses for other networks. Table 2 is the specific plan.

Table 2. Network planning scheme

Network Type	VLAN	Physical interface	IP address planning
Business Private Network	6	eth0	192.168.111.0/24
Management Network	5	eth2	192.168.1.0/24
Storage Network	4	eth3	192.168.0.0/24
Deployment Network	N/A	eth0	10.20.0.0/24
External network/floating network	100	eth1	172.16.0.0/24

3   Enterprise-level JCOS Cloud Platform Deployment Plan

The following conditions must be prepared for the deployment of the JCOS platform on the fuel deployment side.

- The node server and the deployment server are connected in the same Layer 2 network.
- The node server sets PXE priority to start.
- The node server must enable hardware virtualization in the BIOS settings.

3.1   Opensatack Environment

Choose the deployment mode "HA multi-node" mode. In this mode, an odd number of controller nodes need to be deployed. The basic services of the cluster have high availability guarantee in this mode.

If you deploy OpenStack on a physical machine, select "KVM". If you are testing OpenStack in a virtual machine, select "QEMU". This deployment scheme runs on hardware, so we select "KVM".

3.2   Node Allocation

After entering the main interface of the cloud platform, we turn on the power of the node server to automatically obtain the IP address and load the operating system. After the node is discovered, the discovered node will be displayed in the unallocated node pool.

3.3   Assigning Roles

Select the node that needs to be allocated from the unallocated node pool, and assign the corresponding role according to the demand. If it is only a single device, you need to assign all roles to the node.

4   Result Analysis

After the deployment of the JCOS cloud platform, the Windows cloud host and the Cent OS cloud host are created. Under the same host configuration, the CPU uses 2 cores and the memory uses 2G. The efficiency of the JCOS cloud platform is more than 2 times more optimized than the time of VM virtualization, and the result is shown in Fig. 5.

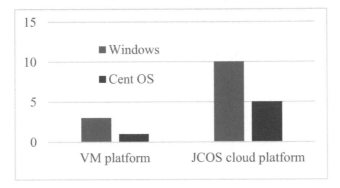

Fig. 5. Comparison of results

5   Conclusion

As the first truly enterprise-level Openstack cloud management platform in China, JCOS has been widely used in education, healthcare, government, IDC, operators and other industries. It has the advantages of high performance, stable operation, wide compatibility, and quick deployment. Platform monitoring and management, and log information maintenance are important features of platform operation and maintenance. After an enterprise deploys a private cloud, the maintenance and update of its system becomes faster, and the task of network administrators becomes relatively easy.

Acknowledgement. Supported by the high-end training project for teachers of higher vocational colleges in Jiangsu Province. Qinglan Project of Jiangsu Province. The 13th Five-Year Plan of Jiangsu Province :Research and Practice of the New Generation Information Technology Talent Training Model under the "1+X" Certificate System (D/2020/03/20).

References

1. Faizi, S.M., Rahman, S.M.: Secured cloud for enterprise computing. In: Proceedings of 34th International Conference on Computers and Their Applications, CATA 2019, pp. 356–367, 13 March 2019

2. Li, W., Zhao, Y., Han, J., Zhang, Z., Yu, H.: Research on construction of innovation cloud service platform in power enterprise. In: E3S Web of Conferences, vol. 136, 9 December 2019
3. Khoo, B.K.: Enterprise information systems in the cloud: implications for risk management. Wireless Telecommunications Symposium, April 2020
4. Longbin, C., De Jana, R.: Sharing enterprise cloud securely at IBM. IT Professional, **23**(1), 67–71 (2021)
5. Shao, M., Li, X.: An empirical study of impact factors on the alignment of cloud computing and enterprise. Inf. Sci. Commun. (CTISC), 70–74 (2020)

Research on Campus Card and Virtual Card Information System

Shuai Cheng[✉]

Faculty of Information Center, University of Electronic Science and Technology of China, No. 2006, Xiyuan Avenue, West Hi-Tech Zone, Chengdu, Sichuan, People's Republic of China
scheng@uestc.edu.cn

Abstract. Campus card system is the core business and platform of University Information System. After more than ten years of development, it covers all aspects of campus life: learning, teaching and research. This paper explains the current situation of campus card system from the perspective of card, account and billing, descripts system design and account model. Based on current system, this paper analyses the development of virtual campus card and describes a Data docking method in Information System.

Keywords: Campus card system · Virtual campus card · Data docking · Information system

1 Background

With the continuous construction of our country's informatization and the continuous popularization of network technology, Internet technology is widely used in society, and new technologies and concepts are constantly emerging, such as: IPv6, 5G, face recognition, biotechnology, drones, block Chain, big data, virtualization, edge computing, etc. [1, 2]. The rapid development of informatization has promoted the construction of informatization in universities and promoted the rapid development of all aspects of informatization in universities. The campus all-in-one card system, which is one of the foundations and core platforms of university informatization, has developed from only solving the problems of canteen catering, shower hot water, supermarket shopping, etc., to covering almost all the campus life: studying, teaching and research by teachers and students. The continuous increasing of business requirement and information system requirement in university has brought higher requirements for third-party docking in campus information system [12, 14]. While exploring the construction of the campus all-in-one card, this article explores the physical card, virtual card, third-party business system docking, and electronic campus card identity data docking. Provide solutions for the construction of a new generation of campus card.

2 Campus Cards and System

At present, the campus all-in-one card system forms an informatized closed-loop management of cards, accounts, and accounts based on service programs, databases, network

Z. Qian et al. (Eds.): WCNA 2021, LNEE 942, pp. 1167–1175, 2022.
https://doi.org/10.1007/978-981-19-2456-9_117

technology, and terminal equipment. At the same time, it is integrated and linked with other systems through docking. All teachers, students and staff of the school only need to hold one campus card, which replaces all the previous certificates, including student ID, teacher ID, library ID, dining card, student medical ID, boarding card, access card, etc. The campus all-in-one card system is the main framework for supporting and running information-based campus applications [7, 11]. Most of it adopts C/S architecture [3]. In the same time, we are talking about another system architecture which uses front-end server or docking server to be compatible with third-party systems and equipment to realize the campus information system. The system business covers all aspects of the teachers and students in the school. The business scope includes: data business, card business, finance, consumer business, water control business, electronic control business, vehicle business, access control business, storage subsidy business, secret key business, etc. [4, 10].

Recently, most campus cards use radio frequency contactless IC cards, and the main card model is the Mifare1 series (M1 card for short) produced by NXP. At the same time, some colleges and universities have adopted CPU chip cards, and most college users use the FM series of Shanghai Fudan Microelectronics (such as FM1208 card, FM1208M01 card, FM1280M-JAVA card). In terms of card security, the CPU card has a central memory (CPU), storage units (ROM, RAM and EEPROM) and a card operating system (COS). The CPU card is not just a single contactless card, but a COS application platform of the system. The CPU card equipped with COS not only has the function of data storage, but also has the functions of command processing, calculation and data encryption. The characteristics of the card surface of the CPU card and the security technology of COS provide a double security guarantee, which can realize the true meaning of one card and multiple applications. Each application is independent of each other and controlled by its own key management system, and storage large capacity. The dynamic password is used by the CPU card, and it is the same card with one password, each time the card is swiped, the authentication password is different, which can effectively prevent security vulnerabilities such as duplicate cards, copy cards, malicious modification of the data on the card, and effectively improve the entire system security. Compare these types of current campus cards as follows (Table 1).

Table 1. Types of current campus cards examples.

Types	Mifare1 card series	CPU card FM series		
Exa	M1	FM1208M01	FM1208	FM1280M-JAVA (as JAVA card)
Cap	8 KB	7 KB + 1 KB mode, compatible with M1	8 KB	80 KB capacity, built-in multiple PBOC applications, independent of each application COS, support multiple authentication methods
Mode	Sector mode	File mode		
COS	Without COS system	With COS system		
Enc	without Hardware encryption	Support hardware DES operation module		
Auth	Fixed key. no SAM authentication	Dynamic key. Using SAM card encryption and authentication to ensure safety		

3 Campus Card Data

The campus card data is the management of cards, accounts, and bills. In the management of accounts, there are different groups of people in different universities, but they all have similar problems and difficulties: data comes from different business departments and systems; there is a lack of system docking between systems, data is isolated, and the systems cannot be linked; data quality is not high due to sparse management; coupled with changes in departmental business and other reasons, it has caused a variety of data and accumulation of historical data. In this paper, the data has been cleaned up, mainly according to the management of the cards and accounts to sort out teachers, students, and other users, and sort out five categories and 28 sub categories of personnel. At the same time, it is connected to the business system, and based on this, we combined with the business and department to screen and clear the data, to solve the problems of management, data, and users in the campus card system. Getting through the business systems of various business departments has played a key step in the future data linkage and data sharing.

4 Physical Card and Virtual Card

After more than ten years of development, physical cards, as the main carrier of identity recognition and campus consumption, have become an indispensable part of the campus all-in-one card system. The main advantages of using physical cards are: easy to carry, high reliability, gradual improvement in security, and convenient to use; but at the same time, physical cards also have many shortcomings: recharge problems, card replacement problems, lost and forgotten problems, card-not-equal-database problems, etc. [15].

With the rapid development of mobile Internet technology and information technology, based on the physical card, the concept of a virtual campus card is proposed. In essence, the virtual campus card is an extension of the mobile Internet service on the existing one-card system [2, 5]. The virtual campus card is a kind of virtual card that is bound to the physical card and can replace the physical card for identity recognition and campus consumption. Teachers and students can use this virtual card to realize consumption and identification at any time. The main advantages of the virtual campus card are: convenient management and function expansion, there is no management cost of the physical card, the virtual card does not have the problem of loss, there is no problem of replacing the card, it can cover most all the campus scene. Of course, the virtual campus card also has some shortcomings: the usage of water control problem, can't be identity cards, data losing problem, high dependence on the network, and the security problem that breaks the closed environment of the private network.

The virtual campus card system adopts Internet technology, mobile application technology, payment technology, etc., unified data management, cards can use multiple carriers, and expand payment methods. The current carriers include: handset terminals with NFC, QR (Quick Response) code, biometrics, web account and passwords, etc. Scanning the QR code is the most common way to realize the virtual campus card. We divide the scanning code into two ways: the Scan and the Scanned. The Scan: The device held by the consumer (user) scans the device or the QR code of the payee (merchant). The Scanned: The QR code generated by the consumer is scanned by the payee.

The process of Scan is:

1) The machine adopts a static QR code that has been generated or a dynamic QR code generated after entering the amount.
2) The consumer scans the QR code and obtains the information, and then applies to the payment platform.
3) The payment platform and the all-in-one card backend perform data verification and conduct transaction processing.
4) The transaction result is returned to the machine tool and the consumer.

The process of Scanned is:

1) The consumer generates a dynamic QR code on the APP or webpage on the handset device.
2) The machine scans the consumer's QR code, enters the amount, and asks the background for data verification. And initiate a transaction request.
3) The payment platform and the all-in-one card background complete data verification and complete the transaction.
4) The transaction result is returned to the machine tool and the consumer.

The following figure is a simplified diagram of the virtual campus card usage (Fig. 1):

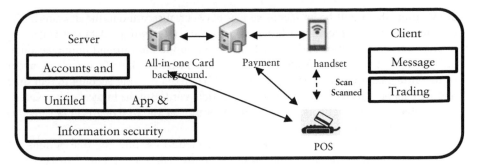

Fig. 1. Virtual campus card usage

5   System Design and Account Modes

The design of the system was implemented in four-layers architecture: Interface layer, Application service layer, Data access layer and Bus service layer. The service content that the platform providing: data access service, security service, infrastructure service, management service, development service, resource management service.

Data access service: Responsible for providing services such as the definition, storage and query of data resources, realizing centralized management of data, and ensuring the legality and integrity of data resources.

Security Service: Responsible for protecting every layer and network from unnecessary threats. Responsible for protecting the legality, integrity and security of data interaction and data communication between each layer of the architecture.

Infrastructure services: Provide efficient use of resources, ensure a complete operating environment, balance workloads to meet service requirement, isolate workload to avoid interference, perform maintenance, secure access, trusted business and data processes, simplify overall system management.

Management services: Provide management tools to monitor service flow, underlying system status, resource utilization, service target realization, management strategy execution, and failure recovery.

Development Service: Provide a complete set of development tools for system expansion.

Resource management service: A service that manages application services registered and running under the architecture.

The most important thing in the design of the above campus card system is to solve the accounting problem. At present, the usual account models are divided into the following types:

- *Offline mode:* transactions are carried out based on the card electronic wallet. This mode is not affected by the factors such as: the network and background, and can be used for offline consumption. However, offline consumption data cannot be uploaded in time, resulting in inconsistency between the balance on the card and the amount of the back-end account (data-base); if the card is dropped and the card is replaced at this time, there will be an inconsistency between the card and the amount in the data-base. If the equipment was broken at this time, there will be data loss Case.

- *Online mode:* transactions are carried out based on the background online account, and the card is for the identification. This model is the realization of the account model of the virtual campus card. The recharge will be credited to the account in real time and will not be affected by the loss of the physical card. But the biggest disadvantage is the reliance on the network. If the network or the background platform fails, it will affect business processing.
- *Offline mode with online allowed*: When connected to the Internet, transactions are carried out based on the back-end online account. The transaction is successfully written into the card electronic wallet. When the terminal is not connected to the Internet, the card electronic wallet shall prevail. The biggest advantage of this mode is that it can have the advantages of the online mode when the network is fine, and can handle the business in the offline mode when the network is blocked. But this mode also has the disadvantages of the offline mode.
- *Online account with electronic wallet separation mode*: one user has two accounts, online account and offline wallet, the two accounts are independent of each other. This mode is a fusion of offline mode and online mode. There are advantages of these two modes as well as disadvantages of these two modes. There are two accounts for users at the same time, which may cause confusion for users.

The above account model analyses several existing account methods, and each university will choose a different method according to its own situation. At present, physical cards mainly use offline mode, while virtual campus cards mostly use online mode. Different account models can also be selected according to different requirements to facilitate the management of system reconciliation.

6 Data Docking

The realization of the virtual campus card can be based on the existing all-in-one card system to expand payment methods. Currently, the methods include: Alipay payment, WeChat payment, Integration payment and so on. Use APP, Web, WeChat, Alipay, etc. However, it's difficult to expand the market of the APP. And it's easy to use the H5 webpage method for multi-party connection. On the other hand, with the expanding of the mobile Internet, the WeChat and Alipay method has also been widely used. Alipay has an Alipay electronic campus card, WeChat has a Tencent WeiXiao electronic campus card, and the Integration payment party also has its own electronic campus card. We use Alipay as an example to explain the identity authentication and consumption of the electronic campus card.

The Alipay electronic campus card mainly uses the interface to identify the identity of people, so it does not affect the existing data access and business processing of the original campus system. All accounting and transactions are completed in Alipay system. Users only need to apply for an electronic campus card. When users receive the electronic campus card in the Alipay card package, they need to initiate an identity authentication request to the background to confirm whether the user has the authentication. Only the person who have passed the certification can receive the electronic campus card. The application for e-campus card is as follows (Fig. 2):

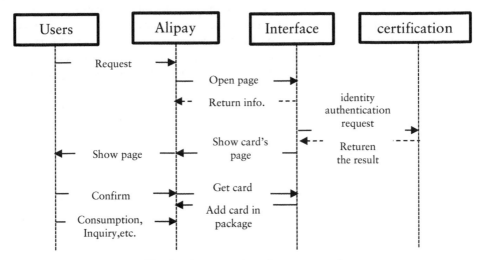

Fig. 2. The processing of e-campus card

The campus all-in-one card database stores identity data. In order to reduce the access pressure to the campus card system and security considerations, a data cache server is added between the campus card database and Alipay APP. The campus card database regularly pushes data to the cache server, and Alipay accesses the data cache server to verify user's identity.

For information security concerning:

1) The campus card identity database only needs to periodically synchronize the latest identity data with the data cache server, which does not affect the existing business of the campus card system.
2) The data cache server is stored in the machinery room to reduce the risk of data leakage.
3) The data cache server opens the firewall, and only opens the public network access permissions for certain necessary ports.
4) Set the access IP whitelist and only allow Alipay server access.
5) When accessing data, a strict encryption and signature mechanism is used to ensure communication security.

At present, this method has been used for identity authentication and consumption in some schools. With the continuous expansion in the later period, it can be extended to other all-in-one cards scenarios.

7 Summary

With the exploration of campus all-in-one card construction, we can see a development trend from physical cards to virtual cards. Comparing the physical cards and virtual cards, we can see that from the saving money, facilitating management, and improving user experience, virtual cards have brought more convenience to schools, but from the current development, virtual cards cannot completely replace the physical cards. At the same time, the virtual cards also need to rely on the current campus card system. There are also defects in the usage of virtual cards, such as the using water control. Due to the dependence of virtual cards on handset terminals, there will be inconveniences when using water control. Of course, there are other solutions that can be found, such as the express delivery method, using temporary digital string generation.

In general, virtual cards and physical cards will co-exist in the campus all-in-one card field, and virtual cards will be a direction for the development of all-in-one cards. With the advancement of technology and practice, the campus all-in-one card will much more focus on users. Based on the existing all-in-one card platform, it is believed that more user-friendly forms and methods will be adopted and used.

References

1. Ye, Y., Xu, F., Cheng, Y.: Design of a new generation campus card payment system based on electronic accounts. J. East China Norm. Univ. (Nat. Sci. Edn.) 536–540 (2015)
2. Wang, T.: Construction and research of virtual campus card system in colleges and universities. Sci. Technol. Commun. **11**, 93–94 (2019)
3. Xu, W., Xu, B., Zhu, X., Wu, W.: Campus card and digital campus. Educ. Inf. **11**, 93–94 (2019)
4. Shao, Y., Mao, X.: Analysis and research on university campus card data in the big data era. Digit. Technol. Appl. 52–53 (2019)
5. Luan, S., Leng, F., Xu, J.: Research and application of virtual campus card system. Inf. Commun. 115–116 (2019)
6. Wu, H.: Analysis of the development prospects of the "all-in-one card" in colleges and universities in the era of cardlessness. China Manag. Inf. Technol. 203–204 (2019)
7. Yang, C.: Analysis of the status quo and trend discussion of the construction of the all-in-one card in colleges and universities. CaiZhi, 249 (2019)
8. Cai, J.: The application and thinking of artificial intelligence in the financial field. Acc. Learn. 56–57 (2019)
9. Wei, N.: Analysis of the internal control activities of the campus all-in-one card under the third-party payment mode. Knowl. Econ. 46–47 (2018)
10. Su, D., Liu, X., Jiang, T., Li, Z.: Research on the application of data mining technology in campus card system. Int. Conf. Smart City Syst. Eng. **2017**, 199–201 (2017)
11. Du, S., Meng, F., Gao, B.: Research on the application system of smart campus in the context of smart city. In: 2016 8th International Conference on Information Technology in Medicine and Education (ITME), pp. 714–718 (2016)
12. Wang, L.: Campus one-card system design. In: Zhang, Z., Zhang, R., Zhang, J. (eds.) LISS 2012, pp. 1105–1116. Springer, Heidelberg (2013). https://doi.org/10.1007/978-3-642-32054-5_156
13. Yang, W.: Network mining of students' friend relationships based on consumption data of campus card. IOP Conf. Ser. Earth Environ. Sci. **632**(5), 052067 (2021)

14. Yang, J.N., Lin, K.: Campus card information query system design and implementation. Appl. Mech. Mater. **467**, 574–577 (2013)
15. Feng, L.: Application and analysis of virtual campus card in college card system. Value Eng. (2018)

A Novel Approach for Surface Topography Simulation Considering the Elastic-Plastic Deformation of a Material During a High-precision Grinding Process

Huiqun Chen[1] and Fenpin Jin[2(✉)]

[1] Faculty of Public Curriculum, Shenzhen Institute of Information Technology, 2188 Longxiang Avenue, Shenzhen, China
[2] Bao'an Maternal and Child Health Hospital, 56 Yulv Road, Shenzhen, China
chqleo@163.com

Abstract. A novel simulation approach for 3D surface topography that considers the elastic-plastic deformation of workpiece material during a high-precision grinding process is presented in this paper. First, according to the kinematics analysis for the abrasive grain during the grinding process, the motion trajectory of the abrasive grain can be calculated. Second, the kinematic interaction between the workpiece and the abrasive grains can be established, which integrates the elastic-plastic deformation effect on the workpiece material with the topography, the simulation results are more realistic, and the simulation precision is much higher. Finally, based on an improved surface applied to the grinding wheel, the surface topography of the workpiece is formed by continuously iterating overall motion trajectories from all active abrasive-grains in the process of high-precision grinding. Both the surface topography and the simulated roughness value of this work are found to agree well with those obtained in the experiment. Based on the novel simulation method in this paper, a brand-new approach to predict the quality of the grinding surface by providing machining parameters, selecting effective machining parameters, and further optimizing parameters for the actual plane grinding process, is provided.

Keywords: Surface topography · High-precision grinding · Abrasive grain · Elastic-plastic deformation · Simulation

1 Introduction

There are two important factors affecting the surface quality of the machined workpiece during the high-precision grinding: the abrasive grains (grinding tools) and the debris formation process. In a traditional grinding process, the machining dimension of the parts and the 3D model of the machined surface are obtained by instrument detection after grinding [1–3]. If the processing parameters are selected improperly, the parts will

© The Author(s) 2022
Z. Qian et al. (Eds.): WCNA 2021, LNEE 942, pp. 1176–1193, 2022.
https://doi.org/10.1007/978-981-19-2456-9_118

not meet the technical requirements, which will result in wasting money and resources [4].

With the development of computer technology, the 3D surface of machined parts has been digitally simulated with the help of computers, and this process is usually called virtual manufacturing. Virtual manufacturing is one of the main development directions of modern manufacturing [5–7].

Many researchers have made significant attempts to study the generation mechanism of workpiece surface during grinding process. Malkin [8] described motion trajectory of any abrasive grain and investigated the relationship between the chip thickness and the grinding parameters. A mathematical model to describe the kinematics of the dressed wheel topography and to reflect the ground workpiece surface texture was established by Liu and his co-authors [9]. Kunz and his co-author [10] utilized a machine vision method to survey the wheel topography of a diamond micro-grinding wheel. Nguyen et al. [11] proposed a kinematic simulation model for the grinding operation, in which the complex interaction relationship between the wheel and the workpiece was taken into account during the creation process of the machined surface. The surface topography of the grinding wheel can affect the surface integrity of grinding workpiece. Chen and his co-authors [12] focused on the modeling for grinding workpiece surface founded on the real grinding-wheel surface topography. Cao and his co-authors [13] investigated the influences of the grinding parameters and the grinding mechanism on surface topography of the workpiece, and a novel topography simulation model considered the relative vibration between the grinding-wheel and the workpiece was proposed, concurrently, the wheel working surface topography was taken into account in this model. Nguyen and Butler [14] described a numerical procedure according to a random field transformation for effectively generating the grinding wheel topography. The correlation between the grinding wheel surface topography and its performance was investigated by Nguyen and Butler in another study [15], which was characterized by using 3D surface characterisation parameters. Li and Rong [16] established the micro interference model of single abrasive grain taking the shape and the size properties of the abrasive grain accompanying the crush between the binder and the grain into account. Because of self-excited vibration, surface grinding processes are bound to be chatter. Sun et al. [17] developed a dynamic model with time-delay and two degrees of freedom feature to reveal the correlation of the dynamic system characteristic and the workpiece topography. Liu and his co-authors [18] took the gear grinding as the research object and revealed the chatter effect on the machined surface topography. The grinding operations under different machining states and surface topographies of gears in each process were discussed comprehensively. Jiang et al. [19] established the kinematics model of machining surface topography of workpiece taking the factors of grinding parameters and vibrational features into account.

However, the machined workpiece materials in the above literatures were assumed that they were non-deformed (under ideal conditions), and all of these researches did not take the influence of workpiece material's elastic-plastic deformation on workpiece surface into account. The simulating precision of the above discussed studies lags behind that of the actual machined surface. How to synthetically consider workpiece material's elastic-plastic deformation during the grinding process and the kinematic prediction for the grinding process proves to be our research emphasis.

In this paper, the abrasive-grain motion trajectory of a plane grinding process is analysed and studied. First, the trajectory equations of abrasive-grain are proposed based on the grinding kinematics. Second, the kinematic interaction relationship between the machining workpiece and the abrasive-grains can be established, a novel approach for surface topography simulation taking the elastic-plastic deformation of a material during a grinding process into account is also developed. Finally, based on the an improved Gaussian surface applied to the grinding wheel, the workpiece surface topography can be formed by continuously iterating overall motion trajectories from all active abrasive-grains in the process of high-precision grinding, and the MATLAB programming method is used to simulate and predict the 3D grinding surface of workpiece.

2 Grinding Kinematics

In the high-precision grinding process, there are two movements: the rotation of the grinding-wheel and the translational movement of the machining workpiece [20, 21].

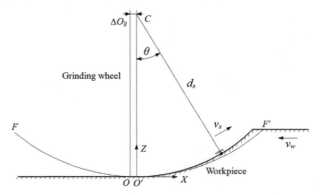

Fig. 1. The motion trajectory of a working abrasive-grain.

In Fig. 1, the coordinates system $O'XYZ$ can be established following the rules that its origin point O' is fixed on the workpiece and coinciding with the abrasive-grain at the lowest point position, and the machining trochoid path $FO'F'$ is formed on the surface of workpiece. This trochoid is synthesized with two motions: abrasive grain rotating around the wheel centre and workpiece translation [11]. The mathematical description of this trochoid is given by Eqs. (1) and (2) [22]:

$$x = \frac{d_s}{2} \sin\theta \pm \frac{d_s v_w}{2v_s}\theta \tag{1}$$

$$z = d_s(1 - \cos\theta) \tag{2}$$

where x and z are the trajectory coordinates of the abrasive-grain, v_w represents the workpiece movement velocity, v_s represents the linear velocity of the grinding-wheel,

θ represents the rotation angle of the grinding-wheel, due to the small of angle θ here, $\sin\theta \approx \theta$, and d_s represents the nominal diameter of the grinding-wheel.

t represents the time required for the abrasive-grain rotating counter-clockwise with an angle θ from the lowest position point O', and $t = \frac{d_s\theta}{2v_s}$. The process in which the linear velocity direction of an abrasive grain revolving around the wheel axis is opposite to that of the workpiece movement is referred to as up-grinding, and the symbol $\pm$ is replaced by $+$ in Eq. (1). Otherwise, when down-grinding occurs, $\pm$ is replaced by $-$.

Because θ is very small here, $\sin\theta \approx \theta$, the trochoid can be simplified to a parabola:

$$z = \frac{x^2}{d_s\left(1 \pm \frac{v_w}{v_s}\right)^2} \tag{3}$$

Due to the workpiece translation, when abrasive-grains cut the workpiece surface, the coordinate origin of each cutting parabola on the workpiece is different. The distance value ΔO_{ij} from the coordinate origin to the initial cutting position can be expressed as

$$\Delta O_{ij} = \frac{\Delta L_{ij}v_w}{v_s} \tag{4}$$

where ΔL_{ij} is the arc length that the initial position of the abrasive grain turns, $\Delta L_{ij} = \pi(n-1)\Delta d_s + l_{ij}$, l_{ij} represents the arc length from the grain on the grinding-wheel surface to the initial point, and n represents the rotation cycle of the grinding-wheel.

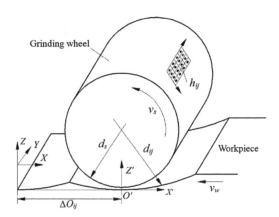

Fig. 2. Cutting model of a single abrasive-grain on a grinding-wheel.

Taking the coordinate system translation and the distance from the abrasive grain to the wheel-axis into account (in Fig. 2), thus the trajectory equation of a single abrasive-grain acting on a grinding wheel surface can be obtained again:

$$z = \frac{(x - \Delta O_{ij})^2}{d_{ij}\left(1 \pm \frac{v_w}{v_s}\right)^2} + h_{max} - h_{ij} \tag{5}$$

where d_{ij} represents the actual distance from the wheel centre to the top point among the cutting points, $d_{ij} = d_s + h_{ij}$, h_{max} represents the maximum coordinate value among cutting points for all abrasive-grains, $h_{max} = max\{h_{ij}\}$, and h_{ij} is the actual radial height of grain cutting points on the wheel surface.

3 Interaction Mechanism of Abrasive-Grains and Workpiece Material in the Grinding Contact Zone

The force acting on a single abrasive grain normal is regarded similar to the stress condition when testing the Brinell-hardness [23]. The deformation condition can be confined as an elastic-plastic deformation. When the spherical grain moves horizontally (along the direction of linear velocity), the plastic-deformation region on the sphere begins tilting, and the material of grinding workpiece is stacked up and torn from the surface of workpiece to generate debris [24].

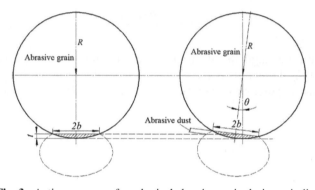

Fig. 3. Action process of a spherical abrasive grain during grinding.

This process is shown in Fig. 3. Where the force R of a single abrasive-grain is derived using the test method of Brinell hardness.

$$R = \frac{\pi}{3}b^2HC'$$ (6)

In the contact zone, C' represents the ratio (the mean pressure is divided by the axial stress), where, generally, $C' = 3$, b is equal to half of the grinded workpiece width, H is the Brinell hardness of the workpiece material, and R represents the normal force acting on abrasive-grain.

The grinding-wheel is a porous body that is composed of abrasive-grains, binder, and pores. The abrasive-grains are elastically supported with the binder. During the actual grinding process, due to the movement of the abrasive-grain centre under the cutting force action, it directly causes the actual interference/contact curve between the wheel and the workpiece to be higher than the theoretical one. Meanwhile, the workpiece surface will attain elastic-recovery when finish grinding, therefore, the final curve formed

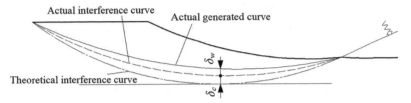

Fig. 4. Action curve of the abrasive-grain.

by the surface formation is higher than the actual interference/contact curve between the grinding-wheel and the machining workpiece (in Fig. 4).

The actual forming curve is realized by attaching the change δ_c of the grain centre, and the elastic-recovery δ_w of the grinding material to the basis of the theoretical interference curve. After discretizing the workpiece surface, the coordinate matrix Z_i^n can be obtained as Eq. (7):

$$Z_i^n = \min\left(z_i^n + \delta_{ci} + \delta_{wi}, Z_i^{n-1}\right) \tag{7}$$

where Z_i^n represents the coordinate matrix of workpiece surface when finish cutting of the $n-th$ abrasive-grain, z_i^n represents the theoretical coordinate matrix of the workpiece surface after machining of the n-th abrasive-grain, Z_i^{n-1} means the coordinate matrix of workpiece surface after machining of the $(n-1)$-th abrasive-grain, and δ_{ci}, δ_{wi} are two types of deformation values at point i, and their expressions are as follows:

$$\delta_c = C(Rcos\theta)^{2/3} \tag{8}$$

$$\delta_w = Rcos\theta/k \tag{9}$$

where C is a constant value that ranges from 0.08 to 0.25 with an average value of 0.15 [25] and k is the stiffness of the workpiece.

In the grinding process, only the undeformed material is removed by the abrasive-grains, while the remaining unresected material undergoes plastic deformation and is stacked on two sides of abrasive-grains, therefore, the grinding efficiency β is utilized here, which is equal to the ratio of the material volume that is undeformed but removed from workpiece surface to the total volume machined by the abrasive-grain in this zone where the abrasive grain has cut. Then, the area A_p that accumulates on both sides of the abrasive grain due to the plastic deformation can be written as

$$A_p = A(1 - \beta)/2 \tag{10}$$

The shape of the material that accumulates on both sides of the abrasive grain can be approximated by a parabola (in Fig. 5).

$$z = \frac{(2a - x)xh}{a^2} \tag{11}$$

The workpiece material is stacked on two sides of the orientation of angle α; then, the stacked material area can be obtained from the stacked material curve:

$$A_p = 4ah/3 \tag{12}$$

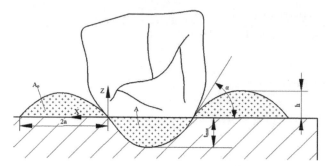

Fig. 5. Plastic accumulation caused by a plough.

Then,

$$a = \sqrt{\frac{3A_p}{2\tan\alpha}} \tag{13}$$

$$h = \sqrt{\frac{3A_p\tan\alpha}{8}} \tag{14}$$

where $\tan\alpha = \frac{t_{max}}{b}$.

4   Simulation of the Workpiece Grinding Surface

During the computer simulation process of the high-precision grinding, such surface parameters of the grinding-wheel can be obtained in two ways. One method is to obtain a height matrix describing the shape of the surface by measuring. This approach, however, takes a lot of time, and computer simulations require massive piece of the wheel-surface. The other method is to randomly generate the position matrix of the abrasive-grains distributed on the grinding-wheel using a computer. Generally, the abrasive-grains are simplified as spheres ignoring the complexity of their shape [26–29]. From a mathematical viewpoint, these abrasive-grains are a set of points with an average distribution in the two-dimensional direction of the wheel surface, and the distances between the grains obey an even distribution [30] in the radial direction. The protrusion-heights of

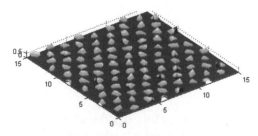

Fig. 6. An improved Gaussian surface of the grinding wheel simulated by the authors of this paper.

these abrasive-grains are described with a distribution [31], furthermore, the size of the abrasive-grain is approximately equal to the number of grains.

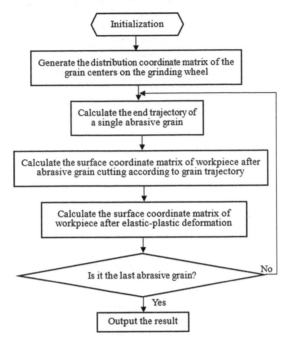

Fig. 7. Computer simulation flow chart.

An improved surface for the grinding-wheel simulated by authors of this paper is shown in Fig. 6. In the computer simulation process, the surface of cutting workpiece can be obtained with the interaction between the abrasive grains and machining workpiece. The trajectory equation of the grain cutting on the workpiece can be obtained from the grinding kinematics model. The machined surface model without elastic-plastic deformation is calculated by the grinding trajectory. The cross sections of the interference formed by the workpiece surface and those abrasive-grains is obtained using the interaction model between the abrasive-grains with the grinding workpiece. The ultimate workpiece surface model is then computed by the cross sectional shapes generated by these interference.

Figure 7 shows the whole simulation process, the flow chart for the axial and circumferential coordinate matrices generation of the abrasive-grain distributed on grinding-wheel surface is shown in Fig. 8. Figure 9 shows the coordinate matrix when finish calculating the elastic-plastic deformation for the workpiece surface.

In the simulation experiment, the material is quenched steel of 45#, and the grinding-wheel is GB70RAP400. The data from the abrasive-grains distributed on the surface of grinding-wheel are as follows: the average gap between two adjacent abrasive-grains in circumferential and axial directions is 0.236 mm, and the variation range is ±0.15 mm.

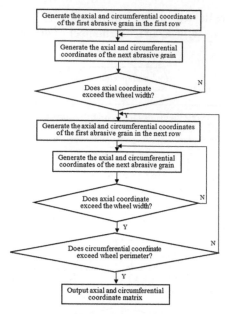

Fig. 8. Flow chart for generating the grain axial deformation of the workpiece material.

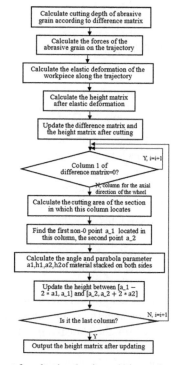

Fig. 9. Flow chart for elastic-plastic and circumferential coordinates.

Table 1. Parametric values of the grinding simulation.

Parameters	Values
Linear velocity of abrasive grains (v_s)	30000 mm/s
Velocity of workpiece translation (v_w)	500/60 mm/s
Nominal diameter of grinding wheel (d_s)	500 mm
Theoretical given cutting depth (a_p)	0.04 mm
Hardness of workpiece material (H)	45HRC (convert to Brinell Hardness when solving)
Coefficient related to the system stiffness of grinding wheel (C)	0.16
Cutting efficiency (β)	0.8
Stiffness of workpiece (k)	320 kg/mm

For these abrasive-grains, the average diameter is 0.125 mm, and the variation range is ±0.11 mm. Table 1 shows the cutting parameters of simulation.

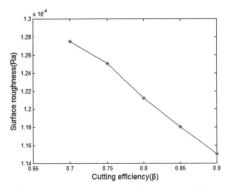

Fig. 10. Relationship between the cutting efficiency and the surface roughness.

When the other parameters are kept unchanged, the surface roughness changes with the cutting efficiency of the workpiece material, which is shown in Fig. 10.

From Fig. 10, a greater cutting efficiency of the workpiece material results in a reduced surface roughness and a better surface quality is obtained, which is the condition under which the other parameters are unchanged. The grinding-wheel surface is meshed (shown in Fig. 11).

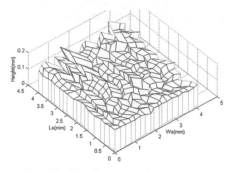

Fig. 11. Simulated wheel topography.

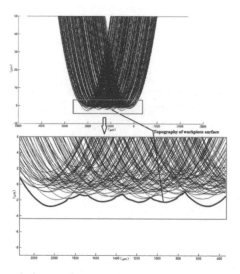

Fig. 12. The workpiece surface topography and the local enlarged drawing.

The workpiece surface topography is formed with continuously iterating the motion trajectories, these motion trajectories are generated by all active abrasive grains in high-precision machining (in Fig. 12). The array of workpiece surface topography needs to be updated

$$[G_{ij}]^k = min\left([G_{ij}]^{k-1}, [g_{ij}]\right) \tag{15}$$

where $[g_{ij}]$ is defined as the initial array, G_{ij} is the protrusion height array of workpiece surface after cutting, the superscript k represents the surface profile index formed by the k-th abrasive-grain. when multi-pass grinding, the preceding simulation for workpiece surface is fed back into the computer program, which is regard as the initial surface texture of the grinding workpiece.

Figure 13 shows a three-dimensional model for the workpiece surface when finish grinding, in which Z represents the height coordinate of the machined workpiece, W_s represents the machined workpiece coordinate in the direction of the grinding-wheel

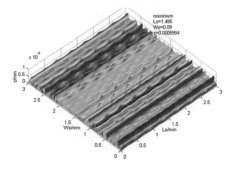

Fig. 13. Simulated surface shape of workpiece.

axis, and L_s is the translational direction coordinate of the workpiece. The labelled values (showing the maximum height and the corresponding position of maximum height) are shown in the upper right corner.

5   Experimental Verification and Analysis

For the sake of verifying the rationality and effectiveness of the algorithm here, comparing the simulation results with the experimental ones is necessary.

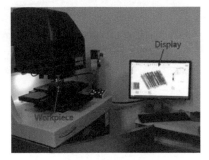

Fig. 14. Yuqing grinder.   **Fig. 15.** 3D optical surface profilometer.

Table 2. Roughness values comparison.

Sample no.	Measured roughness R_a (μm)	Simulated roughness R_a (μm)	Error
1	0.272	0.251	7.7%
2	0.344	0.323	6.1%
3	0.305	0.292	4.3%

Three high-precision grinding experiments were implemented on a multi-function grinder (Model 614S, Taiwan Yuqing Company, as shown in Fig. 14). The grinding surface of all machining parameters was investigated with a 3D optical surface profilometer

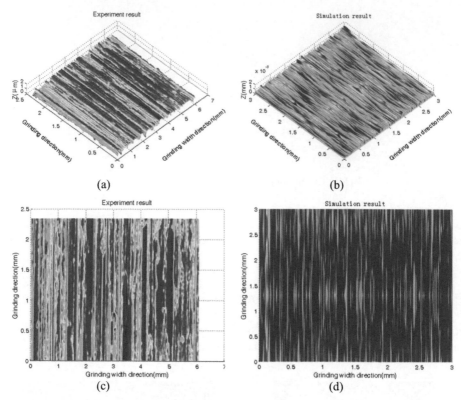

Fig. 16. Comparison of three-dimensional ground surface topography ($v_s = 10$ mm/s, $v_w = 1$ m/min, $a_p = 0.01$ mm)

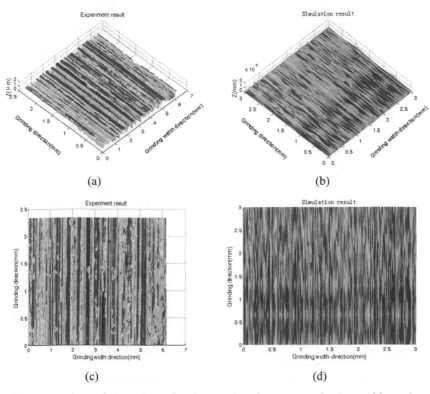

Fig. 17. Comparison of three-dimensional ground surface topography ($v_s = 20$ mm/s, $v_w = 1$ m/min, $a_p = 0.04$ mm).

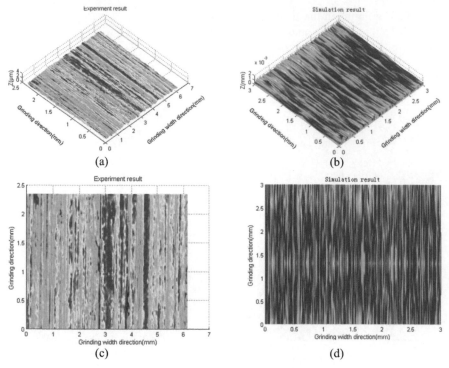

Fig. 18. Comparison of three-dimensional ground surface topography ($v_s = 20$ mm/s, $v_w = 2$ m/min, $a_p = 0.01$ mm).

(ContourGT, American Bruker Company, as shown in Fig. 15). In Figures 16, 17, and 18, the simulated three dimensional surface topography of isometric view figure shown in (b) and top view figure shown in (d), the measured surface topography of isometric view figure shown in (a) and top view figure shown in (c). Table 2 shows that both the measured results and simulation ones have consistent topography features, furthermore, the roughness values also have a small error(less than 8%). In conclusion, it can be said that there is reasonable agreement between the simulated results and the experimental ones.

6 Conclusions

The relationship among the grain parameters, the grinding parameters and the workpiece surface shape is established according to the kinematic model of high-precision grinding and the interaction model between the abrasive-grains and the machined workpiece. The effects of the workpiece material's elastic-plastic deformation are integrated into the kinematic interaction model, the simulation results are more realistic, and the simulation precision is much higher. By using the MATLAB programming environment, based on the an improved Gaussian surface applied to the grinding wheel, the workpiece surface topography can be formed by continuously iterating overall motion trajectories from all active abrasive-grains in the process of high-precision grinding. When comparing the simulated roughness value and the surface topography of this grinding work, under the same machining conditions, both of them are consistent with the measured workpiece surface. The comparison between the simulations and the measurements shows that the accuracy of the presented model is high enough, and both the measured and simulation results have basically consistent topography features and the roughness values also have a small error which is less than 8%. The 3D surface model of the grinding workpiece can be predicted using a computer simulation test, which can provide a basis for selecting machining parameters and further optimizing the parameters.

Acknowledgements. This work was partially supported by the Guangdong Recognized Scientific Research Project in Colleges and Universities of China (grant number 2020KTSCX299), and the Teaching and Research Program for Guangdong Provincial of China (grant number 2020GXSZ165).

References

1. Godino, L., Pombo, I., Sanchez, J.A., Izquierdo, B.: Characterization of vitrified alumina grinding wheel topography using three dimensional roughness parameters: influence of the crystalline structure of abrasive grains. Int. J. Adv. Manuf. Technol. **113**, 1673–1684 (2021)
2. Kang, M.X., Zhang, L., Tang, W.C.: Study on 3D topography modelling of the grinding wheel with image processing techniques. Int. J. Mech. Sci. **167** (2020)
3. Huang, Y., et al.: Study on the surface topography of the vibration assisted belt grinding of the pump gear. Int. J. Adv. Manuf. Technol. **106**, 719–729 (2020)
4. Stalinskii, D.V., Rudyuk, A.S., Solenyi, V.K.: Topography of surface and sub surface layers of grinding balls operating in dry and wet grinding models. J. Steel Trans. **51**, 135–143 (2021)
5. Bellalouna, F.: New approach for industrial training using virtual reality technology. Proc. Cirp. **93**, 262–267 (2020)
6. Bellalouna, F.: Industrial case studies for digital transformation of engineering processes using virtual reality technology. Proc. Cirp. **90**, 636–641 (2020)
7. Mohamed, A.-M.O., Warkentin, A., Bauer, R.: Prediction of workpiece surface texture using circumferentially grooved grinding wheels. Int. J. Adv. Manuf. Technol. **89**(1–4), 1149–1160 (2016)
8. Malkin, S.: Grinding Technology: Theory and Applications of Machining with Abrasives. Ellis Horwood, Chichester (1989)
9. Liu, Y.M., Gong, S., Li, J., Cao, J.G.: Effects of dressed wheel topography on the patterned surface texture and grinding force. Int. J. Adv. Manuf. Technol. **93**, 1751–1760 (2017)
10. Kunz, J.A., Mayor, J.R.: Stochastic characteristics in the micro-grinding wheel static topography. J. Micro. Nano. Manuf. **2**, 29–38 (2014)
11. Nguyen, A.T., Butler, D.L.: Simulation of surface grinding process, part II: interaction of abrasive grain with workpiece. Int. J. Mach. Tools Manuf. **45**, 1329–1336 (2005)
12. Chen, C., Tang, J., Chen, H., Zhu, C.C.: Research about modelling of grinding workpiece surface topography based on the real topography of the grinding wheel. Int. J. Adv. Manuf. Technol. **93**, 1–11 (2017)
13. Cao, Y.L., Guan, J.Y., Li, B., Chen, X., Yang, J., Gan, C.: Modelling and simulation of grinding surface topography considering the wheel vibration. Int. J. Adv. Manuf. Technol. **66**, 937–945 (2013)

14. Nguyen, A.T., Butler, D.L.: Simulation of surface grinding process, part I: generation of grinding wheel surface. Int. J. Mach. Tools Manuf. **45**, 1321–1328 (2005)
15. Nguyen, A.T., Butler, D.L.: Correlation of the grinding wheel topography and the grinding performance: a study from a viewpoint of 3D surface characterization. J. Mater. Process Technol. **208**, 14–23 (2008)
16. Li, X., Rong, Y.: Framework of the grinding process modelling and simulation based on the micro-scopic interaction analysis. Robot. Comput. Integr. Manuf. **27**, 471–478 (2011)
17. Sun, C., Niu, Y.J., Liu, Z., Wang, Y.S., Xiu, S.: Study on surface topography considering grinding chatter based on dynamics and reliabilities. Int. J. Adv. Manuf. Technol. **92**, 1–14 (2017)
18. Liu, Y., Wang, X.F., Lin, J., Zhao, W.: Experimental investigation into effect of chatter on the surface microtopography of gears in grinding. J. Mech. Eng. Sci. **231**, 294–308 (2017)
19. Jiang, J.L., Sun, S., Wang, D.X., Yang, Y., Liu, X.: Surface texture formation mechanism based on ultrasonic vibration assisted grinding process. Int. J. Mach. Tools Manuf. **156** (2020)
20. Pan, J.S., Zhang, X., Yan, Q.S., Chen, S.K.: Experimental study of the surface performance of mono-crystalline 6H-SiC substrates in plane-grinding with a metal-bonded diamond wheel. Int. J. Adv. Manuf. Technol. **89**, 619–627 (2017)
21. Priarone, P.C.: Quality conscious optimization of the energy consumption in a grinding process applying sustainability indicators. Int. J. Adv. Manuf. Technol. **86**, 2107–2117 (2016)
22. Uhlmann, E., Koprowski, S., Weingaertner, W., Rolon, D.: Modelling and simulation of grinding processes with mounted points—part 1 of 2-grinding tool surface characterization. Proc. Cirp. **46**, 599–602 (2016)
23. Li, H.N., Yu, T., Wang, Z.X., Zhu, L.D., Wang, W.: Detailed modelling of cutting forces in the grinding process considering variable stages of grain-workpiece micro-interactions. Int. J. Mech. Sci. **126**, 1–45 (2016)
24. Siebrecht, T., et al.: Simulation of grinding processes using FEA and geometric simulation of individual grains. Prod. Eng. **8**, 345–353 (2014)
25. Brinksmeier, E., Heinzel, C., Bleil, N.: Super-finishing and grind strengthening with the elastic bonding system. J. Mater. Process. Technol. **209**, 6117–6123 (2009)
26. Meng, P.: Micro-structure and performance of mono-layer brazed grinding wheel with poly-crystalline diamond grains. Int. J. Adv. Manuf. Technol. **83**, 441–447 (2016)
27. Tahvilian, A.M., Liu, Z., Champliaud, H., Hazel, B., Lagacé, M.: Characterization of the grinding wheel grain topography under different robotic grinding conditions using the confocal microscope. Int. J. Adv. Manuf. Technol. **80**, 1159–1171 (2015)
28. Wang, J.W., Yu, T.Y., Ding, W., Fu, Y., Bastawros, A.: Wear evolution and stress distribution of the single CBN super-abrasive grain in high speed grinding. Precis. Eng. **54**, 70–80 (2018)
29. Zhou, L., Ebina, Y., Wu, K., Shimizu, J., Onuki, T., Ojima, H.: Theoretical analysis on effects of the grain size variation. Precis. Eng. **50**, 27–31 (2017)
30. Palmer, J., Ghadbeigi, H., Novovic, D., Curtis, D.: An experimental study of the effects of dressing parameters on topography of grinding wheels during the roller dressing. J. Manuf. Process. **31**, 348–355 (2018)
31. Xiu, S.C., Sun, C., Duan, J.C., Lan, D.X., Li, Q.L.: Study on surface topography in consideration of dynamic grinding hardening process. Int. J. Adv. Manuf. Technol. **100**, 209–223 (2019)

A Private Cloud Platform Supporting Chinese Software and Hardware

Man Li[1], Zhiqiang Wang[1(✉)], Jinyang Zhao[2], Haiwen Ou[3], and Yaping Chi[1]

[1] Department of Cyberspace Security, Beijing Electronic Science and Technology Institute, No. 7 Fufeng Road, Fengtai District, Beijing, China
wangzq@besti.edu.cn

[2] Beijing Baidu T2Cloud Technology Co., Ltd., 15A#-2nd Floor, En ji xi yuan, Haidian District, Beijing, China

[3] Department of Cryptography and Technology, Beijing Electronic Science and Technology Institute, No. 7 Fufeng Road, Fengtai District, Beijing, China

Abstract. This paper designs and implements a private cloud platform deployed on an office system that supports domestic software and hardware. With the rapid development of cloud computing, more and more enterprises and users choose cloud platform as a vital Internet resource. At present, most private cloud technologies rely on mature foreign commercial applications and frameworks, and it isn't easy to achieve compatibility between Chinese software and hardware. Therefore, it is urgent to design a private cloud platform that supports Chinese software and hardware. The key private cloud technology of the cloud platform designed in this paper is the key technology of private cloud that supports independent and controllable Chinese software and hardware. The cloud platform uses virtual computing, virtual storage, virtual network, and other technologies to complete the virtualization of computing resources, storage resources, and network resources. Users can centrally schedule and manage virtual resources.

Keywords: Private cloud platform · Virtualization · Cloud computing

1 Introduction

The rapid development and innovation of the Internet have made traditional IT infrastructure platforms increasingly bloated, leading to longer deployment cycles, making it more and more challenging to adapt to business changes. In recent years, as a new type of IT infrastructure platform deployment architecture, cloud computing has frequently appeared in the public's field of vision. Traditional IT platforms have long deployment cycles, high system failure rates, and later operation and maintenance difficulties. The cloud platform attracts more and more people's attention through its low IT cost investment, efficient resource utilization, flexible system adjustment, and low business integration difficulty [1].

Nowadays, with the continuous development and popularization of cloud computing technology and related products, more and more companies and individuals have

Z. Qian et al. (Eds.): WCNA 2021, LNEE 942, pp. 1194–1201, 2022.
https://doi.org/10.1007/978-981-19-2456-9_119

adopted the cloud computing platform as the primary choice for using IT resources [2]. Many excellent features of the cloud platform make it widely used in people's livelihood, finance, military, and business [3]. Many countries have included cloud computing in their national key development plans. Under the current international background, the localization of cutting-edge technology industries is safe and controllable. At present, most of the Chinese cloud platform technologies and solutions are based on mature foreign commercial applications or open-source frameworks, and it is challenging to be perfectly compatible with Chinese office software. Therefore, it is necessary to actively carry out relevant research on cloud platforms that adapt to Chinese software and hardware.

The key technology of private cloud involved in the private cloud platform designed in this paper is the key technology to realize the autonomous and controllable Chinese software and hardware, which provides strong cloud support for Chinese office systems.

The structure of this paper is as follows: first, introduce the research status of the cloud platform; then raise the cloud platform system architecture in more detail; then analyze the system function and performance test results; finally, summarize the paper.

2   Research Status

In 2006, Amazon launched the first batch of cloud products for Amazon Web Services, followed by a series of AWS cloud services. Users can deploy applications with the help of Amazon Elastic Container and perform a series of application extensions as needed [4, 5]. In 2008, Google launched the Google App Engine (GAE) cloud computing service platform [6]. Microsoft released the Microsoft Azure Platform public cloud platform in the same year.

3   Architecture Design of Cloud Platform

3.1   Overall Design

This system uses virtual computing, virtual storage, and virtual networks to complete the virtualization of computing resources, storage resources, and network resources. Through the user portal and administrator portal, users use platform-as-a-service (PaaS) and infrastructure-as-a-service (IaaS) related applications to centrally schedule and manage virtual resources, thereby reducing business operating costs and ensuring system security and reliability.

3.2   Overall Architecture

The cloud platform designed in this paper draws on the best practices of mainstream cloud platforms to provide standard cloud services. The main content of this cloud platform is deployment and application to the cloud, forward-looking planning for operations, and reference to the three-level protection requirements for security. Realize the unified management of traditional IT equipment and resources and the current popular open-source technology on a cloud platform. The overall architecture design of the cloud platform is shown in Fig. 1.

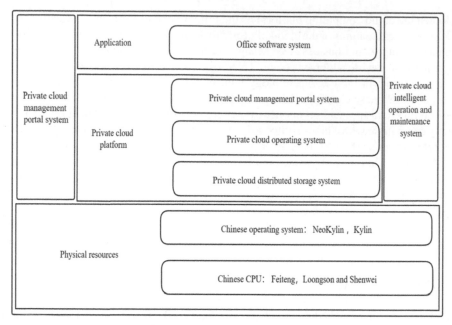

Fig. 1. The overall architecture.

The private cloud platform mainly includes (1) Private cloud management portal system (2) Private cloud operating system (3) Private cloud distributed storage system (4) Private cloud security protection system (5) Private cloud intelligent operation and maintenance system. This cloud platform is compatible with Chinese software and hardware, supports Chinese office software systems in terms of software, adapts Chinese operating systems such as the NeoKylin and Kylin in terms of hardware, and supports Chinese CPUs as Feiteng, Loongson, and Shenwei.

3.3 Technology Architecture

The cloud platform comprises five parts: infrastructure layer, platform service layer, cloud management center, security, and operation and maintenance. Through the collaboration of multiple components, the core service capabilities of the cloud platform are realized.

Infrastructure Layer Design. The infrastructure layer uses virtualization technology to organically combine resources such as computing, storage, and network. The overall IT environment has higher applicability, availability, and efficiency than separate physical hardware resources. It meets the demands of enterprises for cost reduction, simplified management, improved safety, and agile support. Provide core virtualization technology and capabilities for the migration of key businesses of enterprises to the cloud computing environment and the construction of enterprise cloud data centers [7]. The overall structure of the infrastructure layer is shown in Fig. 2.

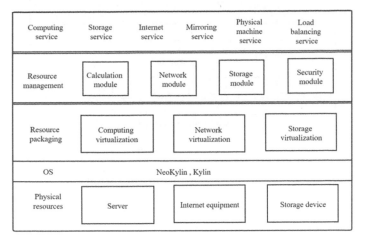

Fig. 2. The infrastructure.

The infrastructure layer includes three layers: physical resources, resource packaging, and resource management. Physical resources mainly include servers, network equipment, and storage devices. The resource encapsulation layer realizes the pooling of different types of physical resources through different virtualization technologies. In addition to driving the resource encapsulation layer, the resource management layer is also responsible for managing various kinds of resources. Finally, the resource management layer provides computing services, storage services, network services, container services, mirroring services, physical machine services, load balancing services, and other service interfaces to the cloud management platform [8].

Platform Service Layer Design. The platform service layer provides information system development and runtime platform environments by creating standard templates and interface packaging to help improve the deployment efficiency of development, testing, and production environments. End users directly develop application system functions and complete configuration and deployment on the platform service layer. The platform service layer includes eight key components of microservice governance, machine learning, integrated middleware as a service, process as a service, message as a service, application middleware as a service, database as a service, and big data as a service.

Software Service Layer Design. SaaS usually positions application software programs developed by PaaS as shared cloud services, which are provided as "products" or available tools [9]. Manufacturers uniformly deploy application software on their own servers. Users can order the required application software services from the manufacturers through the Internet according to their actual needs, pay the manufacturers according to the number of services ordered and the length of time, and obtain the manufacturer's provision through the Internet Service. Users can access through the client interface on various devices, such as a browser. Users do not need to manage or control any cloud computing infrastructure, including networks, servers, operating systems, storage (Fig. 3).

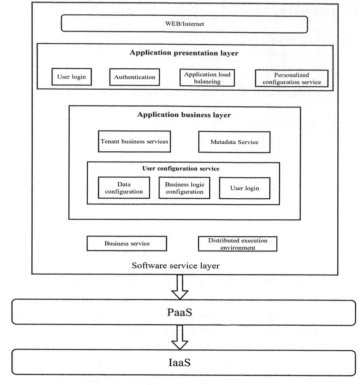

Fig. 3. Software service layer design.

Automation Capability Design. Flexible strategies can provide users with resources and services. Users can increase and decrease the scale of IT infrastructure resources according to system parameter settings to meet business development needs in real-time and save costs. The flexible strategy function supports snapshots and mirroring as templates to create cloud hosts. Users can set the threshold according to the average load of the CPU. When the average load of the cluster reaches the threshold, the system will allocate the resource elastically according to the rules. Elastic distribution is divided into flexible expansion and elastic contraction. When the average cluster CPU load is greater than the threshold, the system expands resources elastically. When the average cluster CPU load is less than the threshold, resources elastically shrink.

Cloud host failover. The system performs periodic detection. When a physical server failure causes a virtual machine failure, the system will migrate the cloud host to other physical servers to quickly recover the cloud host. On the corresponding page, the user can choose whether to support the HA function.

3.4 Security Technology Architecture

Network and Communication Security. Network and communication security ensure the security of the network environment through means such as regional isolation, boundary protection, and traffic identification.

- Deploy an intrusion prevention system.
- Set up Virtual Private Network (VPN).
- TAP replication shunt access platform.
- Perform network system security performance testing.

Equipment and Computing Security. Equipment and computing security adopt measures and technical means such as identity authentication, access control, security audit, intrusion prevention, malicious code prevention, resource control [10].

4 Function Test and Performance Test

4.1 Test Environment

The cloud platform test environment is mainly composed of four server nodes and a test machine. The network topology of the test environment is shown in Fig. 4.

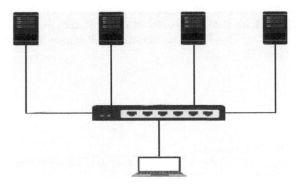

Fig. 4. Test network topology diagram.

The node server used for the test uses the Galaxy Kirin V4.0 operating system, the CPU model is FT1500a@16c CPU 1.5 GHz, the server memory is 64 GB, and the hard disk capacity is 1.5 TB. The software is configured with T2OS cloud operating system V4.0, MariaDB V10.3, and RabbitMQ V3.6.5.

The client used in this test is a Thinkpad T420 laptop, using the Windows 7 flagship operating system. The CPU model is Intel Core i5-2450M 2.50 GHz, the memory is 4 GB, the hard disk capacity is 500 GB, and the client configuration software is Google Chrome 52.0.2743.116.

4.2 Test Results

The cloud platform system designed in this paper realizes the cloud host management and high availability of the virtualized cloud platform. Cloud host management realizes the creation, login, migration, snapshot management, security group management, and other functions of cloud hosts. High availability realizes resource cluster HA capability and master node high availability.

Creating a single cloud host takes an average of 38.8 s; deleting a single cloud host takes an average of 2.2 s; creating a single cloud disk (10 GB) takes an average of 1.0 s. It takes an average of 7.9 s to start a single cloud host.

5 Conclusion

This cloud platform has successfully realized the creation and management of cloud hosts in the cloud platform. It is a unified management platform and has high operating efficiency. This cloud platform realizes a comprehensive high-availability design from business to IT resources, supports on-demand allocation of virtual resources, supports multiple operating systems, uses QoS technology to ensure various resources, and supports multiple hardware devices. This cloud platform's successful research and development provide better and strong cloud support for Chinese office systems. A series of private cloud key technologies have been adapted and optimized in the Chinese software and hardware environment.

Acknowledgments. This research was financially supported by the National Key RD Program of China (2018YFB1004100), China Postdoctoral Science Foundation-funded project (2019M650606), and the First-class Discipline Construction Project of Beijing Electronic Science and Technology Institute (3201012).

References

1. Aleem, A., Sprott, C.R.J.: Let me in the cloud: analysis of the benefit and risk assessment of cloud platform. J. Financ. Crime **20**(1), 6–24 (2012)
2. Chen, Q., Deng, Q.N.: Cloud computing and its key technologies. J. Comput. Appl. **9**, 254–259 (2009)
3. Merin, R., Vaquer, L., Caron, E.: Building safe paas clouds: a survey on security in multitenant software platforms. Comput. Secur. **31**(1), 96–108 (2012)
4. Duan, W.X., Hu, M., Zhou, Q.: Overview of reliability research of cloud computing system. J. Comput. Res. Develop. **57**(1), 102–123 (2020)
5. Pail, C.: Containerization and the Paas Cloud. IEEE Cloud Comput. **2**(3), 24–31 (2015)
6. Magalhes, G., Roloff, E., Maillard, N.: Developing on Google App Engine (2012)
7. Bhardwaj, S., Jain, L., Jain, S.: Cloud computing: a study of Infrastructure As A Service (IAAS). Int. J. Inf. Technol. Web Eng. **2**(1), 60–63 (2010)
8. Wu, L., Garg, S.K., Buyya, R.: SLA-based resource allocation for software as a service provider (SaaS). In: Cloud Computing Environments, pp. 195–204. IEEE (2011)

9. Zhang, Z.H., Zhang, X.J.: A customizable and adaptive load balancing mechanism based on ant colony and complex network theory in open cloud computing federation. In: Proceedings of 2010 2nd International Conference on Intellectual Technology in Industrial Practice (ITIP2010), vol. 2, pp. 45–50 (2010)
10. Aljohani, A.M.: Issues of cloud computing security and data privacy. J. Res. Sci. Eng. **3**(4) (2021)

An Improved Time Series Network Model Based on Multitrack Music Generation

Junchuan Zhao[✉]

International School, Beijing University of Posts and Telecommunications, Beijing, China
zhaojc_music_ai@163.com

Abstract. Deep learning architecture has become a cutting-edge method for automatic music generation, but there are still problems such as loss of music style and music structure. This paper presents an improved network structure of time series model based on multi-track music. A context generator is added to the traditional architecture. The context generator is responsible for generating cross-track contextual music features between tracks. The purpose is to better generate single-track and multi-track music features and tunes in time and space. A modified mapping model was further added to further modify the prediction results. Experiments show that compared with traditional methods, the proposed will partially improve the objective music evaluation index results.

Keywords: Time series model · GAN · Symbolic music generation · Multitrack music generation

1 Introduction

Deep learning is rapid developed technology in the field of AI. From the sky to the ocean, from drones to unmanned vehicles, deep learning is playing its huge potential and capabilities. In the medical field, the machine's disease recognition rate of lung photos has surpassed that of humans; the images and music generated by GAN technology can be fake and real; in the commercial field, micropayments can already be made through human faces; AlphaGo has defeated the real Go master in the official competition.

On the other hand, in my opinion, music is an art that conveys emotions and emotions through sound. It is a way of human self-expression. The creation of music can help people entertain and express their feelings. It is feasible to use deep learning to imitate the patterns and behaviors of existing songs, and to create music content that is real music to human ears. There have been many researchers and research results in the field of music generation based on artificial intelligence and deep learning.

Multi-track music composing [1] requires professional knowledge and a command of the interfaces of digital music software. Besides, few have focused on multi-track composing with emotion great human involvement. According to these, the author presents platform using our life elements. The system can be roughly split into three main parts.

An end-to-end generation framework called XiaoIce Band was proposed [2], which generates a track with several tracks. The CRMCG model utilizes the encoder-decoder

© The Author(s) 2022
Z. Qian et al. (Eds.): WCNA 2021, LNEE 942, pp. 1202–1208, 2022.
https://doi.org/10.1007/978-981-19-2456-9_120

framework to generate both rhythm and melody. For rhythm generation, in order to make generated rhythm in harmony with existing part of music, they take previous generation of music (previous melody and rhythm) into consideration. For melody generation, they take previous melody, currently generated rhythm and corresponding chord to generate melody sequence. Since rhythm is closely related to melody, the loss function of rhythm generation only updates parameters related with rhythm loss, whereas the loss function of melody generation updates all parameters by melody loss. The MICA model is used to solve task, it treats the melody sequence as the input of encoder and the multiple sequences as outputs of decoder. The designed between the hidden layers to learn the relationships and keep the harmony between different tracks.

The Attention Cell is used to capture the relevant parts of other tasks for current task. The author conducted melody generation and arrangement generation tasks to evaluate the effectiveness of the CRMCG and MICA. For melody generation task, they choose the Magenta and GANMidi as baseline methods, meanwhile, chord progression analysis and rest analysis are used to evaluate the CRMCG model. For arrangement generation task, they choose HRNN as baseline methods, meanwhile, harmony analysis and arrangement analysis are used to evaluate the CRMCG model.

The paper [3] proposed a method to generate multiple chord music using GAN. This model will process a transformation from MIDI files and chord music to multiple bass, piano, drum, and guitar tracks and piano rolls., And its dimension is K. After standard preprocessing of the MIDI file, all music is divided into more than one hundred parts according to the beat and the pitch is changed to a certain range. At this time, the dimension is $[K * 5* 192 * 84]$. The model given in the article contains a generator and a discriminator of the convolutional neural network architecture. The structures of the two are symmetrical and opposite. Finally, the activation function sigmoid is used to separate the data. Since the music data is not discrete, and there are often multiple chords pronounced at the same time, the convolution part adopts a full-channel architecture, which helps the network to converge quickly. ReLU + tanh is used in the former, LeakyReLU is used in the latter to deal with the gradient problem, and finally Adam is used to complete the optimization.

Although there are many music generation technologies, the existing music generation methods are still unsatisfactory. Most of the music and songs generated by the music generation technology can be easily distinguished from the real music and songs by the human ear. There are many reasons for this. For example, due to the lack of "alignment" data [4], different styles are used for the same song, leading to the main music style conversion can only use unsupervised methods. The loss of using GAN (RaGAN) during training leads to the inability to guarantee that the original music structure will be retained after conversion [5].

This paper proposes an improved time-series model network structure based on multi-track music MuseGAN, and adds a correction mapping model after the generators to bind the predicted results to the correct results. Experiments on standard data sets show that the method proposed in this paper can further improve subjective and objective evaluation indicators such as Qualified Rhythm frequency.

2   Symbolic Music Generation and Genre Transfer

Furthermore, when style conversion and classification are required, style alignment is first required, with the goal of realizing VAE and style classification in a shared space [6]. While switching the style of music data, this method can also change the types of musical instruments, such as piano to violin, and can also change auditory characteristics such as pitch. This model has a wide range of applications, such as music mixing, music and song mixing, music insertion, and so on. Each data file is in MIDI format, with style tags, that is, specific style tags. By extracting these information from the file and converting them, such as pitch, gauge, and speed. This kind of VAE comes with hyper parameter evaluation Kullback-Leibler to judge the cross entropy loss. In order to obtain the joint distribution of the overall data, three codecs are used to form a shared space.

Another model of musical style conversion is called ycleGAN [7], and the structure of its generator/discriminator is shown in Fig. 1. In order to perform style transfer while retaining the tune and structure of the original music itself, a discriminator is needed to balance the intensity difference between input and output. The generator extracts from the original data and can also input noise, but this method can only handle the transformation of two parts. The goal of the generator is to learn a variety of high-level features, so the discriminator is required to be able to distinguish between the source data and the generated data. The loss function part is measured by consistent loss, which helps to retain more overall information for two-way conversion, the output data can be a true form. When experimenting on the data set, the LeakyReLU + normalization method is used, and the final output is a classifier with a distribution.

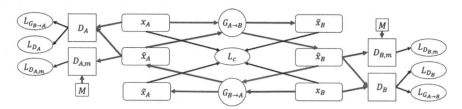

Fig. 1. Architecture of CycleGAN model.

The music generation, especially rhythm patterns of electronic dance music with novel rhythms and interesting patterns, which were not found in the training dataset, could be generated by using deep learning. They extend the framework GAN and encourage inherent distributions by additional classifiers [8]. The author proposes two methods in this paper (Fig. 2).

3   Improved Time Series Model Network on Multitrack Music

The paper [9] proposed the GAN, the quantitative measure estimating the interpretability of a set of generated examples and apply the method to a state-of-the-art deep audio classification model that predicts singing voice activity in music excerpts. Their method

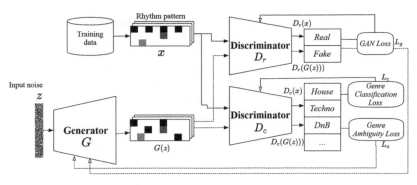

Fig. 2. GAN with genre ambiguity loss.

is designed to provide examples that activate a given neuron activation pattern ("classifier response"), where a generator is trained to map a noise vector drawn from a known noise distribution to a generated example. To optimize the prior weight and optimization parameters as well as the number of update steps, a novel, automatic metric for quickly evaluating a set of generated explanations is introduced. For the generator, they choose a standard normal likelihood. For AM optimization, is performed. The melody composition method could enhance the original GAN based on individual [10].

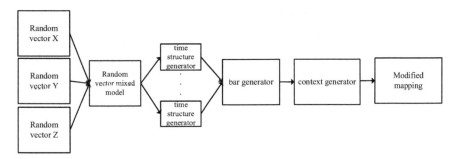

Fig. 3. An improved time series model with multi-generator.

The INCO-GAN [11] is designed to mainly address two problems: 1) cannot judge when to end the generation by itself; 2) no apparent time relationship between the notes or bars. The automatic music generation is two phases: training and generation. The three training steps: Preprocessing, CVG training, and conditional GAN training. CVG provides the conditional vector required for music generation for the generator. It consists of two parts: one part is utilized to generate the relative position vector to represent the generation process, and the other part can predict whether the generation is to end. In the training phase, the CVG training and conditional GAN training are independent of each other. The generation phase comprises three steps: CVG executing, phrase generation, and postprocessing. To evaluate the generated music, the pitch frequency of the music generated by the proposed model was compared with human composer's music.

In summary, these music generation technologies described above are all deep learning technologies. The deep network learns features from a large number of music samples, and generates an effective function approximation method based on the original music sample distribution, and finally generates new music sample data. Since music is a kind of time series data like speech and text, it can be generated by a variety of deep neural networks used to capture long dependencies in the sequence.

This paper proposes an improved time series model network structure based on multitrack music MuseGAN. The sub-network of generators is adhesion on the MuseGAN architecture: in addition to the time structure generator and the bar generator, a context generator is added. After these generators, a modified mapping model was added to further modify the prediction results. The architecture of the improved network model proposed is shown in Fig. 3. The time structure generator is used to characterize the unique time-based architecture of music; the bar generator is responsible for generating a single bar in different tracks, and the timing relationship between bar and bar comes from structures such as Scratch; the context generator is responsible for The music features that are context-sensitive across tracks are generated between tracks. The combination of these three generators can better generate single-track and multi-track music features and tunes in time and space.

4   Experiments and Results

The automatic music generation is divided into two phases: training and generation [11]. The training phase consists of three training steps: Preprocessing, CVG training, and conditional GAN training. CVG provides the conditional vector required for music generation for the generator. It consists of two parts: one part is utilized to generate the relative position vector to represent the generation process, and the other part can predict whether the generation is to end. In the training phase, the CVG and conditional GAN training are independent each other. The generation phase comprises three steps: CVG executing, phrase generation, and post processing. To evaluate the generated music, the pitch frequency of the music generated by the proposed model was compared with human composer's music. The paper [3] uses two sets of programs to track the experimental results.

Table 1. The average score of each model on each indicator of Qualified Rhythm Frequency.

QRF	Traditional model with two generators	Improved time series model
Corpus	0.91	0.93
Duration	0.82	0.87
Beat	0.90	0.89

In this paper, we generate more than 1000 music sequences with the method of each model, and then use some subjective and objective indicators (Qualified Rhythm frequency and Consecutive Pitch Repetitions) to evaluate the performance of each model

[12]. It can be seen from Table 1 that the improved is better than traditional with two generators on the two indicators of the Qualified Rhythm frequency, and worse than the Traditional model with two generators on the Beat indicator. The reason may be that the context generator is in the influence on Beat has the opposite effect.

Table 2. The average score of each model on each indicator of Consecutive Pitch Repetitions.

CPR	Traditional model with two generators	Improved time series model
Corpus	0.01	0.01
Duration	0.08	0.10
Beat	0.05	0.05

It can be seen from Table 2 that the improved is better than traditional with two generators on the two indicators of Consecutive Pitch Repetitions, and is still worse than the Traditional model with two generators on the Beat indicator. The reason may still be the influence of the context generator on Beat.

5 Conclusion

Music generation technology based on deep learning has been widely used, but it still was affected by problems such as loss of music structure during training. This paper proposes an improved time series model network structure, adding a context generator to the traditional architecture, and adding a modified mapping model to further modify the prediction results. Our experiments implied our method proposed can partially improve the index results of Qualified Rhythm Frequency and Consecutive Pitch Repetitions.

References

1. Qiu, Z., et al.: Mind band: a crossmedia AI music composing platform. In: Proceedings of the 27th ACM International Conference on Multimedia, pp. 2231–2233, October 2019
2. Zhu, H., et al.: XiaoIce band: a melody and arrangement generation framework for pop music. In: Proceedings of the 24th ACM SIGKDD International Conference on Knowledge Discovery and Data Mining, pp. 2837–2846, July 2018
3. Chen, H., Xiao, Q., Yin, X.: Generating music algorithm with deep convolutional generative adversarial networks. In: 2019 IEEE 2nd International Conference on Electronics Technology (ICET), pp. 576–580. IEEE, May 2019
4. Cífka, O., Şimşekli, U., Richard, G.: Supervised symbolic music style translation using synthetic data. arXiv preprint arXiv:1907.02265 (2019)
5. Lu, C.Y., Xue, M.X., Chang, C.C., Lee, C.R., Su, L.: Play as you like: timbre-enhanced multi-modal music style transfer. In: Proceedings of the AAAI Conference on Artificial Intelligence, vol. 33, no. 01, pp. 1061–1068, July 2019
6. Brunner, G., Konrad, A., Wang, Y., Wattenhofer, R.: MIDI-VAE: Modeling dynamics and instrumentation of music with applications to style transfer. arXiv preprint arXiv:1809.07600 (2018)

7. Brunner, G., Wang, Y., Wattenhofer, R., Zhao, S.: Symbolic music genre transfer with Cycle-GAN. In: 2018 IEEE 30th International Conference on Tools with Artificial Intelligence (ICTAI), pp. 786–793. IEEE, November 2018

8. Tokui, N.: Can GAN originate new electronic dance music genres?--Generating novel rhythm patterns using GAN with Genre Ambiguity Loss. arXiv preprint arXiv:2011.13062 (2020)

9. Mishra, S., Stoller, D., Benetos, E., Sturm, B.L., Dixon, S.: GAN-based generation and automatic selection of explanations for neural networks. arXiv preprint arXiv:1904.09533 (2019)

10. Li, S., Jang, S., Sung, Y.: Automatic melody composition using enhanced GAN. Mathematics **7**(10), 883 (2019)

11. Li, S., Sung, Y.: INCO-GAN: variable-length music generation method based on inception model-based conditional GAN. Mathematics **9**(4), 387 (2021)

12. Trieu, N., Keller, R.: JazzGAN: improvising with generative adversarial networks. In: MUME Workshop, June 2018

Research on EEG Feature Extraction and Recognition Method of Lower Limb Motor Imagery

Dong Li and Xiaobo Peng[✉]

College of Mechatronics and Control Engineering, Shenzhen University, Shenzhen 518060, Guangdong Province, People's Republic of China
pengxb@szu.edu.cn

Abstract. Aiming at the problems of difficult signal acquisition, low signal-to-noise ratio and poor classification accuracy of BCI technology, based on the theory of EEG, this paper designs a leg raising EEG experiment of lower limb motor imagery and collects EEG signal data from 20 subjects to improve the accuracy of classification and recognition The process of feature extraction and classification recognition is explored, and a multi domain fusion method is proposed for EEG signal feature extraction from time domain, frequency domain, time-frequency domain and spatial domain. At the same time, bagging and gradient boosting ensemble learning algorithms are applied to EEG signal classification and recognition, and multi domain fusion features are tested by constructing different classifiers, The final classification accuracy reaches 87.8% and 93%, which is better than the traditional SVM classification method.

Keywords: Brain computer interface · Motor imagination · Integrated learning · Multi domain fusion · Support vector machine

1 Introduction

Brain is the senior commander of human body, which controls all kinds of information communication between human body and external environment through peripheral nerve and muscle channels. However, with the emergence of global aging problem, a variety of brain diseases are also increasing, such as stroke, epilepsy, depression and so on, which seriously endanger the life safety of patients; In addition, the rapid development of science and technology has greatly changed people's way of travel. While people get convenient transportation, there are also many traffic accidents, such as brain and nervous system damage of drivers, amputation and other problems caused by traffic accidents, which lead to the loss of the ability of human body to control its own muscles [3]. Although these diseases or accidents cut off the channel of information communication between the human brain and the external environment, the brain of the victims can produce consciousness or thinking. Therefore, researchers at home and abroad are trying to help the victims recover and improve their quality of life by using external auxiliary equipment.

© The Author(s) 2022
Z. Qian et al. (Eds.): WCNA 2021, LNEE 942, pp. 1209–1218, 2022.
https://doi.org/10.1007/978-981-19-2456-9_121

In recent years, with the continuous development of computer technology, more and more scientists are committed to the field of brain science. They study the interactive method of combining computer and human brain, and reflect the real intention of patients by recording their EEG signals, so as to carry out rehabilitation treatment, which effectively promotes the brain computer interface, BCI) [5] technology development. Brain computer interface technology refers to a control system that does not rely on human muscle tissue and neural pathways to create channels between the human brain and external devices, so as to realize the communication between the brain and the external environment. As shown in Fig. 1, BCI technology is used to build an external pathway between the brain and the legs, so as to realize the control of the brain over the legs. Brain computer interface technology is not only widely used in biomedicine and neural rehabilitation, but also has significant advantages in education, military, entertainment and so on. BCI was first formed in the 1970s and grew rapidly in the late 1990s. Until now, researchers at home and abroad have never stopped exploring BCI. In recent years, with the in-depth development of artificial intelligence technology, it has opened up a new way for the research of BCI technology. For example, Li [9] proposed the algorithm of using multi-core learning mode to optimize support vector machine, which can quickly classify and recognize EEG with cognitive ability; Hajinoroozi et al. [10] used the method of convolutional neural networks (CNN) to study the EEG of drivers, so as to predict and regress their cognitive ability; Qiao [11] et al. Established a spatiotemporal convolution model to classify and recognize motor imagery EEG signals.

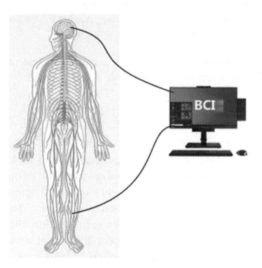

Fig. 1. BCI channel

Motor imaging (MI) refers to the rehearsal of a behavior that is about to be triggered by the brain after receiving external stimulation [12]. At this time, the brain only has the intention to imagine the action, but not the real behavior. When the brain imagines a specific behavior, the related motor areas become active due to stimulation, which enhances the discharge process of neurons and leads to the change of their potential, resulting in event-related changes, and ultimately achieve the purpose of motor control. By collecting the motor imagery EEG signal at the time of brain discharge, and analyzing and processing the signal, different classification algorithms are used to identify the data to obtain the motor imagery intention. Finally, the external device completes the execution of related actions by judging the imported signal [13], and successfully analyzes people's action intention. Motor imagery is widely used in BCI system, sports training, rehabilitation training of lower limb patients and other fields [14]. It is an important tool to study the brain activation, neural network function and psychological process of human body under external stimulation. It is of great significance to the research of medicine and biological brain science.

Based on the theory of EEG, this paper designs EEG experiments of lower limb motor imagery to collect EEG data from 20 subjects. Aiming at the problems of nonstationarity, difficulty in feature extraction and low classification accuracy of motor imagery EEG signal, a multi domain fusion method of feature extraction of EEG signal from time domain, frequency domain, time-frequency domain and spatial domain is proposed, At the same time, the ensemble learning algorithm is used to classify and recognize the fused features, and two kinds of EEG signal classifiers, bagging and gradient boosting, are constructed for experiments. The final classification accuracy reaches 87.8% and 93%, which is better than the traditional SVM EEG signal classification method.

2 Experiment

In this paper, through the construction of the experimental platform of motor imagination, we use the real person leg raising video to stimulate the subjects' motor imagination, which can efficiently and accurately obtain the EEG characteristics of the subjects, and the EEG signal extraction of the subjects uses the safe and convenient non-invasive method, During the experiment, the subjects need to wear a 64 lead quick cap EEG acquisition cap that meets the international 10–20 electrode positioning standard. The EEG signal collected is transmitted to the signal processor through Weaver EEG paste, and then the EEG signal is amplified by a certain proportion through the brain amp amplifier. The experimental paradigm is designed by using E-Prime software to realize synchronous communication.

In this study, a total of 20 college students, male and female, aged 18–26 years old and healthy, without other diseases, were invited. The design of this experiment is based on the motor imagination experiment of resting state and task state under visual stimulation. The human leg raising video is used to induce and stimulate the subjects, and the five electrode channels (FC1, FC2, C1, C2, CZ) of the subjects are explored, as shown in Fig. 2. Before the experiment, each subject is required to carry out a week of motor imagination training to improve the motor imagination ability. At the same time, the whole experimental process and precautions are introduced to the subjects in detail to ensure that the subjects have a clear understanding of the experimental content. In order to ensure that the subjects have a good mental state, they are required to fall asleep before 22 o'clock one day before the experiment; One hour before the experiment, the hair was washed and dried with a hair dryer to ensure a smaller impedance; During the experiment, the subjects are required to blink as little as possible, reduce the number of eye movements and swallowing saliva and other behaviors that affect the effect of the experiment.

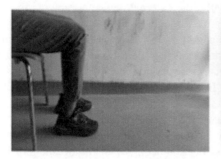

Fig. 2. Video capture of human body in resting state and task state

During the experiment, each person collected 5 groups of experiments, 40 times in each group, 20 times in the resting state and 20 times in the task state, 10 s each time. Before the beginning of each experiment, the screen will display the experiment instructions. After the subjects are ready, they press the keyboard "Q" key to start the experiment. A red "+" will appear in the center of the screen in 0–1 s to remind the subjects to prepare for the experiment; 1–3 s, the screen does not show any content, so that the subjects can relax physically and mentally; In 3–7 s, sit in or leg up videos were randomly displayed on the screen. When the leg up videos appeared, the subjects imagined the movement. When the sit in videos appeared, the subjects only needed to keep their mind blank and did not do any imaginary actions; The rest time is 7–10 s, and the subjects will not be disturbed by the EMG signal generated by fatigue. The experimental process is shown in Fig. 3.

Fig. 3. Flow chart of single experiment

3   Methods

3.1   Data Preprocessing

The original EEG signal collected through the experiment contains a lot of interference noise, such as eye movement, head movement, ECG and 50 Hz power frequency interference. Therefore, before feature extraction of EEG signal, it is often necessary to carry out data preprocessing to effectively filter the noise, as shown in Fig. 4.

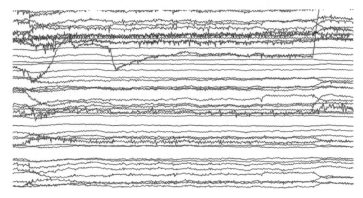

Fig. 4. Original EEG map

The data preprocessing of EEG signal mainly includes: electrode location, removal of useless electrode, re reference, filtering, segmentation, replacement of bad segment, blind source separation and removal of artifacts, among which filtering and blind source separation are particularly important. Because most of the EEG signals of motor imagery of lower limbs are of the same waveform α Wave and β Therefore, the 0.1–40 Hz EEG signal is selected as the band of interest, and the band-pass (low-pass, high pass and sag filter) filter is used for filtering. After filtering, the EEG signal is analyzed by independent component analysis, and different EEG components are separated. The artifact identification and elimination operation are carried out on the separated EEG signal by using the adjust artifact elimination method. As shown in Fig. 5, the EEG signal after preprocessing is shown, and the noise component is significantly reduced, and the signal-to-noise ratio is also greatly improved.

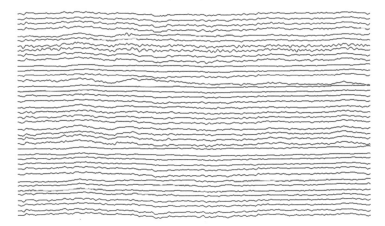

Fig. 5. EEG signal after pretreatment

3.2 Feature Extraction

After preprocessing the collected EEG signals, some electrodes need to be selected to extract their features. Feature extraction is to represent the imagination intention of the brain by using as few feature vectors as possible. It is the basis and basis of classification and recognition in the later stage, and is a necessary part of EEG signal processing. This paper explores the ERD/ERS phenomenon in the brain of the subjects during the experiment, determines the most obvious frequency band and time period of the right leg motor imagination, and represents the two information in the time domain, frequency domain, time-frequency domain and spatial domain respectively. Finally, it is fused into the form of multi domain feature vector, which effectively overcomes the limitations of single feature.

Because of the complexity and non stationarity of EEG signal, the time-domain feature is often abandoned by researchers. It is the characterization of the amplitude of EEG signal at different times, mainly including the maximum, minimum and average of the amplitude of EEG signal. These three common time-domain feature information include all the time information data of EEG signal, which has a strong intuitive feature selection of EEG signal. Frequency domain feature is the change of EEG signal amplitude with frequency. It can identify the correlation of different EEG signals by depicting the spectral feature information of EEG signals in different frequency bands. Power spectral density (PSD) is a common method to study the frequency domain characteristics of EEG signal, which takes frequency as an independent variable to reflect the power value of a specific frequency component. In this paper, the kurtosis, skewness, standard deviation and average power of EEG signal are selected as frequency domain characteristic information by increasing the characteristic number of power spectral density. The feature of time-frequency domain is the dimension reflecting the change of EEG signal frequency with time. By using the method of short-time Fourier transform and introducing the time window function, the non-stationary EEG signal can be effectively extracted, but the time window function cannot meet the local change of time and frequency. Therefore, the processed EEG signal is decomposed and reconstructed by using the method of discrete wavelet transform, Simple and stable time-frequency characteristic information can be obtained. Spatial domain feature extraction is mainly to construct spatial filter for task state and resting state data, and to maximize the covariance difference between the two types of data by using matrix diagonalization and variance scaling method, so as to show the feature vector with high discrimination, as shown in Fig. 6, which is the spatial domain feature map of electrode channel C1 and CZ. The multi domain fusion matrix is obtained by fusing the feature information of time domain, frequency domain, time-frequency domain and spatial domain of the above EEG signal features, which solves the problem of difficult feature extraction caused by the high non-stationary of EEG signal, and brings convenience for the subsequent classification and recognition.

3.3 Classification and Identification

Different classification algorithms are used to classify and identify the extracted feature information, which can help patients to control the external equipment. Compared with the traditional SVM method, this paper proposes an integrated learning algorithm of

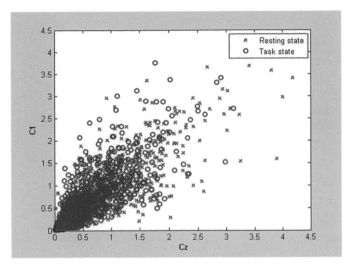

Fig. 6. C1 and CZ airspace characteristic map

bagging and gradient boosting to analyze EEG information, and verifies the advantages and disadvantages of the classification method by comparing its classification accuracy.

Bagging algorithm is one of the integrated learning algorithms, which is characterized by independent sub learners, and its dependence is not strong, and can be generated synchronously [15]. It selects the classification tree in decision tree as weak classifier. After integrating m weak classifiers, bagging can reduce the variance of training set and increase deviation, so that bagging will not show the fitting phenomenon on the training set. Therefore, when using bagging algorithm to classify EEG signals after feature extraction, it can randomly sample and obtain the subset and generate the base classifier after training, The accuracy of EEG signal classification is greatly improved, up to 87.8%. As shown in Fig. 7, the accuracy of multi domain fusion feature classification is shown when using bagging algorithm to iterate for 50 times.

Boosting algorithm is an ensemble learning algorithm that combines multiple weak classifiers into strong classifiers according to the weight. Its principle is to randomly extract samples, add the same initial weight to each sample, observe the performance of weak classifiers after each training round, and increase the proportion of wrong samples, so that such samples can get more attention in the next round, Until m weak classifiers are trained and combined into strong classifiers according to weight, the accuracy of weak classification algorithm can be effectively improved [16]. The gradient boosting algorithm is the optimization of boosting algorithm. It constructs a weak classifier which can reduce the classification error rate along the steepest direction of the gradient by gradient lifting [17]. It can solve the problem of second classification of EEG signal and effectively improve the anti noise ability of the model, with the highest accuracy of 93%, Fig. 8 shows the classification accuracy of multi domain fusion features when the gradient boosting algorithm is used for 50 iterations.

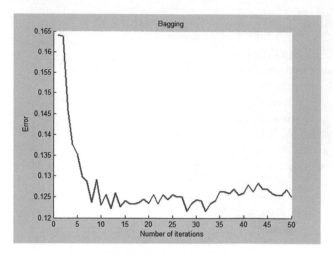

Fig. 7. Bagging classifier

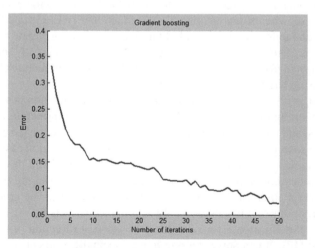

Fig. 8. Gradient boosting classifier

4   Conclusion

In this paper, the EEG data of 20 subjects are collected and explored by building a lower limb motor imagery EEG experimental platform. The multi domain (time domain, frequency domain, time-frequency domain and spatial domain) feature fusion method is used to effectively extract the feature information of complex and high-dimensional EEG signals. At the same time, the ensemble learning algorithm bagging and gradient boosting are used as classifiers, The classification accuracy of EEG signal is greatly improved, but the EEG signal data collected in this experiment is still small data samples, and the experimental objects are normal people. The generalization ability of the classifier model to the EEG signal data of real patients is poor. In the later stage, the EEG signal data of real patients will be collected and the sample size will be expanded to improve the universality of the classifier model.

References

1. Lou, X.: Research on active rehabilitation of stroke patients based on coherence of EEG and EMG, pp. 33–37. Zhejiang University, Zhejiang (2012)
2. Chen, S., Yuanqi, Z., et al.: The method of EEG epilepsy detection based on multiple characteristics. J. Biomed. Eng. **32**(3), 279–283 (2013)
3. Wang, Z.: Introduction to brain and cognitive science. 1st edn. Beijing University of Posts and Telecommunications Press, Beijing, pp. 46–50 (2011)
4. Bigdely-Shamlo, N., Touryan, J., Ojeda, A., et al.: Automated EEG mega-analysis II: cognitive aspects of event related features. Neuroimage **207**, 116054 (2020)
5. Jin, J., Chen, Z., Xu, R., et al.: Developing a novel tactile P300 brain-computer interface with a cheeks-stim paradigm. IEEE Trans. Biomed. Eng. **67**(9), 2585–2593 (2020)
6. Xu, T., Zhou, Y., Wang, Z., et al.: Learning emotions EEG-based recognition and brain activity: a survey study on BCI for intelligent tutoring system. Procedia Comput. Sci. **130**, 376–382 (2018)
7. Friedl, K.E.: Military applications of soldier physiological monitoring. J. Sci. Med. Sport **21**(11), 1147–1153 (2018)
8. Taherisadr, M., Dehzangi, O.: EEG-based driver distraction detection via game-theoretic-based channel selection. In: Fortino, G., Wang, Z. (eds.) Advances in Body Area Networks I. Internet of Things, pp. 93–105. Springer, Cham (2019). https://doi.org/10.1007/978-3-030-02819-0_8
9. Li, X., Chen, X., Yan, Y., et al.: Classification of EEG signals using a multiple kernel learning support vector machine. Sensors **14**(7), 12784–12802 (2014)
10. Hajinoroozi, M., Mao, Z., Jung, T.P., et al.: EEG-based prediction of driver's cognitive performance by deep convolutional neural network. Sig. Process. Image Commun. **47**, 549–555 (2016)
11. Qiao, W., Bi, X.: Deep spatial-temporal neural network for classification of EEG-based motor imagery. In: Proceedings of the 2019 International Conference on Artificial Intelligence and Computer Science, pp. 265–272 (2019)
12. Munzert, J., Lorey B., et al.: Cognitive motor processes: the role of motor imagery in the study of motor representations. Brain Res. Rev. **60**(2), 306–326 (2009)
13. Pfurtscheller, G., Neuper, C.: Motor imagery and direct brain-computer communication. Proc. IEEE **89**(7), 1123–1134 (2002)

14. Xu, F.: Research on brain computer interface algorithm based on motor imagination, pp. 21–25. Shandong University, Shandong (2014)
15. Yueru, W., Xin, L., Honghong, L., et al.: Feature extraction of motor imagery EEG based on time frequency spatial domain. J. Biomed. Eng. **31**(05), 955–961 (2014)
16. Li, W., Yang, X., Huang, L., et al.: Power spectrum and clinical data analysis of sonogram. J. Nanyang Inst. Technol. **4**(4), 31–35 (2012)
17. Liu, L., Li, S.: EEG signal denoising based on fast independent component analysis. Comput. Meas. Control **22**(11), 67–75 (2014)

Author Index

© The Editor(s) (if applicable) and The Author(s) 2022
Z. Qian et al. (Eds.): WCNA 2021, LNEE 942, pp. 1219–1223, 2022.
https://doi.org/10.1007/978-981-19-2456-9

Printed by Printforce, the Netherlands